REVIEWS in MINERALOGY
Volume 7

(PYROXENES)

CHARLES T. PREWITT, Editor

The Authors:

Charles T. Prewitt
James J. Papike
Donald H. Lindsley
Tibor Gasparik

Dept. of Earth & Space Sciences
State University of New York
Stony Brook, NY 11794

Peter R. Buseck
David R. Veblen

Dept. of Geology
Arizona State University
Tempe, AZ 85281

Gordon L. Nord, Jr.
J. Stephen Huebner

U.S. Geological Survey
959 National Center
Reston. VA 22092

Maryellen Cameron

Board of Earth Sciences
University of California
Santa Cruz, CA 95064

George R. Rossman

Division of Geological &
Planetary Science
California Institute of
Technology
Pasadena, CA 91125

John E. Grover

Dept. of Geology
University of Cincinnati
Cincinnati, OH 45221

Peter Robinson

Dept. of Geology
University of Massachusetts
Amherst, MA 01003

Series Editor   Paul H. Ribbe

Department of Geological Sciences
Virginia Polytechnic Institute and State University
Blacksburg, Virginia 24061

MINERALOGICAL SOCIETY OF AMERICA

REVIEWS in MINERALOGY

(Formerly: SHORT COURSE NOTES)

ISSN 0275-0279

VOLUME 7:   PYROXENES

ISBN 0-939950-07-3

Additional copies of this volume as well as those
listed below may be obtained at moderate cost from

Mineralogical Society of America
2000 Florida Avenue, NW
Washington, D.C.   20009

PYROXENES

TABLE of CONTENTS

CHAPTER 4. SUBSOLIDUS PHENOMENA IN PYROXENES

Peter R. Buseck, Gordon L. Nord, Jr. & David R. Veblen

CHAPTER 5. PYROXENE PHASE EQUILIBRIA AT LOW PRESSURE

J. Stephen Huebner

CHAPTER 5, Assemblages of naturally occurring quadrilateral
      pyroxenes, continued

CHAPTER 6.   PHASE EQUILIBRIA OF PYROXENES AT PRESSURES > 1 ATMOSPHERE

Donald H. Lindsley

CHAPTER 9.   THE COMPOSITION SPACE OF TERRESTRIAL PYROXENES - INTERNAL
         AND EXTERNAL LIMITS   Peter Robinson

CHAPTER 9, continued

CHAPTER 10. PYROXENE MINERALOGY OF THE MOON AND METEORITES

James J. Papike

Chapter 1

INTRODUCTION   Charles T. Prewitt

Fourteen years ago the American Geological Institute sponsored a
Short Course on Chain Silicates. At that time, a substantial amount was
known about the crystal chemistry and phase equilibria of pyroxenes, and
this knowledge has been of fundamental importance in guiding research on
pyroxenes in the years following the A.G.I. Short Course. In 1966,
single-crystal x-ray diffractometry was well advanced and good crystal
structure refinements were available for jadeite, spodumene, hypersthene,
clinoferrosilite, orthoferrosilite, and omphacite; the distinction between the clinoenstatite (pigeonite) and diopside (augite) structures
had been established, and the structure of protoenstatite was known, although some doubt existed about the space group of protoenstatite.
Phase diagrams for several joins in the pyroxene quadrilateral had been
published, but often equilibrium had not been established in the experiments and not enough was known about the effects of pressure, oxygen
fugacity, and non-quad elements such as aluminum on the phase equilibria.
Also, inversion relations of Ca-poor pyroxenes were not well understood,
and petrologists had just become aware of the effect of stress on ortho-
to-clinopyroxene transitions.

In 1966 few of us would have guessed how much new data and new analytical results would become available in the next fourteen years. Although most, if not all, of the important instrumental techniques we use
today were available in 1966, the truly spectacular development and application of these techniques did not take place until the Apollo 11
samples and the attendant funding from NASA became available. Pyroxene
research has profited immensely from the application of Mössbauer, optical, and infrared spectroscopy, x-ray and electron diffraction,
transmission electron microscopy, automated electron microprobes, and
digital computers. During these years experimentalists extended the
capabilities of their equipment to examine the behavior of pyroxenes
under conditions of controlled oxygen fugacity, pressure, and temperature, conditions more nearly like those under which pyroxenes crystallize in natural systems.

Looking back, one remembers the excitement of seeing the first

1

lunar samples. We were surprised at the large amounts of pigeonite and the quality of crystals unaffected by water or the presence of sodium. The influence of the lunar program on pyroxene research was extraordinary, and our understanding of pyroxene relationships in terrestrial occurrences benefited tremendously because the lunar pyroxenes provided a basis for comparison with the more complex chemical and structural behavior of terrestrial environments. Probably the most impressive development in the early lunar sample studies was the application of transmission electron microscopy to mineralogy. We were able to see exsolution and other textural features in crystals that looked homogeneous in the optical microscope, thus opening up a wide range of research possibilities that had not existed previously. Advanced crystal growth experiments, detailed phase equilibria, x-ray diffraction at high temperatures, and statistical analyses of microprobe data were all applied to lunar pyroxenes and then extended to terrestrial and meteorite investigations, making this period one of the most productive in history.

In the compilation of this volume, an attempt has been made to review the essential aspects of pyroxene research, primarily those of the last ten or fifteen years. Although the largest fraction of pyroxene research has been performed in the U.S.A., significant advances have been made in other countries, particularly in Europe, Japan, Canada, and Australia, with interest and activity in these countries probably growing at a faster rate than in the United States. Recently, Deer, Howie and Zussman published a second edition of their volume in the *Rock-Forming Minerals* series, *Single-Chain Silicates*, Vol. 2A (John Wiley, New York, 1978). The present volume is intended to be complementary to DHZ and to provide material covered lightly or not at all in DHZ, such as electron microscopy, spectroscopy, and detailed thermodynamic treatments. However, because the range of pyroxene research has grown so much in recent years, there still are important areas not covered comprehensively in either of these volumes. Some of these areas are kinetics, diffusion, crystal defects, deformation, and non-silicate pyroxene crystal chemistry. Because of these omissions and because this volume is intended for use with the M.S.A. Short Course on Pyroxenes to be held at Emory University in conjunction with the

November, 1980 meeting of the Society, a Symposium on Pyroxenes was organized by J. Stephen Huebner for the meeting that is designed to present the latest research results on several different topics, including those above. With DHZ, this volume, and publications from the Symposium, the student of pyroxenes should be well-equipped to advance our knowledge of pyroxenes in the decades ahead.

Finally, it should be noted that the contributions in this volume were written under very strict time limitations so that the volume would be ready to distribute at the Short Course. Because of this, there was not time for the careful manuscript review and editions that normally are associated with scholarly publications. However, the Editor hopes that timely publication will more than offset problems associated with any errors that may remain.

Chapter 2
CRYSTAL CHEMISTRY of SILICATE PYROXENES
Maryellen Cameron & James J. Papike

INTRODUCTION

The general structure of pyroxenes has been known since 1928, when
Warren and Bragg showed that the monoclinic $C2/c$ diopside structure is
characterized by single tetrahedral chains that occur in layers parallel
to (100). A few years later, Warren and Modell (1930) solved the structure of orthorhombic $Pbca$ hypersthene by noting the close correspondence,
except for the doubled a axis, between its unit cell parameters and those
of diopside. The monoclinic $P2_1/c$ structure type was predicted by Ito
(1950), who postulated that $Pbca$ orthopyroxene was a "space-group twin"
of clinopyroxene. He reasoned that the orthopyroxene structure is produced by regular and repeated twinning of the unit cell of clinopyroxene
by operation of a b-glide parallel to (100). Subsequently, Morimoto et
$al.$ (1960) published the first descriptions of $P2_1/c$ pyroxene (pigeonite
and clinoenstatite) structures. Details of another orthorhombic structure type with $Pbcn$ symmetry was first described by Smith (1959). This
space group is restricted to compositions near $MgSiO_3$, and includes a
nonquenchable polymorph of enstatite that exists only at elevated temperatures ($\sim 1000°C$). Pyroxenes with other space groups such as $C2$
(spodumene; Clark et $al.$, 1969), $P2$ (omphacite; Clark and Papike, 1968),
$P2/n$ (omphacite; Matsumoto et $al.$, 1975, and Curtis et $al.$, 1975), and
$P2_1ca$ (lunar orthopyroxenes; Smyth, 1974a, and Steele, 1975) have also
been reported, but in general such occurrences are limited. The latter
space groups are symmetrical subgroups of either $C2/c$ or $Pbca$ (e.g.,
Ohashi and Finger, 1974a; Matsumoto, 1974; W. Brown, 1972). The structure of all of the pyroxenes mentioned above can be described in terms
of alternating tetrahedral and octahedral layers that lie parallel to
the (100) plane. Within the tetrahedral layer, each T tetrahedron shares
two corners with adjacent tetrahedra to form infinite chains parallel to
the c axis (Fig. 1). The base of each tetrahedron lies approximately in
the (100) plane and the repeat unit in each chain consists of two tetrahedra with the formula $(TO_3)^{-2}$. The octahedral layer contains 6- to 8-co-
ordinated M cations. The different symmetries among the various structure
types are a result of different stacking sequences of the octahedral
layers and/or of symmetrically-distinct tetrahedral chains.

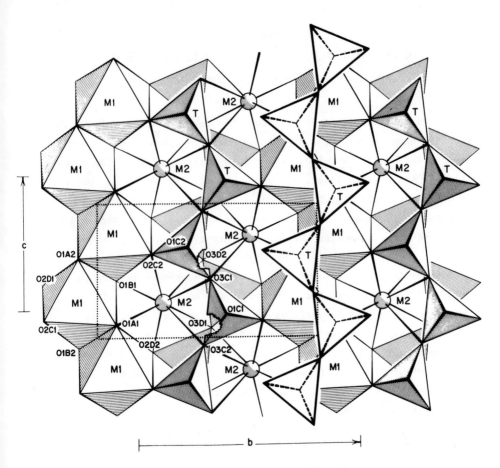

Figure 1. The crystal structure of $C2/c$ clinopyroxene (diopside) projected onto the (100) plane (after Cameron and Papike, 1981). Atom nomenclature is that proposed by Burnham *et al.* (1967).

Since the initial descriptions of the four basic structure types appeared, scores of refinements of both end-member and disordered pyroxenes have provided excellent, detailed data on structural variations as a function of composition, temperature, and pressure. Within the last 15 years, approximately 80 high-quality, three-dimensional structure refinements of terrestrial, lunar, and meteoritic pyroxenes were published. The majority of the refinements involve $C2/c$ clinopyroxenes and $Pbca$ orthopyroxenes, but $P2_1/c$ structures are also well represented. Few data are available on the $Pbcn$ pyroxenes that constitute the fourth major structure type. In addition, within the last two to seven years,

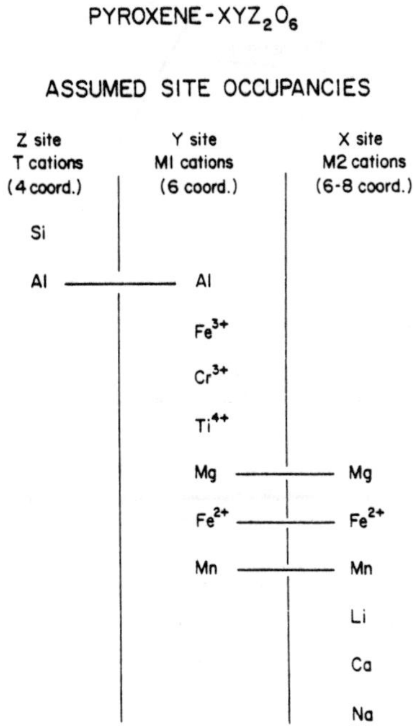

PYROXENE-XYZ_2O_6

ASSUMED SITE OCCUPANCIES

Z site T cations (4 coord.)	Y site MI cations (6 coord.)	X site M2 cations (6-8 coord.)
Si		
Al ————————	Al	
	Fe^{3+}	
	Cr^{3+}	
	Ti^{4+}	
	Mg ————————	Mg
	Fe^{2+} ————————	Fe^{2+}
	Mn ————————	Mn
		Li
		Ca
		Na

Figure 2. Distribution of common cations in the rock-forming pyroxenes.

structure refinements using data collected at temperatures up to 1000°C and pressures up to 53 kbar have become available. In the discussion that follows, we plot, compile, and summarize much of the data (exclusive of abstracts) that appears in these papers. Other reviews of the chemical and structural variations in pyroxenes were presented by Appleman *et al.* (1966), Zussman (1968), Smith (1969), G. M. Brown (1972), Morimoto (1974), Deer *et al.* (1978), and Cameron and Papike (1981).

CHEMICAL CLASSIFICATION AND NOMENCLATURE

The general formula for pyroxene can be expressed as XYZ_2O_6, where X represents Na, Ca, Mn^{2+}, Fe^{2+}, Mg, and Li in the distorted 6- to 8-coordinated M2 site; Y represents Mn^{2+}, Fe^{2+}, Mg, Fe^{3+}, Al, Cr, and Ti in the octahedral M1 site; and Z represents Si and Al in the tetrahedral site (Fig. 2). Ferric iron also enters the tetrahedral site under certain bulk composition, temperature, pressure, and oxygen fugacity conditions (e.g., Huckenholz *et al.*, 1969). Chromium usually occurs as Cr^{3+} and titanium as Ti^{4+}, but under the reducing conditions that obtained on the moon and in meteorites Cr^{2+} and Ti^{3+} may occur. The cations mentioned above are the most common ones in the rock-forming pyroxenes; however, others do occur in trace amounts or as major constituents in synthetic pyroxenes.

The pyroxene nomenclature that we use generally follows the scheme adopted by Deer *et al.* (1978). The major chemical divisions, which are based on occupancy of the M2 site, are given in Table 1. Multiple space groups listed after some of the entires in Table 1 indicate different polymorphs; for example, enstatite has three polymorphs with symmetries *Pbca*, $P2_1/c$, and *Pbcn*. Structural details are given below, but for the P-T conditions of synthesis and relationships among the various polymorphs

7

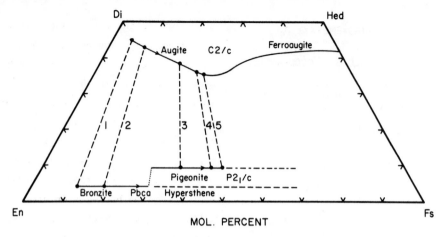

Figure 3. The pyroxene quadrilateral (after Brown, 1967). Dashed lines 1-5 represent tie-lines joining equilibrium pairs of high-calcium and low-calcium pyroxenes.

Table 1. Major Chemical Subdivisions of Pyroxenes
(After Deer et al., 1978)

1. Magnesium - Iron Pyroxenes

Enstatite	$Mg_2Si_2O_6$	Pbca, P2$_1$/c, Pbcn**
Ferrosilite	$Fe_2^{2+}Si_2O_6$	Pbca, P2$_1$/c
Orthopyroxene	$(Mg,Fe^{2+})_2Si_2O_6$	Pbca
Pigeonite	$(Mg,Fe^{2+},Ca)_2Si_2O_6$	P2$_1$/c, C2/c

2. Calcium Pyroxenes

Augite	$(Ca,R^{2+*})(R^{2+},R^{3+},Ti^{4+})(Si,Al)_2O_6$	C2/c
Diopside	$CaMgSi_2O_6$	C2/c
Hedenbergite	$CaFe^{2+}Si_2O_6$	C2/c
Johannsenite	$CaMnSi_2O_6$	C2/c

3. Calcium - Sodium Pyroxenes

Omphacite	$(Ca,Na)(R^{2+},Al)Si_2O_6$	C2/c, P2/n, P2
Aegirine-Augite	$(Ca,Na)(R^{2+},Fe^{3+})Si_2O_6$	C2/c

4. Sodium Pyroxenes

Jadeite	$NaAlSi_2O_6$	C2/c
Acmite	$NaFe^{3+}Si_2O_6$	C2/c
Ureyite	$NaCr^{3+}Si_2O_6$	C2/c

5. Lithium Pyroxenes

Spodumene	$LiAlSi_2O_6$	C2 ($\sim$C2/c)

*R^{2+} = Mn^{2+}, Fe^{2+}, Mg; R^{3+} = Fe^{3+}, Cr^{3+}, Al
**Multiple entries indicate polymorphs having identical composition.

8

the reader is referred to Papike and Cameron (1976), Iijima and Buseck (1975), Buseck and Iijima (1975), Smith (1969), and Burnham (1965).

Most naturally-occurring pyroxenes plot in the $Mg_2Si_2O_6$(En)-$Fe_2Si_2O_6$(Fs)-$Ca_2Si_2O_6$(Wo) system and can be depicted schematically in the pyroxene quadrilateral shown in Figure 3. The four end-members are diopside $CaMgSi_2O_6$, hedenbergite $CaFe^{2+}Si_2O_6$, enstatite $Mg_2Si_2O_6$, and ferrosilite $Fe_2^{2+}Si_2O_6$. Although the quadrilateral pyroxenes can be characrerized chemically by (Wo, En, Fs) percentages, we use the more detailed nomenclature proposed by Poldervaart and Hess (1951), and adopted by Deer *et al.* (1978), when referring to specific pyroxenes described in the literature. Intermediate members of the solid solution series near the diopside-hedenbergite join are termed augite or ferroaugite. All calcium pyroxenes along or near this join have *C2/c* symmetry. Magnesium-iron pyroxenes (i.e., those with compositions near the base of the quadrilateral) that have *Pbca* space group symmetry are referred to as orthopyroxenes whereas those that have $P2_1/c$ symmetry are referred to as pigeonite, clinohypersthene, etc. The reader is referred to Deer *et al.* (1978) for a discussion of frequently-used pyroxene names (e.g., the varieties of orthopyroxene-bronzite, hypersthene, etc., or varieties of Ca pyroxenes--fassaite, salite, titanaugite, etc.). The nomenclature and end-member components for the chemically-complex omphacitic pyroxenes are given in Figure 4 (after Clark and Papike, 1968).

PYROXENE TOPOLOGY

The schematic I-beam diagrams (Fig. 5) of Papike *et al.* (1973) summarize the topologic differences among the four principal structure types of pyroxenes. These diagrams, which are based on the "ideal" models presented by Thompson (1970), depict the pyroxene structure as tetrahedral-octahedral-tetrahedral "I-beam" units whose infinite dimension lies parallel to *c*. In each I-beam, two tetrahedral units point inward and are cross-linked by octahedrally-coordinated cations. These tetrahedral-octahedral-tetrahedral units are highly stylized, and the correspondence between them and a real pyroxene structure is shown in Figure 6.

The symbols within the I-beam units provide information on the symmetry and orientation of individual coordination polyhedra. The A's and B's of the tetrahedral layers refer to two symmetrically-distinct chains:

9

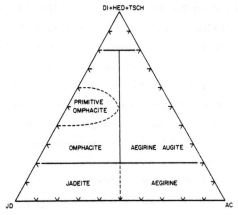

Figure 4. End-member components for omphacitic pyroxenes (after Clark and Papike,
1968). JD = jadeite $NaAlSi_2O_6$, AC = acmite
$NaFe^{3+}Si_2O_6$, DI = diopside $CaMgSi_2O_6$, HED =
hedenbergite $CaFe^{2+}Si_2O_6$, and TSCH = Ca
Tschermak's pyroxene $CaAlSiAlO_6$.

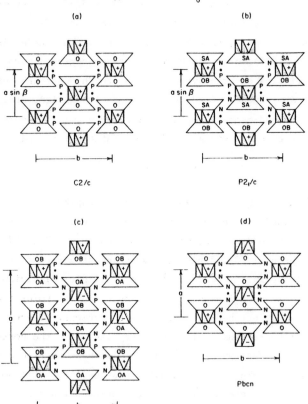

Figure 5. I-beam diagrams of four pyroxene
structure types (after Cameron and Papike,
1981). See text for explanation of symbols.

10

that is, chains that are kinked by different amounts and/or those whose tetrahedra are distorted differently. The absence of this notation indicates that the chains in adjacent layers within one I-beam unit are symmetrically equivalent and are related by 2-fold axes of rotation parallel to b. The O or S notation within the tetrahedral layers refers to the rotational aspect of the chains, as originally defined by Thompson (1970) for regular polyhedra. The completely rotated (i.e., O3-O3-O3 = 120°) O and S configurations shown in Figure 7 are based on a close-packed arrangement of oxygen atoms with a tetrahedral to octahedral edge ratio of 1:1. Cubic close-packing of oxygen atoms (ABCABC...) produces a tetrahedral-octahedral configuration referred to by Thompson (1970) as an O-rotation, whereas hexagonal close-packing produces an S rotation. In an O-rotation, the basal triangular faces of the tetrahedra (those approximately parallel to the bc plane) have an orientation opposite to the triangular faces of the octahedral strip to which they are linked through apical O1 oxygen atoms. In an S-rotation, the triangular faces of the octahedra and tetrahedra that are joined through O1 have the same orientation. The completely-rotated O and S configurations represent the geometric extremes produced by rotating tetrahedra in the chains in opposite directions about imaginary lines passing through oxygen O1 and perpendicular to the (100) layer. Fully extended chains (O3-O3-O3 = 180°; Fig. 7c) are possible only in an ideal structure with a tetrahedral to octahedral edge ratio of $\sqrt{3}$:2. In the structures of real silicate pyroxenes, the tetrahedral chains approach and achieve full extension (O3-O3-O3 = 180°), but they are never kinked by an amount as extreme as 120° (Fig. 8; Table A1[1]).

The positive (+) and negative (-) symbols in the octahedral layers refer to the "skew," "tilt," or direction of stagger of the layer with respect to a right-handed set of crystallographic axes. Within a single layer, each octahedron has a pair of triangular faces that lie approximately parallel to (100). The apices of the upper and lower face of each pair are oriented in an opposite sense, but all triangular faces on one side of an octahedral layer point in the same direction (Figs. 1, 7). In a positive (+) octahedral strip, the apices of the *upper* triangular faces

----

[1]The prefix "A" indicates that the table is located in the Appendix.

11

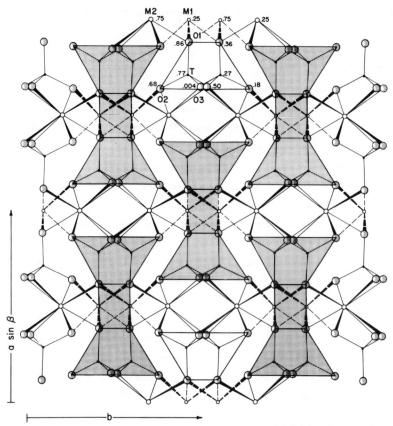

Figure 6. The crystal structure of $C2/c$ pyroxene projected down [001] (after Cameron and Papike, 1981). Shaded areas outline I-beam units depicted in Figure 5.

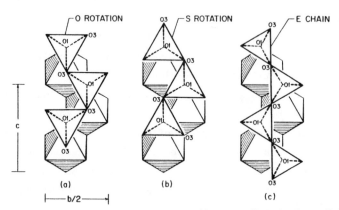

Figure 7. Complete O and S rotations and extended (E) chain configuration for an ideal pyroxene structure (after Cameron and Papike, 1981). Note difference in size of tetrahedra in (a) and (b) vs (c).

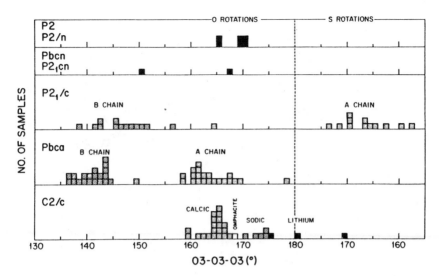

Figure 8. Histogram showing the range in 03-03-03 tetrahedral chain angles for the major structure types of *silicate* pyroxenes.

point in a $+c$ direction, i.e., away from the eye of the reader in Figure 5; in a negative (-) strip, the apices point in a $-c$ direction, i.e., toward the eye of the reader in Figure 5.

Thompson (1970) pointed out in his parity rule that because of geometrical considerations, there are only certain ways that tetrahedral layers can be combined in a structure. His statement follows: "This rule derives from the regularity of the polyhedra and affects the nature of the rotations of adjacent tetrahedral strips in a given tetrahedral layer. If two such strips are both rotated in the same sense then the two octahedral strips (one above and one below the tetrahedral layer) to which they are joined across (100) must both have a 'tilt' or skew of the same sense. If the rotations are in opposite senses then the tilts must be in opposite senses."

Geometrical models for ideal, completely-rotated pyroxene structures (Fig. 9) show that violation of the parity rule results in a mismatch between tetrahedral and octahedral layers. All of the tetrahedral layers in the *Pbcn* pyroxene structure and the A layers in the *Pbca* pyroxene structure contain parity violations. Examination of the tetrahedra within an A layer of the *Pbca* structure shows that all have the same sense of rotation (O) even though the octahedral layers on either side of it have skews of the opposite sense. Despite the parity violation in real *Pbcn* and *Pbca*

13

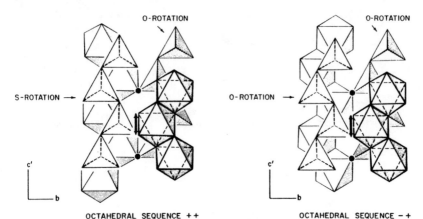

OCTAHEDRAL SEQUENCE + +   OCTAHEDRAL SEQUENCE − +

Figure 9. Stacking sequences for ideal, completely-rotated pyroxene structures. Arrows in both figures show the degree of mismatch between tetrahedra and octahedra in hypothetical structures with parity violations (after Papike *et al.*, 1973).

(a)   (b)   (c)

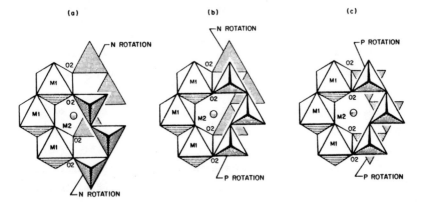

Figure 10. Configuration of tetrahedral chains around the M2 site in an ideal pyroxene structure (after Sueno *et al.*, 1976). See text for discussion of N and P rotations.

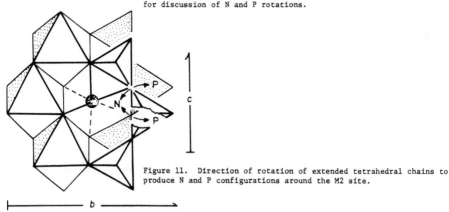

Figure 11. Direction of rotation of extended tetrahedral chains to produce N and P configurations around the M2 site.

14

pyroxene structures, linkage between adjacent octahedral and tetrahedral layers is achieved largely by extension or straightening of the A tetrahedral chains relative to the B chains and by distortion of the cation polyhedra. Papike and Ross (1970) suggested a similar mechanism for the amphibole gedrite. Veblen and Burnham (1978) also discussed in considerable detail the tetrahedral misfit in various pyriboles (including pyroxenes).

The P and N symbols (after Sueno *et al.*, 1976) between the I-beams describe the structural configuration around the M2 site (Fig. 10). They refer to the relative orientations of triangular faces parallel to (100) of octahedra and tetrahedra that are joined laterally through O2 oxygen atoms. (Recall that for O and S rotations, the relative orientations of triangular faces of tetrahedra and octahedra joined through O1 oxygen atoms were important.) In an N configuration, the basal triangles of the lateral tetrahedral chain point in a direction opposite that of the octahedral faces to which they are joined through O2 atoms. In a P configuration, the triangular faces of the tetrahedra and octahedra joined through O2 are similarly oriented. Figure 11 shows the direction of rotation required to produce N and P configurations starting with E-chains. This N-P notation is identical to the U·D notation presented by Papike *et al.* (1973). For example, a D·U or U·D combination in a horizontal row is equivalent to an N symbol, whereas a U·U or D·D combination in a horizontal row produces a P symbol. The N-P symbol thus describes the relative orientations between an octahedral layer and the tetrahedral chains above and below it. It also provides information on the number of shared edges and size of the M2 polyhedron. The complete symbol for each M2 site includes two letters and a dot, which represents the position of the M2 cation. Each N indicates an edge shared between the M2 octahedron and a tetrahedron. The M2 octahedron in those structures with an $\frac{N}{N}$ configuration is relatively small because it shares two edges with tetrahedra. The $\frac{P}{P}$ configuration produces the largest (most open) M2 coordination polyhedron and appears to be the most stable arrangement because no polyhedral edges are shared. In ideal close-packed structures that exhibit no parity violations, only combinations of O with P and S with N are possible.

Referring again to Figure 5 (and Table 2), we can now examine systematically the differences among the four pyroxene structure types.

Table 2. Summary of Characteristics of Silicate Pyroxene Structure Types.

Symmetry (example)	Monoclinic C2/c (diopside)	Monoclinic P2₁/c (pigeonite)	Monoclinic P2/n (omphacite)	Monoclinic P2 (omphacite)	Monoclinic C2 (spodumene)[a]	Orthorhombic Pbca (hypersthene)	Orthorhombic Pbcn (protopyroxene)
Number of different tetrahedral chains	1	2	1	2	1	2	1
Number of different tetrahedra in each chain	1	1	2	2	2	1	1
Total number of different tetrahedra	1	2	2	4	2	2	1
Number of tetrahedral layers	1	2	1	1	1	2	1
Observed tetrahedral chain rotations	all O	A chain = S[b] B chain = O	all O	all O	all O	A chain = O B chain = O	all O
Number of different M1-O octahedral chains	1	1	1	2	1	1	1
Number of different M1 octahedra in each chain	1	1	2	2	2	1	1
Total number of different M1 octahedra	1	1	2	4	2	1	1
Coordination of M2 cation	8	6	8	8	6	6	6
Number of octahedral layers	1	1	1	2	1	1	1
Skew of octahedra	+++...	+++...	+++...	+++...	+++...	++--++...	+-+-...
Octahedral bands	linked	separate	linked	linked	separate	separate	separate

[a]Published refinement of spodumene (e.g., Clark et al., 1969) was calculated using space group C2/c.

[b]One exception is the CaFe pyroxene studied by Ohashi et al. (1975).

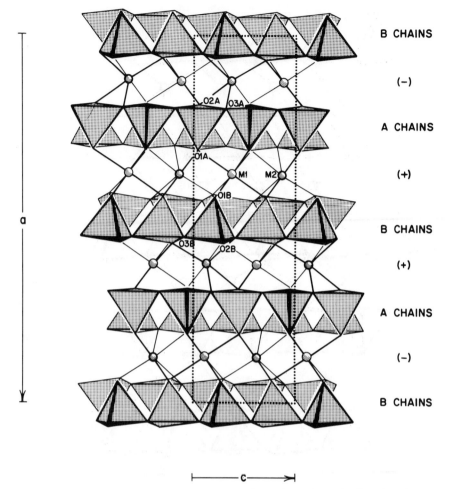

B CHAINS

(−)

A CHAINS

(+)

B CHAINS

(+)

A CHAINS

(−)

B CHAINS

a

c

Figure 12. The crystal structure of *Pbca* orthopyroxene (orthoferrosilite) projected onto the (010) plane (after Burnham, 1966). Note symmetrically-distinct chains in adjacent tetrahedral layers and different skews of the octahedral layers.

The monoclinic $C2/c$ pyroxene structure has octahedral stacking sequence $(+c/3)(+c/3)(+c/3)$... and all O (or all S) rotations of the tetrahedral chains. There is only one type of chain, and those in adjacent tetrahedral layers are related by a 2-fold axis of rotation parallel to b. In this ideal model, the tetrahedra and M2 octahedra share no edges. The orthorhombic *Pbca* structure has octahedral stacking sequence $(+c/3)(+c/3)(-c/3)(-c/3)(+c/3)$..., which produces zero displacement parallel to c for each four octahedral layers (Fig. 12). There are two symmetrically distinct tetrahedral chains, and those in the A layer,

17

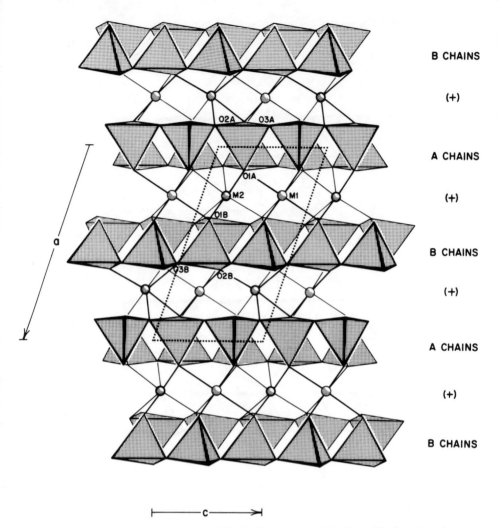

B CHAINS

(+)

A CHAINS

(+)

B CHAINS

(+)

A CHAINS

(+)

B CHAINS

Figure 13. The crystal structure of $P2_1/c$ clinopyroxene (clinoferrosilite) projected onto the (010) plane (after Burnham, 1966). Note symmetrically-distinct chains in adjacent tetrahedral layers.

which is located between octahedral layers with different skews, exhibit parity violations. The M2 octahedron shares one edge with adjoining tetrahedra. The monoclinic $P2_1/c$ structure has octahedral stacking sequence $(+c/3)(+c/3)(+c/3)\ldots$ (Fig. 13). It contains symmetrically-distinct chains in adjacent tetrahedral layers: The A chain is S-rotated and the B chain is O-rotated. No parity violations are present, and the M2 octahedron shares one edge with a tetrahedron as in the *Pbca* structure. The orthorhombic *Pbcn* space group has the stacking sequence

18

$(+c/3)(-c/3)(+c/3)(-c/3)$... and one type of tetrahedral chain, which is
O-rotated. Because the octahedral skew changes in adjacent layers, all
of the tetrahedral layers exhibit parity violations. The M2 octahedron
shares two edges with adjacent tetrahedra.

The stacking arrangements and possible space groups for pyroxenes
were also examined by Pannhorst (1979) and Law and Whittaker (1980).
Pannhorst (1979) presented a new classification in which pyroxene structures are described in terms of various stacking sequences of layer-like
subunits, the smallest of which are termed O layers (oxygen layers).
Law and Whittaker (1980) examined further Thompson's (1970) models for
both pyroxenes and amphiboles.

STRUCTURAL DETAILS

The coordination polyhedra in all pyroxene structures include 4-
coordinated tetrahedra that contain the T cations and 6-8-coordinated
polyhedra that contain the M cations. In the four principal structure
types there are two types of M sites, labelled M1 and M2 (Table A2). In
$C2/c$ and $Pbcn$ space groups, these sites occur in special positions on 2-
fold axes of rotation parallel to b, whereas in the $Pbca$ and $P2_1/c$ structures they occupy general positions. In the $P2/n$ space group reported
for omphacite, there are four symmetrically-distinct M sites, each of
which occupies a special position on a 2-fold axis of rotation parallel
to b. The T cations occupy general positions in all five space groups.
There is only one symmetrically distinct T site in the $C2/c$ and $Pbcn$
structures, whereas there are two in the other three space groups
(Table 2). In the $Pbca$ and $P2_1/c$ pyroxene structures, the more extended chain with the smaller tetrahedra is referred to as the A chain,
and the more kinked chain with the larger tetrahedra is referred to as
the B. chain. Except for the $P2/n$ pyroxenes, there is only one type of
tetrahedron within a given tetrahedral layer and adjacent tetrahedra
along the chain are related by a c glide. The $P2/n$ pyroxene structure
has only one type of chain, but adjacent tetrahedra within each chain
are not equivalent. Characteristics of the $P2$ and $C2$ pyroxene structures are also given in Table 2.

The basic coordination of both anions and cations in the four
principal opacc groups varies only slightly. Each symmetrically

19

Table 3. Average Interatomic Distances in Ordered Silicate Pyroxenes. Crystal Data Given in Table A1.

No.	Pyroxene Name	Space Group	Mean T-O[a]			Mean M1-O			Mean M2-O			Reference
			T cation	Distance (Å)	Q.E.[b]	M cation & Coordination	Distance (Å)	Q.E.	M cation & Coordination	Distance (Å)	Q.E.	
1.	$CaNiSi_2O_6$	C2/c	Si	1.634	—	Ni(6 oxy)	2.072	—	Ca(8 oxy)	2.494	—	Schlenker et al. (1977)[c]
2.	diopside	C2/c	Si	1.634	1.0067	Mg(6 oxy)	2.077	1.0052	Ca(8 oxy)	2.498	—	Clark et al. (1969)
3.	$CaCoSi_2O_6$	C2/c	Si	1.634	—	Co(6 oxy)	2.102	—	Ca(8 oxy)	2.506	—	Schlenker et al. (1977)[c]
4.	hedenbergite	C2/c	Si	1.635	1.0059	Fe^{2+}(6 oxy)	2.130	1.0045	Ca(8 oxy)	2.511	—	Cameron et al. (1973)
5.	johannsenite	C2/c	Si	1.644	1.0055	Mn^{2+}(6 oxy)	2.173	1.0052	Ca(8 oxy)	2.530	—	Freed and Peacor (1967)
6.	jadeite	C2/c	Si	1.623	1.0055	Al(6 oxy)	1.928	1.0151	Na(8 oxy)	2.469	—	Prewitt and Burnham (1966)
7.	ureyite	C2/c	Si	1.624	1.0037	Cr(6 oxy)	1.998	1.0088	Na(8 oxy)	2.489	—	Clark et al. (1969)
8.	acmite	C2/c	Si	1.628	1.0033	Fe^{3+}(6 oxy)	2.025	1.0135	Na(8 oxy)	2.518	—	Clark et al. (1969)
9.	$NaScSi_2O_6$	C2/c	Si	1.632	1.0036	Sc(6 oxy)	2.102	1.0110	Na(8 oxy)	2.564	—	Hawthorne and Grundy (1973)
10.	$NaInSi_2O_6$	C2/c	Si	1.632	1.0026	In(6 oxy)	2.141	1.0106	Na(8 oxy)	2.568	—	Hawthorne and Grundy (1974)
11.	spodumene[d,e]	C2/c	Si	1.618	1.0047	Al(6 oxy)	1.919	1.0149	Li(6 oxy)	2.211	1.2177	Clark et al. (1969)
12.	$LiFe^{3+}Si_2O_6$	C2/c	Si	1.620	1.0030	Fe^{3+}(6 oxy)	2.031	1.0135	Li(6 oxy)	2.249	1.231	Clark et al. (1969)
13.	$LiScSi_2O_6$	C2/c	Si	1.624	1.0030	Sc(6 oxy)	2.107	1.0132	Li(6 oxy)	2.289	1.2399	Hawthorne and Grundy (1977)
14.	$Zn_2Si_2O_6$	C2/c	Si	1.618	1.0029	Zn(6 oxy)	2.145	1.0204	Zn(4 oxy)	1.982	—	Morimoto et al. (1975)
15.	Ca tschermakite[d]	C2/c	½Si, ½Al	1.686	1.0089	Al (6 oxy)	1.947	1.0144	Ca(8 oxy)	2.461	—	Okamura et al. (1974)

Table 3. Cont'd. (page 2 of 2 pages).

No.	Pyroxene Name	Space Group	Mean T-O[a]			Mean M1-O			Mean M2-O			Reference
			T cation	Distance (Å)	Q.E.[b]	M cation & Coordination	Distance (Å)	Q.E.	M cation & Coordination	Distance (Å)	Q.E.	
30.	orthoenstatite	Pbca	SiA SiB	1.628 1.640	1.0099 1.0052	Mg(6 oxy)	2.076	1.0089	Mg(6 oxy)	2.149	1.0488	Hawthorne and Ito (1977)
31.	$Zn_2Si_2O_6$	Pbca	SiA SiB	1.621 1.629	1.0037 1.0040	Zn(6 oxy)	2.128	1.0181	Zn(6 oxy)	2.265	--	Morimoto et al. (1975)
32.	orthoferrosilite	Pbca	SiA SiB	1.626 1.637	1.0071 1.0046	Fe^{2+}(6 oxy)	2.135	1.0087	Fe^{2+}(6 oxy)	2.228	1.0706	Sueno et al. (1976)
57.	clinoenstatite	$P2_1/c$	SiA SiB	1.628 1.638	1.0080 1.0050	Mg(6 oxy)	2.078	1.0086	Mg(6 oxy)	2.142	1.0430	Ohashi and Finger (1976)
55.	clinoferrosilite	$P2_1/c$	SiA SiB	1.623 1.635	1.0058 1.0051	Fe^{2+}(6 oxy)	2.137	1.0083	Fe^{2+}(6 oxy)	2.224	1.0641	Burnham (1966)
68b.	$Mn_2Si_2O_6$	$P2_1/c$	SiA SiB	1.632 1.637	1.0057 1.0039	Mn(6 oxy)	2.177	1.0088	Mn(6 oxy)	2.281	1.0746	Tokonami et al. (1979)
77.	protopyroxene	Pbcn[f]	Si	1.622	1.0073	Mg(6 oxy)[f]	2.096	1.0106	Mg(6 oxy)[f]	2.186	1.142	Smyth and Ito (1977)

[a] Interatomic distances refer to room temperature structures except for protoenstatite.
[b] Quadratic elongation parameter of Robinson et al. (1971).
[c] From Ribbe and Prunier (1977)
[d] Space group may be C2.
[e] True formula for crystal studied is $(Li_{0.952}Fe^{3+}_{0.048})(Fe^{3+}_{0.952}Fe^{2+}_{0.048})Si_2O_6$
[f] M1 contains Sc and M2 contains Li. See Table A5 for site occupancies.

21

distinct T cation is coordinated by one O1, one O2, and two O3 oxygen
atoms (Fig. 1; Table A3). The O1 anions are referred to as apical
oxygen atoms and the O3 anions as bridging oxygen atoms because they
are shared between adjacent tetrahedra in the chains. The M1 cation
is coordinated by four O1 and two O2 anions that have a fairly regular
octahedral configuration. The coordination of M2 varies from 6 to 8
and depends on the size of the cation occupying the site. It ranges
from eight when M2 is occupied by the large Ca and Na atoms ($C2/c$
structures) to seven (some $P2_1/c$ pigeonites) and six (orthorhombic
$Pbca$ pyroxenes or Li-rich $C2/c$ or $C2$ pyroxenes) when it is occupied by
smaller Fe and Mg (or Li) atoms. Although there are minor variations
among the different space groups, in the $C2/c$ pyroxenes the O1 oxygen
atom is coordinated by two M1, one M2, and one T cation; O2 by one M1,
one M2, and one T cation; and O3 by two M2 and two T cations in the Ca
and Na series and by one M2 and two T cations in the Li series. In
terms of classical Pauling bond strengths, O1 is approximately charge-
balanced, O2 is highly underbonded, and O3 is highly overbonded (Table
A3). The apparent charge imbalances are largely eliminated by variations in cation-oxygen bond distances: For example, bonds to the
underbonded O2 oxygen atoms are typically shortened (strengthened)
whereas those to the overbonded O3 atoms are lengthened (weakened).
The bond strengths and valency sums reported for pyroxenes by Clark
et $al.$ (1968, 1969), Ferguson (1974), and Hawthorne and Ito (1977) are
in accord with these observations. The calculations in each of these
three studies included a modification of Pauling's original electro-
neutrality principle in which variations in cation-anion distances were
also taken into account.

The tetrahedral layer

In the end-member pyroxene structures, the mean T-O bond length of
Si-bearing tetrahedra ranges from 1.618 Å to 1.644 Å (Table 3). The
shortest of the four tetrahedral distances (1.585 Å - 1.612 Å) is usually
Si-O2 whereas the longest involves the bridging O3 anions, with the exception of CaTs and the $C2/c$ Li pyroxenes. The mean of the bridging
bond distances is larger than that of the non-bridging distances, and
in general the difference between the two means (Δ) is smallest for

22

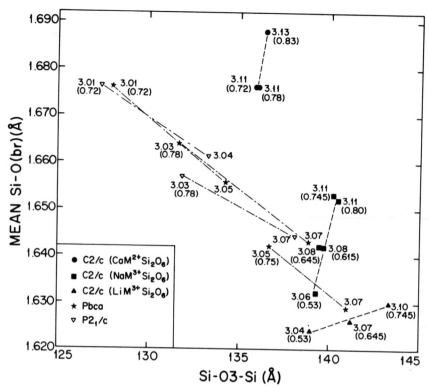

Figure 14. Variation of mean Si-0(br) bonds with Si-03-Si angles (after Cameron and Papike, 1981). Samples plotted include 1-14, 30-32, 55, and 57 (Table A1). Dashed lines are regression curves for $C2/c$ pyroxenes with M2 = Li (triangles), M2 = Na (squares), and M2 = Ca (solid circles). Dot-dashed lines connect data points for A and B chains in $Pbca$ pyroxenes (stars) and $P2_1/c$ pyroxenes (inverted triangles). Numbers associated with each symbol are distances (Å) between adjacent Si cations within the same tetrahedral chain. Numbers in parentheses are ionic radii for M1 cations in $C2/c$ pyroxenes and M1 + M2 in $Pbca$ and $P2_1/c$ pyroxenes.

the $C2/c$ Li pyroxenes with monovalent M2 cations and largest for the $C2/c$ Ca pyroxenes with divalent M2 cations. The Δ's for the $Pbca$ and $P2_1/c$ structures are intermediate between these two groups. Among the various pyroxene structures, shorter bridging distances are associated with larger Si-03-Si (Fig. 14) and 03-03-03 angles. The Si-Si distances within the chains vary between 3.01 Å and 3.13 Å, with the nearest approaches occurring in the highly kinked B chains of clino- and orthoenstatite (Fig. 14). The 3.01 Å distances are close to the lower limit of the Si-Si non-bonded contacts discussed by Hill and Gibbs (1979). In the paragraphs that follow, we discuss some of the factors that are believed to contribute to the distortion and bond length variations in silicate tetrahedra. Specifically, we consider the effects of shared polyhedral edges, of the size and electronegativity of nontetrahedral cations, and of the coordination and local electrostatic environment of the oxygen atoms. In general, the relative importance of the various factors cannot be specified, and in many instances the variations are suitably explained either on the basis of a modified version of Pauling's electrostatic valence rule (e.g., Baur, 1971) or on the basis of generally more realistic covalent bonding models (e.g., Gibbs et al., 1972; Hill and Gibbs, 1979).

We used the quadratic elongation parameter of Robinson et al. (1971) to characterize systematically the tetrahedral distortion in the various pyroxenes (Table 3). In the $C2/c$ structures, the tetrahedra of the Ca pyroxenes exhibit the largest distortion whereas those of the Li pyroxenes have the smallest. Within each of the three groups, the quadratic elongation parameter decreases with increasing radius of the M1 cation. In the $Pbca$ and $P2_1/c$ structures, tetrahedra in the iron end-members are slightly less distorted than those in the magnesium end-members (Table 3), and in addition, the A tetrahedra are more distorted than the B. The greater distortion of the A tetrahedron is largely a result of the cation-cation repulsion across the O-O edge shared with the M2 coordination polyhedra. The approach of M2 and SiA is much closer (2.78 Å - 2.85 Å) than that of either M2 and SiB or M2 and Si (∿3.10 Å to ∿3.30 Å). In the $Pbca$ pyroxenes, the increased distortion also reflects the response of the structure to the parity violation in the A tetrahedral layer. This distortion of individual tetrahedra within the chains is one of the mechanisms that promotes linkage between the tetrahedral and octahedral layers (Papike et al., 1973) and is discussed is more detail in a later section.

The variation in Si-O(br) bond lengths can be explained on the basis of local charge balance considerations. Clark et al. (1969) related the increase in bridging distances in the three groups of $C2/c$ pyroxenes directly to the strength of the M2-O bonds. They stated that "Each O3 bridging oxygen coordinates two M2 cations (except when M2 is Li), and assuming the electrostatic attraction between O3 and Ca^{2+} to be greater than that between O3 and Na^+, then the Si-O3 bridging bonds would be expected to lengthen in the Ca clinopyroxenes." Ribbe and Prunier (1977), in addition, noted the positive correlation in $C2/c$ pyroxenes between the length of the Si-O(br) bonds and the sum of classical Pauling bond strengths (s) at the O3 anion. For example, in the Ca group the bond

24

strength at the O3 oxygen = 2.50 valence units because each O3 is bonded to two Si and two Ca atoms. For the Na pyroxenes, s at O3 = 2.25 v.u., and for the Li pyroxenes, s at O3 = 2.17 v.u. From these values we would expect the $C2/c$ Li group to exhibit the shortest Si-O3 bonds and the smallest differences between the average bridging and non-bridging distances, as is observed. In the CaTs pyroxene, the relatively short bridging O3 bonds can be explained as a result of their being bonded to two (Si+Al) rather than to two Si cations.

The variation in Si-O bond lengths can also be rationalized by covalent bonding models such as extended Hückel molecular orbital (EHMO) theory (e.g., Brown et $al.$, 1969; Gibbs et $al.$, 1972; Tossell and Gibbs, 1977). Analysis of the TO_4^{n-} oxyanions by EHMO theory involves computation of Mulliken bond overlap populations, n(Si-O), for individual bond distances. These n(Si-O) terms are related to the electron density between two bonded atoms, and larger overlap populations imply higher electron densities, greater binding forces, and hence are associated with shorter Si-O bond lengths. Because shorter bond lengths necessarily induce larger overlap populations during the calculations, each Si-O distance is usually set to a fixed value close to 1.63 Å. With this induced correlation eliminated, variations in n(Si-O) and therefore bond length are attributed to the effects of geometrical factors such as O-Si-O and Si-O-Si angles. In general, computations showed that larger n(Si-O) and shorter Si-O bonds are associated with larger Si-O-Si angles (e.g., Gibbs et $al.$, 1972), with larger O-Si-O angles (e.g., Louisnathan and Gibbs, 1972a), and with larger $<O-Si-O>_3$, which is the mean of the three O-Si-O angles common to a bond (e.g., Louisnathan and Gibbs, 1972b). The change in bond overlap populations is apparently related to concomitant changes in both the σ- and π-bonding potentials of the atoms involved (Gibbs et $al.$, 1972). In Figure 14, we show the variation of mean Si-O(br) bond distance with Si-O-Si angle for selected end-member pyroxenes. The overall trend is as expected, even though the tetrahedral angles are undoubtedly affected by bonding of the O3 anions to non-tetrahedral cations. Shorter Si-O(br) bonds are associated with wider Si-O-Si angles.

Several researchers discussed the relationship between electronegativity and variations both in the Si-O bond length and in the difference between the Si-O(br) and Si-O(nbr) distances (McDonald and

25

Cruickshank, 1967; Brown and Gibbs, 1969, 1970; Baur, 1971). In 1967 McDonald and Cruickshank suggested that the formation of covalent bonds to non-tetrahedral cations would diminish the pi bonding potential of the silicate ion, and as a consequence the difference between the bridging and non-bridging bonds should decrease as electronegativity of these cations increases. Brown and Gibbs (1969) noted that in $C2/m$ amphiboles longer Si-O(nbr) bonds are associated with the more electro-negative cations, but Baur (1971) concluded that their arguments were not convincing since the comparisons did not involve strictly isostruc-tural compounds. In the same study, Baur compared four pairs of iso-structural pyroxenes, each with different M1 cations, and concluded that no unambiguous trend attributable to electronegativities is present. We examined, in addition, variations in the Si-O(nbr) distances and Δ's for $C2/c$ pyroxenes along the hedenbergite-ferrosilite join (Ohashi et $al.$, 1975) and for the Fe-Mg series of the $Pbca$ and $P2_1/c$ pyroxenes. In each of these groups, the difference between the bridging and non-bridging Si-O bonds decreases with increasing electronegativity of the non-tetrahedral cations. However, the decrease in Δ is affected as much, if not more, by significant decreases associated with the bridging dis-tances rather than increases in the non-bridging Si-O distances (e.g., Figs. 15b,c, 16b,c).

The mean Si-O distances in the $C2/c$, $Pbca$, and $P2_1/c$ space groups are also influenced by the size of the nontetrahedral cations. The variation is controlled largely by the behavior of the Si-O(br) bond lengths because the Si-O(nbr) distances either remain constant or in-crease slightly with increasing radius of the M1 and M2 cations (Figs. 15, 16). With increasing size of the octahedral layer, the mean Si-O distance in the $C2/c$ pyroxenes increases whereas that in the $Pbca$ and $P2_1/c$ pyroxenes decreases slightly (Fig. 17). In 1969, Morimoto and Koto suggested that the relationship between Si-O distances and the Fe/(Fe + Mg) content of orthopyroxenes was linear. However, Burnham et $al.$ (1971) noted that the mean bridging distances of an orthopyroxene of intermediate composition fell on Morimoto and Koto's curves, but the mean of the non-bridging and hence that of all Si-O bonds did not. Figure 17 shows that there is a definite trend of decreasing mean Si-O distances with increasing Fe content of the octahedral layer, but the

26

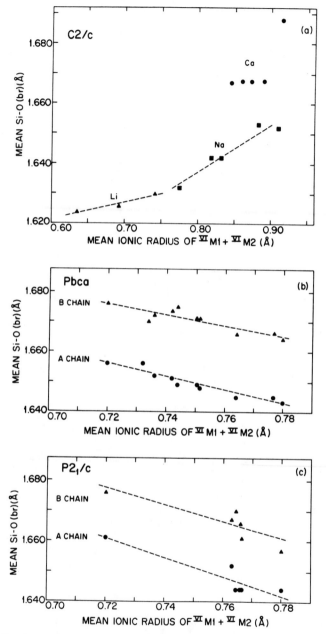

Figure 15. Variation of mean Si-O(br) interatomic distance with mean ionic radius of ($^{VI}M1 + {}^{VI}M2$) (after Cameron and Papike, 1981). Note difference in scale of the abscissa in (a) *vs* (b) and (c). Dashed lines are regression curves. (a) *C2/c* pyroxenes. Triangles represent Li varieties; squares, Na varieties; circles, Ca varieties. Samples plotted include pyroxenes 1-13 (Table A1). (b) *Pbca* pyroxenes. Triangles represent the B chain and circles, the A chain. Samples plotted include pyroxenes 30, 32, 35-38, 41, 43, 44, 46. (c) *P2$_1$/c* pyroxenes. Triangles represent the B chain and circles the A chain. Samples plotted include pyroxenes 55, 57, 61, 64, 66, 67.

27

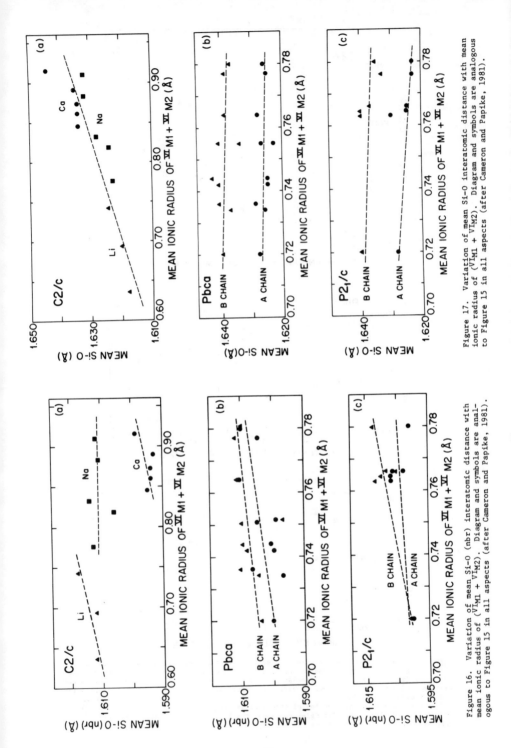

Figure 16. Variation of mean Si-O (nbr) interatomic distance with mean ionic radius of ($^{VI}M1 + ^{VI}M2$). Diagram and symbols are analogous to Figure 15 in all aspects (after Cameron and Papike, 1981).

Figure 17. Variation of mean Si-O interatomic distance with mean ionic radius of ($^{VI}M1 + ^{VI}M2$). Diagram and symbols are analogous to Figure 15 in all aspects (after Cameron and Papike, 1981).

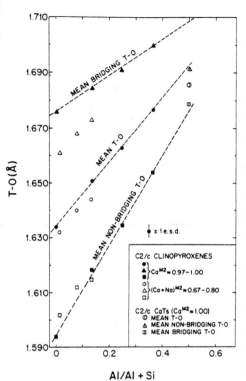

Figure 18. Variation of T-O distances with Al/(Al + Si). Dashed lines are regression curves for the four samples shown in solid symbols. Samples plotted include pyroxenes 2, 25-27 (solid symbols), 19, 20, 24 (open symbols) and 15 (open symbols with vertical lines) from Table A1 (after Cameron and Papike, 1981).

relationship in the *Pbca* pyroxenes may not be linear as originally proposed. A similar, but less well defined, decrease occurs in both chains of the $P2_1/c$ structure. In the $C2/c$ pyroxenes, the increase in Si-O bond lengths is much more pronounced than the variation in either the *Pbca* or $P2_1/c$ structures. The systematic increase in size of the tetrahedra probably reflects both perturbations caused by the increasing size of the octahedral layer and differences in distortion of the tetrahedra. Brown and Shannon (1973) noted that mean bond lengths are approximately a linear function of bond length distortion. The general variations in the $C2/c$ pyroxenes are in accord with this observation; that is, the Ca pyroxenes have the largest and most distorted tetrahedra and the Li pyroxenes have the smallest and least distorted tetrahedra.

The variation in T-O distances *vs* Al/(Al + Si) is linear for $C2/c$ pyroxenes whose M2 sites are filled almost entirely by Ca. In Figure 18, we show only the variation for mean T-O, mean T-O(br), and mean T-O(nbr) distances, even though a linear relationship also exists for T-O2 distances (Clark *et al.*, 1968). The three open symbols associated with each regression curve in Figure 18 show how the relationship changes when there is less than one Ca (≈ 0.75 Ca + Na) per formula unit. The synthetic CaTs ($CaAlSiAlO_6$) studied by Okamura *et al.* (1974) does not lie on two of the three regression lines when it is plotted at Al/(Al + Si) = 0.5, and Hazen and Finger (1977a) suggested that the short range Al-Si ordering reported by Grove and Burnham (1974) might explain its deviation.

In the *Pbca* pyroxenes, aluminum concentrates in the TB tetrahedron (Takeda, 1972b; Kosoi *et al.*, 1974; Brovkin *et al.*, 1975) because TB is

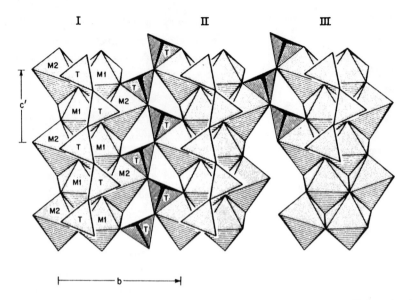

Figure 19. Projection of $P2_1/c$ clinoenstatite onto a plane (∿100) parallel to the chains of octahedra (after Zussman, 1968). Edge-sharing M1 and M2 polyhedra form wide bands (labelled I, II, III) whose long dimension is parallel to c. Note the continuous rift between bands within the octahedral layer.

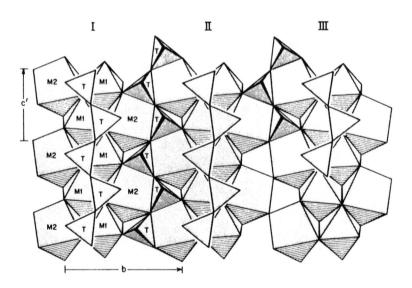

Figure 20. Projection of $C2/c$ diopside onto a plane (∿100) parallel to the chains of octahedra (after Zussman, 1968). The M2 polyhedra associated with different M1 octahedral chains (labelled I, II, III) share edges to form continuous sheets parallel to (100).

inherently larger than TA and because it shares no edges with M2.   This
explanation was first used by Papike and Ross (1970) for the amphibole
gedrite.   The distribution of tetrahedral Al is inferred largely from
the T-O distances.   In those orthopyroxenes that contain tetrahedral Al,
the mean TA-O bond lengths are relatively constant at approximately 1.627
Å, but the mean TB-O bond lengths increase significantly as the amount of
tetrahedral Al in the formula increases.   The relationship between T-O
bond length and Al content is not as well defined as that for the $C2/c$
clinopyroxenes, possibly because of the presence of submicroscopic exsolution lamellae and complex microstructures as suggested by Hawthorne
and Ito (1977).   They concluded that some of the Al represented in chemical
analyses of orthopyroxenes occurs in incoherently diffracting microstructures, and thus does not contribute to the overall diffraction pattern of
the host orthopyroxene.

The octahedral layer

The M1 octahedron and M2 polyhedron share edges to form either
laterally continuous sheets or wide bands of polyhedra that lie parallel
to the c axis within the (100) plane (Fig. 1).   Each M1 octahedron shares
two edges with other M1 octahedra to form zig-zag "chains" whose infinite
dimension is parallel to c.   The M2 polyhedra lie diagonally off to each
side of these edge-sharing M1 octahedra.   Where the M2 cations are small
and 6-coordinated, the edge-sharing polyhedra form wide bands that are
separated by voids or rifts elongated parallel to c (Fig. 19).   The individual bands are connected in the third dimension by tetrahedral chains
in adjacent layers.   Where the M2 cation is large and 8-coordinated, the
M2 polyhedra in adjacent bands share edges and the continuous rifts between the bands disappear (Fig. 20).   Thus, when M2 is 8-coordinated, the
octahedral layer almost has the appearance of a continuous layer, but when
it is 6-coordinated the octahedral layer consists of discrete bands or
strips of polyhedra separated by elongate continuous rifts.

The 6-coordinated M1 site accommodates divalent, trivalent, and
tetravalent cations with ionic radii ranging from ∿0.53 Å (Al) to ∿0.83
Å (Mn).   The variation in mean M1-O with the radii of constituent M1
cations is shown in Figure 21.   The plot includes data for the 15 ordered
$C2/c$ pyroxenes, 10 disordered $C2/c$ pyroxenes (mostly augites), the Fe and

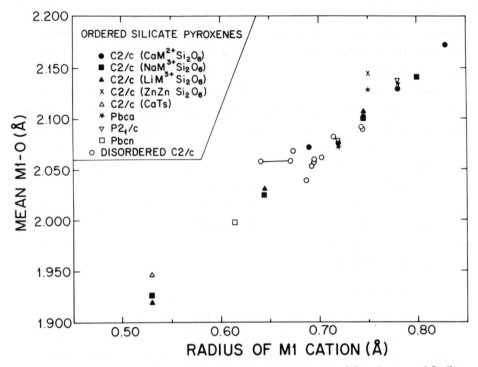

Figure 21. Variation of mean M1-O with radius of the M1 cation (after Cameron and Papike, 1981). Samples plotted include ordered pyroxenes 1-15, 30-32, 54, 55, 69 and disordered pyroxenes 19-27, 29 (Table A1).

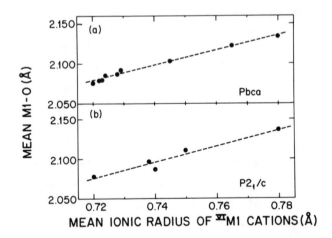

Figure 22. Variation of mean M1-O interatomic distance with mean ionic radius of $^{VI}M1$ cations (after Cameron and Papike, 1981). Pyroxenes shown contain principally (Fe + Mg) in M1 and (Fe + Mg + Ca) in M2. Dashed lines are regression curves. Samples plotted include 30, 32, 36-38, 41, 42, 44, 46 (*Pbca* pyroxenes) and 55, 57, 61, 66, 67 (*P2$_1$/c* pyroxenes). See Table A1.

Mg end-members for the *Pbca* and $P2_1/c$ pyroxenes, and one *Pbcn* proto-pyroxene. This diagram illustrates that the size of the M1 octahedron is dependent upon the radius of the M1 cation (as shown by Ribbe and Prunier (1977) for the *C2/c* pyroxenes) and that it does not vary drastically among the four different structure types. The M1 octahedron appears to be the most important building-block in the pyroxene structure and it has a major effect on the detailed configuration of the M2 polyhedron and on the relative displacement and kinking of the tetrahedral chains linked to it. The trend in Figure 21 can also be used to determine if cation site assignments are grossly in error and if cation valences are properly assigned. For example, the two points connected by the horizontal line represent site occupancies calculated for Ti in two different valence states in a fassaite that is believed to contain considerable trivalent Ti (Dowty and Clark, 1973). The point on the left represents the mean ionic radius for an M1 occupancy of 0.48 Ti + 0.39 Mg + 0.13 Al assuming that all of the Ti is tetravalent. The point on the right, which was calculated with all trivalent Ti, lies closer to the overall trend of the data and appears to be more reasonable. Dowty and Clarke (1973) concluded that this fassaite from the Allende meteorite contains 0.14 atoms of Ti^{4+} and 0.34 atoms of Ti^{3+}. The aberrant point (symbolized by an X), which plots above the curve at a radius = 0.75, represents a synthetic *C2/c* zinc polymorph ($ZnSiO_3$) that has a rather distorted M1 octahedron and a tetrahedrally coordinated M2 cation (Morimoto *et al.*, 1975).

The variation of mean M1-O with mean radius of the M1 cations for the disordered *Pbca* and $P2_1/c$ pyroxenes containing principally Fe and Mg is shown in Figure 22. The scale of the abscissa is expanded relative to Figure 21 in order to show the variation in detail. The regression curves for both structure types are almost parallel and that for the *Pbca* pyroxenes is displaced only very slightly toward lower mean M1-O values at identical cation radii. When more than two cation species are present, the trend for the *Pbca* pyroxenes shows more scatter, but is still well defined. In general, the synthetic Mg, Mn, Co, Zn orthopyroxenes (Hawthorne and Ito, 1977, 1978; Morimoto *et al.*, 1975) plot near the curve for the natural Fe-Mg *Pbca* orthopyroxenes.

Figure 23 is a plot of observed mean M1-O *vs* predicted M1-O distances for the *C2/c* pyroxenes. The predicted M1-O is the sum of Shannon

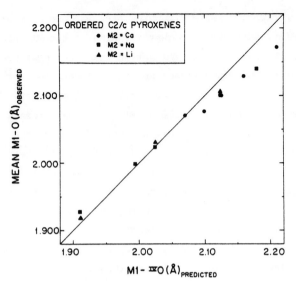

Figure 23. Variation of observed mean M1-O interatomic distance with M1-O predicted from the sum of the ionic radius of the M1 cation and four-coordinated oxygen (after Cameron and Papike, 1981). The line drawn at 45° represents the hypothetical situation in which observed M1-O equals M1-O predicted from the sum of the radii.

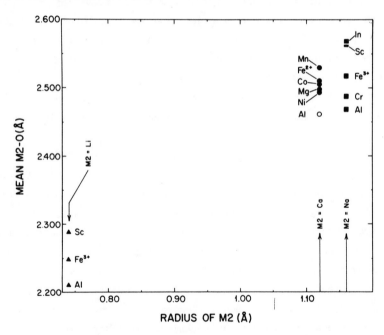

Figure 24. Variation of mean M2-O interatomic distance with radius of the M2 cation in ordered $C2/c$ pyroxenes (after Ribbe and Prunier, 1977). Triangles represent Li pyroxenes; circles, Ca pyroxenes; and squares, Na pyroxenes. Elemental labels associated with each data point indicate M1 occupancy. Samples plotted include 1-13, 15 (Table A1).

and Prewitt's (1969, 1970) ionic radii for the constituent cations and 4-coordinated oxygen. The important feature to note is that the trends for the calcic, sodic, and lithium pyroxenes each cut diagonally across the 45° line representing ideal predicted interatomic distances. The location of the lithium and sodic pyroxenes above the curve at short M1-O distances (= small M1 cations) indicates that the observed values are larger than the predicted, thus implying that the oxygen framework around the M1 site may be held open for small cations. Similarly, at longer M1-O distances, the observed values are smaller than predicted, suggesting that the octahedra are somehow constrained.

The M2 site is 6-, 7-, or 8-coordinated, depending on the size of the cation occupying the site. It is coordinated by four non-bridging (two O2 and two O1) oxygen atoms in all of the structures; however, the number of O3 atoms varies between two and four depending on the relative displacement and kinking of the tetrahedral chains above and below the M2 site. For example, the $C2/c$ structures that contain large M2 cations such as Ca or Na have four O3 oxygen atoms coordinating M2 because tetrahedral chains in adjacent layers are superimposed closely above each other. On the other hand, those $C2/c$ pyroxenes with a small cation such as Li in M2 have only two O3 atoms within the first coordination sphere because the tetrahedral chains in adjacent layers are displaced by relatively large amounts in the c direction. The M2 site in the $Pbca$ structure is 6-coordinated whereas that in the $P2_1/c$ structures have been described as either 6- or 7-coordinated. Morimoto and Güven (1970) considered the M2 site in pigeonite to be 7-coordinated whereas other authors (e.g., Clark et $al.$, 1971; Güven, 1969) described it as 6-coordinated. In many $P2_1/c$ structures (e.g., clinoenstatite, Ohashi and Finger, 1976; pigeonite, Clark et $al.$, 1971; pigeonite, Ohashi and Finger, 1974b), the effective coordination is probably better described as six rather than seven because SiA approaches M2 more closely than does the seventh coordinating oxygen.

The radius of cations that occupy the M2 site in silicate pyroxenes ranges from $\sim$0.72 Å (Mg^{VI}) to $\sim$1.16 Å (Na^{VIII}). In $C2/c$ pyroxenes, the mean M2-O distance increases with increasing radius of both the M2 and M1 cations. For identical M2 occupancy, the mean M2-O distance increases as size of the M1 cation increases (Fig. 24). For example, in the $C2/c$

sodic pyroxenes where M2 = Na, the mean M2-O varies from 2.469 Å (M1 = Al; r_{Al} = 0.53 Å) to 2.568 Å (M1 = In; r_{In} = 0.80 Å). These variations in mean M2-O were also noted by Clark *et al.* (1969) and Ribbe and Prunier (1977), and the latter correlated them to the size of the M1 cation and to the size and charge of the M2 cation. Figure 25 shows the variation of mean M2-O distances with mean radius of cations in M2 of disordered *Pbca* and $P2_1/c$ pyroxenes. Although there is a positive trend for both space groups, the overall relationships are not as good as those for the M1 site. Some of the scatter may be due to the effects of chemically different microstructures in pyroxenes of these space groups, as Hawthorne and Ito (1977) suggested for the orthopyroxenes.

The relationship between mean M-O distances and the Fe-Mg content of orthopyroxene has been discussed repeatedly in recent pyroxene literature. There is general agreement that the size of the M1 and M2 octahedra increases with substitution of Fe for Mg, but the linearity of the relationships is a matter of controversy. Morimoto and Koto (1969) assumed a linear relationship between orthoenstatite and orthoferrosilite and produced two equations that they felt could be used to determine Fe-Mg ratios in the M1 and M2 sites. The hypersthene data of Ghose (1965) did not lie on the curves, but they re-interpreted his site occupancies using their equations. In a footnote at the end of the same paper, Morimoto and Koto stated that Mössbauer work supported the original site occupancies in Ghose's orthopyroxene and that the relationships involving (Mg,Fe)-O distances are probably non-linear. Using a few additional data points, Burnham *et al.* (1971) concluded that the relationship for the M1-O distances was non-linear and that the mean M1-O value was an unreliable indicator of site occupancy. They also noted that the presence of calcium and the extreme distortion of the M2 polyhedron precluded use of mean M2-O distances as anything other than crude indicators of site occupancy. Subsequently, Smyth (1973) stated that the average M1-O distance appeared to vary linearly. Kosoi *et al.* (1974) concluded that the average interatomic distances in both the M1 and M2 octahedra were non-linear whereas Morimoto (1974) described the relationship between Fe-Mg content and mean M1-O and M2-O bond lengths as almost linear. Most recently, Hawthorne and Ito (1977) showed that plots of mean M1-O and M2-O bond lengths *vs* constituent cation radius for five Mg, Fe, Zn,

36

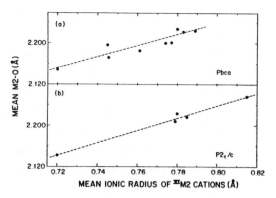

Figure 25. Variation of mean M2-O interatomic distance with mean ionic radius of ^{VI}M2 cations for ordered and disordered pyroxenes of the *Pbca* and *P2₁/c* space groups (after Cameron and Papike, 1981). Dashed lines are regression curves. Samples plotted are the same as those in Figure 22.

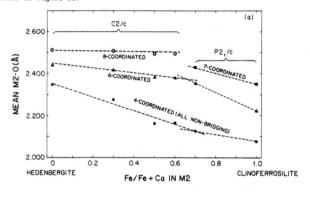

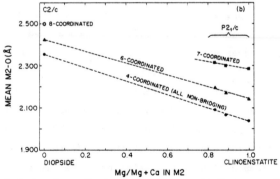

Figure 26. Variation of mean M2-O interatomic distance with M2 occupancy for pyroxenes along the hedenbergite-clinoferrosilite and diopside-clinoenstatite joins (after Cameron and Papike, 1981). In both figures, note that the Ca-rich varieties have space group *C2/c* at room temperature whereas Ca-poor varieties have space group *P2₁/c*. Dashed lines are regression curves. The term non-bridging refers to bonds involving O1 and O2 anions. (a) Variation of mean M2-O with Fe/(Fe + Ca) in M2. Samples plotted include 4, 16-18, 55, 56 (Table A1). (b) Variation of mean M2-O with Mg/(Mg + Ca) in M2. Samples plotted include 2, 57-59.

Cu synthetic orthopyroxenes are extremely non-linear. Their detailed analysis of natural orthopyroxenes indicated that some of the octahedral trivalent cations do not occur in the host orthopyroxenes, but reside in incoherently diffracting microstructures. If these minor trivalent constituents are excluded from the calculation of cation radius, the variation of M1 appears to be linear. Their analysis of bond length variations in the M2 polyhedron was inconclusive because of the difficulty in evaluating precisely the effects of distortion on the bond lengths and, also, because of the ambiguity in assigning calcium to the host crystal vs the incoherently diffracting microstructures.

Ohashi et $al.$ (1975) and Ohashi and Finger (1976) studied at room temperature the structural effects of varying M2 occupancy in a series of synthetic pyroxenes along the $Mg_2Si_2O_6$-$CaMgSi_2O_6$ and $Fe_2Si_2O_6$-$CaFeSi_2O_6$ joins. In both series, M1 occupancy is constant (either all Fe or all Mg), and M2 contains varying proportions of Ca and Fe or Ca and Mg. The Ca end-members of each series have $C2/c$ symmetry whereas the Ca-free end-members have $P2_1/c$ symmetry. In the CaFe-FeFe series, the change in space group at room temperature occurs between $Wo_{15}Fs_{85}$ and $Wo_{20}Fs_{80}$ or where M2 contains approximately 30-40% Ca (Fig. 26a). With decreasing Ca content, the four non-bridging bonds (involving two O1 and two O2 atoms) decrease whereas the four bridging (all O3) bond lengths increase. After the transition to $P2_1/c$ symmetry, the bonds to non-bridging oxygen atoms continue to decrease whereas the bridging oxygens split into two pairs, one of which continues to increase and one of which decreases. Ohashi et $al.$ (1975) also concluded that the geometry of the M2 site does not represent an average structure because variations in the M2 quadrupole splitting indicate that the local configuration around the ferrous ion changes as the bulk chemical composition changes. The variation in mean bond lengths in the CaMg-MgMg series is shown in Figure 26b. Although the data are very limited, Ohashi and Finger (1976) suggested that the $P2_1/c$ to $C2/c$ transition takes place at a composition more Ca-rich than that in the CaFe-FeFe series, assuming that the chain angles vary in a similar manner in both series.

By any measure of distortion, the M1 octahedron in all pyroxene structures is much more regular than the M2 polyhedron. Distortion about

M1 usually involves short interatomic distances to the underbonded O2 oxygen atoms. Among the $C2/c$ pyroxenes, the Ca pyroxenes have the least distorted M1 octahedra but there appears to be no well-defined relationship between size of the cation occupying M1 and distortion as measured by the quadratic elongation parameter (Table 3). The Fe- and Mg-bearing M1 octahedra in the $C2/c$, $Pbca$, and $P2_1/c$ space groups have similar quadratic elongation values, and the octahedra containing Fe^{2+} may be slightly more regular than those containing Mg. In the M2 polyhedra, interatomic distances to the bridging O3 oxygen atoms are usually longer than those to the non-bridging O1 and O2 anions. In the $Pbca$, $P2_1/c$, and $Pbcn$ pyroxene structures, the shortest M2 bonds are those involving O2 anions whereas in the $C2/c$ pyroxenes the shortest distances involve bonds to either O2 or O1 oxygen atoms. Hawthorne and Grundy (1977) discussed in detail the variation in Na and Li clinopyroxenes in terms of size of the cation occupying M1. Quadratic elongation parameters of 6-coordinated M2 octahedra in the $Pbca$ and $P2_1/c$ pyroxene structures are similar for both the Fe and Mg end-members, and the Fe-bearing M2 octahedra are more distorted than the Mg-bearing M2 octahedra. In both structure types, the irregular oxygen configuration around the M2 site relative to M1 is inherent to the pyroxene structure itself and it is augmented only slightly by the presence of Fe.

The distribution of cations between M1 and M2 is influenced by the configuration of anions around each site and by the ionic radius and electronegativity of the cations involved. (The M-site cations are given in Table A4, and individual site occupancies are listed in Table A5.) In all pyroxene structures, the M2 polyhedron is larger and more distorted than M1 (Table 3); thus, we expect it to be enriched, relative to M1, in the larger cations and in those transition metal cations that are stabilized in more distorted environments (Burns, 1970). The preference of Fe^{2+} for M2 over M1 in orthopyroxene was attributed by Ghose (1965) and Burnham et al. (1971) to the slightly enhanced covalent character of the Fe^{2+}-O bonds in the M2 site. In the $C2/c$ pyroxenes, larger cations such as Ca, Na, and Mn are usually assigned to M2 and the trivalent and tetravalent cations to M1 on the basis of size considerations. The same assignment is usually followed for the low-Ca pyroxenes ($P2_1/c$, $Pbca$, $Pbcn$ space groups) even though these cations are usually minor

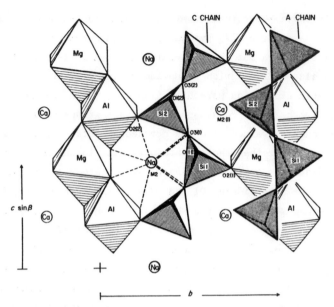

Figure 27. Projection of the $P2$ omphacite structure down the a crystallographic axis (after Clark and Papike, 1968). Note the alternation of Mg and Al within the zig-zag M1 octahedral chains and the presence of symmetrically-distinct tetrahedral chains within one tetrahedral layer.

in abundance. In addition, in these three space groups the general preference of intermediate-sized cations for M2 is Mn > Zn > Fe^{2+} > Co > Mg (Table A5). The distribution of cations in $P2$ omphacite[2] (after Clark and Papike, 1968) is shown in Figure 27. The alternation of Mg($+Fe^{2+}$) and Al($+Fe^{3+}$) along the M1 octahedral chains produces two types of M2 environments. The M2 site that is partially surrounded by two (Al/Fe^{3+}) and one (Mg,Fe^{2+}) octahedra contains approximately two-thirds Na and one-third Ca to maintain local charge balance. The other M2 site, which is surrounded by two (Mg,Fe^{2+}) and one (Al,Fe^{3+}) octahedra, contains approximately two-thirds Ca and one-third Na. A similar distribution was noted in the $P2/n$ omphacite studied by Curtis et al. (1975).

----

[2]There is disagreement over the true space group of ordered omphacite. Clark and Papike (1968) and Clark et al. (1969) refined the structures of two omphacites in space group $P2$. Matsumoto et al. (1975) and Fleet et al. (1978) re-examined these specimens and concluded that the true space group is $P2/n$ rather than $P2$. Matsumoto et al. (1975) and Curtis et al. (1975) both published refinements of additional $P2/n$ omphacites. Subsequently, Carpenter (1978) reported that selected area electron diffraction patterns provided evidence for $P2/n$, $P2/c$, and $P2$ space groups in different parts of ordered omphacite crystals. Regardless of the space group, the basic features of the $P2$ and $P2/n$ structure types are the same.

40

Linkage between the octahedral and tetrahedral layers

The majority of cation substitutions in pyroxenes occur in the octahedral layer. Such substitutions affect the size of the layer and hence
its linkage with the tetrahedral chains. The most obvious means of com-
pensating the misfit simply involves extension or kinking of the O3-O3-O3
chain angle. Additional adjustments include distortion of the tetrahedra
or a change in "out-of-plane" tilting (Cameron *et al.*, 1973). The latter
mechanism involves tilting of the tetrahedra by movement of the O2 atoms
farther from or closer to the plane containing the O3 atoms. At present
there is no published data discussing the *relative* importance of chain
straightening, tetrahedral distortion, and out-of-plane tilting in main-
taining the linkage between the octahedral and tetrahedral layers. Variations involving the first two mechanisms are discussed in detail below,
and Ohashi and Finger (1974c) reported that increased out-of-plane tilting
is highly correlated with increased kinking of the tetrahedral chain.

In the *Pbca* and $P2_1/c$ structures, the chain angle increases significantly as the size of the octahedral layer, as indicated by mean ionic
radius of M1 and M2 cations, increases (Figs. 28, 29). In the *Pbca* orthopyroxenes, the A chain straightens about 10° and the B chain about 5° as
the mean radius of M1 + M2 increases from 0.72 Å to 0.78 Å. The greater
rate of change in the A chain may be related to the increasing mismatch
of the octahedral and tetrahedral layers caused by the parity violation
in the A layers. The B chain appears to have a maximum O3-O3-O3 angle
of ∿145°. In the $P2_1/c$ clinopyroxene structures both the A and B chains
straighten by ∿17° over the same range in ionic radii. The trend for
$P2_1/c$ pyroxenes (Fig. 29) is less well defined as most of the data points
represent samples with Fe-rich compositions. The trends appear to be
real, however, because a similar increase in chain angles occurs with
the substitution of Ca for Mg in $P2_1/c$ structures along the $Mg_2Si_2O_6$-
$CaMgSi_2O_6$ join (Ohashi and Finger, 1976). In both the *Pbca* and $P2_1/c$
structures, straightening of the tetrahedral chains is accompanied by
a reduction in both mean Si-O and bridging Si-O distances, as discussed
previously. The significant reduction in Si-O(br) distances (∿0.12-0.14
Å for *Pbca*; ∿0.12-0.19 Å for $P2_1/c$) is accompanied by small increases in
O3-Si-O3 angles and decreases in O3-O3 distances in both the A and B
chains.

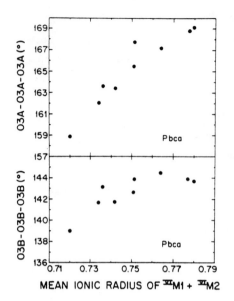

Figure 28. Variation of A and B chain angles with mean ionic radius of $(^{VI}M1 + ^{VI}M2)$ in *Pbca* pyroxenes (after Cameron and Papike, 1981). Samples plotted are the same as those in Figure 22a.

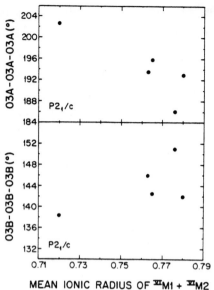

Figure 29. Variation of A and B chain angles with mean ionic radius of $(^{VI}M1 + ^{VI}M2)$ in *P2₁/c* pyroxenes (after Cameron and Papike, 1981). Samples plotted are the same as those in Figure 22b.

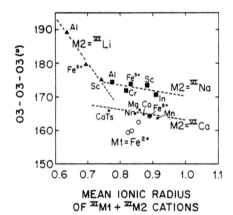

Figure 30. Variation of 03-03-03 chain angle with mean ionic radius of $(^{VI}M1 + ^{VI}M2)$ in *C2/c* pyroxenes. Dashed lines are regression curves for pyroxenes with M2 = Li (triangles), M2 = Na (squares), and M2 = Ca (solid circles). Elemental symbols associated with each data point indicate M1 occupancy. The four open circles refer to a *C2/c* CaFe-FeFe series in which M1 = Fe^{2+} and M2 contains varying amounts of Fe and Ca. Samples plotted include 1-5, 15 (solid circles), 6-10 (squares), 11-13 (triangles), and 4, 16-18 (open circles) from Table A1 (after Cameron and Papike, 1981).

Papike *et al.* (1973) noted a correlation between the mean ionic radius of the M1 and M2 cations and the 03-03-03 tetrahedral chain angle. Using data for eight *C2/c* pyroxenes, they concluded that as the *mean* ionic radius of the M cations decreases, the chain straightens. In Figure 30 we plotted data for 18 *C2/c* pyroxenes. In three series, M2 occupancy is constant (Ca = solid circles, Na = squares, Li = triangles) and in one series M1 occupancy is constant (CaFe-FeFe pyroxenes = open circles). As the size of the M1 cation increases in each of the Ca, Na, and Li series (with the exception of $LiAlSi_2O_6$), the tetrahedral chain kinks. The increased kinking is an obvious result of the increase in size of the M1 octahedron and the O2-O2 octahedral edge (Fig. 1). In a detailed study of the Na and Li pyroxenes, Hawthorne and Grundy (1977) showed that linkage between the tetrahedral and octahedral layers in the Li pyroxenes is accomplished by expansion of the O3-O3 tetrahedral edge and by an increase in the O3-Si-O3 angle, thus producing an increase in chain length without greatly increasing the Si-O(br) distances. With the exception of $NaInSi_2O_6$ (and $?NaCrSi_2O_6$), the tetrahedral chains in the Na pyroxenes straighten by increasing the Si-O(br) distances. In the Ca pyroxenes from M1 = Ni to M1 = Fe^{2+} there is essentially no change in the O3-Si-O3 angles or in Si-O(br) or O3-O3 distances. Between M1 = Fe^{2+} and M1 = Mn, there is a significant increase in both Si-O(br) and O3-O3 distances. An increase in the size of the M2 cation affects the tetrahedral chains in the various $C2/c$ pyroxenes differently. In structures containing divalent M2 cations, such as the CaFe-FeFe series studied by Ohashi et $al.$ (1975), the O3-O3-O3 angle increases as the size of the M2 cation increases. Concomitant with this increase in the chain angle is an increase in the Si-O(br) bonds and a decrease in both the O3-O3 tetrahedral edge and the O3-Si-O3 angle. Among the pyroxenes with monovalent M2 cations such as the pairs LiFe-NaFe and LiSc-NaSc, the O3-O3-O3 angle decreases as the size of M2 increases. Concomitant with this decrease in both pairs is an increase in the Si-O(br) bonds and a decrease in O3-Si-O3. The O3-O3 tetrahedral edge remains approximately constant. As observed for the $Pbca$ and $P2_1/c$ space groups, there is a generally antithetic relationship between O3-O3-O3 tetrahedral chain angle and mean T-O distances. That is, as size of the octahedral layer (= mean radius of cations in M1 + M2) in the $C2/c$ pyroxenes increases, the tetrahedral chain angle decreases and the size of the tetrahedra increase.

Expansion of the octahedral layer in the b direction also requires structural adjustment by the tetrahedral chains. The linkage is accomplished by distortion of the tetrahedra or by a mechanism referred to as c-axis rotation by Hawthorne and Grundy (1977). In a study of the relative importance of these mechanisms for the Li and Na series of $C2/c$ pyroxenes, Hawthorne and Grundy (1977) concluded that tetrahedral edge distortion is more important in the Na pyroxenes whereas c-axis rotation is the mechanism by which octahedral b axis expansion is accommodated in the Li pyroxenes.

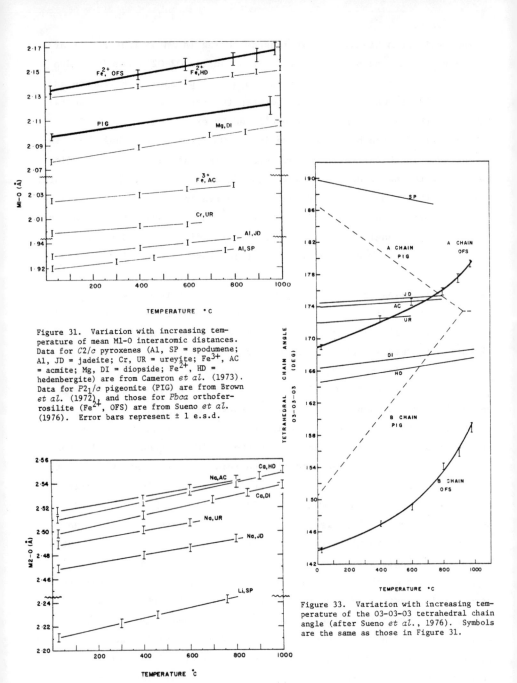

Figure 31. Variation with increasing temperature of mean M1-O interatomic distances. Data for $C2/c$ pyroxenes (Al, SP = spodumene; Al, JD = jadeite; Cr, UR = ureyite; Fe^{3+}, AC = acmite; Mg, DI = diopside; Fe^{2+}, HD = hedenbergite) are from Cameron *et al.* (1973). Data for $P2_1/c$ pigeonite (PIG) are from Brown *et al.* (1972), and those for *Pbca* orthoferrosilite (Fe^{2+}, OFS) are from Sueno *et al.* (1976). Error bars represent ± 1 e.s.d.

Figure 33. Variation with increasing temperature of the O3-O3-O3 tetrahedral chain angle (after Sueno *et al.*, 1976). Symbols are the same as those in Figure 31.

Figure 32. Variation with increasing temperature of mean M2-O interatomic distances in $C2/c$ pyroxenes (after Cameron *et al.*, 1973). Symbols are the same as those in Figure 31. Error bars represent ± 1 e.s.d.

44

Studies of pyroxene structures at elevated temperatures and pressures are necessary to our understanding of such phenomena as exsolution,
solid solution, and phase transitions, which typically do not occur at
room temperature or pressure. To date, six $C2/c$, two $Pbca$, and two $P2_1/c$
pyroxenes have been studied at temperatures up to $\sim$1000°C (Table A6), and
two $C2/c$ pyroxenes have been studied at pressures up to 45 and 53 kbar.

The most important concept to evolve from the high temperature
studies is that of differential polyhedral expansion: with increasing
temperature, polyhedra containing different cations expand at different
rates. In general, smaller mean thermal expansion coefficients are associated with cations having lower coordination number, higher valence,
and higher electronegativity (Cameron et $al.$, 1973). Six $C2/c$ pyroxenes
were examined at a series of temperatures up to 1000°C (Cameron et $al.$,
1973; Finger and Ohashi, 1976). Over the temperature intervals studied,
none of the pyroxenes underwent phase transitions and all exhibited
regular, essentially linear increases in mean M1-O and M2-O interatomic
distances (Figs. 31, 32). The mean Si-O bond lengths in all structures
remained approximately constant with increasing temperature. Mean thermal expansion coefficients for various bonds in the six pyroxenes increase in the following order: Si^{4+}-O < Cr^{3+}-O < Fe^{3+}-O < Al^{3+}-O <
Fe^{2+}-O < Na^{+}-O < Mg^{2+}-O < Ca^{2+}-O < Li^{+}-O. The different expansion rates
of the M and T polyhedra effect additional structural adjustments in
order to maintain linkage between the octahedral and tetrahedral layers
with increasing temperature. Mechanisms for this adjustment include
straightening of the tetrahedral chains ($\sim$2-3° over a maximum of 1000°C;
Fig. 33), small increases in the out-of-plane tilting ($\sim$0.5° for diopside
and ureyite), and very small distortions involving an increase in the
O3-O3 interatomic distance.

Two orthopyroxenes, hypersthene (Smyth, 1973) and orthoferrosilite
(Sueno et $al.$, 1976) were studied at a series of temperatures up to
$\sim$1000°C. With increasing temperature the structural changes in both
pyroxenes were regular and no phase transitions were observed. The
mean Si-O distances showed either essentially no change or perhaps a
slight decrease with increasing temperature. As temperature increased,
both the A and B tetrahedral chains straightened at a significantly

45

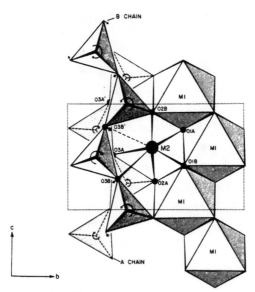

Figure 34. Projection of the $P2_1/c$ low pigeonite structure onto the (100) plane (after Brown *et al.*, 1972). Arrows indicate structural changes involved in the transition to $C2/c$ high pigeonite.

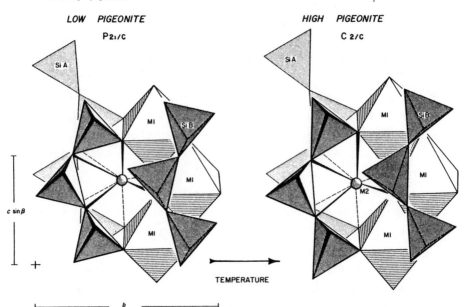

Figure 35. Projection down the a crystallographic axis of the low and high pigeonite structures (after Brown *et al.*, 1972). In low pigeonite, note the kinked configuration of the B tetrahedral chain. In the high pigeonite structure, the A and B tetrahedral chains are symmetrically equivalent.

higher rate than those in the $C2/c$ structures. In orthoferrosilite, the change in 03A–03A–03A was ∿10° whereas 03B–03B–03B changed by ∿15° (Fig. 33). Hypersthene exhibited a change of 5° in the A chain and 15° in the B chain. Maintenance of the linkage between tetrahedral and octahedral layers in these orthopyroxenes principally involves straightening of the tetrahedral chains. The rate of straightening becomes greater at higher temperatures, but at all times the B chain is more kinked than the A chain The pronounced straightening of the tetrahedral chains changes the M2 coordination from six to seven and back to six at the highest temperatures studied.

Two $P2_1/c$ pyroxenes, pigeonite and clinohypersthene, were also examined up to ∿1000°C. Both structures exhibited a low (P) to high (C-centered) transition. Figures 34 and 35 show the structural changes (after Brown *et al.*, 1972) involved with the transition in pigeonite. In Figure 35, note that the B tetrahedral chains in low pigeonite are kinked whereas those in high pigeonite are relatively straight. The pigeonite (En = 39) studied by Brown *et al.* (1972) transformed to $C2/c$ symmetry at ∿960°C whereas the low clinohypersthene studied by Smyth (1974b) exhibited a first-order transition at ∿725°C. The temperature of the transition decreases with increasing Fs component, as noted by Prewitt *et al.* (1971). The clinohypersthene, which was studied at four temperatures below the transition, exhibited a significant decrease in the mean Si–O distances of both the A and B tetrahedra with increasing temperature. As expected, the M1 and M2 polyhedra show much higher rates of expansion relative to the tetrahedra. Over a 700°C temperature interval, the A chain straightened by 7° and the B chain by 8°. In the pigeonite described by Brown *et al.* (1972) the changes in chain angle over 970°C were larger: A∿12° and B∿23° (Fig. 33).

These high temperature studies provide additional insight into the miscibility between the high calcium and low calcium pyroxenes. Refinements of room temperature structures documented significant differences in coordination of the M2 site in augite and pigeonite, and it is generally believed that these different configurations are responsible for the limited miscibility. With increasing temperature, the M2 polyhedra of pigeonite and the calcic pyroxenes (as indicated by the average of the diopside and hedenbergite structural data) become

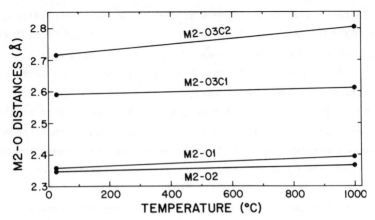

Figure 36. Variation with increasing temperature of M2-O interatomic distances in a hypothetical, average $C2/c$ calcium pyroxene. Data points represent the average of diopside and
hedenbergite M2-O values.

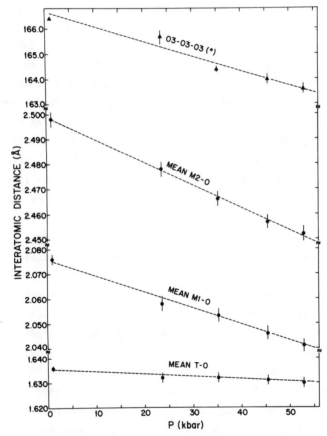

Figure 37. Variation with increasing pressure of mean interatomic distances and O3-O3-O3
chain angle of diopside (data from Levien and Prewitt, in press). Dashed lines are regression curves. Error bars represent ± 1 e.s.d.

much more similar. In the $C2/c$ calcic pyroxenes the change involves the
M2-O3 distances and produces an M2 site that is more 6-coordinated (Fig.
36). The pigeonites, on the other hand, undergo a $P2_1/c$ to $C2/c$ transition with increasing temperature, and specific O3 anions in the first
coordination sphere of M2 are different in the high and low temperature
structures. The structural modifications in both pyroxenes are in a
direction such that solid solution can be accomplished more readily
(Cameron *et al.*, 1973).

Hazen and Finger (1977b) examined the structure of a fassaite at
pressures up to 45 kbar, and Levien and Prewitt (in press) examined
diopside at pressures up to 53 kbar. With increasing pressure, the
mean interatomic distances in both structures decrease regularly (Fig.
37), and the various polyhedra compress differentially. The recent
study by Levien and Prewitt (in press) on end-member diopside is more
precise because of improved experimental methods and because the structure is ordered. Results of this study indicate that the mean compressibility coefficients for interatomic distances increase in the following
order: T-O < M1-O < M2-O.

It is reasonable to assume that increasing pressure and increasing
temperature have generally opposite effects on the crystal structure of
pyroxenes. In addition, topologic changes produced by increasing temperature are expected to be similar to those produced by substitution
of larger cations into a structure. Although these ideas are convenient
generalizations, exceptions are numerous when structural variations are
examined in detail. For example, in the $C2/c$ pyroxenes the mean Si-O
distance remains approximately constant with increasing temperature
whereas it changes significantly with variations in composition of the
octahedral layer. The tetrahedral chain angle in $C2/c$ pyroxenes decreases as the size of cations in the octahedral layer increases, whereas it increases with increasing temperature. Ohashi and Burnham (1973)
pointed out differences in orientation of the strain ellipsoid caused
by increasing temperature and increasing size of constituent cations.
They noted that the direction of largest expansion associated with
addition of Ca to a hedenbergite-ferrosilite series of structures is
about midway between the $+a$ and $+c$. This coincides with the smallest
thermal expansion direction. Levien and Prewitt (in press) showed for

49

diopside that only some of the structural parameters exhibit an inverse relationship with respect to temperature and pressure. Notably, mean M1-O, M2-O, O3-Si-O3, and O3-O3-O3 do, whereas the β cell parameter and mean Si-O do not. The effects of temperature, pressure, and composition on the structures of both silicate and oxide minerals are discussed by Hazen and Prewitt (1977) and Hazen (1977).

UNIT CELL PARAMETERS

The variation of unit cell parameters with chemical composition provides information on both the ideality and the phase relations of a system (e.g., Smith and Finger, 1971; Turnock et al., 1973). Unit cell parameters can also be used to estimate compositions with moderate accuracy

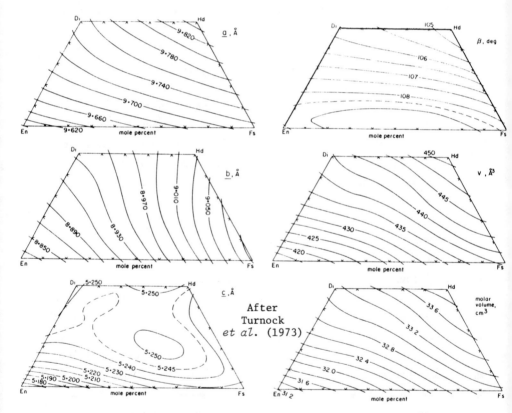

Figure 38. Contours on trend surfaces showing the variation of unit cell parameters for synthetic quadrilateral pyroxenes.

(e.g., 2-5 mole %; see Turnock *et al.*, 1973). The compositional determinations are accurate to the extent that the chemistry of the samples being examined approaches that of the system for which the nomograms were prepared. Thus, the determinative curves for synthetic systems are usually not directly applicable to natural specimens. In addition, the curves for natural pyroxenes should be used only for pyroxenes with a similar paragenesis because cell dimensions reflect both bulk composition and the distribution of atoms among available cation sites (Smith *et al.*, 1969). In Table A7, we have tabulated some of the studies that describe the variation of unit cell parameters with chemical composition. The material in this section is cursory in nature, and the interested reader should consult individual publications for details. Smith *et al.* (1969) summarized and reviewed earlier work on orthopyroxenes, and Ribbe and Prunier (1977) discussed some of the studies on $C2/c$ clinopyroxenes.

Turnock *et al.* (1973) examined the variation of cell parameters for synthetic quadrilateral pyroxenes (Fig. 38). Their study showed that a and c are smallest for clinoenstatite and largest for hedenbergite, that b is sensitive mainly to changes in Fe/Fe + Mg ratios, and that β is affected most by variations in $Ca_2Si_2O_6$ content. Figures 39 and 40 are plots of the β angle *vs* the a and b cell dimensions for ordered, end-member quadrilateral and non-quadrilateral clinopyroxenes. The dotted lines connect pyroxenes with identical M1 occupancy. The Ca, Na, and Li series of the $C2/c$ pyroxenes each have a limited set of β values. As predicted by Whittaker (1960), the magnitude of the angle is dependent upon M2 occupancy, and it is affected little by M1 occupancy. Smith *et al.* (1969) studied the relationship among cell parameters, chemical composition, and site preferences in 30 natural and two synthetic orthopyroxenes. They concluded that an increase in (1) Fe content causes an increase in a, b, and c; (2) Ca content causes a marked increase in a, has little effect on b, and probably increases c; and (3) Al content has little effect on a and c, but causes a strong decrease in b.

Mean thermal expansion coefficients of pyroxene unit cell parameters are given in Table A6. For orthorhombic crystals, the crystallographic axes coincide with the axes of the thermal expansion ellipsoid, and thus the relative values in Table A6 indicate the directions of maximum and minimum thermal expansion of the lattice. In monoclinic crystals, however, the axes of the thermal expansion ellipsoid need not coincide with

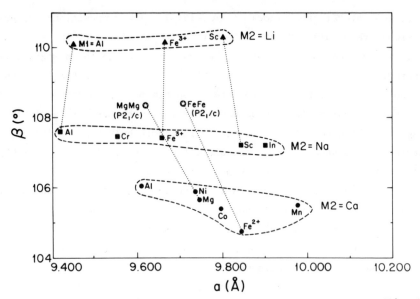

Figure 39. Plot of the β angle *vs* the *a* cell dimension for selected end-member *C*2/*c* and *P*2₁/*c* pyroxenes. Samples plotted include 1-13, 15, 55, 57 (Table A1).

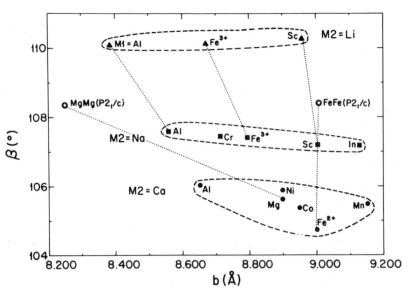

Figure 40. Plot of the β angle *vs* the *b* cell dimension for selected end-member *C*2/*c* and *P*2₁/*c* pyroxenes. Samples plotted are the same as those in Figure 39.

the crystallographic axes. The only constraint is that one axis of the thermal expansion ellipsoid coincide with the b crystallographic axis; the other two ellipsoid axes lie in the (010) plane (similar to an optical indicatrix). Therefore, expansion coefficients for cell parameters in monoclinic crystals may be used to describe relative expansions along the crystallographic axes, but they may not necessarily indicate maximum or minimum expansion within the lattice. Ohashi and Burnham (1973), for example, found that the direction of minimum thermal expansion in Ca-Fe clinopyroxenes was midway between the $+a$ and $+c$ axes in the (010) plane.

SUGGESTIONS FOR FURTHER RESEARCH

The following suggestions for studies are based on our review of recently-published refinements (exclusive of abstracts) of pyroxene structures.

(1) precise structural refinement of clinoferrosilite

(2) precise structural refinement of johannsenite

(3) precise refinement of a series of synthetic orthopyroxenes along the enstatite-ferrosilite join

(4) combined structural and TEM studies on natural pigeonites of varying composition

(5) additional high temperature, structural studies of the enstatite polymorphs

(6) high temperature study of the structural changes accompanying the johannsenite-bustamite transition

(7) high pressure study of additional pyroxene end-members.

ACKNOWLEDGEMENTS

We thank Robin Spencer for her help in preparing the manuscript and in compiling and checking the data tables and references. We are also grateful to Louise Levien and C.T. Prewitt for providing us with some of their high pressure data prior to its publication.

APPENDIX. Tables A1–A7 for Chapter 2, CRYSTAL CHEMISTRY of SILICATE PYROXENES

Table A1. Crystal Data for Pyroxenes Included in Tables and Plotted in Figures.

No.	Name (Composition)[a]	Space Group	Unit Cell Parameters					O3-O3-O3(°)[b]	Source and Locality (Crystal Data)	Reference
			a(Å)	b(Å)	c(Å)	β(°)	V(Å³)			
1.	— $CaNiSi_2O_6$	$C2/c$	9.737	8.899	5.231	105.9	435.9	165.1	synthetic	Schlenker et al. (1977) (in Ribbe and Prunier, 1977)
2.	diopside $CaMgSi_2O_6$	$C2/c$	9.746(4)	8.899(5)	5.251(6)	105.63(6)	438.6(3)	166.38(11)	Gouverneur talc district, Gouverneur, New York	Clark et al. (1969)
3.	— $CaCoSi_2O_6$	$C2/c$	9.797	8.954	5.243	105.4	443.4	164.8	synthetic	Schlenker et al. (1977) (in Ribbe and Prunier, 1977)
4.	hedenbergite $CaFe^{2+}Si_2O_6$	$C2/c$	9.845(1)	9.024(1)	5.245(1)	104.74(1)	450.6(1)	164.5(1)	synthetic	Cameron et al. (1973)
5.	johannsenite $CaMnSi_2O_6$	$C2/c$	9.978(9)	9.156(9)	5.293(5)	105.48(3)	466.0	163.78(47)	U.S.N.M. #R3118 Schio-Vincenti mine, Venetia, northern Italy	Freed and Peacor (1967)
6.	jadeite $NaAlSi_2O_6$	$C2/c$	9.418(1)	8.562(2)	5.219(1)	107.58(1)	401.20(15)	174.7(2)	#184 (H.S. Yoder); veinlets in schist; Santa Rita Peak, New Idria District, California	Prewitt and Burnham (1966)
7.	ureyite $NaCrSi_2O_6$	$C2/c$	9.550(16)	8.712(7)	5.273(8)	107.44(16)	418.6(1.4)	172.1(2)	synthetic	Clark et al. (1969)
8.	acmite $NaFe^{3+}Si_2O_6$	$C2/c$	9.658(2)	8.795(2)	5.294(1)	107.42(2)	429.1(1)	174.0(2)	Green River Formation, Wyoming	Clark et al. (1969)
9.	— $NaSc^{3+}Si_2O_6$	$C2/c$	9.8438(4)	9.0439(4)	5.3540(2)	107.215(2)	455.29	173.6(1)	synthetic	Hawthorne and Grundy (1973)

Table A1. Cont'd. (page 2 of 8 pages).

#										
10.	NaIn³⁺Si₂O₆	$C2/c$	9.9023(4)	9.1307(4)	5.3589(2)	107.200(1)	462.86	170.8(2)	synthetic	Hawthorne and Grundy (1974)
11.	spodumene LiAlSi₂O₆	$C2/c$	9.449(3)	8.386(1)	5.215(2)	110.10(2)	388.1(1)	189.5	Newry, Maine	Clark et al. (1969)
12.	LiFe³⁺Si₂O₆	$C2/c$	9.666(2)	8.669(1)	5.294(2)	110.15(2)	416.4(1)	180(2)	synthetic	Clark et al. (1969)
13.	LiSc³⁺Si₂O₆	$C2/c$	9.8033(7)	8.9581(7)	5.3515(4)	110.281(4)	440.83	175.6(1)	synthetic	Hawthorne and Grundy (1977)
14.	Zn₂Si₂O₆	$C2/c$	9.787(3)	9.161(2)	5.296(1)	111.42(3)	442.0(2)	161.3(1)	synthetic	Morimoto et al. (1975)
15.	Ca-Ts CaAlSiAlO₆	$C2/c$	9.609(3)	8.652(2)	5.274(2)	106.06(2)	421.35(21)	165.93(6)	synthetic	Okamura et al. (1974)
16.	clinopyroxene	$C2/c$	9.812(1)	9.049(1)	5.233(1)	105.34(1)	448.1(1)	162.5(2)	synthetic	Ohashi et al. (1975)
17.	clinopyroxene	$C2/c$	9.781(2)	9.072(2)	5.246(2)	106.55(2)	446.2(3)	159.8(5)	synthetic	Ohashi et al. (1975)
18.	clinopyroxene	$C2/c$	9.760(6)	9.057(8)	5.234(3)	106.28(5)	444.1(6)	159.5(3)	synthetic	Ohashi et al. (1975)
19.	augite	$C2/c$	9.699(1)	8.844(1)	5.272(1)	106.97(2)	432.5(1)	165.8(2)	Kakanui, New Zealand	Clark et al. (1969)
20.	augite	$C2/c$	9.707(2)	8.858(2)	5.274(1)	106.52(3)	434.76	165.47(13)	HK56051801 of Kuno; alkaline olivine basalt; Takasima, North Kyushu, Japan; coexists with #33	Takeda (1972b)
21.	augite	$C2/c$	9.713(2)	8.964(3)	5.266(2)	105.93(2)	440.89	164.26	lamellae in pyroxene grain from lunar (KREEP) basalt 14310,90; cf. pigeonite host #64	Takeda et al. (1974)
22.	augite	$C2/c$	9.704(4)	8.968(5)	5.248(3)	105.65(3)	439.78	—	lamellae in pyroxene	Takeda et al.

Table A1. Cont'd. (page 3 of 8 pages).

No.	Mineral		a	b	c	β	V		Description	Reference
	—								grain from Juvinas eucrite, France; cf. pigeonite host #65	(1974)
23.	augite	$C2/c$	9.726(2)	8.909(3)	5.268(1)	106.82(2)	436.94	164.13	rim from pyroxene grain in lunar basalt 12052; cf. core pigeonite #63	Takeda (1972a)
24.	subcalcic diopside	$C2/c$	9.699(5)	8.871(2)	5.251(3)	107.03(3)	431.9(3)	166.2(3)	#1600E4; pyroxene nodules, Thaba Putsoa pipe, Lesotho	McCallister et al. (1974)
24b.	clinopyroxene	$C2/c$	9.684(2)	8.840(3)	5.266(1)	106.89(3)	431.3(2)	165.57	megacryst; nephelinite breccia; KaKanui, New Zealand	McCallister et al. (1976)
24c.	clinopyroxene	$C2/c$	9.683(3)	8.846(3)	5.264(1)	106.83(2)	431.6(2)	167.15	eclogite; KaKanui, New Zealand	McCallister et al. (1976)
25.	fassaite (titanaugite)	$C2/c$	9.794(5)	8.906(5)	5.319(3)	105.90(3)	446.20	166.0	phenocrysts in nepheline jacupir-angite; Hessereau Hill, Oka, Quebec	Peacor (1967)
26.	fassaite	$C2/c$	9.80(1)	8.85(1)	5.360(5)	105.62(17)	447.70	165.10	Allende meteorite; Chihuahua, Mexico	Dowty and Clark (1973)
27.	fassaite	$C2/c$	9.738(1)	8.874(2)	5.2827(5)	105.89(1)	439.1(1)	165.4(2)	Angra dos Reis achondrite; Rio de Janeiro, Brazil	Hazen and Finger (1977a)
28.	Ca-Ts	$C2/c$	9.780(2)	8.782(2)	5.369(1)	105.78(1)	443.754	—	synthetic	Ghose et al. (1975)
29.	omphacite	$C2/c$	9.646(6)	8.824(5)	5.270(6)	106.59(8)	429.9(5)	168.7(2)	Schmitt #1725; Hareidland, Sunmore, Norway	Clark et al. (1969)
30.	orthoenstatite $Mg_2Si_2O_6$	$Pbca$	18.216(2)	8.813(1)	5.179(1)	—	831.42	A=158.9(1) B=139.0(1)	synthetic	Hawthorne and Ito (1977)
31.	$Zn_2Si_2O_6$	$Pbca$	18.204(5)	9.087(3)	5.278(2)	—	873.1(3)	A=178.3(4) B=149.2(3)	synthetic	Morimoto et al. (1975)

56

Table A1. Cont'd. (page 4 of 8 pages).

#	Mineral								Description	Reference
32.	orthoferrosilite $Fe_2^{2+}Si_2O_6$	Pbca	18.418(2)	9.078(1)	5.2366(4)	—	875.6(1)	A=169.11 B=143.76(24)	synthetic	Sueno et al. (1976)
33.	bronzite	Pbca	18.276(3)	8.819(2)	5.196(1)	—	837.47	A=162.43(12) B=140.89(11)	HK56051802 of Kuno; alkaline olivine basalt; Takasima, North Kyushu, Japan; coexists with #20	Takeda (1972b)
34.	bronzite	Pbca	18.304(3)	8.887(2)	5.215(1)	—	848.31	—	lunar rock (KREEP fragment) 12033, 99	Takeda and Ridley (1972)
35.	bronzite	Pbca	18.301(3)	8.869(2)	5.215(1)	—	846.45	A=164.73 B=142.50	lunar basalt 14310,90	Takeda and Ridley (1972)
36.	bronzite	Pbca	18.233(4)	8.836(3)	5.194(2)	—	836.79	A=162.1 B=141.7	Shaw chondrite; Lincoln County, Colorado	Dodd et al. (1975)
37.	bronzite	Pbca	18.276(7)	8.861(4)	5.201(3)	—	842.27	A=163.41(24) B=141.75(21)	U.S.N.M. #46; Johnstown achondrite; Weld County, Colorado	Miyamoto et al. (1975)
38.	bronzite	Pbca	18.3291(31)	8.8853(11)	5.2212(8)	—	850.32	—	#K7; Luna 20	Ghose and Wan (1973)
39.	aluminous bronzite	Pbca	18.221(2)	8.770(1)	5.192(1)	—	829.67	A=161.88 B=136.80	#2; granulite from Aldan Shield, ?U.S.S.R.	Kosoi et al. (1974)
40.	aluminous bronzite	Pbca	18.224(5)	8.775(5)	5.179(5)	—	828.20	A=161.83 B=137.22	#3; granulite from Aldan Shield, ?U.S.S.R.	Kosoi et al. (1974)
41.	hypersthene	Pbca	18.310(10)	8.927(5)	5.226(5)	—	854.21	A=167.83 B=143.90	#37218 of Ramberg; granulite; Greenland	Ghose (1965)
42.	hypersthene	Pbca	18.295(2)	8.901(1)	5.218(1)	—	849.72	A=165.53 B=142.67	#1; granulite from Aldan Shield, ?U.S.S.R.	Kosoi et al. (1974)
43.	hypersthene	Pbca	18.313(3)	8.912(2)	5.210(1)	—	850.30	--	synthetic	Ghose et al. (1975)

57

Table A1. Cont'd. (page 5 of 8 pages).

#	Name		a	b	c		V	A / B	Locality/Description	Reference
44.	ferrohypersthene	Pbca	18.363(5)	8.990(3)	5.232(4)	—	863.7(5)	A=167.2 B=144.5	B1-9 of Virgo and Hafner (1969); metamorphosed iron formation; Quebec	Smyth (1973)
45.	aluminous hypersthene	Pbca	18.220(5)	8.765(3)	5.188(1)	—	828.51	A=161.65 B=136.39	393/8; Gneiss, Sutam River, Aldan Shield, ?U.S.S.R.	Brovkin et al. (1975)
46.	eulite	Pbca	18.405(1)	9.0338(7)	5.2390(4)	—	871.08(14)	A=168.8(1) B=143.9(1)	XYZ of Ramberg; granulite from Greenland	Burnham et al. (1971)
47.	(Mg, Co) orthopyroxene	Pbca	18.233(7)	8.836(6)	5.188(3)	—	835.82	A=160.1(1) B=139.6(1)	synthetic	Hawthorne and Ito (1978)
48.	(Mg, Mn) orthopyroxene	Pbca	18.270(6)	8.833(6)	5.195(3)	—	838.36	A=160.6(2) B=140.7(2)	synthetic	Hawthorn and Ito (1978)
49.	(Mg, Mn, Co) orthopyroxene	Pbca	18.246(3)	8.839(2)	5.196(1)	—	837.99	A=161.6(2) B=141.4(2)	synthetic	Hawthorne and Ito (1977)
50.	(Zn, Mg) orthopyroxene	Pbca	18.201(5)	8.916(2)	5.209(2)	—	845.32	A=166.9(6) B=143.6(5)	synthetic	Morimoto et al. (1975)
51.	(Zn, Mg) orthopyroxene	Pbca	18.231(4)	8.893(2)	5.209(1)	—	844.53	--	synthetic	Ghose et al. (1975)
52.	(Ni, Mg) orthopyroxene	Pbca	18.203(4)	8.788(2)	5.171(1)	—	827.19	--	synthetic	Ghose et al. (1975)
53.	(Co, Mg) orthopyroxene	Pbca	18.229(4)	8.847(2)	5.182(1)	—	835.71	--	synthetic	Ghose et al. (1975)
53b.	aluminous orthopyroxene	Pbca	18.2248(39)	8.7822(24)	5.1927(12)	—	831.11	A=160.35 B=138.10	#SQ2-70; spinel lherzolite; Baja, California	Ganguly and Ghose (1979)

58

Table A1. Cont'd. (page 6 of 8 pages).

No.	mineral	space group	a	b	c	β	V	A=/B=	description	reference
53c.	aluminous orthopyroxene —	Pbca	18.1991(48)	8.7800(14)	5.1666(12)	—	825.56	A=158.81 B=137.13	synthetic	Ganguly and Ghose (1979)
54.	clinoenstatite $Mg_2Si_2O_6$	$P2_1/c$	9.620(5)	8.825(5)	5.188(5)	108.333(.167)	418.09	A=199.0[c] B=134.0	Bishopville enstatite achondrite; Lee County, S. Carolina	Morimoto et al. (1960)
55.	clinoferrosilite $Fe_2^{2+}Si_2O_6$	$P2_1/c$	9.7085(8)	9.0872(11)	5.2284(6)	108.432(4)	437.6(1)	A=193.0 B=142.0	synthetic	Burnham (1965, 1966)
56.	clinopyroxene —	$P2_1/c$	9.779(1)	9.088(1)	5.258(1)	107.39(1)	445.9(2)	A=164.2(8) B=156.0(7)	synthetic	Ohashi et al. (1975)
57.	enstatite $Mg_2Si_2O_6$	$P2_1/c$	9.605(1)	8.813(1)	5.166(1)	108.46(1)	415.1(1)	A=202.76 B=138.31	synthetic	Ohashi and Finger (1976)
58.	clinopyroxene —	$P2_1/c$	9.639(1)	8.835(1)	5.192(1)	108.39(1)	419.6(1)	A=200.37 B=141.99	synthetic	Ohashi and Finger (1976)
59.	clinopyroxene —	$P2_1/c$	9.657(1)	8.846(2)	5.208(1)	108.34(1)	422.3(1)	A=197.99 B=145.01	synthetic	Ohashi and Finger (1976)
60.	pigeonite —	$P2_1/c$	9.706(2)	8.950(1)	5.246(1)	108.59(1)	431.94	A=190.4 B=148.5(5)	andesite; Isle of Mull, Scotland	Morimoto and Güven (1970)
61.	pigeonite —	$P2_1/c$	9.706(2)	8.950(1)	5.246(1)	108.59(1)	432.0(2)	A=186.1 B=151.0(3)	andesite; Isle of Mull, Scotland	Brown et al. (1972)
62.	pigeonite —	$P2_1/c$	9.678(10)	8.905(10)	5.227(5)	108.71(8)	430.6	A=194.3 B=145.5(5)	lunar basalt 10003,38	Clark et al. (1971)
63.	pigeonite —	$P2_1/c$	9.688(3)	8.890(3)	5.238(2)	108.40(4)	428.07	A=188.85 B=150.05	core from pyroxene grain in lunar basalt 12052; cf. augite rim #23	Takeda (1972a)
64.	pigeonite —	$P2_1/c$	9.715(1)	8.963(1)	5.239(1)	108.64(2)	432.26	A=190.48 B=147.78	host in pyroxene grain from lunar (KREEP) basalt 14310,90; cf. augite lamellae #21	Takeda et al. (1974)
65.	pigeonite —	$P2_1/c$	9.698(2)	8.967(2)	5.223(1)	108.75(2)	430.10	--	host in pyroxene	Takeda et al.

59

Table A1. Cont'd. (page 7 of 8 pages).

No.	Mineral	Space group	a	b	c	β	V	O3-O3-O3 angle	Locality / description	Reference
									grain from Juvinas eucrite; France cf. augite lamellae #22	(1974)
66.	pigeonite	$P2_1/c$	9.683(8)	8.900(7)	5.228(4)	108.50(2)	427.26	A=193.6(2) B=146.0(2)	phenocryst in mare basalt 15476	Ohashi and Finger (1974b)
67.	clinohypersthene	$P2_1/c$	9.691(3)	8.993(3)	5.231(2)	108.61(2)	432.0(2)	A=195.9 B=142.5	B1-9 of Virgo and Hafner (1969); metamorphosed iron formation, Quebec	Smyth (1974b)
68.	(Mn, Mg) clinopyroxene	$P2_1/c$	9.719(2)	8.917(2)	5.248(1)	108.51(2)	431.29	--	synthetic	Ghose et al. (1975)
68b.	MnSiO$_3$	$P2_1/c$	9.864(2)	9.179(2)	5.298(1)	108.22	455.64	A=190.53 B=149.46	synthetic	Tokonami et al. (1979)
69.	protoenstatite Mg$_2$Si$_2$O$_6$	$P2_1cn$	9.304(4)	8.902(4)	5.351(3)	—	443.19	150.12	A.M.N.H. #3846; meteorite; Norton County, Kansas	Smyth (1971)
70.	(Mg,Li,Sc) protopyroxene	$Pbcn$	9.251(2)	8.773(2)	5.377(1)	—	436.39	167.32	synthetic	Smyth and Ito (1977)
71.	omphacite	$P2/n$	9.585(3)	8.776(3)	5.260(3)	106.85(3)	423.5(3)	169.45	eclogite; Iratsu mass; Bessi area, Japan	Matsumoto et al. (1975)
72.	titanian ferro-omphacite	$P2/n$	9.622(2)	8.826(2)	5.279(1)	106.92(2)	428.85	170.24	metamorphosed peralkaline rocks; Red Wine province, Central Labrador	Curtis et al. (1975)
73.	omphacite	$P2$	9.596(5)	8.771(4)	5.265(6)	106.93(7)	423.9(0.4)	A=165.0 C=165.0	#100-RGC-58 Tiburon Peninsula, California	Clark and Papike (1968)
74.	omphacite	$P2$	9.551(8)	8.751(5)	5.254(4)	106.87(8)	420.2(4)	A=170.3 C=169.1	Morgan (1967) #Ca-1059; Puerto Cabello, Venezuela	Clark et al. (1969)

a Compositions are given only for end member pyroxenes.
b O3-O3-O3 angles greater than 180° indicate S-rotations.
c Refinement is based on two-dimensional data.

Table A2. Symmetry Information on Pyroxene Space Groups.

Space Group	Site	Point Symmetry	Multiplicity	Examples
C2/c	M1	2	4	Diopside,
	M2	2	4	Augite,
	T	1	8	Jadeite
	O1	1	8	
	O2	1	8	
	O3	1	8	
P2₁/c	M1	1	4	Pigeonite,
	M2	1	4	Clinoenstatite
	TA	1	4	
	TB	1	4	
	O1A	1	4	
	O1B	1	4	
	O2A	1	4	
	O2B	1	4	
	O3A	1	4	
	O3B	1	4	
Pbcn	M1	2	4	Protoenstatite
	M2	2	4	
	T	1	8	
	O1	1	8	
	O2	1	8	
	O3	1	8	
Pbca	M1	1	8	Hypersthene,
	M2	1	8	Orthoenstatite
	TA	1	8	
	TB	1	8	
	O1A	1	8	
	O1B	1	8	
	O2A	1	8	
	O2B	1	8	
	O3A	1	8	
	O3B	1	8	
P2/n	M1	2	2	Omphacite
	M1(1)	2	2	
	M2	2	2	
	M2(1)	2	2	
	T1	1	4	
	T2	1	4	
	O1(1)	1	4	
	O1(2)	1	4	
	O2(1)	1	4	
	O3(1)	1	4	
	O3(2)	1	4	

Table A3. Oxygen Coordination in Pyroxene Structures.

No.	Space Group Pyroxene (Reference)	Anion	Cations Within 3.4Å	Interatomic Distance (Å)[a]	Cation Charge ÷ Cation Coordination Number
4.	C2/c hedenbergite (Cameron et al., 1973)	01	M1	2.164	2/6 = 0.333
			M1	2.140	2/6 = 0.333
			M2	2.355	2/8 = 0.250
			T	1.601	4/4 = 1.000
					$\overline{1.916}$ = Σ
		02	M1	2.087	2/6 = 0.333
			M2	2.341	2/8 = 0.250
			T	1.585	4/4 = 1.000
			T	3.257	
			T	3.393	
					$\overline{1.583}$ = Σ
		03	M2	2.627	2/8 = 0.250
			M2	2.719	2/8 = 0.250
			T	1.666	4/4 = 1.000
			T	1.686	4/4 = 1.000
					$\overline{2.500}$ = Σ
8.	C2/c acmite (Clark et al., 1969)	01	M1	2.109	3/6 = 0.500
			M1	2.029	3/6 = 0.500
			M2	2.398	1/8 = 0.125
			T	1.629	4/4 = 1.000
					$\overline{2.125}$ = Σ
		02	M1	1.936	3/6 = 0.500
			M2	2.416	1/8 = 0.125
			T	1.598	4/4 = 1.000
			T	3.247	
			T	3.083	
					$\overline{1.625}$ = Σ
		03	M1	3.364	
			M2	2.430	1/8 = 0.125
			M2	2.831	1/8 = 0.125
			T	1.637	4/4 = 1.000
			T	1.646	4/4 = 1.000
			T	3.387	
					$\overline{2.250}$ = Σ
11.	C2/c spodumene (Clark et al., 1969)	01	M1	1.996	3/6 = 0.500
			M1	1.943	3/6 = 0.500
			M2	2.105	1/6 = 0.167
			T	1.637	4/4 = 1.000
					$\overline{2.167}$ = Σ
		02	M1	1.818	3/6 = 0.500
			M2	3.364	
			M2	2.278	1/6 = 0.167
			T	1.586	4/4 = 1.000
			T	3.084	
			T	3.009	
			T	3.366	
					$\overline{1.667}$ = Σ

No.	Space Group Pyroxene (Reference)	Anion	Cations Within 3.4Å	Interatomic Distance (Å)[a]	Cation Charge ÷ Cation Coordination Number
		03	M1	3.317	
			M2	2.251	1/6 = 0.167
			M2	3.144	
			T	1.622	4/4 = 1.000
			T	1.626	4/4 = 1.000
			T	3.230	
					$\overline{2.167}$ = Σ
15.	C2/c Cats (Okamura et al., 1974)	01	M1	2.021	3/6 = 0.500
			M1	1.947	3/6 = 0.500
			M2	2.403	2/8 = 0.250
			T	1.693	(4/4 + 3/4)/2 = 0.875
					$\overline{2.125}$ = Σ
		02	M1	1.872	3/6 = 0.500
			M2	2.420	2/8 = 0.250
			T	1.665	(4/4 + 3/4)/2 = 0.875
			T	3.200	
			T	2.988	
					$\overline{1.625}$ = Σ
		03	M1	3.349	
			M2	2.469	2/8 = 0.250
			M2	2.549	2/8 = 0.250
			T	1.683	(4/4 + 3/4)/2 = 0.875
			T	1.701	(4/4 + 3/4)/2 = 0.875
					$\overline{2.250}$ = Σ
32.	Pbca orthoferrosilite (Sueno et al., 1976)	01A	M1	2.083	2/6 = 0.333
			M1	2.194	2/6 = 0.333
			M2	2.157	2/6 = 0.333
			TA	1.612	4/4 = 1.000
					$\overline{2.000}$ = Σ
		01B	M1	2.194	2/6 = 0.333
			M1	2.123	2/6 = 0.333
			M2	2.128	2/6 = 0.333
			TB	1.619	4/4 = 1.000
					$\overline{2.000}$ = Σ
		02A	M1	2.088	2/6 = 0.333
			M2	2.023	2/6 = 0.333
			TA	1.607	4/4 = 1.000
			TA	3.355	
					$\overline{1.666}$ = Σ
		02B	M1	2.122	2/6 = 0.333
			M2	1.993	2/6 = 0.333
			TB	1.597	4/4 = 1.000
					$\overline{1.666}$ = Σ
		03A	M1	3.380	
			M2	2.459	2/6 = 0.333
			TA	1.632	4/4 = 1.000
			TA	1.652	4/4 = 1.000
			TB	3.269	
					$\overline{2.333}$ = Σ

No.	Space Group Pyroxene (Reference)	Anion	Cations Within 3.4Å	Interatomic Distance (Å)[a]	Cation Charge ÷ Cation Coordination Number
		03B	M1	3.290	
			M2	3.095	
			M2	2.596	2/6 = 0.333
			TA	3.257	
			TB	1.666	4/4 = 1.000
			TB	1.661	4/4 = 1.000
					$\overline{2.333}$ = Σ
57.	P2$_1$/c clinoenstatite (Ohashi and Finger, 1976)	01A	M1	2.142	2/6 = 0.333
			M1	2.030	2/6 = 0.333
			M2	2.090	2/6 = 0.333
			TA	1.610	4/4 = 1.000
					$\overline{2.000}$ = Σ
		01B	M1	2.179	2/6 = 0.333
			M1	2.066	2/6 = 0.333
			M2	2.053	2/6 = 0.333
			TB	1.615	4/4 = 1.000
					$\overline{2.000}$ = Σ
		02A	M1	2.006	2/6 = 0.333
			M2	2.032	2/6 = 0.333
			TA	1.589	4/4 = 1.000
			TA	3.357	
			TA	3.376	
			TA	3.229	
					$\overline{1.666}$ = Σ
		02B	M1	2.044	2/6 = 0.333
			M2	1.985	2/6 = 0.333
			TB	1.586	4/4 = 1.000
					$\overline{1.666}$ = Σ
		03A	M2	2.279	2/6 = 0.333
			TA	1.665	4/4 = 1.000
			TA	1.646	4/4 = 1.000
			TB	3.272	
					$\overline{2.333}$ = Σ
		03B	M1	3.250	
			M2	2.414	2/6 = 0.333
			M2	3.151	
			TA	3.226	
			TB	1.676	4/4 = 1.000
			TB	1.676	4/4 = 1.000
					$\overline{2.333}$ = Σ

No.	Space Group Pyroxene (Reference)	Anion	Cations Within 3.4Å	Interatomic Distance (Å)[a]	Cation Charge ÷ Cation Coordination Number
70.	Pbcn protopyroxene (Smyth and Ito, 1977)	01	M1	2.085	2/6 = 0.333
			M1	2.198	2/6 = 0.333
			M2	2.087	2/6 = 0.333
			T	1.606	4/4 = 1.000
					$\overline{2.000}$ = Σ
		02	M1	2.004	2/6 = 0.333
			M2	2.078	2/6 = 0.333
			T	1.590	4/4 = 1.000
			T	3.374	
					$\overline{1.666}$ = Σ
		03	M1	3.324	
			M2	2.393	2/6 = 0.333
			T	1.631	4/4 = 1.000
			T	1.657	4/4 = 1.000
					$\overline{2.333}$ = Σ
72.	P2/n omphacite (Curtis et al., 1975)	01(1)	M1	2.176	
			M1(1)	1.957	
			M2	2.366	
			T1	1.619	
			T2	3.370	
		01(2)	M1	2.113	
			M1(1)	2.000	
			M2(2)	2.402	
			T1	3.355	
			T2	1.618	
		02(1)	M1(1)	1.886	
			M2	2.373	
			T1	1.605	
			T1	3.082	
			T2	3.217	
		02(2)	M1	2.049	
			M2(2)	3.360	
			M2(2)	2.393	
			T1	3.342	
			T2	1.585	
			T2	3.143	
		03(1)	M1(1)	3.391	
			M2	2.730	
			M2(2)	2.491	
			T1	1.645	
			T1	3.344	
			T2	1.669	
		03(2)	M1	3.306	
			M2	2.456	
			M2(2)	2.838	
			T1	1.655	
			T2	1.648	

[a]Interatomic distances were taken directly from or calculated using the data from references cited in Table A1. Only distances less than 3.4Å are reported.

Table A4. Composition of Pyroxenes Discussed in Text and Included in Tables.[a]

NUMBER / PYROXENE NAME / APPROXIMATE FORMULA (ordered pyroxenes only)	1. -- CaNiSi$_2$O$_6$	2. diopside CaMgSi$_2$O$_6$	3. -- CaCoSi$_2$O$_6$	4. hedenbergite CaFe^{2+}Si$_2$O$_6$	5. johannsenite CaMnSi$_2$O$_6$	6. jadeite NaAlSi$_2$O$_6$	7. ureyite NaCrSi$_2$O$_6$	8. acmite NaFe^{3+}Si$_2$O$_6$
UNIT CELL CONTENTS (based on 6 oxygens)								
Si	2.00	1.97	2.00	2.00	2.045	1.967	1.97	2.00
Al	--	--	--	--	--	0.033	--	--
Fe^{3+}	--	--	--	--	--	--	--	--
Ti^{4+}	--	--	--	--	--	--	--	--
Cr	--	--	--	--	--	--	--	--
Total T Cations	2.00	1.97	2.00	2.00	2.045	2.000	1.97	2.00
Al	--	0.01	--	--	--	0.691	--	0.01
Ti^{3+}	--	--	--	--	--	--	--	--
Ti^{4+}	--	--	--	--	--	--	--	--
Cr^{3+}	--	--	--	--	--	0.002	0.97	--
Fe^{3+}	--	0.01	--	--	--	0.011	--	0.99
Sc	--	--	--	--	--	--	--	--
In	--	--	--	--	--	--	--	--
Ni	1.00	--	--	--	--	--	--	--
Mg	--	1.01	--	--	0.015	0.001	0.04	0.01
Co	--	--	1.00	--	--	--	--	--
Zn	--	--	--	--	--	--	--	--
Fe^{2+}	--	--	--	1.00	0.018	--	--	--
Mn^{2+}	--	--	--	--	1.001	--	0.01	--
Ca	1.00	0.99	1.00	1.00	0.876	0.005	0.05	--
Li	--	--	--	--	--	--	--	--
Na	--	0.02	--	--	--	1.092	1.00	0.95
K	--	--	--	--	--	0.002	--	--
Total M Cations	2.00	2.04	2.00	2.00	1.910	2.074	2.07	1.96
RELEVANT FOOTNOTES	b	d	b	b	g,n	e	d	d
REFERENCE (chemical data)	Schlenker et al. in Ribbe and Prunier (1977)	Clark et al. (1969)	Schlenker et al. in Ribbe and Prunier (1977)	Cameron et al. (1973)	Schaller (1938)	Prewitt and Burnham (1966)	Clark et al. (1969)	Clark et al. (1969)

Table A4. Cont'd. (page 2 of 13 pages).

NUMBER	9.	10.	11. spodumene	12.	13.	14.	15. Ca-Ts
PYROXENE NAME APPROXIMATE FORMULA (ordered pyroxenes only)	$NaSc^{3+}Si_2O_6$	$NaIn^{3+}Si_2O_6$	$LiAlSi_2O_6$	$LiFe^{3+}Si_2O_6$	$LiSc^{3+}Si_2O_6$	$Zn_2Si_2O_6$	$CaAlSiAlO_6$
UNIT CELL CONTENTS (based on 6 oxygens)							
Si	2.00	2.00	1.99	2.00	2.00	2.00	1.034
Al	--	--	--	--	--	--	0.966
Fe^{3+}	--	--	--	--	--	--	--
Ti^{4+}	--	--	--	--	--	--	--
Cr	--	--	--	--	--	--	--
Total T Cations	2.00	2.00	1.99	2.00	2.00	2.00	2.000
Al	--	--	1.00	--	--	--	0.968
Ti^{3+}	--	--	--	--	--	--	--
Ti^{4+}	--	--	--	--	--	--	--
Cr^{3+}	--	--	--	--	--	--	--
Fe^{3+}	--	--	--	0.952	--	--	--
Sc	1.00	--	--	--	1.00	--	--
In	--	1.00	--	--	--	--	--
Ni	--	--	--	--	--	--	--
Mg	--	--	--	--	--	--	--
Co	--	--	--	--	--	--	--
Zn	--	--	--	--	--	2.00	--
Fe^{2+}	--	--	--	0.096	--	--	--
Mn^{2+}	--	--	--	--	--	--	--
Ca	--	--	--	--	--	--	1.030
Li	--	--	1.00	0.952	1.00	--	--
Na	1.00	1.00	1.01	--	--	--	--
K	--	--	--	--	--	--	--
Total M Cations	2.00	2.00	2.01	2.00	2.00	2.00	1.998
RELEVANT FOOTNOTES	b	b	--	b	b	b	b
REFERENCE (chemical data)	Hawthorne and Grundy (1973)	Hawthorne and Grundy (1974)	Clark et al. (1969)	Clark et al. (1969)	Hawthorne and Grundy (1977)	Morimoto et al. (1975)	Okamura et al. (1974)

Table A4. Cont'd. (page 3 of 13 pages).

	16.	17.	18.	19.	20.	21.	22.	23.
NUMBER								
PYROXENE NAME	clinopyroxene	clinopyroxene	clinopyroxene	augite	augite	augite	augite	augite
APPROXIMATE FORMULA (ordered pyroxenes only)	--	--	--	--	--	--	--	--
	--	--	--	--	--	--	--	--
UNIT CELL CONTENTS (based on 6 oxygens)								
Si	2.00	2.00	2.00	1.83	1.731	1.946	1.925	1.786
Al	--	--	--	0.17	0.269	0.054	0.036	0.214
Fe^{3+}	--	--	--	--	--	--	0.002	--
Ti^{4+}	--	--	--	--	--	--	--	--
Cr	--	--	--	--	--	--	--	--
Total T Cations	2.00	2.00	2.00	2.00	2.000	2.000	1.963	2.000
Al	--	--	--	0.16	0.116	0.007	--	0.059
Ti^{3+}	--	--	--	--	--	--	--	--
Ti^{4+}	--	--	--	0.02	0.035	0.033	0.007	0.081
Cr^{3+}	--	--	--	--	0.086	--	0.025	0.036
Fe^{3+}	--	--	--	0.10	--	--	--	--
Sc	--	--	--	--	--	--	--	--
In	--	--	--	--	--	--	--	--
Ni	--	--	--	--	--	--	--	--
Mg	--	--	--	0.90	0.792	0.633	0.657	0.764
Co	--	--	--	--	--	--	--	--
Zn	--	--	--	--	--	--	--	--
Fe^{2+}	1.30	1.50	1.60	0.11	0.177	0.541	0.655	0.453
Mn^{2+}	--	--	--	--	0.006	--	0.021	--
Ca	0.70	0.50	0.40	0.61	0.753	0.773	0.706	0.607
Li	--	--	--	--	--	--	--	--
Na	--	--	--	0.09	0.064	0.005	0.005	--
K	--	--	--	--	0.001	--	--	--
Total M Cations	2.00	2.00	2.00	1.99	2.030	1.992	2.077	2.000
RELEVANT FOOTNOTES	b	b	b	d	c	e	e,r	d,m
REFERENCE (chemical data)	Ohashi et al. (1975)	Ohashi et al. (1975)	Ohashi et al. (1975)	Clark et al. (1969)	Takeda (1972b)	Takeda et al. (1974)	Takeda et al. (1974)	Takeda (1972a)

Table A4. Cont'd. (page 4 of 13 pages).

	24. subcalcic diopside	24b. clinopyroxene	24c. clinopyroxene	25. fassaite (titanaugite)	26. fassaite	27. fassaite	28. Ca-Ts
NUMBER / PYROXENE NAME	subcalcic diopside	clinopyroxene	clinopyroxene	fassaite (titanaugite)	fassaite	fassaite	Ca-Ts
APPROXIMATE FORMULA (ordered pyroxenes only)	--	--	--	--	--	--	--
UNIT CELL CONTENTS (based on 6 oxygens)							
Si	1.973	1.822	1.854	1.506	1.26	1.728	1.00
Al	0.027	0.178	0.146	0.494	0.74	0.272	0.821
Fe^{3+}	--	--	--	--	--	--	0.179
Ti^{4+}	--	--	--	--	--	--	--
Cr	--	--	--	--	--	--	--
Total T Cations	2.000	2.000	2.000	2.000	2.00	2.000	2.000
Al	0.092	0.172	0.170	0.171	0.13	0.161	0.179
Ti^{3+}	--	--	--	--	0.34	--	--
Ti^{4+}	--	0.023	0.024	0.065	0.14	0.059	--
Cr^{3+}	--	0.004	0.004	--	--	0.005	--
Fe^{3+}	0.025	0.048	0.114	0.159	--	--	0.821
Sc	--	--	--	--	--	--	--
In	--	--	--	--	--	--	--
Ni	--	--	--	--	.:	--	--
Mg	1.087	0.885	0.677	0.570	0.38	0.578	--
Co	--	--	--	--	--	--	--
Zn	--	--	--	--	--	--	--
Fe^{2+}	0.125	0.134	0.213	0.063	--	0.223	--
Mn^{2+}	--	0.005	0.007	0.007	--	0.002	--
Ca	0.553	0.634	0.594	0.975	1.01	0.968	1.000
Li	--	--	--	--	--	--	--
Na	0.119	0.093	0.198	0.007	--	0.002	--
K	--	--	--	0.001	--	--	--
Total M Cations	2.001	1.998	2.001	2.018	2.00	1.998	2.000
RELEVANT FOOTNOTES	d,m,p	c	c	e,j	f,s	e,i	b
REFERENCE (chemical data)	McCallister et al. (1974)	McCallister et al. (1976)	McCallister et al. (1976)	Peacor (1967)	Dowty and Clark (1973)	Hazen and Finger (1977a)	Ghose et al. (1975)

Table A4. Cont'd. (page 5 of 13 pages).

	29.	30.	31.	32.	33.	34.	35.	36.
NUMBER PYROXENE NAME	omphacite	orthoenstatite	--	orthoferrosilite	bronzite	bronzite	bronzite	bronzite
APPROXIMATE FORMULA (ordered pyroxenes only)	--	$Mg_2Si_2O_6$	$Zn_2Si_2O_6$	$Fe_2^{2+}Si_2O_6$	--	--	--	--
UNIT CELL CONTENTS (based on 6 oxygens)								
Si	1.995	2.00	2.00	2.00	1.850	1.970	1.941	1.974
Al	0.005	--	--	--	0.150	0.026	0.059	0.004
Fe^{3+}	--	--	--	--	--	--	--	0.002
Ti^{4+}	--	--	--	--	--	--	--	0.020
Cr	--	--	--	--	--	--	--	--
Total T Cations	2.000	2.00	2.00	2.00	2.000	1.996	2.000	2.000
Al	0.233	--	--	--	0.034	0.015	0.015	--
Ti^{3+}	--	--	--	--	--	--	--	--
Ti^{4+}	0.002	--	--	--	0.012	--	0.020	--
Cr^{3+}	--	--	--	--	0.005	--	0.012	0.006
Fe^{3+}	0.123	--	--	--	0.072	--	--	--
Sc	--	--	--	--	--	--	--	--
In	--	--	--	--	--	--	--	--
Ni	--	--	--	--	--	--	--	--
Mg	0.582	2.00	--	--	1.490	1.348	1.403	1.634
Co	--	--	--	--	--	--	--	--
Zn	--	--	2.00	--	--	--	--	--
Fe^{2+}	0.116	--	--	2.00	0.324	0.571	0.448	0.334
Mn^{2+}	--	--	--	--	0.008	0.010	0.007	0.013
Ca	0.583	--	--	--	0.059	0.062	0.093	0.023
Li	--	--	--	--	--	--	--	--
Na	0.325	--	--	--	0.008	0.001	--	0.002
K	--	--	--	--	0.002	--	--	--
Total M Cations	1.964	2.00	2.00	2.00	2.014	2.007	1.998	2.012
RELEVANT FOOTNOTES	d	f,j	b	b	c	g	g	f
REFERENCE (chemical data)	Clark et al. (1969)	Hawthorne and Ito (1977)	Morimoto et al. (1975)	Sueno et al. (1976)	Takeda (1972b)	Takeda and Ridley (1972)	Takeda and Ridley (1972)	Dodd et al. (1975)

70

Table A4. Cont'd. (page 6 of 13 pages).

	37.	38.	39.	40.	41.	42.	43.
PYROXENE NAME	bronzite	bronzite	aluminous bronzite	aluminous bronzite	hypersthene	hypersthene	hypersthene
APPROXIMATE FORMULA (ordered pyroxenes only)	--	--	--	--	--	--	--
UNIT CELL CONTENTS (based on 6 oxygens)							
Si	1.958	1.946	1.752	1.752	1.936	1.942	2.00
Al	0.042	0.054	0.239	0.239	0.064	0.056	--
Fe^{3+}	--	--	--	--	--	--	--
Ti^{4+}	--	--	0.009	0.009	--	0.002	--
Cr	--	--	--	--	--	--	--
Total T Cations	2.000	2.000	2.000	2.000	2.000	2.000	2.00
Al	0.001	--	0.207	0.207	0.035	0.013	--
Ti^{3+}	--	--	--	--	--	--	--
Ti^{4+}	0.003	0.009	--	--	0.011	--	--
Cr^{3+}	0.019	0.022	--	--	--	--	--
Fe^{3+}	--	--	0.028	0.028	0.010	0.039	--
Sc	--	--	--	--	--	--	--
In	--	--	--	--	--	--	--
Ni	--	--	--	--	--	--	--
Mg	1.471	1.389	1.396	1.396	0.885	1.062	1.21
Co	--	--	--	--	--	--	--
Zn	--	--	--	--	--	--	--
Fe^{2+}	0.445	0.501	0.369	0.369	1.000	0.827	0.79
Mn^{2+}	0.015	0.008	--	--	0.010	0.014	--
Ca	0.053	0.083	0.009	0.009	0.049	0.050	--
Li	--	--	--	--	--	--	--
Na	0.001	--	--	--	--	--	--
K	--	--	--	--	--	--	--
Total M Cations	2.008	2.012	2.009	2.009	2.000	2.005	2.00
RELEVANT FOOTNOTES	e	c,k	d	d	f	d	b
REFERENCE (chemical data)	Miyamoto et al. (1975)	Ghose and Wan (1973)	Kosoi et al. (1974)	Kosoi et al. (1974)	Ramberg and DeVore (1951)	Kosoi et al. (1974)	Ghose et al. (1975)

Table A4. Cont'd. (page 7 of 13 pages).

NUMBER PYROXENE NAME APPROXIMATE FORMULA (ordered pyroxenes only)	44. ferrohypersthene --	45. aluminous hypersthene --	46. eulite --	47. (Mg,Co) orthopyroxene --	48. (Mg,Mn) orthopyroxene --
UNIT CELL CONTENTS (based on 6 oxygens)					
Si	1.997	1.83	1.983	2.00	2.00
Al	0.003	0.17	0.017	--	--
Fe^{3+}	--	--	--	--	--
Ti^{4+}	--	--	--	--	--
Cr	--	--	--	--	--
Total T Cations	2.000	2.00	2.000	2.00	2.00
Al	0.002	0.20	0.021	--	--
Ti^{3+}	--	--	--	--	--
Ti^{4+}	0.002	--	--	--	--
Cr^{3+}	--	--	--	--	--
Fe^{3+}	--	0.04	--	--	--
Sc	--	--	--	--	--
In	--	--	--	--	--
Ni	--	--	--	--	--
Mg	0.636	1.46	0.239	1.552	1.850
Co	--	--	--	0.448	--
Zn	--	--	--	--	--
Fe^{2+}	1.327	0.24	1.700	--	--
Mn^{2+}	--	0.01	--	--	0.150
Ca	0.032	0.01	0.038	--	--
Li	--	--	--	--	--
Na	--	0.01	--	--	--
K	--	0.01	--	--	--
Total M Cations	1.999	1.98	1.998	2.000	2.000
RELEVANT FOOTNOTES	c	f,j	f	f,j	f,j
REFERENCE (chemical data)	Smyth (1973)	Brovkin et al. (1975)	Burnham et al. (1971)	Hawthorne and Ito (1978)	Hawthorne and Ito (1978)

72

Table A4. Cont'd. (page 8 of 13 pages).

NUMBER	49.	50.	51.	52.
PYROXENE NAME APPROXIMATE FORMULA (ordered pyroxenes only)	(Mg,Mn,Co) orthopyroxene --	(Zn,Mg) orthopyroxene --	(Zn,Mg) orthopyroxene --	(Ni,Mg) orthopyroxene --
UNIT CELL CONTENTS (based on 6 oxygens)				
Si	2.00	2.00	2.00	2.00
Al	--	--	--	--
Fe^{3+}	--	--	--	--
Ti^{4+}	--	--	--	--
Cr	--	--	--	--
Total T Cations	2.00	2.00	2.00	2.00
Al	--	--	--	--
Ti^{3+}	--	--	--	--
Ti^{4+}	--	--	--	--
Cr^{3+}	--	--	--	--
Fe^{3+}	--	--	--	--
Sc	--	--	--	--
In	--	--	--	--
Ni	--	--	--	0.38
Mg	1.562	1.00	1.55	1.62
Co	0.263	--	--	--
Zn	--	1.00	0.45	--
Fe^{2+}	--	--	--	--
Mn^{2+}	0.175	--	--	--
Ca	--	--	--	--
Li	--	--	--	--
Na	--	--	--	--
K	--	--	--	--
Total M Cations	2.000	2.00	2.00	2.00
RELEVANT FOOTNOTES	f,j	b	b	b
REFERENCE (chemical data)	Hawthorne and Ito (1977)	Morimoto et al. (1975)	Ghose et al. (1975)	Ghose et al. (1975)

73

Table A4. Cont'd. (page 9 of 13 pages).

NUMBER	53.	53b.	53c.	54.	55.
PYROXENE NAME	(Co,Mg) orthopyroxene	aluminous orthopyroxene	aluminous orthopyroxene	clinoenstatite	clinoferrosilite
APPROXIMATE FORMULA (ordered pyroxenes only)	--	--	--	$Mg_2Si_2O_6$	$Fe_2^{2+}Si_2O_6$
UNIT CELL CONTENTS (based on 6 oxygens)					
Si	2.00	1.857	1.915	2.011	2.00
Al	--	0.143	0.085	--	--
Fe^{3+}	--	--	--	--	--
Ti^{4+}	--	--	--	--	--
Cr	--	--	--	--	--
Total T Cations	2.00	2.000	2.000	2.011	2.00
Al	--	0.140	0.093	--	--
Ti^{3+}	--	--	--	--	--
Ti^{4+}	--	--	--	--	--
Cr^{3+}	--	0.002	--	--	--
Fe^{3+}	--	0.008	--	--	--
Sc	--	--	--	--	--
In	--	--	--	--	--
Ni	--	--	--	--	--
Mg	1.26	1.628	1.903	1.966	--
Co	0.74	--	--	--	--
Zn	--	--	--	--	--
Fe^{2+}	--	0.193	--	0.011	2.00
Mn^{2+}	--	0.004	--	--	--
Ca	--	0.021	--	--	--
Li	--	--	--	--	--
Na	--	--	--	--	--
K	--	--	--	--	--
Total M Cations	2.00	1.996	1.996	1.997	2.00
RELEVANT FOOTNOTES	b	g	g	g	b
REFERENCE (chemical data)	Ghose et al. (1975)	Ganguly and Ghose (1979)	Ganguly and Ghose (1979)	Morimoto et al. (1960)	Burnham (1965, 1966)

74

Table A4. Cont'd. (page 10 of 13 pages).

NUMBER	56.	57.	58.	59.	60.	61.	62.	63.
PYROXENE NAME	clinopyroxene	enstatite	clinopyroxene	clinopyroxene	pigeonite	pigeonite	pigeonite	pigeonite
APPROXIMATE FORMULA (ordered pyroxenes only)	--	$Mg_2Si_2O_6$	--	--	--	--	--	--
UNIT CELL CONTENTS (based on 6 oxygens)								
Si	2.00	2.00	2.00	2.00	2.032	2.032	1.983	1.909
Al	--	--	--	--	--	--	0.017	0.091
Fe^{3+}	--	--	--	--	--	--	--	--
Ti^{4+}	--	--	--	--	--	--	--	--
Cr	--	--	--	--	--	--	--	--
Total T Cations	2.00	2.00	2.00	2.00	2.032	2.032	2.000	2.000
Al	--	--	--	--	0.043	0.043	0.043	0.008
Ti^{3+}	--	--	--	--	--	--	--	--
Ti^{4+}	--	--	--	--	0.026	0.026	0.026	0.020
Cr^{3+}	--	--	--	--	--	--	--	0.032
Fe^{3+}	--	--	--	--	--	--	--	--
Sc	--	--	--	--	--	--	--	--
In	--	--	--	--	--	--	--	--
Ni	--	--	--	--	--	--	--	--
Mg	--	2.00	1.888	1.832	0.773	0.773	1.050	1.217
Co	--	--	--	--	--	--	--	--
Zn	--	--	--	--	--	--	--	--
Fe^{2+}	1.70	--	--	--	1.002	1.002	0.718	0.545
Mn^{2+}	--	--	--	--	0.034	0.034	--	--
Ca	0.30	--	0.112	0.168	--	--	0.163	0.183
Li	--	--	--	--	--	--	--	--
Na	--	--	--	--	0.018	0.018	--	--
K	--	--	--	--	--	--	--	--
Total M Cations	2.00	2.00	2.000	2.000	1.896	1.896	2.000	2.005
RELEVANT FOOTNOTES	b	b	b	b	g,k	g,k	h,i	d,k
REFERENCE	Ohashi et al. (1975)	Ohashi and Finger (1976)	Ohashi and Finger (1976)	Ohashi and Finger (1976)	Hallimond (1914)	Hallimond (1914)	Clark et al. (1971)	Takeda (1972a)

75

Table A4. Cont'd. (page 11 of 13 pages).

NUMBER	64.	65.	66.	67.	68.	68b.	69.
PYROXENE NAME	pigeonite	pigeonite	pigeonite	clinohypersthene	clinopyroxene (Mn,Mg)	--	protoenstatite
APPROXIMATE FORMULA (ordered pyroxenes only)	--	--	--	--	--	$MnSiO_3$	$Mg_2Si_2O_6$
UNIT CELL CONTENTS (based on 6 oxygens)							
Si	1.993	1.985	2.00	1.997	2.00	2.00	2.00
Al	0.007	0.015	--	0.003	--	--	--
Fe^{3+}	--	--	--	--	--	--	--
Ti^{4+}	--	--	--	--	--	--	--
Cr	--	--	--	--	--	--	--
Total T Cations	2.000	2.000	2.00	2.000	2.00	2.00	2.00
Al	0.005	0.004	--	0.002	--	--	--
Ti^{3+}	--	--	--	--	--	--	--
Ti^{4+}	0.015	0.009	--	--	--	--	--
Cr^{3+}	--	0.005	--	0.002	--	--	--
Fe^{3+}	--	--	--	--	--	--	--
Sc	--	--	--	--	--	--	--
In	--	--	--	--	t	--	--
Ni	--	--	--	--	--	--	--
Mg	0.767	0.717	1.36	0.636	1.10	--	2.00
Co	--	--	--	--	--	--	--
Zn	--	--	--	--	--	--	--
Fe^{2+}	1.092	1.183	0.52	1.327	--	--	--
Mn^{2+}	--	0.039	--	0.032	0.90	2.00	--
Ca	0.106	0.036	0.12	--	--	--	--
Li	--	--	--	--	--	--	--
Na	0.001	0.001	--	--	--	--	--
K	--	--	--	--	--	--	--
Total M Cations	1.986	1.994	2.00	1.999	2.00	2.00	2.00
RELEVANT FOOTNOTES	e	e,g	b	c	b	b	b
REFERENCE (chemical data)	Takeda et al. (1974)	Takeda et al. (1974)	Ohashi and Finger (1974b)	Smyth (1974b)	Ghose et al. (1975)	Tokonami et al. (1979)	Smyth (1971)

76

Table A4. Cont'd. (page 12 of 13 pages).

NUMBER PYROXENE NAME APPROXIMATE FORMULA (ordered pyroxenes only)	70. protopyroxene (Mg,Li,Sc) --	71. omphacite --	72. titanian ferro-omphacite --	73. omphacite --	74. omphacite --
UNIT CELL CONTENTS (based on 6 oxygens)					
Si	2.00	1.918	1.952	1.96	1.98
Al	--	0.082	0.048	0.04	0.02
Fe^{3+}	--	--	--	--	--
Ti^{4+}	--	--	--	--	--
Cr	--	--	--	--	--
Total T Cations	2.00	2.000	2.000	2.00	2.00
Al	--	0.398	0.399	0.39	0.51
Ti^{3+}	--	--	--	--	--
Ti^{4+}	--	0.005	0.073	--	--
Cr^{3+}	--	--	--	--	0.01
Fe^{3+}	--	0.137	0.141	0.10	0.02
Sc	0.30	--	--	--	--
In	--	--	--	--	--
Ni	--	--	--	--	--
Mg	1.40	0.392	0.053	0.44	0.42
Co	--	--	--	--	--
Zn	--	--	--	--	--
Fe^{2+}	--	0.077	0.334	0.10	0.05
Mn^{2+}	--	--	0.018	--	--
Ca	--	0.516	0.426	0.51	0.47
Li	0.30	--	--	--	--
Na	--	0.484	0.548	0.48	0.48
K	--	--	0.002	--	--
Total M Cations	2.000	2.009	1.999	2.02	1.96
RELEVANT FOOTNOTES	d	f,j	f	f	f
REFERENCE (chemical data)	Smyth and Ito (1977)	Matsumoto et al. (1975)	Curtis et al. (1975)	Clark and Papike (1968)	Clark et al. (1969)

77

Table A4. Cont'd. (page 13 of 13 pages).

[a]See cited publication for details. In some instances the cited publication reports an analysis originally described in an earlier paper.

[b]Only formula given in reference(s) cited. No oxides presented. Unit cells contents and/or site assignments by present authors.

[c]Only unit cell contents given in reference(s) cited. No oxides presented. Site assignments by present authors.

[d]Only unit cell contents and site assignments given in reference(s) cited. No oxides presented.

[e]Oxides and unit cell contents given in reference(s) cited. Site assignments by present authors.

[f]Oxides, unit cell contents, and site assignments given in reference(s) cited.

[g]Only oxides presented. Unit cell contents calculated and site assignments by present authors.

[h]Oxide analysis of entire grain includes ≈10% augite. Formula presented here is for host only.

[i]Unit cell contents calculated on basis of four cations.

[j]Analysis also shows minor amounts of Li_2O, V_2O_5, MoO_3, and/or P_2O_5. Unit cell contents calculated excluding these components.

[k]Contains augite lamellae.

[m]Contains pigeonite lamellae. Analysis has been adjusted to remove pigeonite component.

[n]Analysis presented was derived by subtracting impurities from an original bulk analysis.

[p]Fe^{2+} includes Mn, Ti, Cr.

[r]Fe^{3+} calculated by present authors.

[s]Ti partitioned so that cations sum to 4.0.

Table A5. Site Occupancies of Disordered Silicate Pyroxenes. Crystal Data Given in Table A1.

No.	Pyroxene	Space Group	Site	Occupancy	Relevant Footnotes	Reference
19.	Augite	$C2/c$	M1 M2 T	$0.715Mg + 0.182Al + 0.103Fe$ $0.616Ca + 0.187Mg + 0.107Fe + 0.090Na$ $0.91Si + 0.08Al + 0.01Ti$	a	Clark et al. (1969)
20.	Augite	$C2/c$	M1 M2 T	$0.68Mg + 0.11Al + 0.03Ti + 0.09Fe^{2+} + 0.09Fe^{3+}$ $0.74Ca + 0.06Na + 0.11Mg + 0.09Fe^{2+}$ $0.865Si + 0.135Al$	a,c,g	Takeda (1972b)
21.	Augite	$C2/c$	M1 M2 T	$0.61Mg + 0.39Fe$ $0.83Ca + 0.13Fe + 0.04Mg$ $0.973Si + 0.027Al$	a,c,g	Takeda et al. (1974)
22.	Augite	$C2/c$	M1 M2 T	$0.63Mg + 0.37Fe$ $0.91Ca + 0.06Fe + 0.03Mg$ $0.963Si + 0.018Al + 0.003Ti + 0.016Fe$	a,c,g	Takeda et al. (1974)
23.	Augite	$C2/c$	M1 M2 T	$0.64Mg + 0.22Fe + 0.08Ti^{4+} + 0.06Al$ $0.61Ca + 0.27Fe + 0.12Mg$ $0.893Si + 0.107Al$	a,c	Takeda (1972a)
24.	Subcalcic diopside	$C2/c$	M1 M2 T	$0.854Mg + 0.029Fe^{2+} + 0.025Fe^{3+} + 0.092Al$ $0.553Ca + 0.233Mg + 0.118Na + 0.096Fe^{2+}$ $0.987Si + 0.013Al$	a,c	McCallister et al. (1974)
24b.	Clinopyroxene	$C2/c$	M1 M2 T	$0.684Mg + 0.172Al + 0.121Fe + 0.023Ti$ $0.634Ca + 0.201Mg + 0.093Na + 0.072Fe$ $0.911Si + 0.089Al$	a,c	McCallister et al. (1976)
24c.	Clinopyroxene	$C2/c$	M1 M2 T	$0.604Mg + 0.202Fe + 0.170Al + 0.024Ti$ $0.594Ca + 0.198Na + 0.136Fe + 0.072Mg$ $0.927Si + 0.073Al.$	a,c	McCallister et al. (1976)
25.	Fassaite (titanaugite)	$C2/c$	M1 M2 T	$0.555Mg + 0.166Al + 0.155Fe^{3+} + 0.063Ti^{4+} + 0.061Fe^{2+}$ $0.986Ca + 0.007Na + 0.007Mn$ $0.753Si + 0.247Al$	a,m	Peacor (1967)
26.	Fassaite	$C2/c$	M1 M2 T	$0.48Ti + 0.39Mg + 0.13Al$ $1.00Ca$ $0.645Si + 0.365Al$	u	Dowty and Clark (1973)

Table A5.   Cont'd.

#	Mineral	Space group	Site	Formula	Codes	Reference
27.	Fassaite	C2/c	M1	$0.568Mg + 0.210Fe + 0.161Al + 0.059Ti$	a,c	Hazen and Finger (1977a)
			M2	$0.968Ca + 0.020Fe + 0.010Mg$		
			T	$0.864Si + 0.136Al$		
28.	Ca tschermakite's pyroxene	C2/c	M1	$0.821Fe^{3+} + 0.179Al$	a,h	Ghose et al. (1975)
			M2	$1.00Ca$		
			T	$1.00Si + 0.821Al + 0.179Fe^{3+}$		
29.	Omphacite	C2/c	M1	$0.543Mg + 0.240Al + 0.217Fe$	a,e	Clark et al. (1969)
			M2	$0.590Ca + 0.320Na + 0.057Mg + 0.033Fe$		
			T	$0.998Si + 0.002Al$		
33.	Bronzite	Pbca	M1	$0.86Mg + 0.07Fe^{3+} + 0.03Al + 0.03Fe^{2+} + 0.01Ti$	a,c,i	Takeda (1972b)
			M2	$0.64Mg + 0.29Fe + 0.06Ca + 0.01Na$		
			TA	$0.925Si + 0.075Al$		
			TB	$0.925Si + 0.075Al$		
34.	Bronzite	Pbca	M1	$0.91Mg + 0.07Fe + 0.02(Ti + Al + Cr + Mn)$	a,i	Takeda and Ridley (1972)
			M2	$0.50Fe + 0.44Mg + 0.06Ca$		
			TA	$0.985Si + 0.015Al$		
			TB	$0.985Si + 0.015Al$		
35.	Bronzite	Pbca	M1	$0.89Mg + 0.064Fe + 0.05(Ti + Al + Cr + Mn)$	a,i	Takeda and Ridley (1972)
			M2	$0.52Mg + 0.39Fe + 0.09Ca$		
			TA	$0.97Si + 0.03Al$		
			TB	$0.97Si + 0.03Al$		
36.	Bronzite	Pbca	M1	$0.967Mg + 0.033Fe^{2+}$	a,c,i	Dodd et al. (1975)
			M2	$0.660Mg + 0.317Fe^{2+} + 0.023Ca$		
			TA	$0.987Si + 0.010Cr + 0.002Al + 0.001Ti$		
			TB	$0.987Si + 0.010Cr + 0.002Al + 0.001Ti$		
37.	Bronzite	Pbca	M1	$0.956Mg + 0.044Fe$	a,c,i	Miyamoto et al. (1975)
			M2	$0.511Mg + 0.435Fe + 0.054Ca$		
			TA	$0.979Si + 0.021Al$		
			TB	$0.979Si + 0.021Al$		
38.	Bronzite	Pbca	M1	$0.871Mg + 0.129Fe$	a,d,i	Ghose and Wan (1973)
			M2	$0.593Mg + 0.407Fe$		
			TA	$0.973Si + 0.027Al$		
			TB	$0.973Si + 0.027Al$		
39.	Aluminous bronzite	Pbca	M1	$0.764Mg + 0.200Al + 0.036Fe$	a,e,j	Kosoi et al. (1974)
			M2	$0.631Mg + 0.359Fe + 0.010Ca$		

80

Table A5. Cont'd.

No.	Mineral	Space Group	Site	Composition	Refs	Reference
			TA	$0.996Si + 0.004Ti$	a,e,j	Kosoi et al. (1974)
			TB	$0.756Si + 0.240Al + 0.004Ti$		
40.	Aluminous bronzite	Pbca	M1	$0.700Mg + 0.210Al + 0.090Fe$	a,d,i	Ghose (1965)
			M2	$0.685Mg + 0.305Fe + 0.010Ca$		
			TA	$0.996Si + 0.004Ti$		
			TB	$0.756Si + 0.240Al + 0.004Ti$		
41.	Hypersthene	Pbca	M1	$0.85Mg + 0.15Fe^{2+}$	a,e,i	Kosoi et al. (1974)
			M2	$0.90Fe^{2+} + 0.10Mg$		
			TA	$0.97Si + 0.03Al$		
			TB	$0.97Si + 0.03Al$		
42.	Hypersthene	Pbca	M1	$0.931Mg + 0.069Fe$	a,h	Ghose et al. (1975)
			M2	$0.811Fe + 0.159Mg + 0.030Ca$		
			TA	$0.971Si + 0.028Al + 0.001Ti$		
			TB	$0.971Si + 0.028Al + 0.001Ti$		
43.	Hypersthene	Pbca	M1	$0.810Mg + 0.190Fe^{2+}$	a,i	Smyth (1973)
			M2	$0.396Mg + 0.604Fe^{2+}$		
			TA	$1.00Si$		
			TB	$1.00Si$		
44.	Ferrohypersthene	Pbca	M1	$0.574Mg + 0.425Fe$	k,j,r,s	Brovkin et al. (1975)
			M2	$0.906Fe + 0.062Mg + 0.032Ca$		
			TA	$0.999Si + 0.001Al$		
			TB	$0.999Si + 0.001Al$		
45.	Aluminous hypersthene	Pbca	M1	$0.88Mg + 0.10Al + 0.02Fe^{2+}$	a,d,f,i	Burnham et al. (1971)
			M2	$0.58Mg + 0.25Fe^{2+} + 0.13Al + 0.04Fe^{3+}$		
			TA	$1.00Si$		
			TB	$0.87Si + 0.13Al$		
46.	Eulite	Pbca	M1	$0.75Fe^{2+} + 0.25Mg$	b,h	Hawthorne and Ito (1978)
			M2	$0.96Fe^{2+} + 0.04Ca$		
			TA	$0.992Si + 0.008Al$		
			TB	$0.992Si + 0.008Al$		
47.	(Mg, Co) orthopyroxene	Pbca	M1	$0.871Mg + 0.129Co$	b,h	Hawthorne and Ito (1978)
			M2	$0.681Mg + 0.319Co$		
			TA	$1.00Si$		
			TB	$1.00Si$		
48.	(Mg, Mn) orthopyroxene	Pbca	M1	$0.977Mg + 0.023Mn$		
			M2	$0.873Mg + 0.127Mn$		

Table A5. Cont'd.

No.	Mineral	Space group	Site	Composition	Notes	Reference
			TA	1.00Si		
			TB	1.00Si		
49.	(Mg, Mn, Co) orthopyroxene	$\underline{Pbca}$	M1	$0.904Mg + 0.065Co^{2+} + 0.031Mn^{2+}$	b,h,k	Hawthorne and Ito (1977)
			M2	$0.658Mg + 0.198Co^{2+} + 0.144Mn^{2+}$		
			TA	1.00Si		
			TB	1.00Si		
50.	(Zn, Mg) orthopyroxene	$\underline{Pbca}$	M1	$0.64Mg + 0.36Zn$	v,h	Morimoto et al. (1975)
			M2	$0.64Zn + 0.36Mg$		
			TA	1.00Si		
			TB	1.00Si		
51.	(Zn, Mg) orthopyroxene	$\underline{Pbca}$	M1	$0.933Mg + 0.067Zn$	b,h	Ghose et al. (1975)
			M2	$0.617Mg + 0.383Zn$		
			TA	1.00Si		
			TB	1.00Si		
52.	(Ni, Mg) orthopyroxene	$\underline{Pbca}$	M1	$0.789Mg + 0.211Ni$	b,h	Ghose et al. (1975)
			M2	$0.831Mg + 0.169Ni$		
			TA	1.00Si		
			TB	1.00Si		
53.	(Co, Mg) orthopyroxene	$\underline{Pbca}$	M1	$0.735Mg + 0.265Co$	b,h	Ghose et al. (1975)
			M2	$0.525Mg + 0.475Co$		
			TA	1.00Si		
			TB	1.00Si		
60.	Pigeonite	$P2_1/\underline{c}$	M1	$0.720Mg + 0.280Fe$	a,h	Morimoto and Güven (1970)
			M2	$0.760Fe + 0.18Ca + 0.060Mg$		
			TA	1.00Si		
			TB	1.00Si		
61.	Pigeonite	$P2_1/\underline{c}$	M1	$0.702Mg + 0.298Fe$	a,h	Brown et al. (1972)
			M2	$0.742Fe + 0.180Ca + 0.078Mg$		
			TA	1.00Si		
			TB	1.00Si		
62.	Pigeonite	$P2_1/\underline{c}$	M1	$0.78Mg + 0.15Fe^{2+} + 0.03Ti + 0.04Al$	a,i	Clark et al. (1971)
			M2	$0.57Fe^{2+} + 0.27Mg + 0.16Ca$		
			TA	$0.991Si + 0.009Al$		
			TB	$0.991Si + 0.009Al$		
63.	Pigeonite	$P2_1/\underline{c}$	M1	$0.86Mg + 0.12Fe + 0.01Al + 0.02Ti^{4+}$	a,c,i	Takeda (1972a)

82

Table A5. Cont'd.

No.	Mineral	Space group	Site	Composition	Code	Reference
64.	Pigeonite	$P2_1/c$	M2	0.46Fe + 0.35Mg + 0.18Ca	a,i	Takeda et al. (1974)
			TA	0.955Si + 0.045Al		
			TB	0.955Si + 0.045Al		
65.	Pigeonite	$P2_1/c$	M1	0.62Mg + 0.38Fe	a,c,i	Takeda et al. (1974)
			M2	0.69Fe + 0.20Mg + 0.106Ca		
			TA	0.997Si + 0.003Al		
			TB	0.997Si + 0.003Al		
66.	Pigeonite	$P2_1/c$	M1	0.66Mg + 0.34Fe	a,h	Ohashi and Finger (1974b)
			M2	0.90Fe + 0.06Mg + 0.04Ca		
			TA	0.996Si + 0.003Al + 0.001Ti		
			TB	0.996Si + 0.003Al + 0.001Ti		
			M1	0.930Mg + 0.070Fe		
			M2	0.450Fe + 0.430Mg + 0.120Ca		
			TA	1.00Si		
			TB	1.00Si		
67.	Clinohypersthene	$P2_1/c$	M1	0.503Fe + 0.497Mg	a,i	Smyth (1974b)
			M2	0.834Fe + 0.134Mg + 0.032Ca		
			TA	0.999Si + 0.001Al		
			TB	0.999Si + 0.001Al		
68.	(Mn, Mg) clinopyroxene	$P2_1/c$	M1	0.882Mg + 0.118Mn	b,h	Ghose et al. (1975)
			M2	0.218Mg + 0.782Mn		
			TA	1.00Si		
			TB	1.00Si		
70.	(Mg,Li,Sc) protopyroxene	Pbcn	M1	0.70 Mg + 0.30Sc	b	Smyth and Ito (1977)
			M2	0.70 Mg + 0.30Li		
			T	1.00Si		
71.	Omphacite	$P2/n$	M1	0.815Mg + 0.185Fe	a,i	Matsumoto et al. (1975)
			M1(1)	0.868Al + 0.132Fe		
			M2	0.686Na + 0.314Ca		
			M2(1)	0.716Ca + 0.284Na		
			T1	0.959Si + 0.041Al		
			T2	0.959Si + 0.041Al		
72.	Titanian ferro-omphacite	$P2/n$	M1	0.786Fe + 0.214Al	a,i	Curtis et al. (1975)
			M1(1)	0.749Al + 0.251Fe		
			M2	0.740Na + 0.260Ca		
			M2(1)	0.672Ca + 0.328Na		
			T1	0.976Si + 0.024Al		
			T2	0.976Si + 0.024Al		

83

Table A5. Cont'd.

73.	Omphacite	P2	M1	0.81Mg + 0.19Fe^{2+}	a,i,t	Clark and Papike (1968)
			M1(1)	0.95Al + 0.05Fe^{3+}		
			M1H	0.82Al + 0.18Fe^{3+}		
			M1(1)H	0.80Mg + 0.20Fe^{2+}		
			M2	0.64Na + 0.36Ca		
			M2(1)	0.64Ca + 0.36Na		
			M2H	0.97Ca + 0.03Na		
			M2(1)H	0.64Na + 0.36Ca		
			T1A	0.98Si + 0.02Al		
			T2A	0.98Si + 0.02Al		
			T1C	0.98Si + 0.02Al		
			T2C	0.98Si + 0.02Al		
74.	Omphacite	P2	M1	0.82Mg + 0.10Fe + 0.08Al	a,i,t	Clark et al. (1969)
			M1(1)	1.00Al		
			M1H	0.92Al + 0.08Fe		
			M1(1)H	0.90Mg + 0.10Fe		
			M2	0.76Na + 0.24Ca		
			M2(1)	0.75Ca + 0.25Na		
			M2H	0.69Ca + 0.31Na		
			M2(1)H	0.68Na + 0.32Ca		
			T1A	0.99Si + 0.01Al		
			T2A	0.99Si + 0.01Al		
			T1C	0.99Si + 0.01Al		
			T2C	0.99Si + 0.01Al		

a Assignment of Fe vs Mg in M sites based on site occupancy refinement and/or adjustment of temperature factors, form factors, etc. Remainder of cations assigned to M1 and M2 using crystal chemical principles.

b Assignment of cations to M sites based on site occupancy refinement.

c Minor Ti and/or Cr and/or Mn included with Fe in site occupancy refinement.

d Minor Al and/or Cr and/or Ti and/or Mn ignored in site occupancy refinement.

e Assignment of minor Mn and/or Cr and/or Na and/or Ti not reported.

f Minor ^{VI}Al included with Mg.

g Assignment of cations to T sites done by present authors using crystal chemical principles.

h Tetrahedral site occupancy fixed at 1.0 Si based on general formula in the reference cited.

i No site refinement of Al and Si in T sites attempted and identical site occupancies were assigned by present authors. See original papers for details because bond distances in some crystals indicate a concentration of Al in TB.

j Assignment of Si vs Al in T sites based on site occupancy refinement and/or adjustment of temperature factors, etc.

k T site occupancies adjusted on basis of interatomic distances.

m Cation fractions normalized by present authors so that total occupancy of each site equals 1.0.

n Equal amounts of Ti assigned to TA and TB.

r Al in M1 and M2 not constrained by known composition of sample.

s Authors stated that most of Al occupies the TB position, but in another sentence they state that 13 per cent of the Al atoms are in this position. The discrepancy may have arisen during the translation from Russian and the 13 per cent may indicate 0.13 Al atoms of the 0.17 Al atoms occupy TB.

t Site labels are from cited reference

u Cations assigned to M and T sites using crystal chemical principles and interactomic distances.

v Assignment of cations to M sites using Fourier syntheses.

Table A6. Pyroxene Structures Studied at Elevated Temperatures.

No.	Pyroxene[a]	Space Group	Temperatures (°C) Studied at One Atmosphere Pressure	Mean Thermal Expansion Coefficient ($°C^{-1} \times 10^{-5}$)[b]							
				α_a	$\alpha_{d_{100}}$	α_b	α_c	α_V	α Mean T-O	α Mean M1-O	α Mean M2-O
2.	diopside[c]	C2/c	24,400,700,850,1000	0.779[d]	0.606	2.05	0.646	3.33	0.099	1.44	1.64
4.	hedenbergite	C2/c	24,400,600,800,900,1000	0.724	0.483	1.76	0.597	2.98	0.026	1.05	1.61
6.	jadeite[c]	$\bar{C}2/c$	24,400,600,800	0.850	0.817	1.00	0.631	2.47	0.156	0.947	1.28
7.	ureyite[c]	C2/c	24,400,600	0.585	0.691	0.946	0.390	2.04	0.529	0.633	1.26
8.	acmite[c]	C2/c	24,400,600,800	0.727	0.804	1.20	0.450	2.47	0.182	0.781	1.28
11.	spodumene[c]	$\bar{C}2/c$	24,300,460,760	0.380	0.600	1.11	0.475	2.22	0.160	1.06	1.97
32.	orthoferrosilite	Pbca	24,400,600,800,900,980	0.112	--	0.109	0.168	0.393	-0.350(TA) / -0.736(TB)	1.59	4.59
44.	ferrohypersthene	Pbca	20,175,280,500,700,850	0.135	--	0.145	0.154	0.438	-0.285(TA) / -0.192(TB)	2.26	4.23
61.	pigeonite	$P2_1/c$[e]	24,960	--	--	--	--	--		--	--
67.	clinohypersthene	$\bar{P}2_1/\bar{c}$[f]	20,200,400,600,700	0.162	0.083	0.104	0.138	0.327	-0.212(TA) / -2.88(TB)	1.476	4.70

[a] References for data used in calculations: No. 2-11 from Cameron et al. (1973), No. 32 from Sueno et al. (1976), No. 44 from Smyth (1973), No. 61 from Brown et al. (1972), No. 67 from Smyth (1974b).

[b] Mean thermal expansion coefficients are computed from the equation $\alpha_x = (1/X_{24}) \cdot (X_T - X_{24})/(T-24)$, where the slope of the regression equation is used for the term $(X_T - X_{24})$.

[c] Crystals used in this study were obtained from samples previously used by Prewitt and Burnham (1966; jadeite) and Clark et al. (1969; spodumene, acmite, diopside, and ureyite).

[d] For 0.779, read $0.779 \times 10^{-5} °C^{-1}$.

[e] Transformed to C2/c above $\approx 960°C$.

[f] Transformed to $\bar{C}2/c$ above $\approx 725°C$.

85

Table A7. Studies (post-1962) Describing Unit Cell Variations in Pyroxenes.

Pyroxenes / Join / System	Type of Specimens	Reference
Ca-Mg-Fe quadrilateral pyroxenes	synthetic	Turnock et al. (1973)
quadrilateral clinopyroxenes	natural	Viswanathan (1966)
~$(Fe,Mg)_2Si_2O_6$ orthopyroxenes	natural + synthetic	Smith et al. (1969)
$(Fe,Mg)_2Si_2O_6$; $(Co,Mg)_2Si_2O_6$; $(Ni,Mg)_2Si_2O_6$ orthopyroxenes	synthetic	Matsui et al. (1968)
aluminous orthorhombic enstatites	synthetic	Skinner and Boyd (1964)
~$(Fe,Mg)_2Si_2O_6$ orthopyroxenes	natural	Howie (1963)
C2/c clinopyroxenes	natural + synthetic	Ribbe and Prunier (1977)
$NaAlSi_2O_6-NaCrSi_2O_6$	synthetic	Abs-Wurmbach and Neuhaus (1976)
$LiM^{3+}Si_2O_6$ series	synthetic	Drysdale (1975)
$CaFe^{3+}SiFe^{3+}O_6-CaAlSiAlO_6$	synthetic	Huckenholz et al. (1974)
$CaMgSi_2O_6-NaCrSi_2O_6$	synthetic	Ikeda and Yagi (1972)
$LiM^{3+}Si_2O_6$ and $NaM^{3+}Si_2O_6$ series	synthetic	Brown (1971)
$Fs_{85}En_{15}-Ca_2Si_2O_6$ join (within quadrilateral)	synthetic	Smith and Finger (1971)
$NaAlSi_2O_6-CaAl_2Si_2O_8$ (anorthite)	natural	Mao (1970)
omphacites and related pyroxenes	synthetic	Edgar et al. (1969)
$CaMgSi_2O_6-CaFe^{3+}SiFe^{3+}O_6$	synthetic	Huckenholz et al. (1969)
$CaMgSi_2O_6-NaAlSi_3O_8$ (albite)	synthetic	Kushiro (1969)
$CaFe_2^{2+}Si_2O_6-Fe_2^{2+}Si_2O_6$	synthetic	Lindsley et al. (1969)
$CaMgSi_2O_6-CaFe^{2+}Si_2O_6-NaFe^{3+}Si_2O_6$	natural + synthetic	Nolan (1969)
$CaMgSi_2O_6-CaTiAl_2O_6$ and $CaMgSi_2O_6-CaMgTi_2O_6$	synthetic	Onuma et al. (1968)
$CaMgSi_2O_6-CaFe^{2+}Si_2O_6$	synthetic	Rutstein and Yund (1968)
salites and augites	natural	Lewis (1967)
$NaAlSi_2O_6-NaFe^{3+}Si_2O_6$	synthetic	Gilbert (1967)
$NaFe^{3+}Si_2O_6-CaMgSi_2O_6$	synthetic	Nolan and Edgar (1963)
various diopsides	synthetic	Coleman (1962)
$CaMgSi_2O_6-CaAlSiAlO_6$	synthetic	Clark et al. (1962)

Abs-Wurmbach, I. and Neuhaus, A. (1976) The system $NaAlSi_2O_6$ (jadeite)-$NaCrSi_2O_6$ (kosmochlor) in the pressure range 1 bar to 25 kbars at $800°C$. N. Jahrb. Mineral. Abh., 127, 213-241.

Appleman, D. E., Boyd, Jr., F. R., Brown, G. M., Ernst, W. G., Gibbs, G. V., and Smith, J. V. (1966) *AGI Short Course on Chain Silicates.*

Baur, W. H. (1971) The prediction of bond length variations in silicon-oxygen bonds. Am. Mineral., 56, 1573-1599.

Brovkin, A. A., Novoselov, Yu. M., and Kitsul, V. I. (1975) Distribution of aluminum in the crystal structure of high-alumina hypersthene. Doklady Akad. Nauk SSSR, 223, 148-150.

Brown, G. E. and Gibbs, G. V. (1969) Oxygen coordination and the Si-O bond. Am. Mineral., 54, 1528-1539.

_____ and _____ (1970) Stereochemistry and ordering in the tetrahedral portion of silicates. Am. Mineral., 55, 1587-1607.

_____, _____, and Ribbe, P. H. (1969) The nature and variation in length of the Si-O and Al-O bonds in framework silicates. Am. Mineral., 54, 1044-1061.

_____, Prewitt, C. T., Papike, J. J., and Sueno, S. (1972) A comparison of the structures of low and high pigeonite. J. Geophys. Res., 77, 5778-5789.

Brown, G. M. (1967) Mineralogy of basaltic rocks. *In* H. H. Hess and A. Poldervaart, Eds., *Basalts,* 1, p. 103-162. John Wiley, New York.

_____ (1972) Pigeonitic pyroxenes: a review. Geol. Soc. Am. Memoir, 132, 523-534.

Brown, I. D. and Shannon, R. D. (1973) Empirical bond-strength-bond-length curves for oxides. Acta Crystallogr., A29, 266-282.

Brown, W. L. (1971) On lithium and sodium trivalent-metal pyroxenes and crystal-field effects. Mineral. Mag., 38, 43-48.

_____ (1972) La symétrie et les solutions solides des clinopyroxènes. Bull. Soc. franc. minéral. cristallogr., 95, 574-582.

Burnham, C. W. (1965) Ferrosilite. Carnegie Inst. Washington Year Book, 64, 202-204.

_____ (1966) Ferrosilite. Carnegie Inst. Washington Year Book, 65, 285-290.

_____, Clark, J. R., Papike, J. J., and Prewitt, C. T. (1967) A proposed crystallographic nomenclature for clinopyroxene structures. Z. Kristallogr., 125, 1-6.

_____, Ohashi, Y., Hafner, S. S., and Virgo, D. (1971) Cation distribution and atomic thermal vibrations in an iron-rich orthopyroxene. Am. Mineral., 56, 850-876.

Burns, R. G. (1970) *Mineralogical Applications of Crystal Field Theory.* Cambridge University Press, Cambridge, England.

Buseck, P. R. and Iijima, S. (1975) High resolution electron microscopy of enstatite. II. Geological application. Am. Mineral., 60, 771-784.

Cameron, M., Sueno, S., Prewitt, C. T., and Papike, J. J. (1973) High-temperature crystal chemistry of acmite, diopside, hedenbergite, jadeite, spodumene and ureyite. Am. Mineral., 58, 594-618.

_____ and Papike, J. J. (1981) Structural and chemical variations in pyroxenes. Am. Mineral., in press.

Carpenter, M. A. (1978) Kinetic control of ordering and exsolution in omphacite. Contrib. Mineral. Petrol., 67, 17-24.

Clark, J. R. and Papike, J. J. (1968) Crystal-chemical characterization of omphacites. Am. Mineral., 53, 840-868.

_____, Appleman, D. E., and Papike, J. J. (1968) Bonding in eight ordered clinopyroxenes isostructural with diopside. Contrib. Mineral. Petrol., 20, 81-85.

_____, _____, and _____ (1969) Crystal-chemical characterization of clinopyroxenes based on eight new structure refinements. Mineral. Soc. Am. Spec. Paper, 2, 31-50.

_____, Ross, M., and Appleman, D. E. (1971) Crystal chemistry of a lunar pigeonite. Am. Mineral., 56, 888-908.

87

Clark, Jr., S. P., Schairer, J. F., and de Neufville, F. (1962) Phase relations in the system $CaMgSi_2O_6-CaAl_2SiO_6-SiO_2$ at low and high pressure. Carnegie Inst. Washington Year Book, 61, 59-68.

Coleman, L. C. (1962) Effect of ionic substitution on the unit-cell dimensions of synthetic diopside. *In* A. E. J. Engel *et al.*, Eds., *Petrologic Studies: A Volume to Honor A. F. Buddington*, p. 429-446. Geological Society of America.

Curtis, L., Gittins, J., Kocman, V., Rucklidge, J. C., Hawthorne, F. C., and Ferguson, R. B. (1975) Two crystal structure refinements of a $P2/n$ titanian ferro-omphacite. Canadian Mineral., 13, 62-67.

Deer, W. A., Howie, R. A., and Zussman, J. (1978) *Rock-Forming Minerals*, Vol. 2A, Second Edition, *Single Chain Silicates*. John Wiley, New York.

Dodd, R. T., Grover, J. E., and Brown, G. E. (1975) Pyroxenes in the Shaw (L-7) chondrite. Geochim. Cosmochim. Acta, 39, 1585-1594.

Dowty, E. and Clark, J. R. (1973) Crystal structure refinement and optical properties of a Ti^{3+} fassaite from the Allende meteorite. Am. Mineral., 58, 230-242.

Drysdale, D. J. (1975) Hydrothermal synthesis of various spodumenes. Am. Mineral., 60, 105-110.

Edgar, A. D., Mottana, A., and Macrae, N. D. (1969) The chemistry and cell parameters of omphacites and related pyroxenes. Mineral. Mag., 37, 61-74.

Ferguson, R. B. (1974) A cation-anion distance-dependent method for evaluating valence-bond distributions in ionic structures and results for some olivines and pyroxenes. Acta Crystallogr., B30, 2527-2539.

Finger, L.W. and Ohashi, Y. (1976) The thermal expansion of diopside to 800°C and a refinement of the crystal structure at 700°C. Am. Mineral., 61, 303-310.

Fleet, M. E., Herzberg, C. T., Bancroft, G. M., and Aldridge, L. P. (1978) Omphacite studies, I. The $P2/n \rightarrow C2/c$ transformation. Am. Mineral., 63, 1100-1106.

Freed, R. L. and Peacor, D. R. (1967) Refinement of the crystal structure of johannsenite. Am. Mineral., 52, 709-720.

Ganguly, J. and Ghose, S. (1979) Aluminous orthopyroxene: order-disorder, thermodynamic properties, and petrologic implications. Contrib. Mineral. Petrol., 69, 375-385.

Ghose, S. (1965) $Mg^{2+}-Fe^{2+}$ order in an orthopyroxene, $Mg_{0.93}Fe_{1.07}Si_2O_6$. Z. Kristallogr., 122, 81-99.

_____ and Wan, C. (1973) Luna 20 pyroxenes: evidence for a complex thermal history. Proc. 4th Lunar Science Conf., 1, 901-907.

_____, _____, and Okamura, F. P. (1975) Site preference and crystal chemistry of transition metal ions in pyroxenes and olivines (abst.). Acta Crystallogr., A31, S76.

Gibbs, G. V., Hamil, M. M., Louisnathan, S. J., Bartell, L. S., and Yow, H. (1972) Correlations between Si-O bond length, Si-O-Si angle and bond overlap populations calculated using extended Hückel molecular orbital theory. Am. Mineral., 57, 1578-1613.

Gilbert, M. C. (1967) X-ray properties of jadeite-acmite pyroxenes. Carnegie Inst. Washington Year Book, 66, 374-375.

Grove, T. L. and Burnham, C. W. (1974) Al-Si disorder in calcium Tschermak's pyroxene, $CaAl_2SiO_6$. EOS, Trans. Am. Geophys. Union, 55, 1202.

Güven, N. (1969) Nature of the coordination polyhedra around M2 cations in pigeonite. Contrib. Mineral. Petrol., 24, 268-274.

Hallimond, A. F. (1914) Optically uniaxial augite from Mull. Mineral. Mag., 17, 97-99.

Hawthorne, F. C. and Grundy, H. D. (1973) Refinement of the crystal structure of $NaScSi_2O_6$. Acta Crystallogr., B29, 2615-2616.

_____ and _____ (1974) Refinement of the crystal structure of $NaInSi_2O_6$. Acta Crystallogr., B30, 1882-1884.

_____ and _____ (1977) Refinement of the crystal structure of $LiScSi_2O_6$ and structural variations in alkali pyroxenes. Canadian Mineral., 15, 50-58.

_____ and Ito, J. (1977) Synthesis and crystal-structure refinement of transition-metal orthopyroxenes I: Orthoenstatite and (Mg,Mn,Co) orthopyroxene. Canadian Mineral., 15, 321-338.

_____ and _____ (1978) Refinement of the crystal structures of $(Mg_{0.776}Co_{0.224})SiO_3$ and $(Mg_{0.925}Mn_{0.075})SiO_3$. Acta Crystallogr., B34, 891-893.

Hazen, R. M. (1977) Temperature, pressure and composition: structurally analogous variables. Phys. Chem. Minerals, 1, 83-94.

_____ and Finger, L. W. (1977a) Crystal structure and compositional variation of Angra dos Reis fassaite. Earth Planet. Sci. Letters, 35, 357-362.

_____ and _____ (1977b) Compressibility and crystal structure of Angra dos Reis fassaite to 52 kbar. Carnegie Inst. Washington Year Book, 76, 512-515.

_____ and Prewitt, C. T. (1977) Effects of temperature and pressure on interatomic distances in oxygen-based minerals. Am. Mineral., 62, 309-315.

Hill, R. J. and Gibbs, G. V. (1979) Variation in d(T-O),d(T...T) and ∠TOT in silica and silicate minerals, phosphates and aluminates. Acta Crystallogr., B35, 25-30.

Howie, R. A. (1963) Cell parameters of orthopyroxenes. Mineral. Soc. Am. Spec. Paper, 1, 213-222.

Huckenholz, H. G., Schairer, J. F., and Yoder, H. S., Jr. (1969) Synthesis and stability of ferri-diopside. Mineral. Soc. Am. Spec. Paper, 2, 163-177.

_____, Lindhuber, W., and Springer, J. (1974) The join $CaSiO_3-Al_2O_3-Fe_2O_3$ of the $CaO-Al_2O_3-Fe_2O_3-SiO_2$ quaternary system and its bearing on the formation of granditic garnets and fassaitic pyroxenes. N. Jahrb. Mineral. Abh., 121, 160-207.

Iijima, S. and Buseck, P. R. (1975) High resolution electron microscopy of enstatite. I: Twinning, polymorphism, and polytypism. Am. Mineral., 60, 758-770.

Ikeda, K. and Yagi, K. (1972) Synthesis of kosmochlor and phase equilibria in the join $CaMgSi_2O_6-NaCrSi_2O_6$. Contrib. Mineral. Petrol., 36, 63-72.

Ito, T. (1950) *X-Ray Studies on Polymorphism*, Maruzen, Tokyo, 231 pp.

Kosoi, A. L., Malkova, L. A., and Frank-Kamenetskii, V. A. (1974) Crystal-chemical characteristics of rhombic pyroxenes. Kristallografiya, 19, 282-288 (transl. Soviet Phys. Crystallogr., 19, 171-174, 1974).

Kushiro, I. (1969) Clinopyroxene solid solutions formed by reactions between diopside and plagioclase at high pressures. Mineral. Soc. Am. Spec. Paper, 2, 179-191.

Law, A. D. and Whittaker, E. J. W. (1980) Rotated and extended model structures in amphiboles and pyroxenes. Mineral. Mag., 43, 565-574.

Levien, L. and Prewitt, C. T. (1980) High-pressure structural study of diopside. Am. Mineral., in press.

Lewis, J. F. (1967) Unit-cell dimensions of some aluminous natural clinopyroxenes. Am. Mineral., 52, 42-54.

Lindsley, D. H., Munoz, J. L., and Finger, L. W. (1969) Unit-cell parameters of clinopyroxenes along the join hedenbergite-ferrosilite. Carnegie Inst. Washington Year Book, 68, 91-92.

Louisnathan, S. J. and Gibbs, G. V. (1972a) The effect of tetrahedral angles on Si-O bond overlap populations for isolated tetrahedra. Am. Mineral., 57, 1614-1642.

_____ and _____ (1972b) Variation of Si-O distances in olivines, sodamelilite and sodium metasilicate as predicted by semi-empirical molecular orbital calculations. Am. Mineral., 57, 1643-1663.

Mao, H. K. (1970) The system jadeite $(NaAlSi_2O_6)$-anorthite $(CaAl_2Si_2O_8)$ at high pressures. Carnegie Inst. Washington Year Book, 69, 163-168.

Matsui, Y., Syono, Y., Akimoto, S.-I., and Kitayama, K. (1968) Unit cell dimensions of some synthetic orthopyroxene group solid solutions. Geochem. J., 2, 61-70.

Matsumoto, T. (1974) Possible structure types derived from *Pbca*-orthopyroxene. Mineral. Mag., 7, 374-383.

_____, Tokonami, M., and Morimoto, N. (1975) The crystal structure of omphacite. Am. Mineral., 60, 634-641.

McCallister, R. H., Finger, L. W., and Ohashi, Y. (1974) Refinement of the crystal structure of a subcalcic diopside. Carnegie Inst. Washington Year Book, 73, 518-522.

_____, _____, and _____ (1976) Intracrystalline Fe^{2+}-Mg equilibria in three natural Ca-rich clinopyroxenes. Am. Mineral., 61, 671-676.

McDonald, W. S. and Cruickshank, D. W. J. (1967) A reinvestigation of the structure of sodium metasilicate, Na_2SiO_3. Acta Crystallogr., 22, 37-43.

Miyamoto, M., Takeda, H., and Takano, Y. (1975) Crystallographic studies of a bronzite in the Johnstown achondrite, Fortschr. Mineral., 52, 389-397.

Morgan, B. A. (1967) *Geology of the Valencia Area, Carabobo, Venuzuela.* Ph.D. Thesis, Princeton University, Princeton, New Jersey.

Morimoto, N. (1974) Crystal structure and fine texture of pyroxenes. Fortschr. Mineral., 52, 52-80.

———— and Güven, N. (1970) Refinement of the crystal structure of pigeonite. Am. Mineral., 55, 1195-1209.

———— and Koto, K. (1969) The crystal structure of orthoenstatite. Z. Kristallogr., 129, 65-83.

————, Appleman, D. E., and Evans, Jr., H. T. (1960) The crystal structures of clino-enstatite and pigeonite. Z. Kristallogr., 114, 120-147.

————, Nakajima, Y., Syono, Y., Akimoto, S., and Matsui, Y. (1975) Crystal structures of pyroxene-type $ZnSiO_3$ and $ZnMgSi_2O_6$. Acta Crystallogr., B31, 1041-1049.

Nolan, J. (1969) Physical properties of synthetic and natural pyroxenes in the system diopside-hedenbergite-acmite. Mineral. Mag., 37, 216-229.

———— and Edgar, A. D. (1963) An X-ray investigation of synthetic pyroxenes in the system acmite-diopside-water at 100 kg/cm^2 water-vapour pressure. Mineral. Mag., 33, 625-634.

Ohashi, Y. and Burnham, C. W. (1973) Clinopyroxene lattice deformations: the roles of chemical substitution and temperature. Am. Mineral., 58, 843-849.

———— and Finger, L. W. (1974a) Symmetry reduction and twinning relationships in clino- and orthopyroxenes. Carnegie Inst. Washington Year Book, 73, 531-535.

———— and ———— (1974b) A lunar pigeonite: crystal structure of primitive-cell domains. Carnegie Inst. Washington Year Book, 73, 525-531.

———— and ———— (1974c) The effects of cation substitution on the symmetry and the tetrahedral chain configuration in pyroxenes. Carnegie Inst. Washington Year Book, 73, 522-525.

———— and ———— (1976) The effect of Ca substitution on the structure of clino-enstatite. Carnegie Inst. Washington Year Book, 75, 743-746.

————, Burnham, C. W., and Finger, L. W. (1975) The effect of Ca-Fe substitution on the clinopyroxene crystal structure. Am. Mineral., 60, 423-434.

Okamura, F. P., Ghose, S., and Ohashi, H. (1974) Structure and crystal chemistry of calcium Tschermak's pyroxene, $CaAlAlSiO_6$. Am. Mineral., 59, 549-557.

Onuma, K., Hijikata, K., and Yagi, K. (1968) Unit-cell dimensions of synthetic titan-bearing clinopyroxenes. J. Faculty of Science, Hokkaido Univ., Series IV, Geol. Mineral., XIV, 111-121.

Pannhorst, W. (1979) Structural relationships between pyroxenes. N. Jahrb. Mineral. Abh., 135, 1-17.

Papike, J. J. and Cameron, M. (1976) Crystal chemistry of silicate minerals of geophysical interest. Rev. Geophys. Space Physics, 14, 37-80.

———— and Ross, M. (1970) Gedrites: crystal structures and intracrystalline cation distributions. Am. Mineral., 55, 1945-1972.

————, Prewitt, C. T., Sueno, S., and Cameron, M. (1973) Pyroxenes: comparisons of real and ideal structural topologies. Z. Kristallogr., 138, 254-273.

Peacor, D. R. (1967) Refinement of the crystal structure of a pyroxene of formula $M_I M_{II}$ $(Si_{1.5}Al_{0.5})O_6$. Am. Mineral., 52, 31-41.

Poldervaart, A. and Hess, H. H. (1951) Pyroxenes in the crystallization of basaltic magma. J. Geol., 59, 472-489.

Prewitt, C. T. and Burnham, C. W. (1966) The crystal structure of jadeite, $NaAlSi_2O_6$. Am. Mineral., 51, 956-975.

————, Brown, G. E., and Papike, J. J. (1971) Apollo 12 clinopyroxenes: high temperature x-ray diffraction studies. Proc. 2nd Lunar Sci. Conf., 1, 59-68.

Ramberg, H. and DeVore, G. (1951) The distribution of Fe^{++} and Mg^{++} in coexisting olivines and pyroxenes. J. Geol., 59, 193-210.

Ribbe, P. H. and Prunier, Jr., A. R. (1977) Stereochemical systematics of ordered $C2/c$ silicate pyroxenes. Am. Mineral., 62, 710-720.

Robinson, K., Gibbs, G. V., and Ribbe, P. H. (1971) Quadratic elongation: a quantitative measure of distortion in coordination polyhedra. Science, 172, 567-570.

Rutstein, M. S. and Yund, R. A. (1968) Unit-cell parameters of synthetic diopside-hedenbergite solid solutions. Am. Mineral., 54, 238-245.

Schaller, W. T. (1938) Johannsenite, a new manganese pyroxene. Am. Mineral., 23, 575-582.

Shannon, R. D. and Prewitt, C. T. (1969) Effective ionic radii in oxides and fluorides. Acta Crystallogr., B25, 925-946.

_____ and _____ (1970) Revised values of effective ionic radii. Acta Crystallogr., B26, 1046-1048.

Skinner, B. J. and Boyd, F. R. (1964) Aluminous enstatites. Carnegie Inst. Washington Year Book, 63, 163-165.

Smith, D. and Finger, L. W. (1971) Unit cell parameters of synthetic clinopyroxenes on the join $Fs_{85}En_{15}$-wollastonite in the pyroxene quadrilateral. Carnegie Inst. Washington Year Book, 70, 129-130.

Smith, J. V. (1959) The crystal structure of proto-enstatite, $MgSiO_3$. Acta Crystallogr., 12, 515-519.

_____ (1969) Crystal structure and stability of the $MgSiO_3$ polymorphs; physical properties and phase relations of Mg,Fe pyroxenes. Mineral. Soc. Am. Spec. Paper, 2, 3-29.

_____, Stephenson, D. A., Howie, R. A., and Hey, M. H. (1969) Relations between cell dimensions, chemical composition, and site preference of orthopyroxene. Mineral. Mag., 37, 90-114.

Smyth, J. R. (1971) Protoenstatite: a crystal-structure refinement at 1100°C. Z. Kristallogr., 134, 262-274.

_____ (1973) An orthopyroxene structure up to 850°C. Am. Mineral., 58, 636-648.

_____ (1974a) Low orthopyroxene from a lunar deep crustal rock: a new pyroxene polymorph of space group $P2_1ca$. Geophys. Res. Letters, 1, 27-29.

_____ (1974b) The high temperature crystal chemistry of clinohypersthene. Am. Mineral., 59, 1069-1082.

_____ and Ito, J. (1977) The synthesis and crystal structure of a magnesium-lithium-scandium protopyroxene. Am. Mineral., 62, 1252-1257.

Steele, I. M. (1975) Mineralogy of lunar norite 78235; second lunar occurrence of $P2_1ca$ pyroxene from Apollo 17 soils. Am. Mineral., 60, 1086-1091.

Sueno, S., Cameron, M., and Prewitt, C. T. (1976) Orthoferrosilite: high-temperature crystal chemistry. Am. Mineral., 61, 38-53.

Takeda, H. (1972a) Structural studies of rim augite and core pigeonite from lunar rock 12052. Earth Planet. Sci. Letters, 15, 65-71.

_____ (1972b) Crystallographic studies of coexisting aluminan orthopyroxene and augite of high-pressure origin. J. Geophys. Res., 77, 5798-5811.

_____ and Ridley, W. I. (1972) Crystallography and chemical trends of orthopyroxene-pigeonite from rock 14310 and coarse fine 12033. Proc. 3rd Lunar Sci. Conf., 1, 423-430.

_____, Miyamoto, M., and Reid, A. M. (1974) Crystal chemical control of element partitioning for coexisting chromite-ulvospinel and pigeonite-augite in lunar rocks. Proc. 5th Lunar Sci. Conf., 1, 727-741.

Thompson, Jr., J. B. (1970) Geometrical possiblities for amphibole structures: model biopyriboles. Am. Mineral., 55, 292-293.

Tokonami, M., Horiuchi, H., Nakano, A., Akimoto, S.-I., and Morimoto, N. (1979) The crystal structure of the pyroxene-type $MnSiO_3$. Mineral. J. (Japan), 9, 424-426.

Tossell, J. A. and Gibbs, G. V. (1977) Molecular orbital studies of geometries and spectra of minerals and inorganic compounds. Phys. Chem. Minerals, 2, 21-57.

Turnock, A. C., Lindsley, D. H., and Grover, J. E. (1973) Synthesis and unit cell parameters of Ca-Mg-Fe pyroxenes. Am. Mineral., 58, 50-59.

Veblen, D. R. and Burnham, C. W. (1978) New biopyriboles from Chester, Vermont: II. The crystal chemistry of jimthompsonite, clinojimthompsonite, and chesterite, and the amphibole-mica reaction. Am. Mineral., 63, 1053-1073.

Virgo, D. and Hafner, S. S. (1969) Fe^{2+},Mg order-disorder in heated clinopyroxenes. Mineral. Soc. Am. Spec. Paper, 2, 67-81.

Viswanathan, K. (1966) Unit cell dimensions and ionic substitutions in common clino-pyroxenes. Am. Mineral., 51, 429-442.

Warren, B. and Bragg, W. L. (1928) XII. The structure of diopside, $CaMg(SiO_3)_2$. Z. Kristallogr., 69, 168-193.

_____ and Modell, D. I. (1930) 1. The structure of enstatite $MgSiO_3$. Z. Kristallogr., 75, 1-14.

Whittaker, E. J. W. (1960) Relationships between the crystal chemistry of pyroxenes and amphiboles. Acta Crystallogr., 13, 741-742.

Zussman, J. (1968) The crystal chemistry of pyroxenes and amphiboles, 1. Pyroxenes. Earth Science Rev., 4, 39-67.

Chapter 3

PYROXENE SPECTROSCOPY   George R. Rossman

INTRODUCTION

This chapter reviews spectroscopic studies of pyroxenes conducted
primarily within the past ten years. It includes studies aimed at understanding the origin of the spectroscopic interaction itself, and studies
of broader problems which include spectroscopic approaches. It is neither
comprehensive nor encyclopedic, but is intended to provide a feel for the
applicability of spectroscopic methods in the pursuit of mineralogical
and petrological problems. In many respects, this chapter is looking at
a small subset of a series of larger problems. For instance, the study
of the iron-titanium interaction in sapphire is every bit as important
towards the interpretation of lunar titaniferous pyroxenes as is the
study of the lunar samples themselves.

Five books that review broader areas of mineral spectroscopy are
worth mentioning as a source of a more general perspective: Burns (1970)
presents a review of mineral optical spectroscopy; and Bancroft (1973)
reviews Mössbauer spectroscopy of minerals. The chapters in Karr (1975)
cover a variety of spectroscopic approaches to mineralogical problems.
Lazarev (1972) and Farmer (1974) review the infrared spectroscopy of
minerals.

The three spectroscopies that have been generally used in the study
of pyroxenes are Mössbauer absorption, optical absorption and infrared
absorption. Mössbauer spectroscopy involves measuring the absorption of
gamma rays by ^{57}Fe nuclei. The precise wavelength of absorption is influenced by the oxidation state and coordination environment of iron
atoms in the crystal. Optical spectroscopy measures the absorption of
light in the near-ultraviolet, visible, and near-infrared portions of
the spectrum (300 nm to 3000 nm). It also responds to the oxidation
state and coordination environment of the iron. Interactions with next-
nearest neighbors often are also important. Infrared spectroscopy measures the absorption of photons in the 3 μm to 300 μm region. Absorption
occurs when local units of the crystal are excited into vibration by the
incident energy. It is particularly useful for problems of phase identification.

Mössbauer spectroscopy, although limited to the study of iron in pyroxenes, has the advantage of being readily performed. Typically, a finely ground sample is weighed to provide 20-30 mg of contained iron. The sample is then distributed over a 2.5 cm diameter area and held in place with a binding agent such as wax. All orientations of low symmetry crystals are averaged; special techniques are required to avoid preferred orientation.

Optical spectra require non-cubic crystals to be oriented first along principal optical or crystallographic directions. Crystals are then polished on both sides. With standard spectrophotometers, areas 100 μm in diameter can be measured routinely. With microscope spectrophotometers, areas 10 μm in diameter are practical. Three components usually comprise a pyroxene spectrum. They are measured with plane polarized light vibrating along the three principal axes of the indicatrix and are labeled α, β and γ where the α spectrum is taken with light vibrating in the direction in which the α index of refraction would be measured.

Infrared spectra are commonly run with a 1 mg powdered sample dispersed in 200 mg KBr and pressed to form a coherent 13 mm diameter pellet. With microsampling techniques, 10 μg or less of sample are required.

A most obvious feature of pyroxenes is the wide variety of colors in which they occur. Ultimately, these colors reflect aspects of the crystal chemistry of metal ions in the pyroxene. It is therefore appropriate to begin by considering the origin of the colors.

THE PARTICIPANTS

The first task is to identify which elements contribute to color and optical absorption. Theoretical considerations indicate that only cations with unfilled valence orbitals will directly contribute to the color in the visible region. For most purposes, this means that transition metal cations will be responsible for color. In natural samples, iron, chromium and titanium are most important. The contribution to the color from each transition metal ion could best be determined by examining a suite of synthetic pyroxenes containing all candidate ions in all appropriate oxidation states in the M1, M2 and tetrahedral sites, if possible. Such a comprehensive collection is not available; however, by combining data from natural samples and the synthetics that are available, it will soon be possible to determine the general spectroscopic contributions of all of the first-row transition elements, Ti through Cu.

Iron

Iron, both as Fe^{2+} and Fe^{3+}, is the primary candidate for study by spectroscopic methods. Both optical and Mössbauer spectroscopy are used to determine the oxidation state and site occupancy of iron. Early optical studies were aimed at explaining the origin of pleochroism (Burns, 1966) that derived from the absorption of iron and from the interaction of iron with other cations. As attempts began to determine the M1/M2 distribution of Fe^{2+}, it was quickly realized that Mössbauer spectra could provide the information more conveniently, more rapidly and possibly more accurately than x-ray diffraction, at least for orthopyroxenes.

Orthopyroxenes: Mössbauer spectra, Fe^{2+}. The Mössbauer spectrum of Fe^{2+} in orthopyroxene is composed of two closely overlapping doublets at room temperature (Bancroft *et al.*, 1967; Evans *et al.*, 1967). These doublets are difficult to fit with computer techniques but such studies do indicate that Fe^{2+} prefers the M2 site (Fig. 1). As the figure shows, at liquid nitrogen temperature (77 K), the contributions from M1 and M2 are resolved because of a differential increase of the quadrupole splitting at the two sites (Virgo and Hafner, 1968). This effect has allowed detailed study of the Fe^{2+}/Mg disorder of heated orthopyroxenes. These studies established the strong preference of Fe^{2+} for the M2 site in samples equilibrated at low temperature and obtained distribution equilibrium constants and other thermodynamic data (Virgo and Hafner, 1969, 1970).

Orthopyroxenes: optical spectra. The spectrum of the Bamble bronzite (Fig. 2) displays nearly exclusively M2 Fe^{2+} features. The dominant features are the strong absorption in the α spectrum centered at 900 nm, and the absorption in the β spectrum at 1850 nm. Each of these bands has subsidiary components in the other polarization directions. Neither of these strong absorptions has an effect on the color of the mineral in the thicknesses which are ordinarily encountered.

The orthopyroxene M1 Fe^{2+} features can be best observed in samples that have been quenched from high temperatures. An M1 feature of Fe^{2+}

95

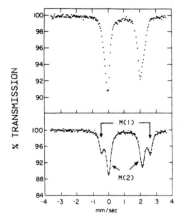

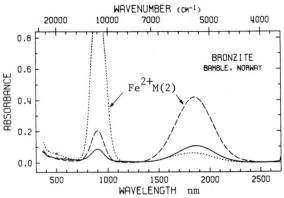

Figure 1. Mössbauer spectrum of an orthopyroxene heated at 1000°C to re-equilibrate Fe^{2+} in the M1 and M2 sites. Top: measured at room temperature. Bottom: measured at 77°K where the M1 and M2 features separate. From Virgo and Hafner (1968).

Figure 2. Optical absorption spectrum of an orthopyroxene showing the M2 Fe^{2+} features and sharp, weak spin-forbidden band at 505 nm. Dotted line: α spectrum; dashed line: β spectrum; solid line: γ spectrum. Crystal 100 μm thick.

at ∿1180 nm is well developed in the γ spectrum of a heated bronzite from Gore Mountain (Fig. 3). The existence of an additional component from M1 near 800 nm is indicated by computer generated difference spectra.

Additional weak Fe^{2+} features known as spin-forbidden bands can be observed in Figures 2 and 3. The most prominent occur as spikes near 505 and 440 nm. Also present in Figure 3, a broad, weak band at 600 nm, called an intervalence charge transfer band, arises from the interaction of Fe^{2+} with minor amounts of Fe^{3+}. At the shortest wavelengths there is a rapid rise in absorption intensity. It begins at about 400 nm in Figure 3 and will continue to rise to over a thousand absorbance units in the ultraviolet portion of the spectrum. This is a charge transfer band that occurs because the incident light causes the transfer of electron density from the oxide ions to iron. There has been very little work in the ultraviolet spectral range because of the necessity for ultrathin specimens (Langer and Aub-Eid, 1977). The color of most pyroxenes is controlled by the combination of the tail of the ultraviolet charge transfer band, the spin-forbidden bands, the interaction of Fe^{2+} and Fe^{3+}, and other impurity ions which may be present.

Clinopyroxenes: optical spectra. Whereas the optical spectra of natural orthopyroxenes are in many ways similar to each other, the spectra of clinopyroxenes show much more diversity. The spectra of two terrestrial

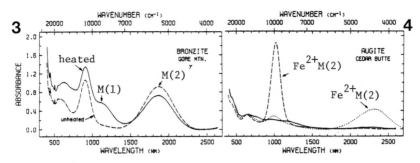

3

WAVENUMBER (cm⁻¹)
20000 10000 7000 5000 4000

heated

BRONZITE
GORE MTN.

M(2) γ

M(1)

unheated

4

WAVENUMBER (cm⁻¹)
20000 10000 7000 5000 4000

$Fe^{2+}M(2)$

AUGITE
CEDAR BUTTE

$Fe^{2+}M(2)$

Figure 3. Optical spectrum in the γ direction of bronzite from Gore Mountain, New York, showing the growth of the M1 Fe^{2+} features and decline of the M2 Fe^{2+} features after being quenched from high temperature. 100 μm thick. From Rossman (1979).

Figure 4. Optical spectrum of augite from Cedar Butte, Oregon, showing the M2 Fe^{3+} features in a clinopyroxene. α = ···; β = --; γ = —. 180 μm thick.

clinopyroxenes illustrate the differences brought about by different ratios of Fe^{2+} in M1 and M2. Figure 4 gives the spectrum of an augite for which the features of Fe^{2+} in M2 dominate the spectrum. It resembles the spectra of the orthopyroxnes but with differences in the wavelength of the M2 features. The M2 bands occur at 1030 nm and 2310 nm. This difference from the orthopyroxene wavelengths is a direct result of the structural differences of the M2 site. The M1 features are at about 970 and 1200 nm. Minor contributions from the Fe^{2+}/Fe^{3+} intervalence charge transfer interaction near 800 nm, and Cr^{3+} near 650 and 450 nm are also present.

The spectrum of the hedenbergite (Fig. 5) shows virtually no contribution from Fe^{2+} in M2 near 2300 nm and only a weak band at 1030 nm in the β spectrum. The M1 Fe^{2+} features are strong because of the necessarily high content of Fe^{2+} in the M1 site. The Fe^{2+}/Fe^{3+} intervalence charge transfer interaction band near 800 nm is the strongest feature in the visible region and controls the color of the mineral.

The lunar samples ultimately provide the best control for identification of Fe^{2+} features in the absence of Fe^{3+} (Bell and Mao, 1972a; Burns *et al.*, 1972a,b, 1973; Rossman, 1977). The spectra of the lunar pyroxenes, as well as terrestrial pyroxenes, contain absorption features from Fe^{2+} in both the M1 and M2 sites, although as the site preference studies would suggest, the M2 features dominate the spectrum in the case of orthopyroxenes. The spectra of two lunar pyroxenes are shown for comparison in Figures 6 and 7. The features of both M1 Fe^{2+} and M2 Fe^{2+}

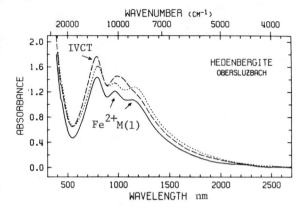

Figure 5. Optical spectrum of hedenbergite from Obersluzbach, Austria, showing the M1 Fe^{2+} features and the Fe^{2+}/Fe^{3+} intervalence charge transfer feature. $\alpha = \cdots$ $\beta = --$ $\gamma = ---$ 250 μm thick.

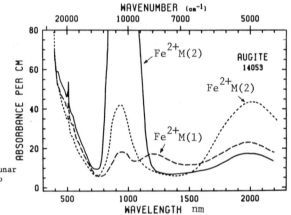

Figure 6. Optical spectrum of a lunar augite showing Fe^{2+} features but no Fe^{3+} absorption. $\alpha = \cdots$ $\beta = --$ $\gamma = ---$ From Bell and Mao (1972).

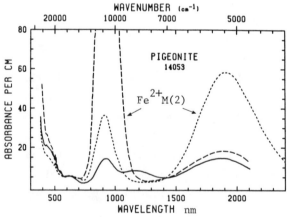

Figure 7. Optical spectrum of a lunar pigeonite showing Fe^{2+} features but not Fe^{3+} features. Weak Cr^{3+} absorption occurs near 630 and 440 nm. $\alpha = \cdots$ $\beta = --$ $\gamma = ---$ From Bell and Mao (1972).

are present but the Fe^{2+}/Fe^{3+} intervalence charge transfer feature is missing, further confirming the lack of Fe^{3+} in the lunar samples. It is somewhat ironic that the optical spectral of many more lunar pyroxenes have been studied than terrestrial samples. Yet, this fact serves to indicate the growth potential of this approach for solving the scientific and applied problems of earth materials.

Compositional trends. The position of the M2 bands near 1000 and 2300 nm are different in Figures 6 and 7. In general, the positions of the 1000 nm and 2000 nm bands vary systematically with pyroxene composition (Adams and McCord, 1972; Adams, 1974; Sung *et al.*, 1977; Hazen *et al.*, 1978). These variations have been of special interest to astronomers who wish to use them in telescopic remote planetary surface composition studies (Adams, 1975; Charette *et al.*, 1976, 1977). The details of these variations have been reported by Hazen *et al.* who find that the smooth variations with composition can be plotted on a pyroxene quadrilateral diagram and used for determinations of compositions (Figs. 8, 9 and 10). Significant deviations from these trends were noted only in the case of chromian diopside. An additional complication for the use of these plots in remote sensing is the effect of temperature upon the position of the Fe^{2+} bands. Sung *et al.* (1977) have demonstrated that the 1000 nm band shifts slightly to larger wavelengths with increasing temperature and the 2000 nm band becomes distorted due to thermal emission. Diagrams such as Figures 8-10 have not been established at temperatures other than room temperature.

Clinopyroxenes: Mössbauer spectra. The interpretation of the Mössbauer spectra of clinopyroxenes is more complicated than that of the orthopyroxenes because of the pervasive problems of exsolution intergrowths and chemical inhomogeneity of the individual grains (Bancroft and Burns, 1967b; Williams *et al.*, 1971). Mössbauer spectra have figured prominently in the study of lunar clinopyroxenes (Hafner and Virgo, 1970; Hafner *et al.*, 1971; Dowty *et al.*, 1972) and were used to demonstrate the absence of Fe^{3+}. Dowty and Lindsley (1973) made a comprehensive study of iron-rich clinopyroxenes and concluded that it is necessary to fit the spectra with multiple M1 doublets that arise from different types of next-nearest neighbor configurations about Fe^{2+} in the M1 site. Even at liquid

99

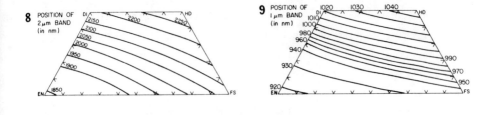

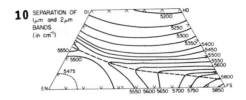

Figure 8. Position of the 2 μm pyroxene absorption maxima vs composition, projected on the pyroxene quadrilateral. From Hazen et al. (1978).

Figure 9. Position of the 1 μm pyroxene absorption maxima vs composition, projected on the quadrilateral. From Hazen et al. (1978).

Figure 10. Difference in energy between the 1 and 2 μm bands. This difference is related to the distortion of the M2 octahedral site. From Hazen et al. (1978).

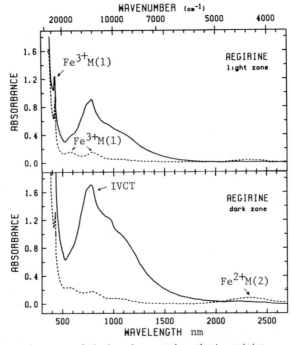

Figure 11. Optical spectra of the hourglass zoned synthetic aegirine taken in the light and dark zones. In both spectra features of Fe^{3+} and M(2) Fe^{2+} and the Fe^{2+}/Fe^{3+} IVCT are seen. The concentration of Fe^{2+} is greater in the dark zone. Consequently, the intensity of the IVCT band which determines the color is more intense in the dark zone. α = ··· β = -- γ = — 138 μm thick.

nitrogen temperature, the overlap cannot be accounted for satisfactorily in computer fits. This means that accurate site distributions probably cannot be obtained for high-calcium clinopyroxenes. Similar types of complexity were also observed in the spectra of pyroxenes in the diopside-hedenbergite series (Bancroft *et al.*, 1971).

Clinopyroxenes: Fe^{3+} in the M1 site. Fe^{3+} manifests its presence in pyroxene spectra in two distinct forms. The first form arises from Fe^{3+} in the M1 site of near-octahedral symmetry. The Mössbauer spectrum of aegerine is typical of octahedral ferric iron: a small isomer shift, 0.39 mm/sec, and a quadrupole splitting, 0.35 mm/sec, much smaller than Fe^{2+}. The positions of the bands in the optical spectrum (Fig. 11) are close to the textbook-perfect example of what theory predicts them to be. They consist of two broad, weak bands at 815 nm and 595 nm and a sharp spike at about 435 nm actually composed of two closely spaced components at 431 and 437 nm. Similar spectra result from Fe^{3+} in jadeite (Rossman, 1974) and spodumene. The sharp 435 nm Fe^{3+} band can often be seen in the spectra of other pyroxenes. It does not significantly modify their color, but does find use in gemological testing of jadeite.

Clinopyroxenes: Fe^{3+} in the tetrahedral site. Ferri-diopside, which shows bright yellow color in thin section, is a special case of Fe^{3+} spectroscopy. Mössbauer studies (Virgo, 1972) have indicated that a substantial fraction of the Fe^{3+} is present in the tetrahedral sites. The isomer shift and quadrupole splitting (Table 1) are different than for octahedral Fe^{3+} and are in the range encountered for Fe^{3+} in tetrahedral sites. Furthermore, the ferri-diopsides that show this spectroscopic behavior have significantly fewer than 2 silicons per 6 oxygens.

The portion of the optical spectrum associated with the Fe^{3+} consists of only a prominent band at $\sim$450 nm (Fig. 12). The multiple absorption bands associated with Fe^{3+} in well-defined tetrahedral sites are not present in the spectra of ferri-diopside. The spectra of tetrahedral Fe^{3+} in iron arsenates, feldspar, $LiAlO_2$, ferri-micas, and sillimanite all have a theoretically predicted complexity consisting of multiple bands in the visible region. Bell and Mao (1972b) suggest that an oxygen to Fe^{3+} charge transfer interaction causes the 450 nm absorption. It is puzzling why a similar charge transfer is not observed in other

101

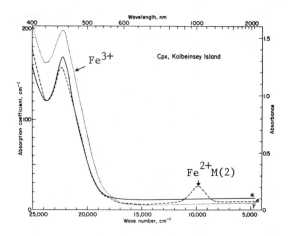

Figure 12. Optical spectrum of ferridiopside from Kolbeinsey Island showing the feature associated with tetrahedral Fe^{3+}. A weak M(2) Fe^{2+} feature is also present near 1000 nm. From Bell and Mao (1972).

Table 1. Selected Mossbauer parameters of pyroxene.

		M1		M2		Tetr.	
		I.S. mm/sec	Q.S. mm/sec	I.S. mm/sec	Q.S. mm/sec	I.S. mm/sec	Q.S. mm/sec
Orthopyroxene							
Fe^{2+}	RT	1.00	2.25 (1)	0.94	1.97 (1)	–	–
	LN₂	1.12	3.11 (1)	1.05	2.04 (1)	–	–
Fe^{3+}	RT	0.47	0.70 (2)	–	–	0.21	1.34 (2)
	LN₂	0.59	0.71 (2)	–	–	0.34	1.41 (2)
Clinopyroxene							
Fe^{2+}	RT	1.10-1.22	2.05-2.76 (3)	1.11-1.15	1.72-1.91 (3)	–	–
	LN₂	1.28-1.31	2.74-3.17 (3)	1.21-1.27	1.80-2.02 (3)	–	–
Fe^{3+}	RT	0.39	0.35 (4)	–	–	0.40.26	1.27-1.62 (5)
	LN₂	0.50	0.37 (4)	–	–		
IVCT	RT	1.31	1.81 (4)				
	LN₂	1.08	2.17 (4)				

This table is not comprehensive. It is intended to illustrate typical values for a few selected examples. All values of the isomer shift are referred to metallic iron.

1) Virgo and Hafner (1968)
2) Annersten et al., (1978)
3) Dowty and Lindsley (1973)
4) Amthauer and Rossman (in preparation)
5) Virgo (1972)

minerals that contain Fe^{3+} in tetrahedral sites. The answer to this puzzle may have to do with interactions between Fe^{3+} in the octahedral and tetrahedral sites. This absorption band is also found in titanaugites at 448 nm (Burns *et al.*, 1976) and likewise attributed to tetrahedral Fe^{3+}. The titanaugites also have indications of tetrahedral Fe^{3+} in their Mössbauer spectrum.

Interactions between Fe^{2+} and Fe^{3+}

Two dissimilar cations occupying adjacent sites can interact to produce properties that are not the sum of the individual constituents. Most relevant to this chapter is the intervalence charge transfer interaction (IVCT) between Fe^{2+} and Fe^{3+}. In the spectrum (Fig. 5) of hedenbergite from Obersluzbach, Austria, a salient spectroscopic feature is a band at 800 nm in all three polarizations. This band varies in intensity in the spectrum of various clinopyroxenes, is nearly absent in the Cedar Butte augite spectrum (Fig. 4), and is absent in the spectra of lunar samples (Figs. 6 and 7). It arises from the exchange of electrons from Fe^{2+} to adjacent Fe^{3+} ions stimulated by the incident radiation. The Fe^{2+}/Fe^{3+} IVCT bands are often more important than the Fe^{2+} bands in determining the color of a pyroxene observed in thin section. IVCT interactions have also been implicated in the spectra of orthopyroxenes (Burns, 1970; Goldman and Rossman, 1977).

In many systems displaying IVCT in the optical spectra, Mössbauer spectra indicate the presence of both Fe^{2+} and Fe^{3+}. Additional features may also be present solely as a result of the intervalance interaction. The study of the Mössbauer spectra of the features unique to the Fe^{2+}/Fe^{3+} interactions in pyroxenes is just beginning. Amthauer and Rossman (1980) have observed that in the Mössbauer spectrum of aegirines with sizable Fe^{2+} concentrations, a broad absorption pattern of intermediate quadrupole splitting is present in addition to the Fe^{3+} and Fe^{2+} doublets. The relative area of this absorption increases as the sample temperature increases suggesting that it originates from Fe^{2+} and Fe^{3+} ions involved in thermally-activated electron delocalization.

Chromium

Chromium produces a bright emerald green color in pyroxenes that is most frequently encountered in the case of chromian-diopside and dark

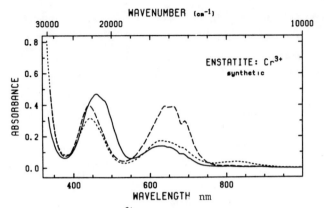

Figure 13. Optical spectrum of Cr^{3+} in an enstatite synthesized by Jun Ito. 0.8% Cr_2O_3. The structure in the 600-700 nm region is characteristic of Cr^{3+}. α = ···; β = --; γ = —. 210 μm thick.

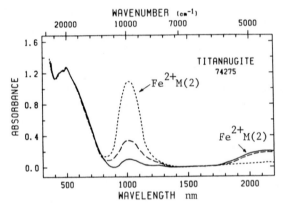

Figure 14. Optical spectrum of a lunar titanaugite showing strong absorption in the 500-700 nm region. Absorption in this region also occurs in terrestrial titanaugites. From Sung et al. (1974).

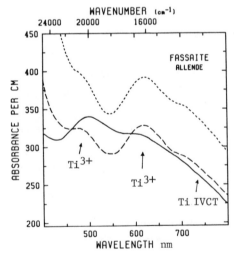

Figure 15. Optical spectrum of a titanium-rich fassaite from the Allende meteorite showing the Ti^{3+} features and the Ti^{3+}/Ti^{4+} IVCT. α = ···; β = --; γ = —. From Mao and Bell (1974).

green precious jade. The spectrum of Cr^{3+} in all pyroxenes consists of two regions of absorption of roughly comparable intensity (Fig. 13). The first, in the 650 nm region has much structure, whereas the second in the 450 nm region has little. The structure of the 650 nm region is distinctive and is the basis for the spectroscopic identification of Cr^{3+} in minerals and for gemological tests. The pattern which is typical of Cr^{3+} in nearly octahedral sites indicates that the Cr^{3+} is in the M1 site.

Other oxidation states of Cr. Two groups have synthesized blue chromian diopside and have presented optical absorption spectra that contain minor Cr^{3+} features plus other features that have been interpreted variously as Cr^{3+} in a tetrahedral site, Cr^{2+}, or Cr^{4+} (Mao *et al.*, 1972; Burns, 1975; Ikeda and Yagi, 1977, 1978; Schreiber, 1977, 1978). The experimental work necessary to resolve this question has not been done. Its importance exists because of need to establish the oxidation state of minor elements such as Cr in mantle minerals and the need to understand elemental partitioning in synthetic systems.

Titanium

Titanaugites have been recognized in thin section by their strong color and distinct violet color in the γ direction. An absorption band centered near 550 nm together with Fe^{3+} features is responsible for the color. The 550 nm band was initially curve-resolved and attributed to Ti^{3+} (Manning and Nickel, 1969). Although the color and absorption band position was analogous to aqueous solutions of Ti^{3+} chemicals, Ti^{3+} was recognized as unlikely to occur in near-surface terrestrial rocks. The reflectance spectrum of synthetic Ti^{3+} pyroxene, $NaTiSi_2O_6$, was unlike the titanaugite spectrum. It had a single absorption band at 640 nm which was attributed to Ti^{3+}/Ti^{4+} IVCT (Prewitt *et al.*, 1972).

Extra-terrestrial samples have had prominent roles in stimulating interest in Ti^{3+} per se and in IVCT interactions involving titanium. Five classes of pyroxenes have figured prominently in the study of Ti. They are (1) the terrestrial titanaugites, (2) the synthetic $NaTiSi_2O_6$, (3) lunar clinopyroxenes, (4) the Angra dos Reis (ADOR) meteoritic fassaite, and (5) the Allende meteoritic fassaite. Their spectra involve Ti^{3+}, Ti^{3+}/Ti^{4+} IVCT, and Fe^{2+}/Ti^{4+} IVCT.

Spectra of lunar titanaugites display strong titanium-related absorption in the 450-700 nm region with bands often resolvable at $\sim$470 nm and $\sim$660 nm (Fig. 14). Weaker features of the same type contribute to the spectra in Figures 6 and 7. These samples instigated a series of articles aimed at resolving the nature of Ti in pyroxenes (Bell and Mao, 1972a; Burns et $al.$, 1972, 1973, 1976; Sung et $al.$, 1974), which proposed interpretations that included Ti^{3+}, Ti^{4+}/Fe^{2+} IVCT, and Ti^{3+}/Ti^{4+} IVCT for the 400-700 nm region.

The interpretation of titanium-containing pyroxenes has been especially aided by the study of meteoritic pyroxenes. The fassaite from the Allende meteorite is unusual in its low iron content (<1% FeO) and high titanium content ($\sim$17% TiO_2). Measurements of this fassaite (Dowty and Clark, 1973a,b; Mao and Bell, 1974) have indicated that the Ti^{3+} absorptions occur near 490 and 620 nm and the Ti^{3+}/Ti^{4+} IVCT occurs at 730 nm (Fig. 15). The work by Mao and Bell utilized what is becoming an important experimental approach in the spectroscopic study of minerals: high pressure techniques that literally drive the ions closer together to facilitate charge transfer and shift absorption bands in predictable fashions. Based in part upon these results, the conclusion was reached by Burns et $al.$ (1976) that Fe^{2+}/Ti^{4+} IVCT occurs in the 500 to 600 nm region in both lunar and terrestrial titanaugites, whereas Ti^{3+}/Ti^{4+} IVCT occurs in the 600 to 750 nm region.

The fassaite from the Angra dos Reis meteorite, containing both iron and titanium as major elements, is notable because of its unusually deep, brownish red color. Mössbauer spectra demonstrated that all of the iron is Fe^{2+} (Mao et $al.$, 1977). The optical spectrum (Fig. 16) is dominated by an intense band at 480 nm (Mao et $al.$, 1977), and the pressure dependence (Hazen et $al.$, 1977) suggests a charge transfer process. Although they did not propose a definite assignment for the 480 nm band, it is likely that it is a Fe^{2+}/Ti^{4+} IVCT such as Dowty and Clark (1973a,b) proposed for the spectra of other pyroxenes. The greater intensity observed by Mao et $al.$ in the E∥c direction suggests that it is an interaction among M1 sites in accord with the predictions of Dowty and Clark.

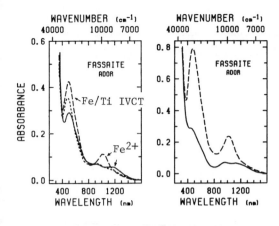

WAVENUMBER (cm⁻¹)
40000   10000   7000

WAVENUMBER (cm⁻¹)
40000   10000   7000

Figure 16. Optical spectrum of fassaite from Angra dos Reis meteorite which has both Fe^{2+} and Ti as major elements. Fe^{2+} features occur from 900 to 1300 nm. The strong band near 480 nm most likely results from an Fe^{2+}/Ti^{4+} interaction. Spectra on the left are in the $\alpha = \cdots$; $\beta = --$; and $\gamma = ---$ directions. Spectrum on the right shows how the 480 nm band can be essentially absent when the incident light is plane polarized in directions which do not correspond to a principal direction of the optical indicatrix. $\beta = --$; $E \perp c$ and $E \perp b = ---$. From sections ~30 μm thick (left) and ~42 μm thick (right).

Other cations

Little has been published about optical spectra of other cations in pyroxenes. Manganese is a prevalent minor element in spodumene. Upon exposure to x-radiation, manganese-containing spodumene will turn from lilac color to bluish green. Schmitz and Lehmann (1976) and Hassan and Labib (1978) have followed this change with optical spectroscopy and conclude that a transformation from Mn^{3+} to Mn^{4+} is associated with this change. Reflectance spectra that indicate band positions have been obtained for Ni^{2+} and Co^{2+} in pyroxenes (White *et al.*, 1971).

QUANTITATIVE CONCENTRATIONS AND APPLICATIONS TO ZONING

One of the ultimate aims of the mineral spectroscopist is to be able to determine the concentration of all the spectroscopically active elements in a mineral. The determination would distinguish among the oxidation states of an element in each crystallographic site. In the case of Fe^{2+} and Fe^{3+} in orthopyroxenes, this type of determination is most conveniently realized through the use of Mössbauer spectra.

For pyroxenes, Mössbauer spectra can be obtained only for the element iron; optical spectra can be used to obtain concentration information although, for elements other than iron, the detailed absorption intensity *vs* concentration plots that would be necessary for accurate determinations are not available. The information available is tabulated in Tables 2 and 3.

Concentration information in Tables 2 and 3 is tabulated in terms of ε-value in analogy to the methods used in solution colorimetric analysis.

Table 2.   Intensities of Absorption – Orthopyroxenes [*]

Element	wavelength (nm)	a = β	b = α	c = γ	site	ref.
Cr^{3+}	442	40	29	49	M1	1
	646	45	18	15		
Mn^{3+}	540	–	≥ 9.0	–	?	1
Fe^{2+}	930	5.2	40.8	2.5	M2	2
	2000	9.6	1.6	2.5		
Fe^{2+}	1160	–	–	4.6	M1	2
Co^{2+}	500	21	33	4.9	(M2+M1)?	1
Ni^{2+}	1406	11.6	4.7	34	M2+M1	3
	770	5.9	12.5	67		
	408	9.0	4.2	10		
Cu^{2+}	820	17	39	5	M2	1
	1373	28	5	24		

[*] All values reported as ε-values (liters per mole per cm)
1) Rossman, G. R., unpublished data
2) Goldman and Rossman (1979)
3) Rossman, Shannon and Warring (in preparation)

Table 3.   Intensities of Absorption – Clinopyroxene [*]

element	site	λ		sample		ε	Ref.
Fe^{2+}	M2	1000 nm	β	pigeonite	140306,6	≥ 9.0	(1)
				pigeonite	12063,79	≥14.5	(1)
				augite	14053	≥28.0	(1)
				pigeonite	14053	≥29.6	(1)
Cr^{3+}	M1	664 nm	α	diopside Serrania de la Ronda		23.6	(2)
		460 nm	α	"	"	22	(2)

[*] All ε values (in liters per mole per cm) are calculated based on total cation. For Fe^{2+} these values are lower limits because some M1 Fe^{2+} is also present in the sample.
(1) Calculated from the data presented in Bell and Mao (1972a)
(2) Calculated from the data presented in Mao et al. (1972)

At sufficiently low concentrations, generally below 10%, metal ions in crystals follow Beer's law, so ε, the molar absorptivity, is a useful analytical constant defined in terms of A, the absorbance read directly from a spectrophotometer; p, the optical path length: namely, the thickness of the crystal in units of cm; and C, the concentration of the absorbing ion in units of moles per liter. The equation is:

$$A = \varepsilon p C \tag{1}$$

Its use is illustrated as follows:

From Figure 2 the absorbance of the band at 1850 nm in β polarization, 0.42, is read directly from the figure. The path length is stated at 0.01 cm. The concentration can be calculated from the density and FeO wt % as follows:

FeO = 9.77%, from electron microprobe analysis assuming all Fe is Fe^{2+}.

Support for this assumption comes from the spectrum itself in view of the absence of Fe^{3+} features.

Density = 3.32, from Figure 74 of Deer *et al.* (1978).

Fe concentration = 3320 g/l x 0.0977 = 324.4 g FeO/liter = 4.51 moles Fe/liter.

Thus from equation 1:

0.42 = ε x 0.01 x 4.51; ε = 9.31.

This value was calculated under the assumption that all the iron is in M2. Corrections for the Fe^{2+} content in M1 obtained from Mössbauer spectra would bring this value into even closer agreement with the value of 9.6 in Table 2.

The aegirine, whose spectrum appears in Figure 11, shows a distinct hourglass sector zonation. The composition of this crystal is close to stoichiometric aegirine in both the dark and the light zones. The problem, to explain the color zoning, is exactly the type that is best suited to a spectroscopic approach. The absorption spectrum (Fig. 11) shows the Fe^{3+} absorption bands characteristic of aegirine, but also shows the M2 Fe^{2+} band in α-polarization at 2310 nm. The Fe^{2+} band is 2.6 times more intense in the dark zone. Because the intensity *vs* concentration curves have not been established for clinopyroxenes, the absolute FeO content cannot be determined. By comparison with the orthopyroxene intensities, a maximum estimate of about two percent seems appropriate. The Fe^{2+} absorption *per se* will not cause the hourglass zonation. However, the intervalence charge transfer interaction of the Fe^{2+} with Fe^{3+} produces an intense absorption centered at 780 nm that strongly influences the color and causes the hourglass zonation. From the one determination of the intensity of an IVCT band in an aegerine (Amthauer and Rossman, 1979), an FeO content of 1.6% is estimated for the dark zone.

INFRARED SPECTRA AND OTHER SPECTROSCOPIC APPROACHES

Infrared spectra have been obtained from a number of pyroxenes (Lazarev, 1972; Estep, *et al.*, 1971; Omori, 1971; Rutstein and White, 1971; Kovach, 1975; Kieffer, 1979). Typical spectra are shown in Figure 17. Much of the infrared work with pyroxenes has been concerned with either establishing compilations of data sets across the range of pyroxene compositions or with the analysis of the spectroscopic data in terms of the structure of the pyroxene. Infrared methods remain especially useful as a method of phase identification and are most widely used for this purpose. Systematic variations with composition have been established along some solid-solution series, but direct chemical analysis remains the preferred method of compositional determination. Cation order-disorder appears to influence the spectra but no quantitative measure of order-disorder by infrared spectra has been established. Infrared spectroscopy is ideally suited for the study of minor amounts

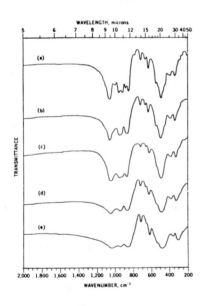

of OH^- in crystals, but very little use of this ability has been applied to pyroxenes. In a crystallographic study, Beran (1976) used infrared spectra to determine orientation of OH^- groups in diopside.

Closely related to infrared spectra are Raman spectra, which also provide information on vibrational frequencies of crystals. The few Raman data which exist for pyroxenes (White, 1975) have been combined with infrared data by Kieffer (1979, in preparation) to calculate the temperature dependence of the heat capacity from only elastic, crystallographic and spectroscopic data.

Other types of spectroscopy have not been widely applied to problems involving pyroxenes. X-ray photoelectron spectroscopy has been used

Figure 17. Infrared spectral comparison of lunar pyroxene with synthetic pigeonites (a) synthetic $Wo_{10}En_{75}Fs_{15}$; (b) synthetic $Wo_{10}En_{60}Fs_{30}$; (c) lunar pyroxene (Fs_{32}) from rock 12018,26,104 dark amber fragments; (d) synthetic $Wo_{10}En_{52}Fs_{38}$; (e) synthetic $Wo_{10}En_{30}Fs_{60}$. From Estep *et al.* (1971).

to determine the oxidation state of V^{3+} in an aegirine of high vanadium content (Nakai *et al.*, 1976). Ghose and Schindler (1969) used electron spin resonance (ESR) to determine that Mn^{2+} occurs in two sites in diopside. They found that the distribution between the sites was sample-dependent, but concluded that before trace Mn^{2+} distribution could be used for geothermometry, more would have to be known about the effects of other cations on the Mn^{2+} distribution. ESR spectra work well for low concentrations of Mn^{2+} and Fe^{3+} (<0.1%) but are unable to determine the oxidation state distribution of Mn in the violet Italian clinopyroxenes ("violan") because the Mn content is too high (Mottana *et al.*, 1979).

THEORETICAL ASPECTS

The electronic spectrum of Fe^{2+} in the M2 site of orthopyroxene has been the subject of several studies aimed at a fundamental understanding of the origin of absorption bands in silicates (White and Keester, 1966, 1967; Bancroft and Burns, 1967a; Runciman *et al.*, 1973; Goldman and Rossman, 1976, 1977). The low symmetry and large size of the M2 site generate extremes of polarization and splitting of electronic states of Fe^{2+} that are manifested in the large energy separations among the absorption bands of Fe^{2+} in pyroxene. The separation between the 1000 nm and 2000 nm M2 Fe^{2+} absorption bands is larger than the separation between the M1 Fe^{2+} bands as a direct result of the greater distortion of the M2 site from octahedral geometry.

The spectra of the ADOR fassaite highlighted a fundamental issue regarding the relationships among the orientation of the indicatrix, directions of maximum absorption, and crystallographic directions. Mao and Bell (1974) observed that even though the α, β and γ directions are all reddish, there exists a direction that does not coincide with a principal direction of the indicatrix in which the sample is nearly colorless. Figure 15 compares the spectrum in this direction with the α, β and γ spectra. Dowty (1973a) has indicated that this property should be a general feature of low symmetry crystals, and will be accentuated in crystals with IVCT. Absorption spectra differing from those obtained in the principal directions of the indicatrix have since been observed in many monoclinic crystals.

CONCLUSIONS

In many applications optical and Mössbauer spectroscopic studies of pyroxenes offer the capability of determining oxidation states of ions and site population analyses with a sensitivity unrivaled by any other method. Detailed quantitative calibrations still must be obtained for iron in clinopyroxenes, and for other elements that may be present as minor components in natural samples. Many of the calibrations await only the synthesis of suitable standards. Our understanding of the processes that generate the spectroscopic interaction has improved to a great extent in the past ten years, although the consequences of interactions among dissimilar ions have not been fully explored experimentally or explained theoretically.

The potential exists for more extensive collaboration between mineralogists and petrologists as petrological problems are reformulated in terms of the spatial variation of oxidation states and site occupancies in zoned crystals and in coexisting phases. The role of water in the form of H_2O and OH^- in the alteration of pyroxenes is virtually unexplored from the spectroscopic point of view. Here again, the potential exists for innovative new science to be accomplished in the fields of low temperature geochemistry and metamorphic petrology. High resolution electron microscopy studies would be an important adjunct to the study of hydrous alteration.

Adams, J. B. (1974) Visible and near infrared diffuse reflectance spectra of pyroxenes as applied to remote sensing of solid objects in the solar system. J. Geophys. Res., 79, 4829-4836.

_____ (1975) Interpretation of visible and near-infrared reflectance spectra of pyroxenes and other rock-forming minerals. *In* C. Karr, Jr., ed., *Infrared and Raman Spectroscopy of Lunar and Terrestrial Minerals*, Academic Press, N. Y., p. 90-116.

_____ and McCord, T. B. (1972) Electronic spectra of pyroxenes and interpretation of telescopic spectral reflectivity curves of the moon. Proc. 3rd Lunar Sci. Conf., 3021-3034.

Amthauer, G. and Rossman, G. R. (1979) Mössbauer electronic absorption studies of mixed valence iron silicates and phosphates. Geol. Soc. Am. Abstr. Programs, 11, 378.

_____ and _____ (1980) Mössbauer optical absorption studies of aegirines (abstr.). Internat. Mineral. Assoc., Orléans, France.

Bancroft, G. M. (1973) *Mössbauer Spectroscopy: An Introduction for Inorganic Chemists and Geochemists*. Halstad Press, N. Y.

_____ and Burns, R. G. (1967a) Interpretation of the electronic spectra of pyroxenes. Am. Mineral., 52, 1278-1287.

_____ and _____ (1967b) Distribution of iron cations in a volcanic pigeonite from Mössbauer spectroscopy. Earth Planet. Sci. Lett., 3, 125-127.

_____, _____, and Howie, R. A. (1967) Determination of the cation distribution in orthopyroxene series by the Mössbauer effect. Nature, 213, 1221-1223.

_____, Williams, P. G. L., and Burns, R. G. (1971) Mössbauer spectra of minerals along the diopside-hedenbergite tie line. Am. Mineral., 56, 1617-1625.

Bell, P. M. and Mao, H. K. (1972a) Crystal-field spectra of lunar samples. Carnegie Inst. Wash. Year Book, 71, 479-489.

_____ and _____ (1972b) Crystal-field determination of Fe^{3+}. Carnegie Inst. Wash. Year Book, 71, 531-534.

Beran, A. (1976) Messung des Ultrarot-Pleochroismus von Mineralen. XIV. Der Pleochroismus der OH⁻ Streckfrequenz in Diopsid. Tschermaks Min. Pet. Mitt., 23, 79-85.

Burns, R. G. (1966) Origin of optical pleochroism in orthopyroxenes. Mineral. Mag., 35, 715-719.

_____ (1970) *Mineralogical Applications of Crystal Field Theory*. Cambridge University Press, Cambridge, England.

_____ (1975) On the occurrence and stability of divalent chromium in olivines included in diamonds. Contrib. Mineral. Petrol., 51, 513-522.

_____, Abu-Eid, R. M., and Huggins, F. E. (1972a) Crystal field spectra of lunar pyroxenes. Proc. 34d Lunar Sci. Conf., 533-543.

_____ and Huggins, F. E. (1973) Visible-region absorption spectra of a Ti^{3+} fassaite from the Allende meteorite: A discussion. Am. Mineral., 58, 955-961.

_____, _____, and Abu-Eid, R. M. (1972b) Polarized absorption spectra of single crystals of lunar pyroxenes and olivines. The Moon, 4, 93-102.

_____, Parkin, K. M., Loeffler, B. M., Leung, I. S., and Abu-Eid, R. M. (1976) Further characterization of spectral features attributable to titanium on the moon. Proc. 7th Lunar Sci. Conf., 2561-2578.

_____, Vaughan, D. J., Abu-Eid, R. M., Witner, M., and Morawski, A. (1973) Spectral evidence of Cr^{3+}, Ti^{3+}, and Fe^{2+} rather than Cr^{2+} and Fe^{3+} in lunar ferromagnesian silicates. Proc. 4th Lunar Sci. Conf., 983-994.

Charette, M. P., Soderblom, L. A., Adams, J. B., Gaffey, M. J., and McCord, T. B. (1976) Age-color relationships in the lunar highlands. Proc. 7th Lunar Sci. Conf., 2579-2592.

_____, Taylor, S. R., Adams, J. B., and McCord, T. B. (1977) The detection of soils of Fra Mauro basalt and anorthositic gabbro composition in the lunar highland by remote spectral reflectance techniques. Proc. 8th Lunar Sci. Conf., 1049-1061.

Deer, W. A., Howie, R. A., and Zussman, J. (1978) *Rock-Forming Minerals, Vol. 2A: Single-Chain Silicates*. 2nd ed., John Wiley and Sons, New York.

Dowty, E. and Clark, J. R. (1973a) Crystal structure refinement and optical properties of a Ti^{3+} fassaite from the Allende meteorite. Am. Mineral., 58, 230-242.

_____ and _____ (1973b) Crystal structure refinement and optical properties of a Ti^{3+} fassaite from the Allende meteorite. Reply. Am. Mineral., 58, 962-964.

_____ and Lindsley, D. H. (1973) Mössbauer spectra of synthetic hedenbergite-ferrosilite pyroxenes. Am. Mineral., 58, 850-868.

_____, Ross, M. and Cuttitta, F. (1972) Fe^{2+}-Mg site distribution in Apollo 12021 pyroxenes. Evidence for bias in Mössbauer measurements and relations of ordering to exsolution. Proc. 3rd Lunar Sci. Conf., Geochim. Cosmochim. Acta, Suppl. 3, Vol. 1, 481-492.

Estep, P. A., Kovach, J. J., and Karr, C. (1971) Infrared vibrational spectroscopic studies of minerals from Apollo 11 and Apollo 12 lunar samples. Proc. 2nd Lunar Sci. Conf., 3, 2137-2151.

Evans, B. J., Ghose, S., and Hafner, S. (1967) Hyperfine splitting of ^{57}Fe and Mg-Fe order-disorder in orthopyroxene ($MgSiO_3$-$FeSiO_3$ solid solutions). J. Geol., 75, 306-322.

Farmer, V. C., ed. (1974) *The Infrared Spectra of Minerals*. Mineralogical Society of London.

Ghose, S. and Schindler, P. (1969) Determination of the distribution of trace amounts of Mn^{2+} in diopsides by electron paramagnetic resonance. Mineral. Soc. Am. Spec. Paper, 2, 51-58.

Goldman, D. S. and Rossman, G. R. (1976) Identification of a mid-infrared electronic band of Fe^{2+} in the distorted M(2) site of orthopyroxene (Mg,Fe)SiO_3. Chem. Phys. Lett., 41, 474-475.

_____ and _____ (1977) The spectra of iron in orthopyroxene revisited: the splitting of the ground state. Am. Mineral., 62, 151-157.

_____ and _____ (1979) Determination of quantitative cation distribution in orthopyroxenes from electronic absorption spectra. Phys. Chem. Minerals, 4, 43-53.

Hafner, S. S. and Virgo, D. (1970) Temperature-dependent cation distributions in lunar and terrestrial pyroxenes. Proc. Apollo 11 Lunar Sci. Conf., Geochim. Cosmochim. Acta, Suppl. 1, Vol. 3, 2183-2198.

_____, _____, and Warburton, D. (1971) Cation distributions and cooling history of clinopyroxenes from Oceanus Procellarum. Geochim. Cosmochim. Acta, Suppl. 2, Vol. 1, 91-108.

Hassan, F. and Labib, M. (1978) Induced color centers in α-spodumene called kunzite. Neues Jahrb. Mineral. Abh., 134, 104-115.

Hazen, R. M., Bell, P. M., and Mao, H. K. (1977) Polarized absorption spectra of Angra dos Reis fassaite to 52 kbar. Carnegie Inst. Year Book, 76, 515-516.

_____, _____, and _____ (1978) Effects of compositional variation on absorption spectra of lunar pyroxenes. Proc. 9th Lunar Planet. Sci. Conf., 2919-2934.

Ikeda, K. and Yagi, K. (1977) Experimental study on the phase equilibria in the join $CaMgSi_2O_6$-$CrSi_2O_6$ with special reference to the blue diopside. Contrib. Mineral. Petrol., 61, 91-106.

_____ and _____ (1978) Reply to H. D. Schreiber. Contrib. Mineral. Petrol., 66, 343-344.

Karr, C., ed. (1975) *Infrared and Raman Spectroscopy of Lunar and Terrestrial Minerals*. Academic Press: New York.

Kieffer, S. W. (1979) Thermodynamics and lattice vibrations of minerals, 2. Vibrational characteristics of silicates. Rev. Geophys., 17, 20-34.

Kovach, J. J., Hiser, A. L., and Karr, C. (1975) Far-infrared spectroscopy of minerals. *In* C. Karr, Jr., ed., *Infrared and Raman Spectroscopy of Lunar and Terrestrial Minerals*, Academic Press: New York, 231-254.

Langer, K. and Abu-Eid, R. M. (1977) Measurement of the polarized absorption spectra of synthetic transition metal-bearing silicate microcrystals in the spectral range 44,000-4000 cm^{-1}. Phys. Chem. Minerals, 1, 273-299.

Lazerev, A. N. (1972) Vibrational spectra and structure of silicates. Consultants Bureau: New York.

114

Manning, P. G. and Nickel, E. H. (1970) A spectral study of the origin of colour and pleochroism of a titanaugite from Kaiserstuhl and of a riebeckite from St. Peter's Dome, Colorado. Canadian Mineral., 10, 71-83.

Mao, H. K. and Bell, P. M. (1974) Crystal-field effects of trivalent titanium in fassaite from the Pueblo de Allende meteorite. Carnegie Inst. Year Book, 73, 488-492.

_____, _____, and Dickey, J. S. (1972) Comparison of the crystal-field spectra of natural and synthetic chrome diopside. Carnegie Inst. Wash. Year Book, 71, 538-541.

_____, _____, and Virgo, D. (1977) Crystal-field spectra of fassaite from the Angra dos Reis meteorite. Earth Planet. Sci. Lett., 35, 353-356.

Mottana, A., Rossi, G., Kracher, A., and Kurat, G. (1979) Violan revisited: Mn-bearing omphacite and diopside. Tschermaks Min. Petr. Mitt., 26, 187-201.

Nakai, I., Hideo, O., Sugitani, Y., and Niwa, Y. (1976) X-ray photoelectron spectroscopic study of vanadium-bearing aegirines. Mineral. J. Japan, 8, 129-134.

Omori, K. (1971) Analysis of the infrared absorption spectrum of diopside. Am. Mineral., 56, 1607-1616.

Prewitt, C. T., Shannon, R. D., and White, W. B. (1972) Synthesis of a pyroxene containing trivalent titanium. Contrib. Mineral. Petrol., 35, 77-82.

Rossman, G. R. (1974) Lavender jade. The optical spectrum of Fe^{3+} and $Fe^{2+}-Fe^{3+}$ inter-valence charge transfer in jadeite from Burma. Am. Mineral., 59, 868-870.

_____ (1977) Optical absorption spectra of major minerals in Luna 24 sample 24170 (abstr.). *In* Papers Presented to the Conference on Luna 24, 156-159. The Lunar Science Institute: Houston, Texas.

Runciman, W. A., Sengupta, D., and Marshall, M. (1973) The polarized spectra of iron in silicates. I. Enstatite. Am. Mineral., 58, 444-450.

Rutstein, M. S. and White, W. B. (1971) Vibrational spectra of high-calcium pyroxenes and pyroxenoids. Am. Mineral., 56, 877-887.

Schreiber, H. D. (1977) On the nature of synthetic blue diopside crystals: the stabilization of tetravalent chromium. Am. Mineral., 62, 522-527.

_____ (1978) Chromium, blue diopside and experimental petrology. Contrib. Mineral. Petrol., 66, 341-342.

Schmitz, B. and Lehmann, G. (1976) Color centers of manganese in natural spodumene. Bunsenges, Phys. Chem., 79, 1044.

Sung, C.-M., Abu-Eid, R. M., and Burns, R. G. (1974) Ti^{3+}/Ti^{4+} ratios in lunar pyroxenes: implications to depth of origin of mare basalt magma. Proc. 5th Lunar Sci. Conf., 717-726.

_____, Singer, R. B., Parkin, K. M., and Burns, R. G. (1977) Temperature dependence of Fe^{2+} crystal-field spectra: implications to mineralogical mapping of planetary surfaces. Proc. 8th Lunar Sci. Conf., 1063-1079.

Virgo, D. (1972) ^{57}Fe Mössbauer analyses of Fe^{3+} clinopyroxenes. Carnegie Inst. Wash. Year Book, 71, 534-538.

_____ and Hafner, S. (1968) Re-evaluation of the cation distribution in orthopyroxene by the Mössbauer effect. Earth Planet. Sci. Lett., 4, 265-269.

_____ and _____ (1969) Fe^{2+},Mg order-disorder in heated orthopyroxenes. Mineral. Soc. Am. Spec. Paper, 2, 67-81.

_____ and _____ (1970) Fe^{2+}-Mg order-disorder in natural orthopyroxenes. Am. Mineral., 55, 201-223.

Williams, P. G. L., Bancroft, G. M., Brown, M. G., and Turnock, A. C. (1971) Anomalous Mössbauer spectra of C2/c clinopyroxenes. Nature, 230, 149-151.

White, W. B. (1975) Structural interpretation of lunar and terrestrial minerals by Raman spectroscopy. *In* C. Karr, Jr., ed., *Infrared and Raman Spectroscopy of Lunar and Terrestrial Minerals*, 325-358. Academic Press: New York.

_____ and Keester, K. L. (1966) Optical absorption spectra of iron in the rock-forming silicates. Am. Mineral., 51, 774-791.

_____ and _____ (1967) Selection rules and site assignments for the spectra of ferrous iron in pyroxenes. Am. Mineral., 52, 1508-1514.

_____, McCarthy, G. J., and Sheetz, D. E. (1971) Optical spectra of chromium, nickel and cobalt-containing pyroxenes. Am. Mineral., 56, 72-89.

Chapter 4

SUBSOLIDUS PHENOMENA in PYROXENES

Peter R. Buseck, Gordon L. Nord, Jr. & David R. Veblen

INTRODUCTION

When pyroxenes cool from the igneous or metamorphic temperatures
at which they form, they undergo structural and chemical changes in
order to remain in equilibrium. Some pyroxenes are also reheated
during metamorphic episodes, requiring further readjustments through
the resulting thermal cycles. In many cases, however, kinetic
factors inhibit such structural and chemical reequilibration. As a
consequence, many pyroxene crystals contain, within their structures,
traces of their high-temperature histories. Crystals that are not
perfectly periodic may result. These crystals can consist of intimately
intergrown regions of pre- and post-reaction material, of transitional,
intermediate types of materials, or of ordered or disordered combinations of the above. It is these subsolidus features that are emphasized
in this chapter.

Our goals are (1) to describe the features that develop in pyroxene
crystals in response to changing temperatures and pressures and (2) to
explore information contained in these features regarding subsolidus
phenomena and the histories of the host pyroxenes. Specific questions
to be addressed include: How can the traces of subsolidus reactions
be studied? What sorts of polymorphic transformations occur among
pyroxenes? What polytypic variations can occur in the pyroxene structure?
What are the ways that exsolution is initiated and what are the several
mechanisms by which it proceeds? How do antiphase domains form in
pigeonite and to what extent can they be used for geothermometry? By
what mechanisms do pyroxenes react to form amphiboles, wide-chain
silicates, and other silicate minerals?

Pyroxenes are either orthorhombic or monoclinic and crystallize in
several space groups. The most obvious structural feature of pyroxenes
is the single chain of corner-sharing tetrahedra having stoichiometry
SiO_3. These chains are articulated to strips of edge-sharing polyhedra
that contain cations in 6- or 8-fold coordination. In most pyroxenes,
there are two symmetrically distinct sites for non-tetrahedral cations,
which are designated M(1) and M(2). Large cations such as calcium or

sodium occupy M(2), whereas smaller octahedrally-coordinated cations such as iron, magnesium, and manganese distribute themselves between M(1) and M(2).

In all of the pyroxene structures, the octahedral strips and tetrahedral chains form layers parallel to (100). These layers can be stacked in various ways, leading to different pyroxene polytypes when the stacking sequence is periodic. Mistakes in the stacking sequence of these layers produce pyroxenes with stacking disorder; the degree of disorder can range from isolated faults occurring in an otherwise ordered polytype to complete stacking disorder, in which case there is no recognizable pattern to the stacking of the layers. Reactions involving changes in the stacking sequence of the polyhedral layers are discussed, under the heading "Polymorphism," in the section "Reconstructive reactions."

Another class of polymorphic reactions involves the geometrical details of the individual silicate chains, rather than the ways in which the (100) pyroxene layers are stacked. Transformations that occur by rotation of tetrahedra within chains are generally rapid, reversible, and nonquenchable. The microstructures that arise during these reactions are described in the section "Displacive reactions."

Reactions that involve the diffusional ordering or disordering of octahedrally-coordinated cations, generally within distances equal to a few bond lengths, comprise another type of polymorphic transformation. Such reactions are of particular importance for omphacites, which undergo space group changes produced by ordering within the M(1) and M(2) sites. These reactions and resulting antiphase features are discussed in the section on "Order-disorder reactions."

In considering pyroxene transformations, we have chosen to distinguish between polymorphic reactions, which do not require long-range compositional changes, and exsolution reactions, which do involve such changes. By long-range we mean distances greater than one or two unit cell translations. Therefore, the formation of a cluster containing a few unit cells of a different composition is considered to be exsolution. The diffusing species are, in general, those that occupy the sites in the octahedral strip of the pyroxene structure. The section on "Exsolution" describes the varied mechanisms by which these reactions can occur and the rich variety of microstructures that can result from them.

118

In addition to pyroxene transformations in which the reaction
products are also pyroxenes, there are a variety of transformations to
related, non-pyroxene minerals, such as amphibole and wider-chain
silicates, sheet silicates like talc and serpentine, and pyroxenoids.
In many such reactions there is chemical exchange between the pyroxene
and its environment, so that the reaction product no longer retains
the original pyroxene composition. These sorts of transformations are
considered in the last section, "Alteration reactions."

All standard mineralogical techniques have been used to study
pyroxene subsolidus features. Many of these features are non-periodic
and occur on a very small scale, so that transmission electron microscopy, which provides great spatial resolution of structural irregularities, has become a powerful and widely used technique for the study of
pyroxene microstructures. For this reason, we give considerable attention
to electron microscopy.

EXPERIMENTAL TECHNIQUES

Over the years, emphasis has shifted on the types of problems and
instrumental techniques that are used for studying pyroxenes. The
initial quantitative studies of the crystal structures and phase
relations of pyroxenes were completed in the 1930's. Emphasis during
the 1940's and 1950's was on determining their optical and bulk chemical
properties. Electron microprobe measurements and experimental studies of
the phase relations made great strides in the 1960's and early 1970's.
A high degree of structural understanding was also provided by single
crystal x-ray and spectroscopic measurements during this time. In
addition, during the last decade, transmission electron microscopy was
first used to help understand the textures and microstructures produced
in pyroxenes by subsolidus reactions.

In this section we provide a brief overview of several of the major
procedures that are used to study the subsolidus behavior of pyroxenes.
In addition to the methods discussed here, spectroscopic techniques
have been used, mainly for the determination of cation ordering among
the several crystallographic sites. These methods are reviewed in the
chapter by G. Rossman (this volume).

Light Optical Microscopy

The petrographic microscope has been and continues to be an instrument of vital importance for the study of pyroxenes. Most pyroxenes can be identified by their optical properties, and there is an extensive literature relating optical properties to chemical composition (Deer *et al.*, 1978).

In a series of classic papers, Hess and his colleagues described the use of light optical methods to define and clarify the distinctions among augite, pigeonite, and orthopyroxene (e.g. Hess, 1941, 1952; Hess and Phillips, 1938, 1940; Poldervaart and Hess, 1951). They also determined many of the details of exsolution and inversion reactions in the major rock-forming pyroxenes and showed that the orientations of exsolution features can be influenced by thermal history. Recent studies using a combination of light microscopy, x-ray methods, and transmission electron microscopy have underlined the complexities of the exsolution features that were first noted in these early papers (Robinson *et al.*, 1971, 1977; Jaffe *et al.*, 1975). These studies have shown that small angular variations in the planes of exsolution are dependent on temperature of formation, as well as on subsequent cooling history.

In general, there has been a trend of decreasing reliance on precise light optical measurements. As indicators of chemical composition, these techniques are indirect; they rely on calibrations using optical properties of crystals of known composition to establish reference curves to be used for pyroxenes of unknown composition. Such calibration curves were of great utility for many years because it is easier to determine the optical properties of a crystal than to perform a complete wet chemical analysis. An important change occurred, however, with the development and subsequent wide availability of electron microprobes. Today, most chemical investigations of pyroxenes rely on chemical analysis with electron beam instruments, rather than on optical measurements.

Electron and Ion Beam Analysis

The electron microprobe and related instruments to a large extent have replaced wet chemical analysis and light optical techniques as the major analytical tools for the chemical study of pyroxenes. The great appeal of electron microprobe analyses is that they are rapid and can be performed on areas as small as 1μm in diameter. Thus, it is possible to study fine-grained intergrowths, zoning relations, and reaction coronas, as well as the compositions of small crystals (e.g. Boyd and Brown, 1968; Ishii and Takeda, 1974). Microprobe analyses are usually accurate to within 2 to 3% for all elements heavier than sodium and are improving for the lighter elements.

Analyses for the lightest elements are precluded with most electron beam techniques. By using a beam of charged ions, however, it is possible to determine elements such as hydrogen, lithium, boron, and beryllium, although there remain many difficulties. However, the ion microprobe will probably be used extensively in the coming years, once some of the instrumental complexities have been mastered. Limited results from ion microprobe measurements of pyroxenes have already appeared. For example, Shimizu et al. (1978) demonstrated the feasibility of quantitative analyses for major and minor elements in several terrestrial and meteoritic pyroxenes, and Steele et al. (1980) showed that reference to a single, well-documented standard can suffice for minor element analyses in low-Ca pyroxenes.

Another instrument that is finding greatly increased use for the chemical analysis of pyroxenes is the scanning transmission electron microscope (STEM) equipped with an x-ray energy analyzer (e.g. Cliff & Lorimer, 1975; Goldstein, 1979; Zaluzec, 1979). While energy analyzers do not have the high resolution of the crystal spectrometers that are used for most x-ray wavelength analyses on electron microprobes, energy analyzers have high collection efficiencies and work well for the low x-ray production encountered with STEMs. Copley el al. (1974) have used the method to study the distribution of phases in an iron-rich pyroxene in order to understand its exsolution mechanism and cooling history, and Nobugai et al. (1978) determined the compositional differences between exsolved and host augite and pigeonite from the Skaergaard complex.

The combination of imaging the interphase boundaries and determining phase compositions provides a powerful approach to the study of pyroxene subsolidus phenomena. The great advantage of such analytical transmission electron microscopes (TEMs) is that they extend the spatial resolution for analyses from the 1μm limit of the electron microprobe

to 500A or less. Furthermore, new dedicated STEMs have electron
beams as small as 5A and thus are theoretically capable of analyses
of areas approaching unit cell dimensions, provided the specimens
are sufficiently thin.

There are definite limitations, however, for the use of the STEM
as an analytical instrument. For specimens of reasonable thickness,
x-ray emissions occur from a volume having a diameter that is
appreciably greater than that of the electron beam. Also, the
yield of emitted x-rays decreases as specimen thickness decreases;
counting statistics therefore become problematical for very thin
specimens, and long counting times are required. Another difficulty
is that the x-ray production varies with thickness. Consequently,
the methods of data reduction used in electron microprobe analysis
are not feasible with the TEM, unless the thicknesses of both unknown
and standard are accurately known. Since this is not generally
possible, a technique using the ratios of x-ray emission peaks in
conjunction with correction factors derived from standards has
been used successfully (Lorimer and Cliff, 1976; Goldstein, 1979).
Instrumental factors, such as mechanical and electronic stability
and specimen contamination rates, also severely affect the spatial
resolution and quality of STEM analyses.

In addition to measuring compositions from emitted x-rays, STEMs
potentially can also be used to determine the concentrations of
lighter elements. When an electron beam is transmitted through a
specimen, some of its electrons interact inelastically with the atoms
of the specimen (electron diffraction is the result of elastic interactions). During this process the transmitted electrons lose energy
(in the range of 0 to 1000 eV) by interaction with the inner atomic
shells of the specimen. The effect is analogous to the K-, L-, and
M-edges seen in x-ray absorption spectra, and under favorable circumstances the technique of electron energy loss spectroscopy (EELS)
can be used to measure the presence and concentrations of a wide range
of elements (Egerton, 1978; Joy *et al.*, 1979).

One of the major advantages of EELS is that it can provide great
spatial resolution; in distinction to emitted x-radiation, the excited
volume is not appreciably larger than the electron beam, so that

features approaching the diameter of the beam can be analyzed. In pyroxenes it may eventually be possible to compositionally image features such as narrow exsolution lamellae, Guinier-Preston (GP) zones, compositional segregation along anti-phase boundaries, and "tweed" textures.

EELS is a relatively new development and does not yet yield reliable quantitative results in complex materials, but it will probably find increasing use in the future for the study of pyroxenes, as well as other minerals.

X-ray Diffraction

Single crystal x-ray diffraction measurements at both standard and elevated temperatures have provided the data for structure determinations and refinements of pyroxenes. Reviews include those by Smith (1969a), Clark *et al.* (1969), Morimoto (1974), Papike and Cameron (1976), Deer *et al.* (1978), and by Cameron and Papike in this volume. The emphasis in such refinements is on the structural state of nearly perfect crystals with low densities of non-periodic features. In comparison to TEM studies, described below, x-ray methods are useful for obtaining a complete structural picture of a given crystal, since one crystal can be placed into any combination of orientations. More accurate cell parameters can be obtained, and these contain compositional as well as structural information. Three-dimensional orientation information in intergrowths is readily obtained through x-ray measurements, as are statically meaningful views of the degree of disorder in given crystals (Nakazawa and Hafner, 1977; Nakajima and Hafner, 1980).

All of the standard x-ray diffraction methods have been applied to pyroxenes, but the most productive instrument has been the computer-controlled single crystal x-ray diffractometer (Clark *et al.*, 1969; Cameron *et al.*, 1973; Finger and Ohashi, 1976). Structure refinements of high accuracy and precision both at room and elevated temperatures have been obtained, and the structures of the rock-forming pyroxenes are now known well. A result is that the number of papers on pyroxene structure refinements has decreased in recent years.

During the last several years, single-crystal x-ray film methods have also produced extremely interesting and important results regarding exsolution processes. For example, Nakazawa and Hafner (1977) used precession and Laue methods to distinguish among four phases within lunar pigeonites and to show that there are two generations of exsolution and two host-exsolution pairs. X-ray precession methods were used by Ross *et al.* (1973) for phase relation studies in order to define a miscibility gap for lunar pyroxenes, and Smyth (1974a) has studied enstatite phase relations with high-temperature single crystal methods. Robinson *et al.* (1971, 1977) and Jaffe *et al.* (1975) used single crystal x-ray photographs to identify exsolved pyroxenes and to determine the precise orientation relationships of host and exsolved phases.

Transmission Electron Microscopy

The transmission electron microscope (TEM) has come into greatly increased use for pyroxene research since the early years of lunar sample analysis. It has rapidly become a standard instrument for mineralogical investigations. As one of the newest techniques for the study of pyroxenes and one that is well suited for studying subsolidus reactions, greater attention will be paid to it here than to the other methods.

The attraction of the TEM for mineralogists is that it provides great spatial resolution together with information on non-periodic features. A somewhat arbitrary distinction is commonly made between standard or conventional transmission electron microscopy (CTEM) and high-resolution transmission microscopy (HRTEM). In CTEM a single diffracted or transmitted beam is generally utilized to form the image, whereas for HRTEM multiple beams and greater magnifications are used. CTEM methods have been widely used for studying processes of exsolution in pyroxenes, antiphase domains, mechanisms of dislocation formation and migration, and the extent of pyroxene twinning. HRTEM has been particularly useful for resolving structural details associated with crystal defects and interfaces.

As with x-rays, electrons can be used to produce diffraction patterns. Electrons have much shorter wavelengths than x-rays (0.037A for 100 kev electrons versus 1.54A for CuKα x-rays), so that Ewald's sphere of reflection, the radius of which is inversely proportional to the wavelength, is quite large for electrons in a TEM. Therefore, many

124

reciprocal lattice points may lie near Ewald's sphere, and numerous diffracted beams can be observed simultaneously. As a consequence, a large section of a reciprocal lattice plane can be sampled and photographed without having to use the moving-crystal methods necessary with x-rays.

Electrons interact strongly with crystals, so that electron diffraction patterns can be photographed in a few seconds with the TEM. In addition, changes in the electron diffraction pattern can be observed instantaneously, while the crystal is being tilted. (Such tilting is required for most electron microscopy in order to obtain the desired crystallographic orientations.) Electron diffraction patterns of pyroxenes commonly contain both space group information and data relating to the microstructural state of the mineral. Recording and interpreting electron diffraction patterns is a standard part of electron microscopy and is one of the ways used to determine whether a given pyroxene contains unusual structural features that warrant further study.

Perhaps the most familiar and striking aspect of the TEM is its ability to yield highly magnified images of crystals. Such images cannot be produced with x-rays, which are not readily focussed by lenses in the same way as light and electrons. However, because electrons are charged, they can be focussed by magnetic lenses, and diffracted electron beams can be combined to produce a greatly magnified image of the crystal. Both periodic and non-periodic features separated by 3 or 4A can be resolved in many modern TEMs.

Many TEM studies of pyroxenes have utilized bright field (BF) microscopy. Another method, dark field (DF) microscopy, is useful for distinguishing between parts of pyroxene crystals that have slightly different structures or orientations from the rest of the crystal; examples are twins, exsolution lamellae, and antiphase boundaries. In BF microscopy, the direct electron beam is allowed to pass through an aperture in the TEM and form the image. Most HRTEM images are multiple-beam BF images, which are formed from a combination of the central beam with several diffracted beams. For DF images, the direct electron beam is excluded and only diffracted beams are used for image formation. As a consequence, only those parts of the crystal that

125

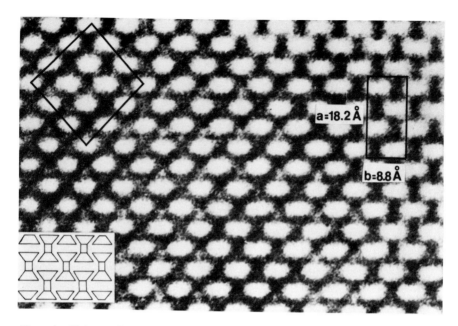

Figure 1. High magnification structure image of an a-b section of orthoenstatite. The white regions lie between the M(2) sites. A unit cell is outlined; the insert shows the I beams in the orthoenstatite structure at the same scale as the image. Possible cleavage surfaces are sketched in the upper left. (Modified from Buseck and Iijima, 1974)

are in proper orientation for diffraction will be visible in the image. Proper tilting of the crystal makes it possible to associate specific diffracted beams (and thus interplanar spacings) with particular microstructural features.

Before discussing structural pertubations, it is necessary to consider the appearance of "ideal" pyroxenes at high magnification. In HRTEM studies, two crystallographic orientations have proven to be the most useful. Projections along the b-axis, yielding images of the a - c plane, display features such as twinning and stacking disorder; c-axis projections, yielding images of the a^{*} - b plane, are ideal for the imaging of the pyroxene "I-beams" and the study of intercalation structures.

Figure 1 (adapted from Buseck and Iijima, 1974) shows an image of enstatite from Bamble, Norway, as viewed down [001] (a c-axis image). The white spots are positions of low charge density and are centered on the positions between the M(2) sites of adjacent cation strips. The pyroxene I-beams lie in the intervening black regions, with the "I"s

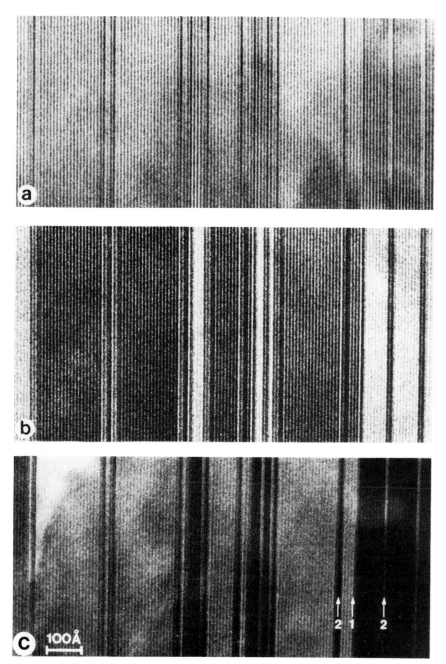

Figure 2. Lattice fringe image of an $a-c$ section of clinoenstatite. The vertical fringes are parallel to c and the silicate chain extensions. Spacings between adjacent fringes are 9A. When the crystal is tilted slightly the contrast in (b) and (c) develops (the crystal was tilted in opposite directions to produce the reversed contrast). Arrow #1 points to the trace of an individual twin plane, whereas #2 points to paired twin planes that are 9A apart. (After Iijima and Buseck, 1976)

positioned vertically. Although the structural details of the I-beams are not resolved, their widths can be observed at a glance by noting the separations of the white spots along the b direction. As can be seen in the illustrations of papers by Veblen and Buseck (1979a,1980), silicates with wider chains have greater separations between the centers of these white regions.

Figure 2a shows an image of artificially heated and quenched Bamble enstatite, as viewed down [010] (a b-axis image). The black vertical lines (fringes) are parallel to the SiO_3 chains and the tetrahedral and octahedral strips in the pyroxene structure, and consequently they indicate the layering parallel to (100). These strips are well illustrated in the crystal structure drawings projected onto the (010) plane in the chapter by Cameron and Papike (this volume). In some of the TEM images along directions near [010], there are irregularities in the contrast and spacings of these fringes (Fig. 3); these variations reflect stacking disorder, discussed in more detail below. This orientation and type of image is also useful for studying pyroxene twinning and polytypism.

POLYMORPHISM

There have been a number of attempts to classify polymorphic reactions in minerals (see Heuer and Nord, 1976, for a review). Classifications have been based on kinetic (i.e., sluggish or rapid), thermodynamic (i.e., first or second order), and structural grounds. For the purposes of this section, we have followed the classification of Buerger (1951), which was a major clarification for mineralogists. Buerger used the premise that the change in internal energy during a transformation reflects the changes in the bonding of the atoms in the structure. A structural classification of transformations should, therefore, be based on these differences in bonding (i.e., coordination changes). Buerger proposed four categories: (1) transformations of second coordination, (2) order-disorder transformations, (3) transformations of first coordination and (4) transformations of bond type; only the first two are important for polymorphic reactions in pyroxenes. Second-coordination transformations were further divided into two types, reconstructive and displacive. A reconstructive transformation

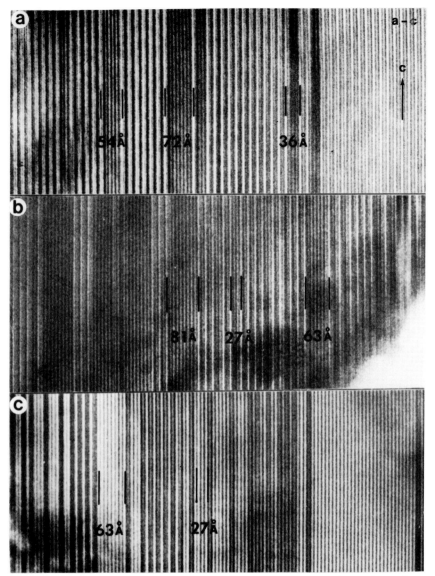

Figure 3. Lattice fringe images of a-c sections of clinoenstatite similar to Figure 2, except that this image shows irregularities in the spacings of the fringes. The crystals in (b) and (c) were heated in the laboratory into the protopyroxene stability field and quenched. (After Buseck and Iijima, 1975)

involves (1) delinking and then relinking of the structure, (2) retention of first-coordination contacts in the new structure, (3) temporary breaking of first coordination bonds during the transformation, and (4) a high activation energy barrier (thermal or stress activated).

Buerger gives the reaction of tridymite to quartz as an example of a reconstructive transformation. A displacive transformation involves (1) distortion of one structure to another, (2) no bond breaking, and (3) a very low activation energy barrier. Buerger gives the reaction of high to low quartz as an example of a displacive transformation. Reconstructive transformations are generally sluggish (although some can be quite rapid), and displacive transformations are generally rapid. We discuss a variety of reconstructive, displacive, and order-disorder reactions in pyroxenes in this section.

Reconstructive Reactions

Reconstructive transformations in pyroxenes are generally polytypic, involving changes in stacking sequences, and are illustrated well by enstatite ($MgSiO_3$). Pyroxenes having the enstatite composition are known to occur in at least four and perhaps five space groups, more than for any other pyroxene composition (Smith, 1969a). The form that is stable at high temperatures, protoenstatite (PEN), crystallizes in orthorhombic space groups $Pbcn$ or possibly $P2_1cn$ (Smyth, 1971), a non-centrosymmetric subgroup of $Pbcn$. Upon cooling, it inverts to either of two forms: low clinoenstatite (CLEN), in monoclinic space group $P2_1/c$ (also the space group of pigeonite), or orthoenstatite (OREN), in orthorhombic space group $Pbca$. Above about 980°C, CLEN inverts reversibly to the non-quenchable high clinoenstatite, $C2/c$. In addition, several orthorhombic pyroxenes have been reported in space group $P2_1ca$ (Smyth, 1974a,b; Steele, 1975; Krstanovic, 1975; Harlow et al., 1979), but it is possible that some of these reports are the result of extra x-ray reflections arising from multiple diffraction or included phases. In the following discussion we will emphasize the geologically most significant forms -- PEN, CLEN, and OREN -- and their structural relationships to one another.

These three enstatite polymorphs have many structural similarities but are topologically distinct, and the inversions among them are reconstructive (Smyth, 1974a). It is probable that the free energy differences at standard temperature and pressure are small among at least some of the structure types, but that activation energies for some of the transformations are significant, thereby accounting for the common coexistence of CLEN and OREN, for example.

Stacking Sequences and the Block Abstraction. In their solution
of the enstatite structure, Warren and Modell (1930) noted that the
orthopyroxene structure can be derived from clinopyroxene by a simple
glide reflection operation. The notion that orthopyroxene can be considered as clinopyroxene twinned on the unit-cell scale has been further
elaborated since the time of this early work (Ito, 1935, 1950; Brown *et
al.*, 1961; Burnham, 1967; Smith 1969a). The solution of the PEN structure
further demonstrated that the pyroxene polymorphs are all closely related, being built from layers of similar structure that are stacked
together in different ways (Smith, 1959, 1969a).

The ways in which pyroxene layers are stacked are referred to as
"stacking sequences." These sequences are frequently discussed in terms
of the sequence of orientations, or "skews," of the M(1) octahedra
(Thompson, 1970; Papike *et al.*, 1973), as discussed in detail by Cameron
and Papike (this volume). For our purposes, it is useful to further
abstract the pyroxene structure types into blocks that are about 4.5A
wide in the $a*$ direction, as shown in Figure 4. The boundaries of
these blocks that are parallel to (100) pass through the layers of
tetrahedral chains that are common to all pyroxenes.

The pyroxene blocks can assume two possible orientations, which
are designated + and -. In PEN (Fig. 4a), there are blocks of both
orientations, arranged in the stacking sequence ...+-+-..., which
results in a unit cell with $a \sim 9A$. In a perfect crystal of CLEN,
however, all of the blocks have the same orientation; this structure
can thus be represented either as ...++++... or as ...----..., depending on the orientation of the blocks with respect to the unit cell
axes. In CLEN, SiO_3 chains that are adjacent to one another along [100]
are alternately more or less extended parallel to c; these symmetrically distinct chains are designated A and B. Since adjacent 4.5A
blocks are not equivalent by translation, two blocks having similar
"sign" are required to define a unit cell, again resulting in a cell
with $a \sim 9A$.

If both block orientations are intergrown in the same crystal,
the + and - orientations are related to each other by a b-glide twinning
operation, with the twin planes parallel to (100). Such a twin plane

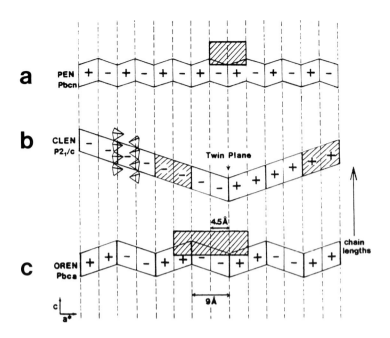

Figure 4. Schematic diagram of the structures of (a) protoenstatite ("PEN") (b) clinoenstatite ("CLEN"), and (c) orthoenstatite ("OREN"), all projected along [010]. The
parallelogram-shaped blocks are abstractions designed to show the differences among the
structure types. In this projection the long pyroxene chains are aligned vertically and
are located at the boundaries of the blocks. The + or - orientations of the blocks
reflect the relative displacements of adjacent chains. In PEN the blocks alternate every
4.5A, producing a unit cell — the basic repeat unit — that is 9A wide; an example is
shaded. CLEN can consist of either + or - type blocks; if both occur the crystal is
"twinned," with the zone of separation defining the twin plane. The repeat units (shaded)
consist of pairs of like blocks and are 9A wide. OREN, like PEN, consists of paired blocks,
but the pairs alternate rigorously along twin planes every 9A resulting in an 18A unit cell
(shaded).

is represented in Figure 4b, which also shows shaded unit cells for
CLEN. The OREN structure is represented in Figure 4c. Here the stacking
sequence is ...++--++--..., and the resulting orthorhombic unit cell has
$a \sim 18A$.

As we can see from these diagrams, the major pyroxene polytypes
can be constructed from blocks having two different orientations. Not
all pyroxene cyrstals, however, consist of these perfectly periodic
structure types. Several types of deviations from perfect periodicity
can occur. For example, there may be isolated mistakes in the ideal
stacking sequences, resulting in stacking faults or twinning.

Alternatively, more than one of the ideal structure types may be found intergrown together in the same crystal. It is even theoretically possible that a pyroxene crystal could be constructed with a completely random stacking sequence, resulting in complete lack of periodicity in the stacking direction. It is possible to study such deviations from perfect periodicity with HRTEM methods, and, in turn, to understand certain aspects of the phase transformations that commonly result in this structural disorder. Iijima and Buseck (1975) used HRTEM to study the partial dislocations, associated strain fields, and atomic offsets that occur at the terminations of OREN lamellae within disordered CLEN.

Transformation of Protoenstatite. PEN is the stable enstatite polymorph above about 980°C (Smyth, 1974a). At lower temperatures it transforms to either orthorhombic or monoclinic enstatite. In either case, the symmetrically equivalent SiO_3 chains of PEN change to alternating A- and B-chains. The particular polymorph that forms depends on the sequence of + and - blocks that develop. In an experimental study of enstatite polymorphism, Smyth (1974a) found that the PEN→CLEN inversion is essentially instantaneous, diffusionless, oriented, and reversible, and it is thus martensitic. Strain energies that arise from differences in cell dimensions cause the reaction to be athermal, so that the extent of the reaction is characteristic of temperature, but does not increase with time.

The transformation between PEN and OREN is more complicated (Smyth, 1974a). In order to go to completion, the OREN→PEN reaction requires several days at 1200°C, 200°C above the lower limit of· PEN stability. Smyth describes the reaction as requiring coordinated motion of alternate pairs of octahedral layers, thereby indicating its complex character. He also suggests that CLEN may be a necessary intermediate phase. In any case, slower cooling appears to be required to produce ordered OREN from PEN than to produce ordered CLEN.

Orthoenstatite-Clinoenstatite Transformation. The stability relations between OREN and CLEN have received much attention but are nevertheless still a matter of uncertainty. Diffraction patterns of many crystals of enstatite show streaking parallel to a^*, indicating an intergrowth of the two polymorphs, and consideration of Figure 4 indicates how the two can be intergrown coherently.

Smith (1969a) has summarized much of the stability data and concluded that OREN is the stable phase at room temperature. Most subsequent authors have tended to agree, although Grover (1972) has suggested that CLEN is the stable form. Smyth (1974a) has performed experiments using a high-temperature single-crystal x-ray precession camera. He found that CLEN slowly inverts to OREN between 650° and 950°C, and he suggests that it is a time-dependent, non-reversible reaction in the absence of externally applied stress. It is still not certain, however, which polymorph is stable at room temperature, because the sluggishness of the reactions has precluded definitive experiments.

There is abundant evidence that external stress favors the formation of clinopyroxene from orthopyroxene. Turner *et al.* (1960) showed experimentally that mechanical deformation can induce the transformation and suggested that it is a martensitic reaction, with slip occurring on (100). Considerable attention has been focussed on the structural mechanism of the transformation. Turner *et al.* (1960) originally suggested that a macroscopic angle of shear through 18.3° parallel to [001] could account for the formation of the clinopyroxene. Alternate transformation mechanisms have been proposed by Brown *et al.* (1961), Sadanaga *et al.* (1969), and Coe (1970). Coe and Müller (1973) and Coe and Kirby (1975), based on CTEM studies, detailed a model that requires a 13.3° shear combined with a dislocation mechanism. The various models have been examined by McLaren and Etheridge (1976) in a careful analysis of the structural transformation mechanisms. They propose that either a 13.3° or a 18.3° shear theoretically could produce the observed transformations, but that the 18.3° mechanism is more likely because it requires only homogeneous shearing.

When clinopyroxene is produced from orthopyroxene, there is a relative displacement of the orthopyroxene structure on opposite sides of the inserted clinopyroxene. This displacement (parallel to *c*) results in features that can be detected by DF tilting experiments, and a number of CTEM studies have been made of both naturally and experimentally deformed pyroxenes. In addition, Iijima and Buseck (1974, 1975) used DF HRTEM methods to detail the displacements and resulting partial dislocations. In the initial CTEM studies of stacking faults on (100), several different fault vectors were proposed (Lally *et al.*, 1972;

Kohlstedt and VanderSande, 1973). McLaren and Etheridge (1976) used DF experiments to determine the number of adjacent unit cells in a given lamella of clinopyroxene. They also demonstrated that these transformations can result from the passage through the crystal of edge dislocations with displacements parallel to [001]. Since the formation of such dislocations occurs as essentially isolated events, the resulting clinopyroxene lamellae are likely to be narrow, although extensive shearing of orthopyroxene would presumably produce extensive regions of clinopyroxene.

Polytypes with Repeats Greater than 18A. Polytypism is essentially one-dimensional polymorphism, where important differences among polytypes occur along only one crystallographic axis. These differences are produced by varying the stacking sequences of structurally and chemically similar units. Dimensional relationships perpendicular to the stacking direction are more or less constant. Because of the layered character of the pyroxene structure parallel to (100), polytypism occurs within pyroxenes. In the preceding sections, we have discussed the stacking sequences that give rise to the well-known orthorhombic and monoclinic pyroxene polytypes. It is also possible that other ordered stacking sequences could occur, producing yet other pyroxene polymorphs.

With the notable exception of a description by Bystrom (1943) of a 36A a repeat, pyroxenes having a dimensions greater than 18A have not been reported from x-ray diffraction measurements. Such ordered pyroxenes with a > 18A apparently do not occur or are extremely rare, at least in sufficient abundances to be observable in x-ray measurements. On the other hand, with HRTEM methods it is possible to observe local regions of crystal that contain repeats that are larger than 18A.

Iijima and Buseck (1975) calculated the possible pyroxene polytypes that can contain up to nine layers in a repeating stacking sequence. They used a 9A basic repeating layer rather than the 4.5A unit that is shown in Figure 4; the 4.5A unit should have been used in order to include all of the possible stacking combinations, so that their Table 1 properly applies to spacings up to 40.5A rather than 81A. In either case, it is clear that the number of possible pyroxene polytypes is large.

Only very few of the possibilities listed by Iijima and Buseck (1975) have been observed. In addition to the 36A enstatite reported by Bystrom (1943), Buseck and Iijima (1975) showed TEM images of sequences having 27, 36, and 54A repeats. However, by applying the statistical test of Veblen and Buseck (1979a), it can be shown that the probability is very high that the illustrated sequences arose by a random process. The existence of statistically significant long-period pyroxene polytypes therefore remains to be demonstrated.

Electron Diffraction Effects and Imaging. X-ray measurements were used for determining the stacking sequences of the macroscopically-ordered enstatite polymorphs that are described above. However, such measurements do not allow a precise determination of the numbers and dimensions of twinned regions, and, as we discuss below, such data can be of geological interest. On the other hand, twinning can be observed and studied on a fine scale with TEM imaging. In most crystals twins occur in small concentrations, but some crystals contain many twinned regions. As a consequence of their different orientations, twins diffract electrons at different angles relative to the primary electron beam. As a result, in certain orientations twins produce regions of dissimilar contrast in TEM images (Iijima and Buseck, 1974, 1976). Such a twinned region is shown in Figure 5 (Iijima and Buseck, 1975). In this example, the white and black regions bear a twin relationship to each other. An interpretation of the image is given for the portion between the arrows. Slabs of OREN four and one unit cell wide (72 and 18A, respectively) can be identified within the CLEN. For the choice of unit cell origins shown, the region of CLEN surrounded by OREN is seven unit cells wide (63A).

Most b-axis images of pyroxene crystals contain fringes that are regularly spaced and of uniform contrast, but some images contain regions of different contrast with widths that are multiples of the predominant 9A spacing, the value of d_{100} for primitive clinopyroxene (Müller, 1974; Buseck and Iijima, 1974, 1975). The above authors have ascribed genetic significance to the widths of these lamellae. However, Boland (1974) and McLaren and Etheridge (1976) have raised questions regarding the determinations of the widths of a given lamella of pyroxene.

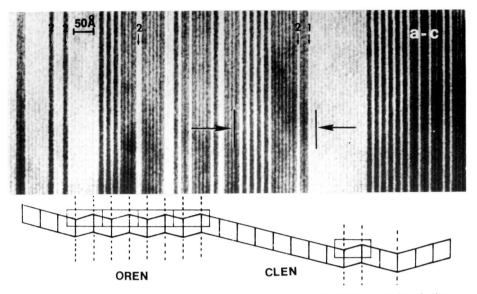

Figure 5. Electron image of an a-c section of heated and quenched Bamble enstatite showing polysynthetic twinning. The vertical arrows indicate similar features to those described in Figure 2. The insert is a schematic representation of the structure in the area between the horizontal arrows. (After Iijima and Buseck, 1975)

Boland, McLaren, and Etheridge point out that there is no necessary correlation between the position of a given fringe on an image and a particular structural feature. It is thus difficult if not impossible to establish the exact positions where a given lamella starts and stops and therefore to determine its width. In fact, depending on the thickness of the crystal fragment that is being imaged, "extra" fringes may appear. Thus, Iijima and Buseck (1975; Fig. 1c) show 4.5A fringes in a very thin fragment of clinopyroxene, whereas most primitive clinopyroxene images show 9A fringes. McLaren and Etheridge (1976) make a similar point regarding 9 and 18A fringes in the photograph of Müller (1974). Here again, the change in spacing is a result of differences in thickness of the crystal that is being imaged.

The questions raised about the interpretation of (100) proxene fringe images are important ones. However, when images obtained under the same experimental conditions, such as crystal thickness and orientation, are compared, a self-consistent interpretation is possible. Thus, for some of the specimens examined by Iijima and Buseck (1975) and Buseck and Iijima (1975), it is possible to observe that the CLEN lamellae are either all even multiples of 9A or all odd multiples wide. Structural consideration of the reactions producing the lamellae further suggests that the interpretation of widths was, in fact, correct, although this has not yet been confirmed experimentally. With the resolutions now available in TEMs, it is possible to determine exactly the stacking sequence in a pyroxene crystal using HRTEM methods (Veblen and Buseck, work in progress). By using calculated images of crystals (O'Keefe, 1973; O'Keefe *et al.*, 1978), it is possible to relate specific stacking sequences to the experimental images to confirm the interpretation. Work on more modern electron microscopes will therefore resolve the problems that have been pointed out by McLaren and Etheridge (1976).

There is apparent confusion in the literature regarding the number of unit cells along $a*$ that occur within a given lamella that has been produced by transformation from orthopyroxene. McLaren and Etheridge (1976) showed that the lamellae contain an odd number of 9A clino-pyroxene unit cells, whereas Müller (1974) and Buseck and Iijima (1975) showed that they contain even multiples. Moreover, both sets of authors obtained internally consistent results.

The apparent difference lies in the choice of unit cell origin for counting, and the reason for the seeming ambiguity is that in the $a-c$ plane the edges of adjacent orthorhombic and monoclinic cells do not coincide. McLaren and Etheridge start counting at the edge of the monoclinic lamella. Buseck and Iijima, on the other hand, observed that the orthorhombic cell edge extends halfway into the first adjacent monoclinic cell (Fig. 5), and thus do not count the first half mono-clinic unit cell. Since a similar "orthorhombic overlap" also occurs at the other side of the slab, Buseck and Iijima count two half-cells less in a given monoclinic lamella than do McLaren and Etheridge. The results of the two sets of authors are thus compatible.

Twin Orientations, Numbers, and Widths: Criteria for Determining how Clinopyroxene Formed. There are a number of distinct ways that CLEN can be produced. Among the possible origins are generation from OREN by shock, as during meteoritic and lunar impacts; by gradual and long-sustained directed stresses, as occurs during tectonic processes; by cooling from PEN, as may occur in lava flows (Dallwitz *et al.*, 1966); and by static transformation from OREN, as proposed by Grover (1972). Buseck and Iijima (1975) proposed that HRTEM can be used to distinguish among these origins. With four proposed formation mechanisms for CLEN,

Table 1. Characteristics of the OREN → CLEN Reaction, Depending
on Origin (revised from Buseck and Iijima, 1975)

	CLEN field widths $(2n)9\overset{\circ}{A}*$	$(2n+1)9\overset{\circ}{A}*$	Twinning present
1. PEN inversion; high temperature	Yes	Yes**	Yes
2. Shearing; moderate to low temperature (a) homogeneous shear	Yes	No	No
(b) inhomogeneous shear (*e.g.*, shock)	Yes	No	Yes
3. Static; moderate to low temperature	Yes	No	Yes

* Where n = any integer; these values depend on the counting
convention, as described in the text.

** The possibility, as yet undemonstrated, exists that some
CLEN lamellae resulting from PEN inversion could have widths
of $(2n+1)4.5\overset{\circ}{A}$.

several criteria are needed. As summarized in Table 1 (revised from
Buseck and Iijima, 1975), these are the widths of CLEN lamellae within
OREN and the presence or absence of twinning in CLEN. In this table,
the lamellar width counting convention of Buseck and Iijima is used.
In order to distinguish between a shock origin and static transformation,
it is necessary to observe other features, such as intense fracturing
and mosaic structure.

CLEN that has inverted from PEN can be in fields that are any
multiple of 9A wide, and in the absence of external shear there are no
constraints against producing abundant twins having either + or −
orientation (using the terminology of the section on Stacking Sequences).
In fact, there is no theoretical reason why CLEN lamellae that are
multiples of 4.5A wide could not occur, although lamellae having such
widths have not yet been observed.

CLEN that is produced from OREN will be in fields that are multiples
of 18A wide. If produced slowly and in the absence of shear, as during
static transformation, the CLEN will probably contain few or no twins.
When twinning does occur, + and − orientations should be present in
approximately equal proportions.

Shearing can produce abundant CLEN from OREN. If homogeneous,
the shear will result in most of the CLEN regions having a common
orientation. Shock, on the other hand, will promote both twin orientations; these result from the strong shock wave interactions that tend
to produce shear couples having a variety of orientations throughout
the shocked sample.

139

In utilizing the pyroxene microstructures as observed by HRTEM to determine the history of a crystal, care must be taken to look at enough of the crystal to assure that the observed features are present in significant amounts. Also, relatively large regions of pyroxene may have to be studied in order to establish whether or not fields of CLEN occur that are odd multiples of 9A wide. With these precautions in mind, the use of structural details observable with HRTEM are potentially of great help for establishing the origin of certain pyroxenes.

Displacive Reactions

Displacive reactions that occur in pyroxenes include the $C2/c$ to $P2_1/c$ transitions in pigeonite, enstatite-ferrosilite clinopyroxenes, and kanoite (MnMg pyroxene). As noted in the chapter by Cameron and Papike, all of the silicate chains in the C-centered clinopyroxenes are symmetrically equivalent. In the primitive clinopyroxenes, on the other hand, there are two symmetrically distinct silicate chains, which are designated as the A and B chains. The A and B chains are also conformationally distinct, having, for example, different chain rotation angles. It has been shown that clinopyroxenes having space group $P2_1/c$ at room temperature undergo rapid, nonquenchable phase transitions to space group $C2/c$ at higher temperatures (Smith, 1969b; Smyth, 1969; Smyth and Burnham, 1972; Brown *et al.*, 1972; Gordon *et al.*, 1980). It has been argued on the basis of high-temperature x-ray data that the $C2/c$-$P2_1/c$ transformation is thermodynamically a rapid first-order transition (Smyth, 1974c), although the possibility that it is of second or higher order has not been completely ruled out.

Morimoto and Tokonami (1969a) first suggested that diffuseness of x-ray reflections with indices $h + k$ = odd in pigeonites resulted from antiphase domains (APD's) having a displacement vector of $1/2(a + b)$. Figure 6a shows the distribution of A and B chains in a c-axis projection of unfaulted primitive clinopyroxene structure. The arrangement of chains in a crystal with antiphase domains is shown in Figure 6b; the light lines represent antiphase boundaries (APB's), which separate the APD's. The existence of such APD's and the $1/2(a + b)$ antiphase vector were confirmed by CTEM DF observations in which an

a
```
A A´A A´A A´A A´
B´B B´B B´B B´B
A A´A A´A A´A A´
B´B B´B B´B B´B
A A´A A´A A´A A´
B´B B´B B´B B´B
```

b
```
A A´A A´|B B|A A´
B´|A A´A A´A|B´B
A|B´B B´|A A´A A´
B|A A´A|B´B B´B
A A´A A´A A´A A´
B´|A A´A A´|B B´B
```

image is formed with one of the $h + k$ = odd diffractions (Bailey *et al.*, 1970; Christie *et al.*, 1971). During the past ten years, numerous observations of APB's have been made by electron microscopists in $P2_1/c$ pyroxenes.

Figure 7 shows APB's in a pyroxene with composition $Wo_4En_{46-48}Fs_{48-50}$. TEM observations on pigeonite have

Figure 6. (a) Schematic representation of the pigeonite structure projected down the c-axis. A and A' represent the positions of the A silicate chains in the structure, and B and B' are the positions of the symmetrically distinct B chains. Primed and unprimed letters refer to chains with apical oxygens pointing up and down, respectively. (b) The same projection showing antiphase domains. Light lines represent the antiphase domain boundaries. (Morimoto and Tokonami, 1969a)

been reviewed by Heuer and Nord (1976) and by Carpenter (1979a).

Although it is possible for APB's to form during crystal growth, it is generally thought that the APB's in $P2_1/c$ pyroxenes form when a crystal passes through the C-centered to primitive transition during cooling, as suggested by Morimoto and Tokonami (1969a). During this phase transition, ordering of the silicate chains into A and B chains takes place simultaneously in many parts of the crystal, forming numerous primitive domains. Where these domains meet, they will either be in registry, or they will be out of registry by $1/2(a + b)$. This vector, of course, corresponds to the C-centering

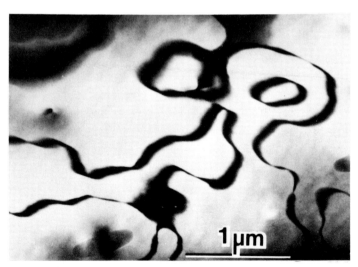

Figure 7. Dark field image, taken with the 102 diffraction, of antiphase domain boundaries in quenched clinohypersthene, $Wo_4En_{46-48}Fs_{48-50}$. The clinohypersthene was produced by J.S. Huebner (USGS) by heating orthopyroxene from Moore County meteorite in platinum foil at 1061°C for 743 hours. The sample was quenched in mercury to room temperature within a few seconds. (G.L. Nord, Jr.)

translational symmetry that was lost when the crystal reacted from $C2/c$ to $P2_1/c$ symmetry.

The mechanisms of APD formation, APB motion, and the textures of APD's in pigeonites and omphacites are discussed by Carpenter (1979a). The domains in $P2_1/c$ pyroxenes form either by nucleation and growth or by a continuous mechanism in which there is no nucleation energy barrier, depending on whether the $C2/c$ to $P2_1/c$ reaction is first order or higher order. Following APD formation, coarsening of the domains can take place by growth of large domains at the expense of small ones. Such coarsening can occur in pigeonites on a time scale of days at temperatures above 700°C (Carpenter, 1979b). APD's in pigeonites and kanoite ($MnMgSi_2O_6$) are typically elongated in the c direction, and the experimental annealing studies have shown that the shape remains relatively constant during annealing. Although the APD's of volcanic pigeonites tend to have irregular morphologies, the boundaries in slowly-cooled pigeonites from near the center of the Palisades Sill (Walker, 1969) are almost perfectly planar and parallel to (100), presumably owing to the annealing process (Veblen, unpublished data). The APD shapes may not be reliable indicators of cooling rate in $P2_1/c$ pyroxenes, however, since morphology may also depend strongly on the pyroxene composition. It has been suggested that APB orientations are related to calcium contents of primitive clinopyroxenes because certain orientations provide more suitable sites for calcium ions, which may be concentrated on the domain boundaries (Carpenter, 1978a).

Complex interactions exist in some cases between antiphase domains and exsolution features in pigeonites. The temperature of the inversion from $C2/c$ to $P2_1/c$ pigeonite varies at least from 500°C for Fe-rich compositions up to 1000°C for Mg-rich pigeonites (Prewitt et al., 1971), and temperatures of nucleation for different generations of exsolution lamella can vary substantially as well. When exsolution precedes APD formation, textures observed in the TEM suggest that APD's nucleate preferentially on the precipitate interfaces, but, conversely, augite lamellae can also nucleate on pre-existing pigeonite APB's. Where exsolution lamellae of augite are closely spaced, they can interfere in a complex fashion with the coarsening of pigeonite APD's during annealing (Lally et al., 1975; Smith, 1978; Carpenter, 1979a).

Numerous authors have used the sizes of APD's in pigeonites as a rough guide to cooling rates; coarsening of APD's in slowly-cooled rocks tends to produce larger APD's than those in rapidly-cooled rocks (Radcliffe *et al.*, 1970; Christie *et al.*, 1971; Ghose *et al.*, 1972; Lally *et al.*, 1975). It has been shown experimentally by Carpenter (1979b), however, that the coarsening of pigeonite APD's is governed by a complex rate law that is in part controlled by segregation of impurity ions on the APB's; these ions are presumably calcium, as first suggested by Morimoto and Tokonami (1969a). In addition, Carpenter suggested that large APD's can be produced either by very slow or by very rapid cooling; he produced small domains with intermediate cooling rates. On the other hand, Grove (1979) has calibrated the scale of the APD's as a function of geologically-reasonable cooling rates in a quartz-normative basalt and found that domain size increases in a regular fashion with slower cooling rates. It is now clear, however, that APD sizes and morphologies in $P2_1/c$ pyroxenes are very complex functions of composition and cooling history. Quantitative use of the scale of the APD microstructures will therefore require extremely careful calibration of the specific system of interest, combined with the application of geological common sense.

Order-Disorder Reactions

It is now generally agreed that omphacites having compositions near $Na_{0.5}Ca_{0.5}Al_{0.5}(Mg,Fe)_{0.5}Si_2O_6$ (jadeite 50%, augite 50%) can occur in either a high-temperature form with $C2/c$ symmetry or in a low-temperature form with $P2/n$ symmetry (Matsumoto *et al.*, 1975; Fleet *et al.*, 1978). There are also indications that low-temperature omphacites with symmetries $P2/c$ and $P2$ may exist (Carpenter, 1978b,c, 1979c, 1980). The primary difference between high-temperature $C2/c$ omphacite and the low-temperature forms lies in the ordering of the octahedral cations; in the structure of Matsumoto *et al.* (1975), Mg and Al are ordered between two different M1 sites, and Na and Ca are partially ordered between two M2 sites. In this section, the microstructures associated with ordering in omphacites with compositions close to Jd50-Aug50 will be considered; the far more complex interactions that can occur between exsolution and ordering processes in other compositions will be discussed in the section "Combined reactions."

143

0.1μm

Figure 8. Dark-field image of omphacite with smooth antiphase boundaries with displacement $\frac{1}{2}(a+b)$. (Champness, 1973)

The C-centered to primitive ordering reaction in omphacites is much more sluggish than are the non-quenchable, displacive transitions discussed in the last section (Fleet et $al.$, 1978). Just as in the $P2_1/c$ clinopyroxenes, however, anti-phase domains having displacement vectors of $1/2(a + b)$ have been observed in $P2/n$ omphacites using DF TEM techniques, as shown in Figure 8 (Champness, 1973; Phakey and Ghose, 1973; Carpenter, 1978b,c, 1979c, 1980). The temperatures of omphacite inversion from $C2/c$ to $P2/n$ have been estimated to be 300-400°C by Yokoyama et $al.$ (1976), and Fleet et $al.$ (1978) have measured a disordering temperature (P to C) of about 725°C. On the basis of this temperature, Fleet et $al.$ argue that omphacites grow as ordered crystals in space group $P2/n$ and that the APD's that have been observed are primary growth domains. On the other hand, Champness (1973) and Carpenter (1978b, 1979c, 1980) have proposed that omphacites grow metastably as disordered crystals in $C2/c$ and that the APD's form during subsequent cation ordering and inversion to a primitive symmetry. Many of the textures described by Carpenter are consistent with this view and difficult to reconcile with the notion that the APD's are primary growth features.

The properties and annealing behavior of APD's in pigeonite and omphacite have been compared recently by Carpenter (1979a). The differences can in general be logically related to the transformation mechanisms of the reactions that produce the domains: in the $P2_1/c$ pyroxenes, APD's are produced by a nondiffusional, displacive trans-formation, whereas the domains in omphacite arise through a cation ordering reaction that requires diffusion to take place. In the absence of effects such as pinning of APB's by segregation of calcium, the domain boundaries in pigeonite can move quite rapidly, since APB motion requires only minor adjustments in atomic positions. In omphacites, however, APB motion requires diffusion of octahedral cations, generally at relatively low temperatures. The coarsening of domains therefore is a very sluggish process in omphacites, and such coarsening is only observed in omphacites that have been annealed on metamorphic time scales. As in pigeonites, APD's in omphacite are observed to coarsen by a homogeneous mechanism, in which there is a continuous increase in average domain size. In this case, the increase in domain size is more or less uniform throughout the crystal. Omphacite domains can, however, coarsen by a second, heterogeneous mechanism as well, in which a single large domain grows and consumes the smaller domains. The omphacite APD morphologies also contrast with those in pigeonite. Unlike the elongated domains with ragged or irregular boundaries in most pigeonite, omphacite typically has equant domains with smooth boundaries. This difference in shape is probably related to domain growth mechanisms or to a higher interface energy associated with the APB's in omphacite, as compared with the APB's in $P2_1/c$ pyroxenes.

EXSOLUTION

Pyroxenes generally contain a variety of microscopic or submicroscopic exsolution features that arise from the attempts of individual crystals to lower their total free-energies upon cooling or decompression. The purpose of this section is to (1) classify and define the mechanisms and kinetics of both exsolution and subsequent microstructural development, (2) illustrate the microstructures produced during exsolution, and (3) evaluate the use of exsolution features as geothermometers and geospeedometers. The following references are particularly useful for exsolution in silicate systems: Champness and Lorimer (1976), Yund and McCallister (1970), and, for detailed reviews of exsolution in solids, Christian (1965, 1975) and the book on Phase Transformations (1970).

When a crystal enters a two-phase field from a single-phase field, the reduction of free-energy is the driving force for exsolution and any subsequent microstructural changes. The total free-energy of a crystal undergoing exsolution can be expressed as a sum of several terms:

$$G_{total} = G_{volume} + G_{interface} + G_{coherency\ strain}$$

The free-energy per unit volume (chemical free-energy) is that of the stress-free material. The interfacial free-energy is the contribution from the chemical change across the host-precipitate interface. The coherency strain term is the contribution of the elastic work necessary to match the lattices of the two phases. Although the main driving force in exsolution reactions is the reduction in volume free-energy achieved by the formation of a two-phase product, the interfacial and strain contributions can be minimized as well. This is commonly accomplished by different exsolution mechanisms, by reducing the interfacial areas, and by preferentially orienting and/or changing the structure of the interface. Each of these phenomena is, in turn, subject to changes in kinetics and diffusion rates with temperature and bulk composition. Understanding the exsolution mechanisms, kinetics, and diffusion rates as a function of bulk composition, therefore, gives us a chance to determine the geologic history of the sample from the microstructure, observed in both real and reciprocal space.

Subsolidus Phase Relations in (Ca,Mg,Fe)SiO$_3$ Pyroxenes

It is useful to begin the discussion of exsolution microstructures in pyroxenes with simplified free-energy versus composition and temperature versus composition diagrams of a pseudobinary section through the pyroxene quadrilateral (Fig. 9). These sections are at approximately

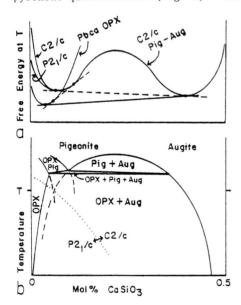

Figure 9. Free-energy versus composition (a) and temperature versus composition (b) diagrams for a pseudobinary section at approximately En/Fs = 2. The free-energy curves are those at temperature T (∼1100°C). The curve $C2/c$ describes high-pigeonite - augite. Low pigeonite, $P2_1/c$, is represented as a curve continuous with $C2/c$, which assumes that the $P2_1/c \rightarrow C2/c$ transition is second order. Orthopyroxene ($Pbca$) is represented as a discrete phase. Common tangent lines are drawn between the equilibrium phases orthopyroxene and augite ("Opx-Aug") (solid line) and possible metastable pairs ("Opx-Pig" and "Pig-Aug") (dashed lines).

146

En/Fs = 2 and for the purpose of illustration contain all of the subsolidus phase relations necessary to discuss the majority of exsolution
features. The sections are pseudobinary, because the two-phase fields
cannot be truly represented on a plane; the tie-lines rotate in and
out of the plane as the temperature changes. We have ignored the
solidus-liquidus relations, since they change radically with minor
changes in composition (Huebner and Turnock, 1980).

For discussing exsolution the most important features of the subsolidus diagrams are the broad two-phase fields of orthopyroxene +
augite and pigeonite + augite. At high temperatures the structures of
pigeonite and augite are similar, with space group $C2/c$. This similarity is shown by the continuous free-energy curve for pigeonite-
augite. The transition in pigeonite from $C2/c$ to $P2_1/c$, discussed
above, is shown as a second-order transition; there is a continuous
change from the high temperature structure to the low temperature
structure. The free-energy curve for $P2_1/c$ pigeonite, therefore, is
continuous with that for $C2/c$ pigeonite-augite. The transition temperature in T-X space decreases with increasing calcium content. The
orthopyroxene structure $Pbca$ is different enough from that of $C2/c$
or $P2_1/c$ so that it has a discrete free-energy curve. Continuous
free-energy curves between two phases, such as that between pigeonite and
augite, indicate that a continuous change of structure is possible.
We shall learn later that this continuity enables clinopyroxenes
to exsolve by spinodal decomposition as well as by nucleation and
growth. The free-energy discontinuity between orthopyroxene and
augite, however, requires a nucleation step before exsolution; spinodal
decomposition is not possible for this pair.

The free-energy curves of Figure 9 are drawn for a temperature of
~1100°C and the compositions of the equilibrium phases (orthopyroxene)
and augite) are given by the intersection of the common tangent tieline
with the free-energy curves. The loci of these points as a function of
temperature delineate a miscibility gap, the limbs of which are known
as solvus curves that define the limits of solubility. Dashed tielines
between orthopyroxene-pigeonite and pigeonite-augite free-energy curves
are also shown, indicating the possible formation of metastable phases
in the event that the equilibrium phases cannot form as a result of

147

kinetic factors. The compositions of the metastable phases are also
indicated in the T-X diagram (dashed lines).

Coherency and its Effect on the Compositions of Exsolved Phases

The structure of the interface between a host and the exsolved
phase plays an important role in microstructural development. Interphase boundaries can be completely coherent, semi-coherent, or incoherent,
and the boundary orientation that occurs is generally the low-energy
configuration (Willaime and Brown, 1974). A completely coherent interface requires that the lattice planes of the host be continuous with,
though not necessarily parallel to, the lattice planes of the precipitate. A semi-coherent boundary contains an array of interface dislocations, but coherency is maintained between the dislocations, whereas
an incoherent boundary has no lattice correspondence across the boundary.
Depending on the amount of strain built up during growth, a lamellar
boundary may start as coherent, become semi-coherent, and perhaps finally
incoherent.

Figure 10a is a schematic drawing of a single pigeonite lamella
parallel to (001) that is coherently constrained by an augite host.
A schematic diagram of the expected b-axis electron diffraction pattern
is shown in Figure 10b. The coherent lamellae in an experimentally
annealed ferroaugite, $Wo_{25}En_{31}Fs_{44}$, and its electron diffraction
pattern are shown in Figures 10c and 10d. In all four figures the
pigeonite lamellae share the (001) plane with augite. In a perfectly
coherent boundary, any plane normal to the boundary (i.e., (010)) has
an identical d-spacing in both phases and crosses the boundary without
deviation. It also follows that any crystallographic direction lying
in the boundary plane is identical for both phases. In Figure 10 the
crystallographic axes a and b lie in the (001) plane of both phases and
are therefore identical. Planes at some angle to the boundary (other
than 90°) are bent upon crossing the interface. In Figure 10a, (100)
is bent 1.83° at the interface. This bending of the (100) plane shows
up in the diffraction pattern as a streak connecting the (h00) spot
of pigeonite with that of augite (inset of the 600 diffraction spots
in Figure 10d). Ghose et $al.$ (1972) and Lally et $al.$ (1975) have
observed this connecting streak in diffraction patterns of lunar clinopyroxenes that have thin (001) exsolution lamellae. The following

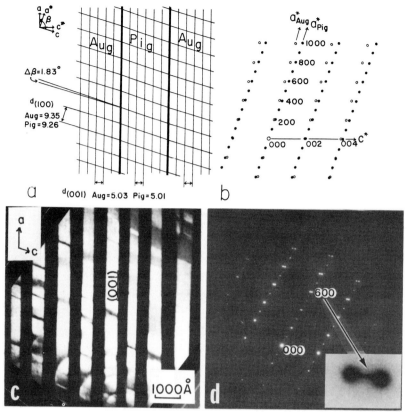

Figure 10. Characteristics of coherent pigeonite lamellae in an augite host. (a) is a drawing of the (100) and (001) lattice planes of a pigeonite lamella that shares (001) with its host augite. The crystallographic β angle of pigeonite is 1.83° larger than that of augite, thereby accounting for the slight bending of (100) across the interface. (b) is a drawing of the expected b-axis electron diffraction pattern for lamellae coherent on (001). The filled circles are diffraction spots for $P2_1/c$ pigeonite, and the open circles are diffraction spots for $C2/c$ augite. (c) is a dark-field electron micrograph of (001) pigeonite-augite lamellae in a synthetic ferroaugite, $Wo_{25}En_{31}Fs_{44}$, annealed at 1000°C for 3886.5 hours. The image was formed using the 102 diffraction ($h + k$ = odd), which arises only from the primitive pigeonite. Pigeonite lamellae are bright, and augite lamellae are dark. Within the pigeonite lamellae, dark boundaries parallel to (100) are antiphase domain boundaries. The actual shape of the domains is cylindrical, with the axis parallel to [001]. (d) is a b-axis electron diffraction pattern of 5000Å diameter area of the lamellar microstructure in (c). Bending of the lattice planes causes continuous diffraction intensity between the pigeonite and augite spots, as shown in the inset.

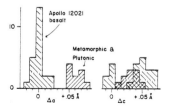

Figure 11. Histograms of the differences in a and c lattice parameters between (001) lamellae and host in pyroxenes from quickly cooled rocks and slowly cooled rocks. The values of Δa cluster about zero for the basaltic pyroxenes, indicating that the lattice parameters are constrained by coherency, and those values for pyroxenes from metamorphic and plutonic environments cluster about 0.6 and 0.7Å, indicating noncoherency. c is not constrained for (001) lamellae in either case, and a wide range of Δc values exist.

table summarizes tests for coherency for (001) and (100) lamellae.

<table>
<tr><td align="center"><u>(001) lamella</u></td><td align="center"><u>(100) lamellae</u></td></tr>
<tr><td align="center">$a_{aug} = a_{pig}$</td><td align="center">$a_{aug} \neq a_{pig}$</td></tr>
<tr><td align="center">$b_{aug} = b_{pig}$</td><td align="center">$b_{aug} = b_{pig}$</td></tr>
<tr><td align="center">$c_{aug} \neq c_{pig}$</td><td align="center">$c_{aug} = c_{pig}$</td></tr>
</table>

Coherent intergrowths are particularly common in the rapidly cooled lunar basalts, and several papers have documented the similarity in lattice parameters between pigeonite and augite. Δa and Δc, calculated from the studies of Ross *et al.* (1973) and Papike *et al.* (1971), for the case of lamellae sharing (001), are shown on the histogram in Figure 11. The values of Δa cluster about zero. Also plotted on the histogram are Δa and Δc from (001) pigeonite lamellae in Bushveld augite (Robinson *et al.*, 1977) and metamorphic augites from New England (Jaffe *et al.*, 1975). The lamellae in these augites have been found to be semi-coherent, which results in the large Δa; Δa is ideally zero for the coherent case. Δc is not constrained in either case for (001) lamellae, as expected for either coherent or semi-coherent lamellae. Δb was found to be zero for both the coherent and semi-coherent cases.

Because of coherency, both host and precipitate are elastically strained. In order to minimize the elastic strain energy, both lattice misfit and stiffness anisotropy for both phases must be minimized (Willaime and Brown, 1974). The addition of elastic strain energy to the total free

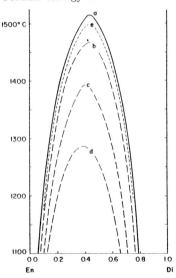

Figure 12. Temperature–composition diagram for iron-free (enstatite–diopside) clino-pyroxenes from Tullis and Yund (1979). Curve a is the hydrostatic (strain-free) solvus, taken from Kushiro (1972) and Boyd and Schairer (1964), and extrapolated into the metastable regions at higher and lower temperatures. Curves b, c, and d are calculated coherent solvi, based on poly-nomial fits to the room temperature cell parameters of Turnock et al. (1973): curve b is for the two ($h0\ell$) orientations of zero compositional strain; curve c is for the (001) orientation; and curve d is for (100) orientation. Curve e is the coherent solvus calculated by a linear fit to the room temperature cell parameters, for the two directions of zero composi-tional strain, (100) and 7° from c toward a. The coherent solvus based on the high temperature cell parameters is essentially indistinguishable from the hydrostatic solvus, so that it is also represented by curve a.

energy of the phases then describes a second solvus, the coherent solvus, which lies within the strain-free solvus of Figure 9. This solvus was calculated by Tullis and Yund (1979) for Fe-free clinopyroxenes. They used the variation of unit-cell parameters and elastic constants as a function of composition, temperature, and pressure to calculate the elastic stresses and to calculate the total elastic strain energy. The

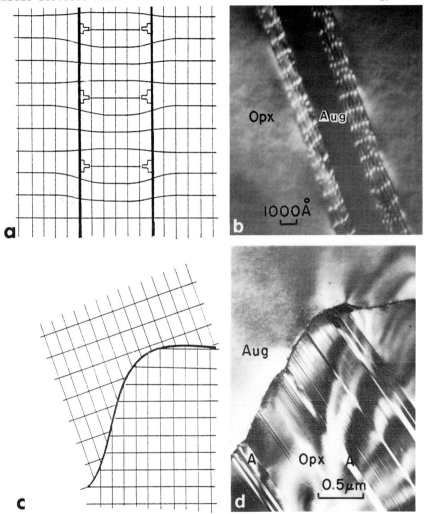

Figure 13. Semi-coherent and incoherent interfaces between augite and orthopyroxene. (a) is a drawing of a semi-coherent interface between host and lamella. Interface dislocations in the lamellar boundary relieve strain that results from the difference in unit cell sizes. (b) is a dark-field electron micrograph (Ross and Huebner, 1979) of interface dislocations on the boundary between orthopyroxene host ("Opx") and augite lamella ("Aug"). (c) is a drawing of an incoherent interface, and (d) is a bright-field electron micrograph of an irregular incoherent interface between an orthopyroxene lamella and augite host from the Bushveld complex (Champness and Copley, 1976). (100) augite lamellae in the orthopyroxene are labeled A.

151

magnitude of this strain energy controls the amount of depression of the coherent solvus from the strain-free solvus. Since the strain energy is a function of boundary orientation, Tullis and Yund, using several sets of lattice parameter values, calculated coherent solvi for orientations with minimum strain energies, as well as (100) and (001) orientations. These solvi are shown in Figure 12 as curves depressed from curve a, the strain-free solvus. Curves b and e are calculated for orientations with zero compositional strain using a polynomial and linear fit respectively, to the room temperature lattice parameters. The curves calculated for the actual orientations (001) and (100) are depressed even further. The coherent solvus calculated using high-temperature lattice parameters, the temperature at which exsolution actually occurs, was only depressed by a few degrees and is therefore nearly identical with the strain-free solvus, curve a.

The elastic strain energy of the host-precipitate interface can also be reduced by an array of interface dislocations; coherency is only maintained between the dislocations. This is known as a semi-coherent interface and is illustrated in Figure 13a. The example in Figure 13b shows a tilted augite lamella in an orthopyroxene host (Ross and Huebner, 1979). At the interface on each side of the lamella there is an array of nearly vertical dislocations approximately 600-900 A apart. Such a semi-coherent interface can arise in two ways. It could have formed originally as a necessary condition for nucleation and growth, as in augite exsolution in orthopyroxene, or an originally coherent lamella could have lost coherency when the elastic strain energy of the interface became too large for continued growth. Coherency loss is accomplished by the attraction or nucleation of dislocations at the interface. As in the completely coherent case, the semi-coherent boundary orientation is controlled by minimum lattice misfit and elastic stiffness.

Boundaries that have no lattice correspondence across the interface are termed incoherent (Fig. 13c). There are no areas of coherency along an incoherent boundary, and structurally the boundary is disordered. The example in Figure 13d shows the irregular nature of an incoherent boundary. The irregularity is possible because the interfacial energy is effectively independent of orientation (Christian, 1965, p. 332).

152

Figure 14. Free-energy versus composition and temperature versus composition diagrams illustrating the exsolution mechanisms of nucleation and growth and of spinodal decomposition. (a) shows free-energy curves g_{Pbca} and $g_{C2/c}$ for the strain-free phases, and $\phi_{C2/c}$ for the strained phases, at temperature T. The compositions of the two coexisting pairs of strain-free phases indicated by the common tangents (labeled strain-free), are "Opx" and "Aug(strain-free)," and "Pig(strain-free)" and "Aug(strain-free)." The compositions of the coexisting pair of coherent phases, indicated by the common tangent (labeled coherent), are given by the position of "Pig" and "Aug." (b) shows a free-energy curve for $C2/c$ phases strained by coherency. (c) shows the pseudobinary phase diagram. The coherent spinodal and chemical spinodal are curves defined by the loci of the inflection points (s), on the free-energy curves $\phi_{C2/c}$ and $g_{C2/c}$, respectively, as a function of temperature. The coherent solvus and strain-free solvus are curves defined by the loci of the common-tangent points of free-energy curves $\phi_{C2/c}$ and $g_{C2/c}$, respectively. The orthopyroxene-augite strain-free solvus (outermost curves) is defined by the common-tangent points on free-energy curves g_{Pbca} and $g_{C2/c}$. T_1, T_2, and T_3 are the temperatures used in reference to Figure 20.

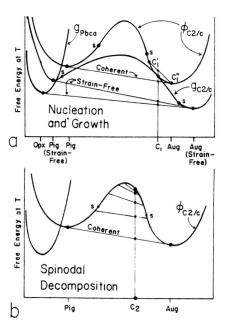

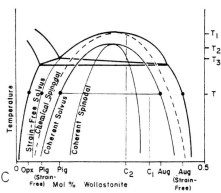

Exsolution Mechanisms in Pyroxenes

Exsolution is initiated by a fluctuation in either the composition or the structure of the high temperature solid solution. In 1875, Gibbs (in Gibbs, 1948) showed that an unstable solid could decompose by forming compositional fluctuations that are large in magnitude and small in extent, a process conventionally known as nucleation and growth, or by forming compositional fluctuations that are large in extent but small in magnitude, a process that, following Cahn (1968), is now known as spinodal decomposition.

The distinction between the two mechanisms is best shown in the free-energy vs composition diagrams in Figure 14. Free-energy curves are shown for both strain-free orthopyroxene (g_{Pbca}) and clinopyroxene ($g_{C2/c}$). An additional curve for clinopyroxene ($\phi_{C2/c}$) describes the free-energy function obtained by including the strain energy due to coherency. In nucleation and growth processes (Fig. 14a), both coherent and strain-free exsolution can occur, so that all three free-energy

153

curves are necessary to describe the resultant microstructures.
Only coherent exsolution occurs via the spinodal mechanism and we need
only consider the curve $\phi_{C2/c}$ (Fig. 14b). On a phase diagram the curve
outlining the compositional limits of nucleation and growth is the solvus
and that outlining the limits of spinodal decomposition is the spinodal.

An augite of composition C_1 (Fig. 14a) is metastable with respect
to either pigeonite-augite or orthopyroxene-augite pairs (the curves
$g_{C2/c}$ and $\phi_{C2/c}$ have the same free-energy value at composition C_1, because there is no strain-energy contribution in the unexsolved, homog-
enous phase; for different compositions, the curve $\phi_{C2/c}$ will be slightly
shifted). A pyroxene of composition C_1 could undergo phase separation
to composition $C_1'-C_1''$, but that would result in a net increase in free
energy. Instead, in order to lower its free energy, the composition of
the calcium-poor phase must differ considerably from C_1. This formation
of a compositional fluctuation that is large in magnitude is known as
nucleation. After nucleation, composition C_1 can lower its free-energy
by separating into two phases, the compositions of which are described
by the law of common tangents. If the phases are coherent, they both
lie on $\phi_{C2/c}$ (i.e., pigeonite-augite). However, if they are strain-free,
they lie on g_{Pbca} and/or $g_{C2/c}$; the possible strain-free pairs are
pigeonite-augite or orthopyroxene-augite. In this diagram orthopyroxene-
augite is the most stable assemblage.

An augite of composition C_2, which lies on the portion of the coherent free-energy curve that is concave downwards (Fig. 14b), can decompose to two phases that have only small compositional differences
and have a total free energy that is lower than that of the homogeneous
solid solution. Decomposition can proceed continuously in this case until the compositions of the common-tangent rule are reached. This mode
of exsolution is called spinodal decomposition, and no nucleation event
is necessary. The compositional range in which spinodal decomposition
can occur lies between the points of inflection (s) on the free-energy
curves where $\frac{\partial^2 g}{\partial c^2} = 0$. The loci of these inflection points as a function
of temperature can be plotted on the pseudobinary phase diagram (Fig.
14c). The curve described by inflection points on the strain-free curve
$(g_{C2/c})$ is called the chemical spinodal, whereas the curve described by
inflection points on the coherent free-energy curve $(\phi_{C2/c})$ is called
the coherent spinodal. The coherent spinodal is depressed from the

154

chemical spinodal because of the strain energy of coherency and is tangent to the coherent solvus at its apex. The coherent spinodal has the same relationship to the coherent solvus as the chemical spinodal has to the strain-free solvus. The chemical spinodal is unimportant in solids because it does not account for coherency strains.

The only experimental determination of the position of the coherent spinodal in pyroxenes has been by McCallister and Yund (1977). They delineated the Ca-rich side of the coherent spinodal in Fe-free clinopyroxenes by defining a kinetic break in their exsolution experiments; runs that exhibited exsolution were considered to be within the spinodal, and runs that did not exhibit exsolution were considered to be outside the spinodal.

It should be pointed out that the coherent spinodal is not a phase boundary or solvus but is only a boundary that delineates the pyroxene compositions that can decompose by the spinodal mechanism. In the phase diagram (Fig. 14c) the compositions of both coherent and strain-free exsolved phases are shown by the black dots at temperature T, defined by the free-energy curves at temperature T in Figures 14a and 14b.

Nucleation. Exsolution by a nucleation mechanism may be divided into three stages: (1) formation of nuclei of the new phase, (2) growth of nuclei until the matrix is chemically depleted in the components of the growing phase and equilibrium compositions are achieved, and (3) coarsening of the new phase. In order to achieve the first stage, a nucleation energy barrier must be overcome.

Figure 15 shows the height of the barrier in terms of the increase in free energy (ΔG) *vs* the radius (r) of the chemical fluctuations. At the critical radius (r_c), a nucleus can decrease its free energy by increasing its size. A nucleus with a radius larger than r_c enters into the second stage of the three stages, that of growth. The height of the barrier (ΔG) and the radius of the critical nucleus (r_c) are both reduced in size as the supersaturation is increased and more fluctuations are able to reach the critical size, thereby increasing the nucleation rate.

The change in free energy necessary to form a nucleus consists of the change in chemical free energy (negative term), the energy of the

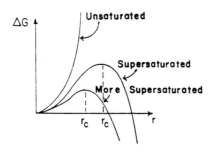

Figure 15. Change in free-energy (ΔG) with the radius of compositional fluctuations (r). The heights of the "supersaturated" and "more supersaturated" curves represent the nucleation barrier, and the radius at that maximum is the critical radius for growth of a nucleus.

surface of the nucleus (positive term), and the strain energy necessary to accommodate the nucleus in the host (positive term). Once the super-saturation is large enough and the fluctuations grow to critical size, nucleation will occur uniformly throughout the crystal. This process is known as homogeneous nucleation, and, if the nuclei are coherent with the host, the resulting host-preci-pitate compositions will lie on the coherent solvus (Fig. 14c). Homog-enous coherent nucleation, therefore, can occur anywhere beneath the co-herent solvus.

The nucleation rate can be enhanced if the strain or surface energy of the nucleus is reduced by its formation on a pre-existing defect, such as a grain boundary or dislocation. This type of nucleation is said to be hetereogeneous, and the exsolved phase is distributed hetero-geneously as well. Particularly common nucleation sites in pyroxenes are grain boundaries and grain boundary dislocations. Grain boundaries are the most effective nucleation sites, since they can relieve strain in several directions depending on their structure. Champness and Lorimer (1973) have described (100) augite lamellae nucleating on grain boundaries in orthopyroxene and a (100) transition phase nucleating on dislocations in subboundaries in orthopyroxene (Fig. 16a). Isolated dislocations are less effective nucleation sites, since they relieve the strain of the nuclei in one direction only. Nord *et al.* (1976) described both (001) and (100) pigeonite lamellae that had nucleated on dislocations in an Apollo 17 augite (Fig. 16b). The (100) pigeonite lamellae nucleated predominantly on subboundary dislocations, whereas the (001) pigeonite lamellae nucleated predominantly on isolated unit dislocations. The preference of a particular orientation for a particular type of dis-location occurs because the strain is minimized when the prinicpal misfit of the precipitate is approximately parallel to the Burgers vector of the dislocation (Nicholson, 1970).

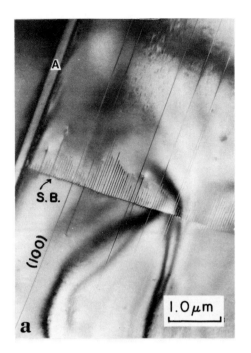

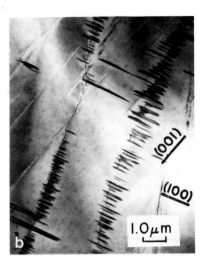

Figure 16. Electron micrographs of heterogeneous nucleation in pyroxenes. (a) shows a thick (100) augite lamella A and many thin transition phase lamellae that have nucleated on a subgrain boundary (s.b.) in Stillwater orthopyroxene (Champness and Lorimer, 1973). (b) shows (100) pigeonite lamellae that have nucleated on dislocations in augite from Apollo 70017 mare basalt. (001) lamellae are also present (Nord *et al.*, 1976). (c) shows fine augite lamellae nucleated at APB's. The dark-field image was formed using an $h \ 1 \ k = $ even reflection. The fringes (N–S parallel lines) arise from the APB's and the fine augite lamellae are parallel to (001) and cut these APB's (Carpenter, 1978a)

Another type of nucleation site is one that is already enriched in solute. Carpenter (1978a) has shown that in pigeonite antiphase domain boundaries (APB's) having orientations approximately parallel to (211) and ($\bar{2}$11) are favorable sites for calcium enrichment. By modeling the structure of an APB in several orientations, Carpenter suggested that APB's cutting the pyroxene chains may allow easier accommodation of large cations (i.e., Ca^{2+}) because of the extended nature of the chains. Figure 16c shows fine augite that has nucleated on segments of pigeonite APB's that are presumably calcium-enriched.

By virtue of the structural defects that supply the heterogeneous nucleation sites, heterogeneously exsolved phases are usually strain free with semi-coherent or incoherent interfaces. In this case, the compositions of the host and exsolved phase lie on the strain-free solvus (Fig. 14c). It follows that heterogeneous nucleation and growth can occur anywhere beneath the strain-free solvus. On the other hand, the example of augite nucleating on APB's in pigeonite points out the possibility of heterogeneous coherent nucleation. Carpenter (1978a) suggests that the lamellae are probably coherent or semi-coherent. If coherent, then the host and lamellae compositions will be constrained to the coherent solvus in this particular case of heterogeneous nucleation.

Spinodal Decomposition. Spinodal decomposition differs from nucleation in both its kinetics and early stages of development. Whereas in nucleation a distinct interface is present at the onset of exsolution, in spinodal decomposition exsolution proceeds by the development of a compositional modulation. Typical wave-like microstructures in clinopyroxenes are shown in Figure 17. These micrographs are of quenched clinopyroxenes from a kimberlite and a lunar basalt and illustrates an early stage of spinodal decomposition. The interface between the crest and trough of a compositional modulation is diffuse at the early stages, but becomesdistinct and eventually sharp at the completion of decomposition.

The contrasting early stages of spinodal decomposition and nucleation are shown schematically in Figure 18 and are easily understood by studying the free-energy curves of Figure 14. A particularly interesting

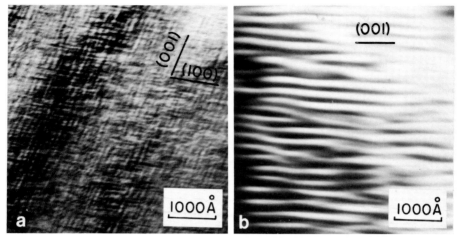

Figure 17. Decomposition microstructures in quickly cooled clinopyroxenes. (a) shows a (001) and (100) "tweed" microstructure in augite from an Apollo 12 mare basalt, 12053 (Nord *et al.*, 1976). (b) shows a modulated microstructure (λ = 200A) along (001) in a subcalcic diopside from the Sekameng kimberlite, Lesotho, Africa. Electron diffraction patterns of this crystal give $\Delta\beta$ (Pig_{ss}-Di_{ss}) = 1°54', which indicates that decomposition is well advanced, although the microstructure is still very fine (McCallister and Nord, in preparation).

aspect of Figure 18 is the direction of movement of the diffusing species, in this case mainly calcium. In the example of augite nucleation, calcium diffuses down a concentration gradient to the nucleus, which grows in thickness and depletes the host of solute. In the example of spinodal decomposition, calcium diffuses up a concentration gradient and builds up a calcium-rich modulation. The major difference between the two mechanisms, therefore, is the direction of diffusion; uphill diffusion is characteristic of spinodal decomposition (Cahn, 1968), downhill for nucleation and growth.

Cahn (1961) described the early stages of spinodal decomposition by modifying Fick's second law of diffusion to include the effects of inter-

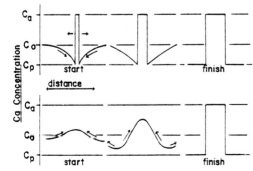

Figure 18. The evolution of decomposition is shown in a schematic diagram for nucleation and growth of an augite lamella (top) and spinodal decomposition (bottom). The composition of the pigeonite, augite and original homogeneous pyroxene are given as C_p, C_a, C_o (Nord *et al.*, 1976, after Cahn, 1968).

159

facial energy on the diffusion coefficient. The derivation is beyond the scope of this chapter, but several useful aspects, especially morphological development, are quantified by Cahn's treatment. In general, Cahn added two terms to the diffusion equation for unstressed solids. The first allows for the effects of coherency strains on the driving force for diffusion, and the second is a gradient-energy term. The gradient-energy term is the energy necessary to maintain the compositional gradient at the interface. The theory predicts that decomposition will proceed by the selective amplification (i.e., change in composition) of sinusoidal compositional fluctuations. The theory further predicts that one particular wavelength (λ_m) will undergo maximum amplification. That maximum is the result of two competing effects: (1) increasing growth rate with decreasing wavelength, because of the reduction in diffusion distance; and (2) reduction in driving force for growth by the increase in interfacial energy associated with the increasingly sharp compositional fluctuations. λ_m is infinite at the temperature of the coherent spinodal, but within one degree Celsius below the spinodal it decreases to several hundred Ångströms. Characteristics of the early stages of spinodal decomposition are summarized by Hilliard (1970).

For isothermal spinodal decomposition:

1. There will be one wavelength, λ_m, that will have the maximum growth rate.

2. λ_m decreases with (1) decreasing temperature, (2) decreasing values of the gradient-energy term, and (3) the shift of the bulk composition towards the center of the spinodal field.

3. Preferential growth will occur in those directions that minimize the coherency strain energy.

4. The rate of decomposition is zero at the temperature of the spinodal and passes through a maximum at some lower temperature (the maximum corresponds to the nose of a time-temperature-transformation curve, discussed below).

For spinodal decomposition during continuous cooling:

1. The structures resulting from decomposition during continuous cooling are similar to those produced isothermally.

2. There is no single wavelength that has a maximum growth rate.

3. For slow cooling rates and complete decomposition, λ_m increases with decreasing cooling rate by the 1/6 power, a very small change from the isothermal case.

4. For more rapid cooling rates and incomplete decomposition, the wavelength is independent of cooling rate and corresponds to that wavelength characteristic of the nose on a time-temperature-transformation curve, discussed below.

These characteristics are consistent with the types of modulated microstructures observed in synthetic clinopyroxenes decomposed by isothermal annealing (McCallister, 1978; Nord and McCallister, 1979) and in natural clinopyroxenes decomposed by rapid continuous cooling (Champness and Lorimer, 1971a,b, 1972, 1976; Champness *et al.*, 1971; Christie *et al.*, 1971; Lally *et al.*, 1975). The isothermal experiments are discussed below, but it is appropriate here to outline the observations on modulated microstructures in natural clinopyroxenes.

Two types of modulated microstructures have been observed, a fine tweed texture (Fig. 17a) composed of modulations oriented close to (100) and (001) and a coarser single orientation (Fig. 17b) composed of modulations oriented close to (001). In the tweed texture the set of modulations close to (001) are generally coarser (greater wavelength) and have more intense diffraction contrast in the image (greater amplitude) than the set of modulations close to (100). The wavelengths of the tweed microstructures range between 100 and 200 A. The amplitudes (compositional differences) of the modulations in the tweed texture are generally small; the electron diffraction patterns only show streaking along a^* and c^* (Lally *et al.*, 1975) or, in one case, a pair of satellites along a^* and c^* (Champness and Lorimer, 1972). There is no evidence for two lattices.

In the tweed microstructure where the (001) set is coarser or has more intense diffraction contrast, splitting of the diffraction spots along c^* indicates the formation of two distinct lattices and the development of a growing (001) compositional fluctuation. The (100) orientation usually does not grow and eventually disappears entirely when the (001) modulations coarsen beyond 200 to 300 A. The difference between the growth behavior of the two sets of modulations is explained by the difference in their elastic strain energies; the modulation with the lower energy orientation will be more likely to grow and coarsen, whereas the modulation with the higher energy orientation will be more likely to decay.

Morimoto and Tokonami (1969b) calculated strain energies of 0.02 and 0.04 Kcal/mole, respectively, for (001) and (100) ferroaugite lamellae in a Moore County ferropigeonite. Tullis and Yund (1979) calculated a molar strain coefficient k for Fe-free clinopyroxenes, where the exact strain energy depends on the compositional difference (amplitude) of the modulations. The value of k for the (001) orientation is 0.650 Kcal/mole and for the (100) orientation is 1.060 Kcal/mole. The lower calculated elastic strain energy of the (001) orientation compared to that of the (100) orientation is consistent with the observation that (001) is the most common orientation.

Later Stages of Spinodal Decomposition. Cahn's derivation of the early stages of decomposition used a linearized diffusion equation that predicts that the compositional modulations will continue to grow indefinitely with time. Introduction of nonlinear terms for the later stages introduces harmonic distortions that, for bulk compositions at the center of the spinodal field, cause a squaring of the composition profile, as in Figure 18. If the bulk compositions are off the center of the spinodal region, the distortions increase the compositional difference of one phase relative to the other, while narrowing its spatial extent, a development that makes the resulting microstructure consistent with the lever rule (Cahn, 1966). It is during these later stages that spinodal decomposition eventually leads to a microstructure similar to that produced by nucleation and growth. Ditchek and Schwartz (1980) recently described amplitude and wavelength changes in Cu-Ni-Sn alloy using the nonlinear theory and showed in detail the later stages of decomposition. In particular, they showed that coarsening (discussed below) can begin when the amplitude (percent of decomposition) is in the range of 75 to 90% of achieving compositions that lie on the coherent solvus. In terms of exsolution in clinopyroxene this result suggests that coarsening can begin before the host-lamellae compositions undergoing spinodal decomposition reach the limbs of the coherent solvus.

Growth. Once nucleation has occurred, the nuclei grow by depleting the matrix of solute (as illustrated in Fig. 18). The rate of growth is commonly controlled by the diffusion rate of the slowest-moving cation. In some instances, however, the transfer of atoms across the interface may be slower than volume diffusion, and the growth rate is then considered to be interface controlled. An incoherent or disordered interface, such as that between orthopyroxene and augite in Figure 13d, presents no structural barrier to growth, and its migration kinetics should depend only on volume diffusion. A number of studies reviewed by Aaronson *et al.* (1970) have shown that the thickening kinetics of precipitates with incoherent interfaces follow a parabolic rate law

$$s = \alpha t^{1/2}$$

where s = half thickness of the precipitate, α = rate constant, and t = time.

When the interface is coherent or semi-coherent the growth rate may be quite different from the incoherent case. The movement of a coherent interface (or coherent regions between interface dislocations in a semi-coherent interface) is energetically more difficult than the movement of an incoherent interface. Aaronson *et al.* (1970) suggest that the reason for this difficulty is that growth of the coherent or semi-coherent interface by means of atom-by-atom diffusion requires that substitutional atoms be inserted in interstitial sites in the interface between the precipitate and the host. In order to overcome this barrier to growth, precipitate interfaces form ledges, which then add material to the interface one layer at a time by lateral movement parallel to the plane of the interface. The large face of a new layer, parallel to the interface, has a coherent or semi-coherent structure, while the moving edge of the layer is incoherent or disordered. This disordered edge presents no barrier to growth and its migration proceeds as rapidly as new material can be brought in by diffusion.

Ledges have been identified in orthopyroxene-augite semi-coherent interfaces (Champness and Lorimer, 1974; Kohlstedt and Vander Sande, 1976) and in augite-pigeonite semi-coherent interfaces (Copley *et al.*, 1974). Kohlstedt and Vander Sande (1976) described in detail, by lattice

Figure 19. Ledges at the (100) interface between orthopyroxene on the top (18A fringes) and augite on the bottom (4.5A fringes) from gem orthopyroxene $Wo_3En_{90}Fs_7$ (Sultan Hamud, Kenya). The inset is a schematic representation of the lattice fringes around these ledges (Kohlstedt and Vander Sande, 1976).

fringe imaging, the ledge structure of an augite-orthopyroxene interface (Fig. 19). (Ledges can also be seen between an amphibole-pyroxene inter-growth, Fig. 34c). They found that the augite grew into the orthopyroxene parallel to (100) by the initial formation of two 9.1 A lattice spacings from an 18.2 A (100) spacing of orthopyroxene, followed by the formation of two 4.5 A (200) lattice spacings, characteristic of augite, from each of these 9.1 A spacings. The 9.1 A spacing is characteristic of the $P2_1/c$ space group of pigeonite. We might speculate that these observa-tions of Kohlstedt and Vander Sande suggest that the ledge mechanism for the growth of augite in orthopyroxene proceeds by the following sequence: low calcium $Pbca$ → intermediate calcium $P2_1/c$ → high calcium $C2/c$. This sequence is consistent with the identification of a calcium-enriched primitive transition phase in orthopyroxene (Champness and Lorimer, 1974). The introduction of a transition phase with an intermediate structure and composition may be energetically more favorable for growth than a direct ledge step of one 18.2 A lattice spacing to four 4.5 A lattice spacings.

Transformation Kinetics. In order to use exsolution microstructures properly as indicators of past geologic conditions, we must examine the effects of transformation kinetics. Time-temperature-extent of transformation (T-T-T) diagrams (Fig. 20)(Nord *et al.*, 1975), which relate cooling histories to the extent of completion and mechanism of subsolidus decomposition, are useful for understanding and interpreting exsolution products. The T-T-T diagram, developed as a guide for the heat treatment of steels (Davenport and Bain, 1930), indicates the extent to which a transformation can proceed at a particular cooling rate for a unique composition. The use of T-T-T diagrams for mineralogical problems is discussed by McConnell (1975) and Champness and Lorimer (1976). Each of the transformation mechanisms is represented as C-shaped "start" and "finish" curves on the diagram. The start curve indicates the time at which transformation is first detectable for a particular annealing temperature, whereas the finish curve indicates the time necessary for the transformation to go to completion at a particular annealing temperature. In the case of nucleation, the C-shape is due to a rapidly increasing nucleation rate as undercooling increases and then a decreasing diffusion rate as temperature decreases. In spinodal decomposition, there is no nucleation event, but the extent of transformation follows a similar C-shaped curve. This is because as the temperature is reduced below the temperature of the spinodal, the maximally amplified wavelength, λ_m, decreases, which decreases the diffusion distance necessary for growth of the microstructure and thus increases the reaction rate. As the temperature is reduced further, the decreasing diffusion rate requires longer times for the start and finish of the transformation. The competition between decreasing diffusion distance and rate therefore results in a transformation curve with a maximum (the nose).

In order to understand the kinetics for the entire pyroxene quadrilateral, many T-T-T diagrams should be determined experimentally. Three regions are shown on the T-T-T diagram in Figure 20, illustrating the transformation kinetics for one pyroxene composition (C_2 in Fig. 14c). The three regions correspond to heterogeneous nucleation and growth (below T_1, on the strain-free solvus, Fig. 14c), homogeneous nucleation and growth (below T_2, on the coherent solvus, Fig. 14c), and spinodal

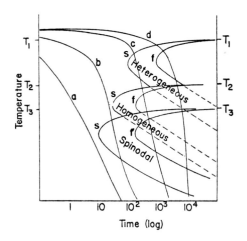

Figure 20. Time-temperature-transformation diagram for composition C_2 (Fig. 14c), showing the start (s) and finish (f) curves for three mechanisms of exsolution in clinopyroxenes: heterogeneous nucleation and growth, homogeneous nucleation and growth, and spinodal decomposition. The temperatures T_1, T_2, T_3 refer to the maximum temperatures at which the three transformation mechanisms can operate for compositon C_2 in Figure 14c. Cooling curves a, b, c and d are calculated for various depths in a lunar basalt flow; the exsolution microstructures representative of different cooling rates are discussed in the text for each cooling curve.

decomposition (below T_3, on the coherent spinodal, Fig. 14c). The relative positions of the curves have not been determined experimentally, but studies of lunar pyroxenes indicate that spinodal decomposition is the dominant mechanism in rapidly-cooled clinopyroxenes, and nucleation and growth is dominant in more slowly-cooled clinopyroxenes.

To illustrate the microstructural development in the lunar samples, four cooling curves calculated from a heat-flow model for an 8 m thick lunar lava flow are shown in Figure 20 (Provost and Bottinga, 1974). Curve (a) corresponds to the immediate crust of the flow, where the pyroxenes are effectively quenched; no decomposition takes place. Curve (d) passes through the field of heterogeneous nucleation and growth and corresponds to a point 100 or more centimeters within the flow.

Two lunar augites of similar composition, $\sim Wo_{35-40}En_{40-50}Fs_{15-25}$, illustrate the microstructures produced by cooling along curves b and c. Curve b intersects the start curve for spinodal decomposition, but not the finish curve, consistent with the 12053 augite microstructure (Fig. 17a). The 12053 augite shows a modulated "tweed" exsolution structure, and the diffraction pattern gives a small $\Delta\beta$, indicating that decomposition was not complete (Lally et al., 1975) (Cameron and Papike, this volume, give the variation of β with composition). In contrast the 70017 augite (Fig. 16b) shows heterogeneously nucleated pigeonite lamellae in a matrix containing finer pigeonite lamellae. The fine

lamellae are interpreted as a product of homogeneous nucleation (Nord
et al., 1976). Curve c intersects the heterogeneous nucleation start-
curve but exits before growth of the lamellae can drain the matrix of
solute and before the transformation is complete. The cooling curve
then enters the homogeneous nucleation field, and those areas still
supersaturated with solute nucleate a second, finer set of pigeonite
lamellae. Denuded zones occur next to the earlier lamellae, where
solute depletion has occurred. Cooling along curve c therefore would
produce a microstructure similar to that of the 70017 augite.

Grove (1979) has studied clinopyroxenes from synthetic Apollo 15
quartz normative basalts crystallized and cooled at rates that varied
from 0.5°C/hour to 150°C/hour. He found homogeneously nucleated lamellae in 0.5°, 4°, and 10°C/hour experiments and heterogeneously nucleated
lamellae in 0.5° and 4°C/hour experiments, suggesting that homogeneous
nucleation occurs at a lower temperature and at an earlier time than
heterogeneous nucleation. This is consistent with the relative positions of the curves in Figure 20. Grove also noted that the lamellar
thickness increases with decreasing cooling rate.

Coarsening. Upon completion of growth in the case of nucleated
lamellae, or completion of decomposition in the case of lamellae formed by
the spinodal mechanism, any further size increase in the microstructure
represents a coarsening phenomenon. If there is a reduction in temperature,
however, the two phases may no longer be in equilibrium and growth will
compete with coarsening. The major driving force for coarsening is the
reduction of the surface free-energy. There is no difference in the
coarsening behavior between spinodally decomposed pyroxenes and those that
exsolved by nucleation and growth, because coarsening is controlled only
by the diffusion kinetics.

The rate law for coarsening of exsolution lamellae has been given
as

$$\lambda = \lambda_0 + Kt^{1/3}$$

where λ is the wavelength of the lamellae at time t, λ_0 is the initial
wavelength of the lamellae after completion of growth or decomposition,
and K is the kinetic constant (Park *et al.*, 1976). (Brady and McCallister,

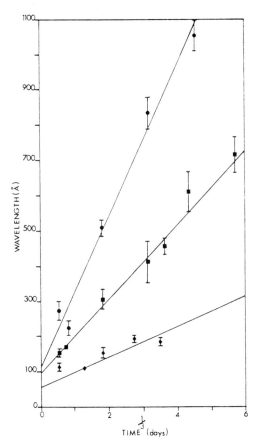

1980, have recently questioned the $t^{1/3}$ rate law as it applies to clinopyroxenes.) The $t^{1/3}$ rate law describes coarsening phenomena of particles with either coherent, semi-coherent, or incoherent interfaces, as long as the character of the interface remains constant during the experiment. A loss of coherency during coarsening would change the value of the kinetic constant because the interfacial free energy of a diffuse coherent interface is less than that of a sharp semi-coherent or incoherent interface.

Coarsening kinetics have been studied in detail for only one pyroxene composition to date. Mc-Callister (1978) determined the isothermal coarsening kinetics of (001) exsolution lamellae in synthetic $Wo_{27}En_{73}Fs_0$ annealed in evacuated silica tubes at temperatures of 1300°, 1200°, and 1100°C for times of up to 190 days.

Figure 21. Wavelength in Ångströms versus (time)$^{1/3}$ (in days) for exsolution microstructure coarsening experiments at 1300°C (dots), 1200°C (squares) and 1100°C (diamonds). The intersection of each line with the ordinate corresponds to λ_0, with values of 109 ± 30A, 89 ± 4A and 53 ± 10A, respectively (McCallister, 1978).

Figure 21 shows a graph of wavelength, λ, vs $t^{1/3}$ in days. A cubic least-squares fit shows that the coarsening is consistent with the $t^{1/3}$ law, and the slope of the fitted line at each temperature yields the kinetic constant, K. McCallister has plotted the K's for each temperature on an Arrhenius plot, ln K vs $\frac{1}{T}$. From the Arrhenius equation

$$\ln K = -\frac{G}{R} \frac{1}{T}$$

where R is the gas constant, he determined an activation energy G = 99 $\pm$ 2 Kcal/mol for coarsening. The activation energy is that energy

168

necessary to surmount the barrier between the initial and final states
of the reaction path, in this instance coarsening. Because coarsening
is mainly dependent upon diffusion, McCallister has suggested that the
99 Kcal/mol activation energy is a measure of the energy barrier associated with diffusion.

The kinetic constant is a function of many factors, one of which
is the diffusion rate. A comparison of the coarsening kinetics determined by McCallister (1978) for the iron-free system and those determined by Nord and McCallister (1979) and Nord (1980a) for composition
$Wo_{25}En_{31}Fs_{44}$ indicates an increase in the value of K by a factor of ~ 10
between the two compositions for similar temperatures. This increase
in coarsening rate is a result of an increase in diffusion rates, and
diffusion controlled processes in pyroxenes will therefore be more rapid
in compositions with greater Fe content. The variation in rates should
be carefully considered when extrapolating kinetic data determined in
the Fe-poor pyroxene system to processes in Fe-rich pyroxenes.

Experimental Verification of Spinodal Decomposition. There have been
a number of electron microscopic observations of modulated microstructures
in clinopyroxenes that suggest that clinopyroxenes can decompose by the
spinodal mechanism. However, proof of the mechanism involves observation
of both the change of wavelength and amplitude of the compositional modulations with time. Laughlin and Cahn (1975) recently confirmed the spinodal
mechanism in Cu-Ti alloys by observing continuous phase separation while
the wavelength remained constant. The increase in the amplitude of the
modulations (phase separation) was observed with the TEM by the continuous
increase in strain contrast associated with the compositional modulations.

Recently, Nord and McCallister (1979, in preparation) used this
"microstructural sequence method" of Laughlin and Cahn to verify the
spinodal mechanism in synthetic $Wo_{25}En_{31}Fs_{44}$ ferroaugite. The ferroaugite was synthesized at 1200°C, 20 Kbar for 24 hours from an oxide gel
and then annealed isothermally in evacuated silica tubes at 800, 900,
1000 and 1050°C for times up to five months. The continuous change in
amplitude was observed in electron diffraction patterns by the increase
in the difference between the β-angles of the two phases as the compositional modulations developed. A constant wavelength was verified at

169

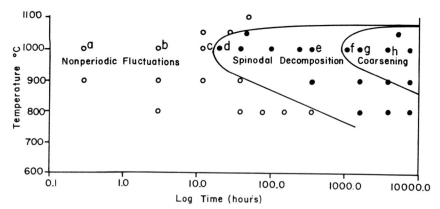

Figure 22. Time-temperature-transformation diagram for synthetic clinopyroxene $Wo_{25}En_{31}Fs_{44}$. Three transformation regions are defined by the start and finish curves: nonperiodic fluctuations, spinodal decomposition, and coarsening (Nord and McCallister, 1979). The solid circles are runs that have undergone detectable decomposition, while the open circles are runs that have not yet reacted.

the same time by measurements of the lamellae on electron micrographs and the distance between satellite reflections in the diffraction patterns. Figure 22 shows the T-T-T diagram determined from the series of experiments. Three distinct kinetic regions are defined on the diagram: (1) nonperiodic fluctuations, where the product has the same microstructural characteristics as the quenched starting material; (2) spinodal decomposition, where the microstructure consists of periodic fluctuations and has constant wavelength while the amplitude increases; and (3) coarsening, where the microstructure increases wavelength while the amplitude remains constant. Figure 23 shows a sequence of electron micrographs of the lamellae developed during the 1000°C runs.

The results of these isothermal decomposition experiments are consistent with the characteristics summarized by Hilliard (1970) for spinodal decomposition. Nord and McCallister found that (1) there is one wavelength (λ_m) that has a maximum growth rate; (2) this wavelength decreases with decreasing temperature (λ_m's measured at the three temperatures of investigation are approximately 85 A at 800°C, 145 A at 900°C and 240 A at 1000°C); (3) preferential growth occurred in those orientations that minimize the coherency strain energy (the growth is along (001), consistent with the calculations of Tullis and Yund (1979) and Fletcher and McCallister (1974)); and (4) the rate of decomposition describes a maximum with decreasing temperature (this maximum is the nose of the T-T-T diagram in Figure 22).

170

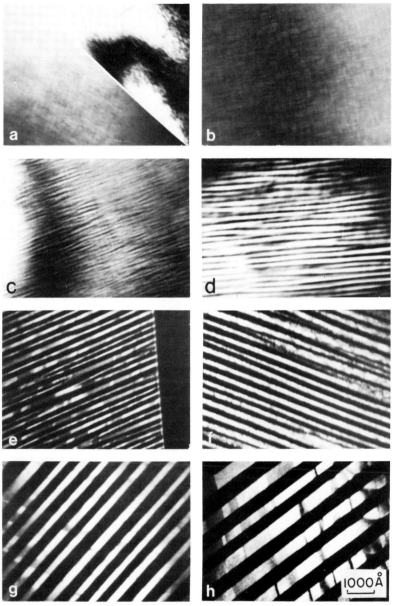

Figure 23. Electron micrographs showing changes in the microstructures with time at a temperature of 1000°C, during spinodal decomposition and coarsening, for a synthetic clinopyroxene, $Wo_{25}En_{31}Fs_{44}$. The earliest stages exhibit a faint tweed microstructure [(a) and (b)], (c) shows a more intense contrast between the lamellae, but the wavelength is still nonperiodic, whereas (d) and (e) show that decomposition is underway; the modulation interfaces become more periodic and then sharper. In micrographs (f), (g), and (h) the spinodal microstructure coarsens to a wavelength approaching 1000A. Throughout the entire sequence of annealing experiments at 1000°C with annealing times up to almost 5.5 months, the exsolution lamellae have remained completely coherent (Nord and McCallister, 1979).

171

Guinier-Preston Zones and Transition Phases. The initial formation
of metastable phases followed by the formation of a stable phase is a
common phenomenon in heat-treated alloys such as Al-Cu (Lorimer and Nicholson, 1970). This sequence occurs because the metastable phases have lower
activation energies and can form more rapidly than the stable phase; they
are more closely related to the host, in structure and/or composition, than
the stable exsolved phase. These metastable phases are called transition
phases when both the structures and compositions are intermediate to those
of the host and the the equilibrium precipitate. When the structures are
identical to those of the host and only the compositions are intermediate
to that of the host and the stable precipitate, the metastable phases are
called Guinier-Preston zones after the co-discoverers of these metastable
zones in Al-Cu. In the case of orthopyroxene, both stable and metastable
exsolution products have been identified (Champness and Lorimer, 1974). The
stable exsolution product is augite; the metastable products are (1) a primitive monoclinic transition phase (tentatively identified as pigeonite) and
(2) small zones of calcium-enrichment that retain the structure of the host
orthopyroxene. These three phases are shown coexisting in an orthopyroxene
host from the Stillwater complex in Figure 24 (Champness and Lorimer, 1974).

Champness and Lorimer (1974) identified the transition phase as
being enriched in calcium, having a primitive lattice, and having cell
dimensions close to those of clinoenstatite or pigeonite. Although the
identification is tentative, the possibility of the metastable precipitation of pigeonite in either orthopyroxene or augite can easily be seen
by the dashed common-tangent lines in Figure 9a. The dashed lines in
the free-energy diagram represent the metastable pairs orthopyroxene-
pigeonite and pigeonite-augite. The exact composition of a metastable
pigeonite phase depends on the shape and position of the free-energy
curves. In Figure 9a the pigeonite minimum is shown with a rather low
curvature, and the metastable pigeonite composition is slightly different
for an orthopyroxene host *vs* an augite host. The exact form of the phase
diagram is unknown.

Small zones of calcium-enrichment like those in Figure 24 have been
found in both lunar and terrestrial orthopyroxenes of plutonic or relatively high-grade metamorphic origin (Champness and Lorimer, 1973; Lorimer

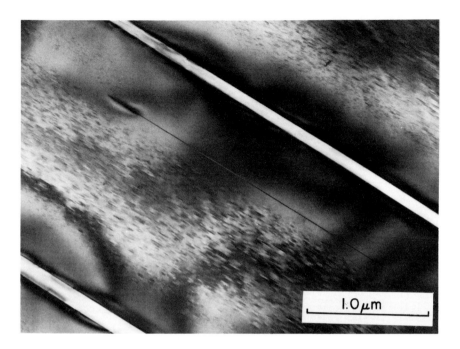

Figure 24. Electron micrograph showing precipitate distribution in Stillwater orthopyroxene. The two large lamellae are augite, one thin lamella terminated by a partial dislocation (showing strain contrast) is a transition phase, and the fine homogeneous precipitates in the background are Guinier-Preston zones. Precipitate-free zones occur adjacent to both types of lamellae (Champness and Lorimer, 1973).

and Champness, 1973; Nord *et al.*, 1977; Nord, 1980b). The zones are completely coherent with the orthopyroxene host, calcium-enriched, disc-shaped, oriented parallel to (100), one to several lattice planes thick, and expanded parallel to a^* relative to the host. Champness and Lorimer (1974) called the zones Guinier-Preston (G.P.) zones, since they have similar characteristics to G.P. zones originally observed in alloys. Figure 25 shows the compositional range of orthopyroxenes in which G.P. zones have been described. Figure 26 shows G.P. zones in a bronzite from the Stillwater complex (Champness and Lorimer, 1974) and larger zones in a bronzite from lunar troctolite, 76535.

The composition, structure, and stability of G.P. zones in lunar and terrestrial orthopyroxene have been described by Nord (1980b). The largest of the lunar G.P. zones were particularly informative because their size (up to 6 unit-cells wide) gave rise to diffraction maxima in electron diffraction patterns. Nord determined that the wollastonite contents of the G.P. zones were ~25 mol % by (1) direct x-ray energy analysis in STEM; (2) comparison of the crystallographic a dimension of the zones with the known increase in cell dimension with wollastonite content; and (3) consideration of the volume percent of the G.P. zones in the orthopyroxene host.

The large lunar G.P. zones also gave rise to a-glide violations in $hk0$ diffraction patterns. Because the G.P. zones retain the c-glide of the pyroxene structure and assuming that their space group is a subgroup of $Pbca$, the space group of the G.P. zones must be $Pbc2_1$ (Matsumoto, 1974). The loss of the a-glide results in four crystallographically distinct silicate chains (A1, A2, B1, B2) and four non-equivalent cation sites (M1A, M1B, M2A, M2B). Figure 27a illustrates the $Pbc2_1$ structure, showing that crystallographically equivalent M2 sites are clustered together in one half of the unit-cell. Since it has been shown previously that the G.P. zones contain $\sim$25% wollastonite component, one-fourth of the divalent cations are calcium. Ordering of calcium onto only the M2A or M2B sites would produce a zone one-half unit cell thick, extended parallel to (100). The half-cell-thick zone would have a composition of Wo_{50} and the structure of orthopyroxene (Fig. 27b). The adjacent one-half unit-cell-thick zone would contain no calcium atoms. The pairs of Wo_{50} and Wo_0 zones would then constitute an 18 A thick G.P. zone of composition Wo_{25}. The addition of more layers by extended annealing would produce the larger five or six unit-cell-thick G.P. zones with a $Pbc2_1$ space group symmetry and Wo_{25} composition.

The stability of the G.P. zones has been investigated by annealing experiments. These experiments showed that the zones could be homogenized into the orthopyroxene host between 950 and 1050°C for times of one week; therefore, G.P. zone formation occurs at relatively lower temperatures than exsolution of the stable phase augite. At these lower temperatures nucleation and growth of the stable phase is inhibited by lower diffusion rates, and thus, in orthopyroxenes that are cooling from plutonic or high-grade metamorphic temperatures, G.P. zones can form in the supersaturated areas between earlier augite or pigeonite lamellae (Fig. 24). For low-grade metamorphic orthopyroxenes and high-temperature orthopyroxenes that are unable to nucleate the stable phase augite, G.P. zone formation may be the only exsolution mechanism available to lower their initial free energy. G.P. zones, therefore, may provide valuable information about the thermal history of orthopyroxenes.

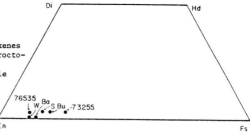

Figure 25. The compositions of orthopyroxenes that contain G.P. zones: (76535) lunar troctolite, (W) Webster-Addie peridotite, North Carolina; (Ba) Bamble orthopyroxene, Bamble area, Norway (Nord, 1980); (73255) lunar norite clast (Nord and McGee, 1979); (S) Stillwater complex; and (Bu) Bushveld complex (Champness and Lorimer, 1974).

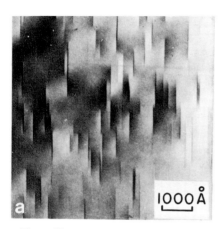
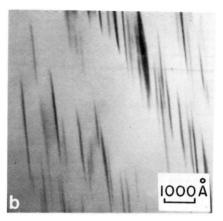

Figure 26. Bright-field electron micrographs of G.P. zones, parallel to (100), in (a) Stillwater orthopyroxene (Champness and Lorimer, Fig. 5a, 1973) and in (b) orthopyroxene from lunar troctolite 76535 (Nord, Fig. 2a, 1980). The G.P. zones are the straight vertical lines.

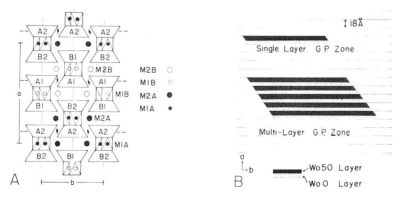

Figure 27. Structural models of G.P. zones. The model in (a) (after Smyth, unpublished) is an ab section of an orthorhombic pyroxene possessing space group symmetry $Pbca2_1$. The tetrahedral Si-O chains are indicated by A1, A2, B1, B2 and the divalent cation sites are M1A, M1B, M2A, M2B. In the orthopyroxene space group $Pbca$, which possesses an a-glide, only A and B chains and M1 and M2 sites are present. The layer model in (b) shows (100) lattice fringes 18 Å apart with single and multiple layer G.P. zones. The single layer G.P. zone consists of a Wo_0 layer and a Wo_{50} layer (Nord, 1980).

175

Cellular Reactions in Pyroxene. Cellular (or duplex) reactions involve the transformation of a supersaturated host into volumes that have completely reacted to the equilibrium, two-phase, assemblage and volumes of unreacted supersaturated host (Christian, 1965). The transformed volume formed by this type of reaction is called a "cell" or duplex product, thus the term cellular or duplex reaction. The cell usually has a lamellar texture, always nucleates at a grain boundary, and grows by the migration of the grain boundary into the untransformed grain; the grain boundary, therefore, serves as a reaction front. The cell takes on the orientation of the grain that shares the grain boundary with the grain into which the cell is growing. This process is similar to recrystallization but involves a compositional change. One condition for cellular growth is that the lamellae maintain a constant spacing, to retain the correct lamellae/host volume ratio. Kinetically, a cellular reaction is more rapid than the nucleation and growth reactions, discussed above. In those reactions, growth is parabolic with respect to time, whereas in a cellular reaction the growth rate is linear.

There are two types of cellular reactions. The first type is cellular precipitation where

$$\alpha_{supersaturated} \rightarrow \alpha_{saturated} + \beta_{saturated}$$

whereas the second type is eutectoidal decomposition where

$$\alpha_{supersaturated} \rightarrow \beta_{saturated} + \gamma_{saturated}.$$

Both types of cellular reactions have been proposed to explain some exsolution microstructures in pyroxenes. Boland and Van Gijlswijk (1980) have suggested that cellular precipitation accounts for a symplectic texture of augite–orthopyroxene in augite ($En_{50}Fs_{15}Wo_{35}$) that occurs in an olivine gabbro from Norra-Storfjall, Norway. The lamellae in those symplectic areas have curved interfaces with a texture similar to cellular reaction textures reported in alloys. Boland and Van Gijlswijk interpret the texture as a cellular precipitation reaction of augite → pigeonite + augite; the pigeonite has subsequently inverted to orthopyroxene. Evidence for the second reaction type, eutectoidal decomposition, is discussed in the section on combined reactions below.

Orientation and Stacking-fault Density of Clinopyroxene Exsolution
Lamellae as Geothermometers

The orientation of clinopyroxene lamellae exsolved from clinopyroxene has been found to be a function of the formation temperature
(Robinson *et al*., 1971; Nakazawa and Hafner, 1977). Jaffe *et al*. (1975)
demonstrated that for metamorphic augites coexisting with orthopyroxenes
ranging in composition from Fs_{15} to Fs_{95}, exsolved pigeonite lamellae were
not exactly parallel to (001) and (100), as had previously been supposed
(Poldervaart and Hess, 1951), but were oriented on planes near (001) and
(100). The largest deviations from (001) and (100) were found in the most
magnesian augites. In augites of igneous origin, several sets of exsolution lamellae of different orientations within a single host crystal have
been described by Robinson *et al*. (1977) and Robinson (Chapter 9, this
volume). The multiple sets formed in igneous pyroxenes that cooled through
a much greater temperature interval than metamorphic ones.

The orientation of an exsolution lamella boundary is determined at
the time and temperature of nucleation to optimize the dimensional fit
of the two lattices, thereby minimizing the total elastic strain energy
between the host and the lamellae. In general, this elastic energy
is a function of the lattice mismatch between the two phases and the
elastic anisotropy of each of the two phases. The relative importance
of the lattice mismatch and the elastic anisotropy on the orientation of
the boundary is also dependent on the type of interfacial boundary in
question. An incoherent boundary, with no lattice correspondence between the two phases, has no preferred orientation (Christian, 1965).
A semi-coherent boundary, however, has segments of coherent boundary
separated by interface dislocations. In the latter case, the lattice
mismatch term predominates because the strain due to the elastic anisotropy is relaxed within a few unit cells of the boundary by the presence
of dislocations in the interface (Aaronson *et al*., 1970). In a completely coherent boundary, however, the elastic anisotropy must be
considered because both phases are elastically strained.

Tullis and Yund (1979) calculated lamellar orientations under the
condition of zero compositional strain by determining the actual principal strain for the particular composition of each coherent lamella.

177

The crystallographic orientations of the directions of zero strain were determined directly from the Mohr circle for strain (Robin, 1977). These orientations are expected for completely coherent lamellae that one might expect to arise from homogeneous nucleation or spinodal decomposition. The calculated orientations for the iron-free clinopyroxenes are 3° from (001) toward c and 7° from (100) toward a. Nord and McCallister (1979) used the TEM to measure lamellar orientations in rapidly cooled synthetic clinopyroxenes of composition $Wo_{25}En_{31}Fs_{44}$, relative to (100) twin planes. The orientations are 0 to 6° from (001) away from c and 6 to 12° from (100) toward a. The differences from Tullis and Yund's calculated orientations presumably reflect the addition of ferrosilite component. Tullis and Yund (1979) have also shown that the elastic anisotropy contribution to the orientation of lamellae in clinopyroxene is small in comparison to the effect of lattice misfit. Therefore, models that neglect elastic aniso-tropy can be used successfully with clinopyroxenes.

Several models have been proposed using lattice misfit to explain observed lamellae orientations. The optimal phase boundary model of Bollman and Nissen (1968) utilizes coincident lattice theory, in which rotation of one phase with respect to the other produces common lattice points. These points describe an O-lattice, and one of the faces of the O-lattice unit cell is the optimal phase boundary orientation. Robin (1977) utilized a knowledge of the lattice parameters of both phases and the Mohr circle for strain to calculate lamellar orientations. Robinson et $al.$ (1977) utilized a numerical method for calculating the lamellar orientations at different exsolution temperatures. This method and the comparison of model orientations with those determined by uni-versal stage methods from augites of the Bushveld complex are given in the chapter by Robinson (this volume). Lamellar orientations can also be determined with the TEM by using (100) twins or stacking faults as crystallographic markers found in pigeonite hosts or lamellae. Three lamellae from the same Bushveld augite grain are shown in Figure 28a,b,c; the boundary orientations correspond to formation temperatures of 1000°, 800° and 560°C using the method of Robinson et $al.$ (1977). These esti-mates of temperature, of course, rely on the accuracy of the lattice parameters and their variations with temperature. The method of Robinson et $al.$ (1977) has been applied to lunar pyroxenes by Nord and James (1978) and Nord and McGee (1979), and to terrestrial samples by

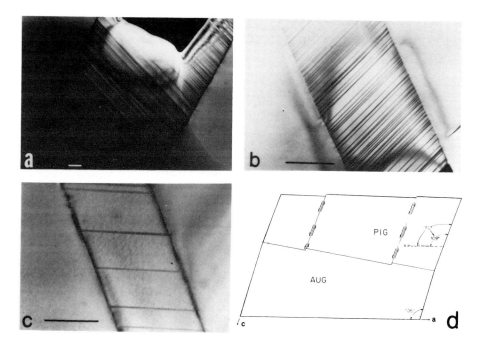

Figure 28. Transmission electron micrographs of pigeonite lamellae in Bushveld augite. (a) shows a 5 µm wide lamella with a high density of (100) stacking faults ["001" interface Λ (100) faults = 107°]. Taken in dark-field, the black area at the top is the edge of the specimen. (b) shows a 1 µm wide lamella with (100) faults ["001" interface Λ (100) faults = 109°]. (c) shows a 0.8 µm wide pigeonite lamella with six (100) stacking faults in the field of view ["001" interface Λ (100) faults = 114°]. Scale bars equal 0.5 µm (Robinson *et al.*, 1977). (d) is a diagrammatic view of a model "001" interface between high-temperature pigeonite lamella and augite host after cooling. A decrease in the pigeonite β from 110° to 109° during cooling has been accommodated by development of a series of (100) stacking faults. Although augite and pigeonite lattices are strained where the faults terminate, β of pigeonite within the areas bounded by the faults can adjust without straining the entire augite lattice, and the *average* angle of the phase boundary remains essentially unchanged. The greater the amount that β decreases in pigeonite, the greater the number of stacking faults that are needed to accommodate it (Robinson *et al.*, 1977).

Nobugai *et al.* (1978), Rietmeijer (1979), Rietmeijer and Champness (1980), and Nobugai and Morimoto (1979).

After nucleation, additional lattice rotation occurs because the β angle of pigeonite decreases more rapidly than the β angle of augite with falling temperature (see Cameron and Papike, this volume). In order to relieve the lattice deformation in the confined pigeonite lamellae, (100) stacking faults propagate across the lamellae (Fig. 28). In effect, these break up the lamellae into segments, and the lattice strain becomes concentrated at the partial dislocations on the lamellar interfaces (Fig. 28d). The greater the change in β, the more faults are necessary to relieve the strain. The high temperature (thicker) lamellae have a greater concentration of stacking faults; thus, the relative concentration of stacking faults provides a "relative" geothermometer.

A secondary effect of the (100) faulting may be to limit the growth of "001" pigeonite lamellae in augite, because the lamellae grow by a ledge mechanism. The presence of partial dislocations along the lamellae-host boundaries with Burgers vectors that are not parallel to those boundaries may impede the movement of the ledges and hence the growth (Robinson et $al.$, 1977). The growth of such a faulted lamella would be interface controlled rather than diffusion controlled.

COMBINED REACTIONS

Transformation of Pigeonite to Orthopyroxene by Competing Mechanisms: Martensitic, Massive, and Eutectoidal Decomposition

At low temperatures pigeonite becomes metastable, relative to the stable phases, orthopyroxene and augite. Pigeonite is, however, usually retained metastably in rapidly cooled rocks and sometimes even in slowly cooled plutonic rocks, presumably because nucleation of orthopyroxene is a sluggish process in clinopyroxene. When pigeonite does transform to orthopyroxene, the resulting product is called "inverted" pigeonite. There may or may not be any crystallographic relationship between the parent and product; both cases have been described. These parent/product crystallographic relationships are known because augite lamellae invariably exsolve prior to or during the transformation, providing "fossil" markers of the original pigeonite orientation. There are numerous papers describing the crystallographic relationships among parent, product, and exsolved phases. Some of the most informative are Poldervaart and Hess (1951), Brown (1957, 1972), Ishii and Takeda (1974), Huebner et $al.$ (1975), Harlow et $al.$ (1979), and Rietmeijer (1979).

Three different transformation mechanisms for the reaction pigeonite → orthopyroxene ± augite have been proposed. Two of these are essentially diffusionless transformations ("massive" and "martensitic") while the third ("eutectoidal decomposition") requires diffusion. With proper consideration of the microstructures induced by these three transformations, it should be possible to differentiate among them.

One feature of some inverted pigeonites is the lack of a common crystallographic orientation between the pigeonite parent and the orthopyroxene product. Smith (1969) suggested that this feature results from orthopyroxene nucleating in a random orientation, and then growing by sweeping an interface through the entire crystal of pigeonite before any other nuclei are able to grow. Bonnichsen (1969) described textures in metamorphic pyroxenes from the Biwabik Iron Formation, Minnesota, that indicate that the first orthopyroxene nucleus to form grows to incorporate several adjacent pigeonite crystals by growing across the pigeonite grain boundaries. The orientations and outlines of the precursor pigeonite grains are defined by the fossil augite lamellae. Champness and Copley (1976) have proposed that this texture is produced by a massive transformation, common in alloys. The massive transformation starts by nucleation and growth of the new phase at a grain boundary and proceeds by a process that involves rapid, non-cooperative transfer of atoms across a relatively high-energy, incoherent interface; there is no change in the overall composition (Massalski, 1970). The only diffusion involved is the atomic jumping necessary at the transformation interface. The transformation is very rapid in alloys, and may be fairly rapid in pyroxenes, because large areas of crystallographically continuous orthopyroxene can form by this mechanism (Bonnichsen, 1969).

In the case where crystallographic continuity is retained between parent (pigeonite) and product (orthopyroxene), the transformation appears to proceed by a martensitic mechanism. Rietmeijer and Champness (1980) have suggested that some inverted pigeonites from Rogaland, Norway, have transformed by this mechanism. Takeda (1973) has shown that orthopyroxene in the Moore County meteorite is only partly transformed from pigeonite and retains the orientation of the host. The character of the partial transformation was verified by x-ray precession photographs that show the presence of streaks parallel to a^*, suggesting disorder, and by TEM micrographs that show 18 A and 9 A layer stacking disorder (Nord and Huebner, unpublished).

One of the most important features in martensitic transformations is a change of crystal shape. In the case of the orthopyroxene to clinopyroxene transformation, it is possible to produce a shape change by translating parts of the structure with respect to each other through a displacement that is not a repeat vector of the lattice. Coe and Kirby (1975) have shown that this displacement can be accomplished by glide on (100) of partial dislocations with Burgers vectors = 0.83[001]. This is effectively an "untwinning" of alternate 9 A subcells of orthopyroxene (see section on Electron Diffraction Effects and Imaging, above). The reverse of this transformation would produce inverted pigeonite and is effectively unit-cell twinning of 9 A clinopyroxene cells to make 18 A orthopyroxene. Coe and Kirby experimentally reversed the ortho → clino transformation and showed that the reverted samples contained remnant (100) stacking faults. This is the type of microstructure that would be expected to be found in a pigeonite which had inverted to orthopyroxene by a martensitic mechanism.

Martensitic transformations are of two types: (1) athermal and (2) isothermal (Christian, 1975). In athermal martensite the amount of transformation is a function of temperature, virtually independent of time, and related to the number of nuclei that can form at any given temperature and to the size to which these nuclei can grow; in addition athermal martensite is non-quenchable. As discussed previously, the transformations between protopyroxene and clinopyroxene are probably examples of non-quenchable, athermal martensitic reactions in pyroxene (Smyth, 1974a).

The characteristics of isothermal martensite are (1) the amount of transformation is a function of time and not temperature; (2) the transformation can be suppressed by rapid cooling; and (3) nucleation rather than growth is the rate-limiting step. The transformation of pigeonite to orthopyroxene is therefore possibly an isothermal martensitic transformation; thus, the extent of reaction is dependent on annealing time, so that the extent of pigeonite inversion may indicate cooling rate. Coe and Kirby (1975), in their reversion experiments, showed experimentally that the clinopyroxene → orthopyroxene transformation is time dependent.

Ishii and Takeda (1974) have suggested that pigeonite decomposes into orthopyroxene and augite at the temperature of the pigeonite eutectoid reaction point to produce blebby augite sharing (100) with the host orthopyroxene. Inverted pigeonite with blebby augite has been called "Kintoki-San type" by Kuno (1966) and has been found in many geologic environments. These environments include plutonic (Skaergaard; Brown, 1957), metamorphic (Biwabik Iron Formation; Bonnichsen (1969), and hypabyssal (basaltic andesite dike of Kintoki-San; Ishii, 1973). Ishii and

Takeda (1974) propose that pigeonite undergoes eutectoidal decomposition if it crystallizes at a temperature and composition above the ortho-pyroxene-pigeonite-augite eutectoid reaction point (this point can be seen in Figure 9 as the peak of the three-phase field opx-pig-aug) (also see Huebner, this volume).

A eutectoidal reaction involves the decomposition of a high-temperature solid phase to two new lower temperature phases. Eutectoidal decomposition must proceed in such a manner that the two new phases are formed simultaneously (Christian, 1965). In alloys the two phases commonly form a characteristic lamellar structure in growing cells (called pearlite in steels). This structure is similar but not identical to the blebby microstructure described for the Kintoki-San type of inverted pigeonites, but subsequent coarsening could have obliterated the lamellar habit.

The three reaction mechanisms suggested for inverted pigeonites compete kinetically. Eutectoidal decomposition is restricted to pigeonites with eutectoid compositions and is most likely limited to relatively high temperatures, where diffusion rates are still significant. The relative positions of the isothermal martensitic and massive reaction curves on a T-T-T diagram are unknown for pigeonite compositions, but in alloys the massive transformation usually occurs at a higher temperature than a martensitic one (see Fig. 1 of Massalski, 1970). Harlow *et al.* (1979), in a study of the Serra de Magè meteorite, suggest that the competition between eutectoidal decomposition and an oriented inversion (martensitic reaction) of the original pigeonite is responsible for the coexistence of "Stillwater" and "Kintoki-San" type inversion textures. They interpret this coexistence as resulting from an originally heterogeneous calcium distribution (now homogenized) that allowed both mechanisms to operate.

Subsolidus Phase Relations and Possible Exsolution Mechanisms in Omphacite Pyroxenes

Omphacite is an important mineral in high pressure, low temperature metamorphic rocks and has potential for use as a cooling rate indicator in such rocks. This potential exists because the transformations in omphacite are diffusion controlled and therefore cooling-rate dependent. Recent work by Carpenter (1978b, 1980) has uncovered a variety of complex

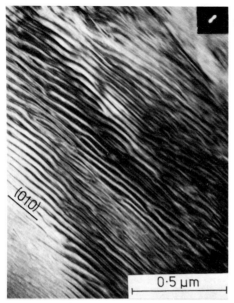

microstructures that result from the interaction of exsolution and ordering processes in omphacites. The ordering reaction itself was discussed above in the section on Order-Disorder Reactions for a composition of jadeite 50%, augite 50%. Cooling of this composition results in a single ordered phase with antiphase domains. Compositions away from the 50:50 composition have been found to have an exsolution microstructure and to consist of two phases, one of which is ordered and the other disordered, or of two phases, both of which are apparently ordered. Selected area electron diffraction patterns suggest the existence of space groups $P2/n$, $P2/c$ and $P2$ in different parts of the ordered crystals. Carpenter suggests that cation ordering occurred in a series of steps controlled by the transformation kinetics of the different ordering schemes. The evidence for this step-wise transformation lies in microstructures that show APB's of one ordering scheme partially replaced by APB's of another ordering scheme.

Carpenter's study of omphacites from Syros and Guatemala showed that the exsolution microstructures were consistent with spinodal decomposition. Figure 29 shows a modulated exsolution texture in impure jadeite consisting of ordered and disordered components approximately parallel to (010) (Carpenter, 1980). This microstructure is similar to the modulated, branching exsolution textures found in spinodally decomposed clinopyroxenes and feldspars (Yund, 1975).

Because the variety of exsolution and ordering microstructures in omphacite appear to have formed by continuous mechanisms, Carpenter (1980) explained the subsolidus phase relations by using free-energy diagrams in which ordered and disordered phases are assumed to transform from one

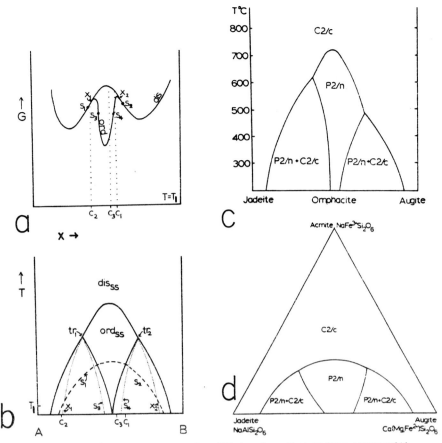

Figure 30. Schematic free-energy curves and equilibrium phase diagrams for a binary solid
solution with limited miscibility and an intermediate ordered phase. (a) shows the free-
energy relations at a temperature, T_1 (as shown in b), with limited miscibility in the
disordered phase and a second order transformation; $\partial G/\partial c$) is continuous between the ordered
("ord") and disordered ("dis") phases. (b) is the equilibrium phase diagram corresponding
to (a). The spinodal for the disordered phase is marked by a broken line, and the conditional
spinodals are marked by dotted lines that represent the loci of X_1, S_3, S_4, and X_2; tr_1
and tr_2 are tricritical points.* (c) is the equilibrium phase diagram proposed for the
system jadeite-augite and (d) is for jadeite-augite-acmite (Carpenter, 1980). The binary
diagram in (c) corresponds to that in (b). The spinodal curves are not indicated in (c)
(Carpenter, 1980). * Note added in proof: The solid curve separating "dis" and "ord"
drawn between tr_1 and tr_2 marks the transition from the single-phase disordered field to the
single-phase ordered field and is not a solvus. The solid curves below tr_1 and tr_2 are solvus
limbs for the "dis" phase ("outer curves") and "ord" phase ("inner curves"); these limbs delineate two two-phase fields. Accordingly, there should be a metastable solvus curve drawn
between the points tr_1 and tr_2 and below the ordering transition curve, completing the closure
of the "dis" miscibility gap.

to the other by a second order, continuous transformation. His free-

energy *vs* composition and temperature *vs* composition diagrams for the

case that best explains the omphacite microstructures are reproduced in

Figure 30. In these diagrams Carpenter introduces two concepts that are

unfamiliar to mineralogists, although similar thermodynamic conditions

185

may be present in many important mineral systems. The thermodynamic condition is that the peak ordering temperature of a second or higher ordering transition lies above the peak of the solvus. The point at which the second order transition curve meets the solvus (first new concept) is called a tricritical point (Allen and Cahn, 1976). There are two tricritical points (tr_1 and tr_2) in the phase diagram (Fig. 30b). Beneath the tricritical point there is a conditional spinodal (second new concept); compositions within the conditional spinodal may decompose to give compositional modulations with ordered and disordered components. A conditional spinodal is conditional in the sense that the ordered phase must be present in order for the spinodal to exist (Carpenter, 1980).

Free-energy curves are shown for the ordered ("ord") and disordered ("dis") phases in Figure 30a. The curves are continuous and both share two minima. The interaction between the two curves has an interesting effect on the equilibrium phase diagram (Fig. 30b). Two single-phase fields are shown ("dis_{ss}" and "ord_{ss}"); the boundary between the two at high temperature is that for the ordering reaction. At lower temperatures the central field of a single ordered phase is bounded by two-phase fields, each of dis_{ss} + ord_{ss}. There are two distinct types of spinodal decomposition possible in this situation. First, there is the possibility of spinodal decomposition of the disordered phase, as indicated by the inflection points S_1 and S_2 on the free-energy curve and the spinodal S_1-S_2 on the phase diagram. Second, there are two conditional spinodals, which are indicated by the inflection points X_1, S_3 and X_2, S_4 in the free-energy diagram and by the spinodal curves X_1, S_3 and X_2, S_4. Omphacites that have compositions within these conditional spinodals can decompose into an ordered and disordered component. Carpenter (1980) observed microstructures consistent with this interpretation.

Several ordering-exsolution paths are possible, depending on the bulk composition; three examples are indicated in Figure 30b. C_1 can decompose spinodally into two disordered phases, each of which can then order, or C_1 can first order and then decompose. C_2 can decompose into two disordered phases, one of which would become ordered. Crystals with a composition such as C_3, which is outside the conditional spinodal, can exsolve a disordered phase only by nucleation.

Carpenter (1980) synthesized the available petrographic and electron optical data on omphacite into equilibrium phase diagrams, based on the above relations. These are shown in Figure 30c,d. The careful study of the microstructures in omphacite by Carpenter point out the value of transmission electron microscopy to mineralogy and petrology, especially in mineral systems with stranded metastable phases.

ALTERATION REACTIONS TO HYDROUS PYRIBOLES,
SHEET SILICATES, AND PYROXENOIDS

In the previous sections, we have considered reactions that do not involve chemical exchange between a pyroxene crystal and its environment. In such reactions, the pyroxenes behave as chemically closed systems. In the present section, we will consider some of the reactions that can occur when a pyroxene crystal behaves as a chemically open system, free to exchange material with its surroundings. We will restrict ourselves to a few examples of solid-solid replacement reactions that occur at relatively high temperatures, thus excluding from consideration the likewise interesting problem of dissolution and other reactions that can occur under weathering conditions (Berner *et al.*, 1980, for example). Specifically, we will focus on three broad categories of alteration reactions that have been studied with TEM and petrographic methods: (1) replacement of pyroxene by hydrous pyriboles (amphibole and wide-chain silicates); (2) replacement of pyroxene by sheet silicates, such as serpentine; and (3) replacement of pyroxene by pyroxenoids having compositions other than the primary pyroxene.

Replacement of Pyroxene by Hydrous Pyriboles

The formation of amphibole pseudomorphs after pyroxene is a common phenomenon in plutonic igneous rocks, in rocks that have undergone retrograde metamorphism, and in pyroxene skarns. The replacement process is often referred to as "uralitization" and was described petrographically and chemically in some detail by Goldschmidt (1911). Pyroxenes that have been completely replaced by amphibole are often referred to as "uralites," which generally form from diopside or augite, or "smaragdites," which may form from omphacites. In addition to such complete replacement, it has been shown that pyroxenes are in some cases only partially replaced by amphibole lamellae or by chain silicates having triple or wider chains. In all of these reactions, the primary and secondary chain silicates share common crystallographic axes, indicating that the reactions are topotactic. This section will focus on recent electron optical study of such lamellar intergrowths and uralites.

Uralites from Ocna de Fier, Romania, and Copper Mountain, Alaska, have been studied by high-resolution TEM methods (Veblen and Buseck, 1979b, in preparation).[*] The latter specimen contains small islands of remnant pyroxene, which is separated from the amphibole by boundaries with irregular orientation (Fig. 31). This texture strongly suggests that the pyroxene → amphibole reaction took place by a bulk replacement mechanism along a reaction front that advanced through the primary pyroxene. In both specimens, the amphibole contains moderate numbers of chain-width errors (Fig. 32); in the Romanian uralite, the errors consist primarily of material with triple, quadruple, and quintuple silicate chains, whereas the Copper Mountain amphibole contains mainly single, triple, and quadruple errors. These chain-width errors probably formed at least in part during the pyroxene → amphibole replacement reaction, possibly as the result of the coalescence and misfit between domains of amphibole representing different nucleation events. Alternatively, some of the chain-width errors may have formed during subsequent alteration of the amphibole. Near fractures, defects with sextuple and wider chains are also present locally; these almost certainly are the result of late-stage amphibole alteration along zones that were readily accessible to the altering fluids. In addition, substantial reaction to sheet silicates has taken place along such fractures, as discussed in the next section. Taken in total, the textures of both these uralites are indicative of an initial stage of pyroxene → amphibole conversion by a bulk mechanism along a reaction front, followed by further hydrothermal alteration of the amphibole along fractures.

The reaction scheme observed in uralites is very different from that which can be inferred from TEM studies of lamellar intergrowths of hydrous pyriboles in pyroxenes. In these lamellar reactions, the transformation from pyroxene to amphibole or wide-chain silicate takes place by nucleation and growth of lamellae that may be only one or two silicate chains in width. Although it is quite clear that some intergrowths of this type

_ _ _ _ _
* Read: Veblen, D.R. and Buseck, P.R. (1981) Am. Mineral. 66, 1107-1133 wherever "in preparation" appears on pp. 188-198.

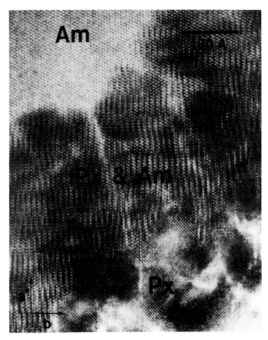

Figure 31. Amphibole ("Am") and a remnant patch of pyroxene ("Px") in a uralite (Copper Mountain, Alaska). The HRTEM image suggests that pyroxene was altered to amphibole by a bulk mechanism along a reaction front. The amphibole-pyroxene interface is tilted, so that the two structures overlap in part of the image. (Veblen and Buseck, in preparation).

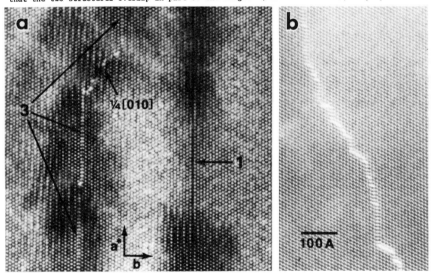

Figure 32. Chain-width errors in amphibole of Copper Mountain uralite. (a) Single-chain ("1") and triple-chain ("3") errors parallel to (010). The triple-chain error is offset by displacive faults having projected displacements of 1/4[010]. (b) A boundary between two amphibole domains. The boundary consists of quadruple-chain structure combined with displacive faults having projected displacements of 1/2[100]. (Veblen and Buseck, in preparation).

189

result from hydrothermal alteration involving chemical exchange with the surroundings of the pyroxene crystal, it is still possible that in some specimens the lamellae grew by exsolution from pyroxene that originally had some degree of solid solution toward amphibole composition, as discussed below. Some lamellar intergrowths of hydrous pyriboles in pyroxene can be seen in petrographic thin section, but others are on a fine enough scale that they are observable only with TEM methods.

Intergrowths of pyroxene and amphibole have been observed in augites from basic intrusions, such as the Skaergaard, the Bushveld, and the Palisades Sill; in omphacites from eclogites; in jadeites from glaucophane schists; in orthopyroxenes from anorthosites; and in augites, orthopyroxenes, and diopsides from peridotites, lherzolites, and harzburgites (Bown and Gay, 1959; Papike *et al.*, 1969; Desnoyers, 1975; Francis, 1976; Smith, 1977; Yamaguchi *et al.*, 1978; Nakajima and Ribbe, 1980; Veblen and Buseck, 1980, in preparation; David L. Bish and Charles W. Burnham, pers. comm.). Amphibole lamellae in such intergrowths are typically oriented parallel to either (010) or (100), but amphibole can also occur intergrown in pyroxene as blebs, rods, or "flames" (Francis, 1976; Veblen and Buseck, in preparation). Triple-chain silicate and material with wider chains have also been described in intergrowths with pyroxene (Veblen and Buseck, 1977, 1980, in preparation; Buseck and Veblen, 1978; Nakajima and Ribbe, 1980).

One example of amphibole lamellae intergrown with augite from a harzburgite is shown in Figure 33 (Smith, 1977). The amphibole appears dark, and the lamellae are parallel to (010). Altered (100) orthopyroxene lamellae are also present normal to the amphibole. The diffraction pattern from the augite-amphibole intergrowth exhibits sharp pyroxene diffractions, streaking parallel to b^*, and diffuse maxima corresponding to amphibole. Amphibole lamellae are also found in some of the orthopyroxene crystals of this specimen (Fig. 34), which have been studied with HRTEM (Veblen and Buseck, in preparation). With very rare exceptions, the amphibole lamellae in the orthopyroxene contain even numbers of silicate chains and are typically 2, 4, 6, 8, or 10 chains wide. Growth of the lamellae apparently takes place by nucleation and growth of narrow lamellae two chains wide (Fig. 34b). Once a lamella has formed, it can expand by nucleation and migration of ledges, again two amphibole

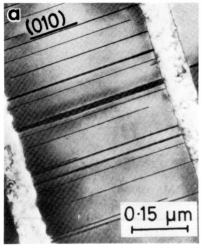 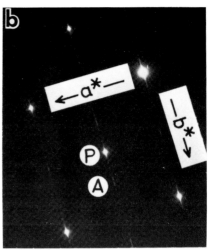

Figure 33. (a) CTEM image of amphibole lamellae (dark) parallel to (010) of an augite from a harzburgite. The two wide bands normal to the amphibole lamellae are (100) orthopyroxene exsolution lamellae that have been altered to sheet silicates. (b) Electron diffraction patterns of the augite-amphibole intergrowth. Pyroxene diffractions ("P") are relatively sharp, whereas the amphibole diffractions ("A") are diffuse and accompanied by streaking parallel to b^*, as a result of the narrowness of the amphibole lamellae. (Smith, 1977).

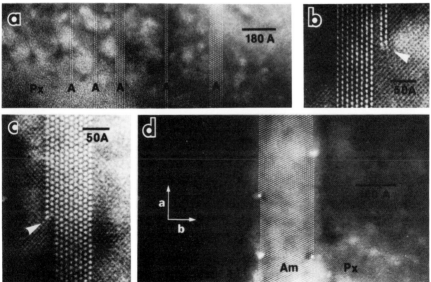

Figure 34. HRTEM Images of amphibole lamellae in an orthopyroxene from a harzburgite. (A) The pyroxene contains narrow amphibole lamellae ("A") that nearly always contain an even number of silicate chains. The lamellae in this image are two, six, and eight chains wide. (b) An amphibole lamella two chains wide terminating beside a lamella eight chains wide. (c) A ledge two chains wide in an amphibole lamella. Lamellae presumably can widen by the nucleation and growth of such ledges. (d) An unusually wide lamella containing an odd number of silicate chains (27). Ledges in the pyroxene-amphibole interfaces that are one amphibole chain wide can be seen by viewing at a low angle. These ledges are accompanied by strain contrast. (Veblen and Buseck, in preparation).

191

chains wide (Fig. 34c). In this example, structural considerations have constrained the lamellar widths to contain an even number of silicate chains, in order to minimize strain at the terminations of growing lamellae (Veblen and Buseck, 1980, in preparation). The lamella in Figure 34d is unusual for this specimen because it contains an odd number of silicate chains (27) in its width and because it exhibits ledges in the pyroxene-amphibole interface that are only one chain wide and are accompanied by strain contrast. This lamella is the widest observed in this orthopyroxene; clearly, no amphibole is observable with optical petrographic methods.

Although it is clear from the extent of reaction that uralites form by alteration involving chemical exchange between a pyroxene and its surroundings, the question of the formation mechanism of thin amphibole lamellae in pyroxene has not yet been fully resolved. Papike *et al.* (1969) suggested several possible modes of formation for such intergrowths. In more recent years, two of these mechanisms have been propounded by electron microscopists: (1) exsolution of amphibole from a pyroxene host having minor solid solution toward amphibole composition; and (2) alteration (hydration reaction) of an initially stoichiometric pyroxene. Unfortunately, the *structural* mechanisms for both exsolution and alteration reactions can be identical, so that many of the microstructures produced in both cases will be the same (Veblen and Buseck, in preparation). Differentiation of the mode of reaction must therefore be based on chemical and textural grounds.

Compositions of amphibole lamellae and host pyroxenes have been analyzed by Desnoyers (1975), Smith (1977), and Yamaguchi *et al.* (1978) using electron microprobe and analytical TEM methods. Desnoyers and Smith have suggested that the differences in chemistry between the amphibole and pyroxene are too great to be the result of an alteration reaction. Yamaguchi *et al.* suggest that element depletions near the amphibole lamellae are indicative of exsolution, rather than alteration. The structural details at the terminations of growing amphibole lamellae and ledges are, however, conducive to rapid diffusion in and out of the reacting crystal, and it has been alternatively argued that all of the available chemical data are also completely consistent with an alteration mechanism for the formation of the amphibole lamellae (Veblen and Buseck, in preparation).

192

AMPHIBOLE LAMELLAE (vertical)

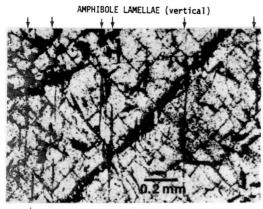

Figure 35. Visible light micrograph of petrographically resolvable
amphibole lamellae in an augite.
The amphibole lamellae, which appear
dark, are concentrated along a
pyroxene cleavage trace that runs
from northeast to southwest. Plane
polarized light. (Veblen and Buseck,
in preparation).

Textural arguments have also been used in support of an exsolution

mechanism for the growth of amphibole lamellae in pyroxenes. Smith (1977)

showed that in certain augites with orthopyroxene lamellae, the amphibole

lamellae are restricted to the augite, whereas alteration might be ex-

pected to affect the orthopyroxene as well. In addition, in some speci-

mens, the development of amphibole lamellae does not seem to be related

to fractures or cleavages (Smith, 1977). This does not, however, rule

out a pervasive alteration reaction. In fact, in an augite specimen

investigated by Veblen and Buseck (in preparation), there is pervasive

reaction to (010) amphibole lamellae, but the degree of reaction is

clearly enhanced along a pyroxene cleavage trace in one place (Fig. 35).

This texture suggests strongly that the reaction is, in fact, the result

of alteration. Other textures indicative of a hydration reaction mecha-

nism for the growth of amphibole in pyroxenes are given by Veblen and

Buseck (in preparation). It is possible that in some specimens amphibole

lamellae form by an exsolution reaction, whereas in others they form by

an alteration process. It is worth noting, however, that nearly all

specimens from which amphibole lamellae have been reported exhibit at

least minor alteration in thin section. Furthermore, all of the textures

that have been reported to date are at least consistent with an alteration

origin for such lamellae, and the textures in a number of specimens ob-

viously are not the result of exsolution.

It has been shown that hydration reactions involving replacement of

amphibole by talc can occur by several different mechanisms of varying

complexity (Veblen and Buseck, 1980). Likewise, TEM evidence indicates

that the alteration of pyroxenes to hydrous biopyriboles such as amphi-

bole, triple-chain silicate, and talc can also proceed by more than one

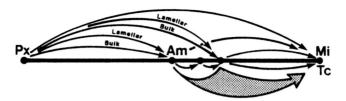

Figure 36. Hydration reaction paths for pyriboles inferred from TEM studies. Paths for pyroxene alteration are shown above the composition line, and paths for amphibole alteration are below the line. Pyroxene can alter to amphibole or triple-chain silicate by either a bulk transformation mechanism or by the growth of narrow lamellae. Sheet silicates ("Mi" or "Tc") can be produced either by direct replacement or pyroxene, of by reaction of intermediate amphibole or triple-chain silicate. (Veblen and Buseck, in preparation).

structural mechanism (Veblen and Buseck, in preparation). For example, pyroxene can react to amphibole either by a bulk replacement mechanism along a reaction front or by a mechanism involving the nucleation, growth, and widening of narrow amphibole lamellae. Similarly, it appears that replacement of pyroxene directly by triple-chain silicate or hydrous pyribole having a disordered chain sequence can occur by either a bulk or lamellar reaction path. Figure 36 shows some of the hydration reaction paths for pyroxene (paths above composition line) and amphibole (below line) that have been inferred from TEM studies (Veblen and Buseck, in preparation). This diagram emphasizes the complexities of biopyribole hydration reactions. Still, it is a simplification of the alteration schemes observed in nature, completely omitting, for example, replacement of pyroxene by non-biopyriboles such as serpentine minerals and pyroxenoids, as discussed in the next sections.

Replacement of Pyroxene by Sheet Silicates

Observations in hand specimen and in thin section show that pyroxenes can be replaced by a number of different sheet silicates, including talc, chlorite, serpentine minerals, and clay minerals. In recent years, studies using TEM and x-ray microbeam camera techniques have suggested that some aspects of these transformations echo the complexities of the pyroxene → hydrous pyribole reactions described in the previous section. In particular, there appear to be several different ways in which some of the pyroxene → sheet silicate reactions can take place.

Just as in the replacement of pyroxene by amphibole, the replacement of pyroxene by some sheet silicates can be "topotactic," so that the crystallographic orientation of the product mineral is dictated by the orientation of the primary pyroxene. A striking example of such a

194

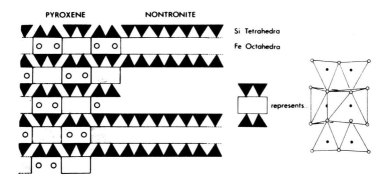

PYROXENE   NONTRONITE

Si Tetrahedra
Fe Octahedra

represents

Figure 37. The structural relationship between the clay mineral nontronite and pyroxene. Because the two structures can intergrow coherently, the reaction of hedenbergite to nontronite is topotactic, and the crystallographic orientation of the nontronite is controlled by the pyroxene orientation. (Eggleton, 1975).

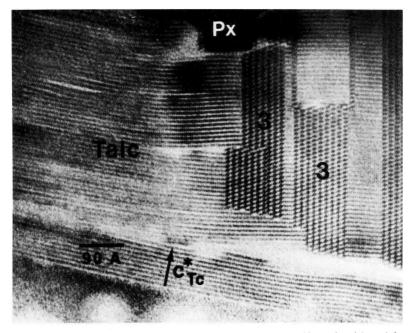

Figure 38. HRTEM image of an augite crystal that has been partially replaced by triple-chain silicate ("3"), very wide chain silicate, and talc. The orientation of the triple-chain silicate is rigorously controlled by the pyroxene orientation, but the talc layers are only parallel to the (100) plane of the pyriboles in the neighborhood of the interface between talc and chain silicate. (Veblen and Buseck, in preparation).

reaction is the oriented replacement of hedenbergite by nontronite (an iron-rich montmorillonite), which was explored by Eggleton (1975). It was proposed by Eggleton that large blocks of the parent structure may be inherited by the product structure in such a topotactic transformation. The structural relationship between the two structures can be seen in

195

Figure 37; the silicate sheets of the nontronite form parallel to (100) of the pyroxene.

Talc is another sheet silicate that can topotactically replace pyroxenes. In this case, again, the talc sheets are typically parallel to the pyroxene (100) planes, with the b-axes of the sheet and chain silicate parallel to each other (Boland and Eggleton, 1978; Veblen and Buseck, in preparation). On the other hand, talc can also replace pyroxene in a non-oriented fashion. Boland and Eggleton (1978) describe a particularly complex reaction scheme in which orthopyroxene first alters to a mixture of talc, smectite, and serpentine; this initial mixed reaction product is then converted to talc that for the most part bears no regular orientation relationship to the parent pyroxene. HRTEM studies have also shown that pyroxene can react to mixtures of talc and wide-chain silicate (Fig. 38) (Veblen and Buseck, in preparation).

The above observations point out some of the complexities of pyroxene hydration reactions involving sheet silicate products. First, the reactions may or may not be topotactic. Second, the reaction product need not be a single, homogeneous phase, but may instead by a complex, nonequilibrium mixture of different structures. Finally, the alteration reactions need not take place in a single step. Instead, there may be intermediate steps in the reaction, and a single specimen may consist of texturally complicated intergrowths of the primary pyroxene, phases representing intermediate steps in the reaction, and final reaction products.

These characteristics are particularly well exhibited in HRTEM studies of uralites (Veblen and Buseck, 1979b, in preparation). As described in the preceding section on pyroxene → hydrous pyribole reactions, uralites are primarily oriented pseudomorphs of amphibole after pyroxene. In the cases that have been studied with TEM methods, however, there is evidence of substantial further reaction of the amphibole to form sheet silicates. These minerals can include chlorite, talc, serpentine minerals, and mixed-layer intergrowths of two or more different types of silicate layers. In this stepwise reaction from pyroxene → amphibole → sheet silicate, the initial replacement by amphibole is a topotactic reaction, whereas the second stage of replacement is not generally a strictly oriented reaction. However, the sheet silicates generally tend to form in such a way that their layers are parallel to the pyribole *c*-axis. The end result of these stepwise reactions is a texturally complicated, nonequilibrium mixture of chain and sheet silicates. Some of the observed textures and sheet silicate microstructures are shown in Figures 39 and 40.

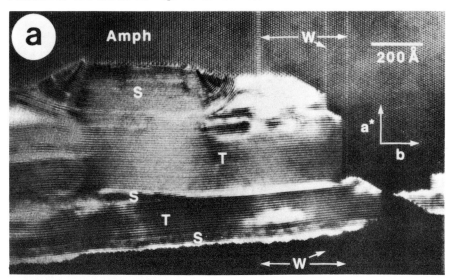

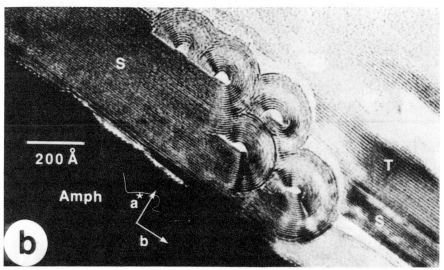

Figure 39. Sheet silicates in a uralite. Amphibole has replaced pyroxene and is in turn being altered to interlayered serpentine ("S") and talc ("T"). (a) The amphibole contains chain-width errors ("W"), and part of the serpentine consists of combined planar (lizardite) and curled (chrysotile) forms. (b) Complex serpentine structures, with planar layers ending in rolls of curved layers. In one of the rolls, the direction of curvature reverses, producing a hairpin structure. (Veblen and Buseck, 1979).

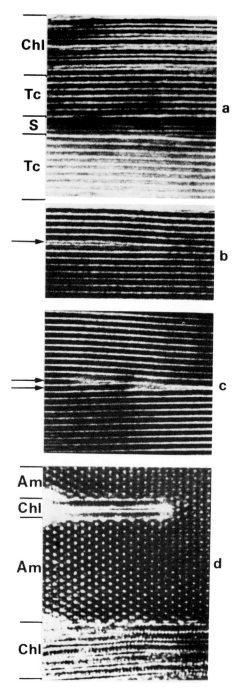

One of the most common pyroxene alteration products is serpentine. Pyroxenes that have been completely replaced by serpentine minerals are called "bastites." In the past decade, the textures and microstructures of bastites have been explored in some detail with petrographic, electron microprobe, x-ray microbeam camera (Wicks and Zussman, 1975), and TEM methods. Serpentinization can occur both in orthopyroxenes ("opx-bastites") and, less easily, in clino-pyroxenes ("cpx-bastites"), and the replacement can be either retrograde or prograde (Wicks and Whittaker, 1977). The alteration products can be lizardite, chrysotile, or anti-gorite, or mixtures of these three serpentine minerals with or without brucite and magnetite. A classifi-cation of different alteration types based on mineralogical and textural criteria has been presented by Wicks and Whittaker (1977) and amplified by Wicks and Plant (1979).

Studies with the x-ray micro-beam camera have provided a valuable

Figure 40. Microstructures in sheet silicates in a uralite. (a) Intimate intergrowth of chlorite ("Chl"), talc ("Tc"), and serpentine (S") structures. (b) Intergrowth and termination of a 5A brucite-like layer (arrowed) in talc. The structure produced by this extra layer has a chlorite configuration. (c) The termination of two layers (arrowed) in talc. (d) Intergrowth of chlorite and amphibole. A slab of chlorite structure consisting of two talc layers and an interleaved brucite sheet can be seen to terminate in the upper part of the figure. (Veblen and Buseck, in preparation).

198

look at the average mineralogical features of bastites (Wicks and Whittaker, 1977; Wicks and Plant, 1979). Both opx- and cpx-bastites in most cases contain lizardite as their predominant serpentine mineral, although cylindrical chrysotile and Povlen-type chrysotile,which has a polygonal structure, have also been observed with x-rays. Further mineralogical complexities, such as replacement by antigorite, can arise during continued alteration. In some cases, preferred orientation relationships suggest that the replacement of pyroxene by serpentine minerals may be, at least in part, a topotactic reaction (Dungan, 1979; Wicks and Plant, 1979). Such replacement would be structurally analogous to the partially oriented replacement of amphiboles by lizardite, chrysotile, and antigorite that has been observed with HRTEM (Veblen and Buseck, 1979; Veblen, 1980).

A number of opx- and cpx-bastites have been studied by CTEM techniques by Cressey (1979). These results provide little evidence for an oriented transformation from pyroxene to serpentine. Instead, the TEM observations suggest that coarse, well-crystallized plates of lizardite are usually separated from remnant pyroxene by a zone of extremely fine-grained, randomly-oriented serpentine. Cressey (1979) interprets this texture as indicating that the initial product in the serpentinization reaction bears no strict orientation relationship to the pyroxene structure, and that the coarser lizardite forms by recrystallization of the fine-grained material. If this interpretation is correct, then the replacement reaction is clearly not topotactic in the parts of the specimens examined with TEM. In some cases, coarse serpentine clearly does replace finer material, but an alternative interpretation of the texture described by Cressey might be that coarse lizardite is a primary, early reaction product and that the fine-grained, randomly-oriented serpentine is produced only during the very latest stages of serpentinization. In a few places, Cressey (1979) found that there is a rough orientation relationship between lizardite and pyroxene, and in some instances the layers of cylindrical and Povlen-type chrysotile are aligned with the silicate chains of the primary pyroxene. Thus, there is a limited influence of the parent structure on the orientation of the reaction product in some cases, as suggested by Dungan (1979) and the studies of Wicks. Differences in the results obtained with TEM and x-ray methods

Figure 41. A bastite from a harzburgite. Remnant (100) lamellae of clinopyroxene contain (010) amphibole lamellae (dark vertical lines); the clinopyroxene has been serpentinized on the right side of the figure. Between the clinopyroxene lamellae are (100) lamellae of orthopyroxene that have been completely serpentinized, indicating easier serpentinization of the orthopyroxene. Part of the serpentine replacing the clinopyroxene is Povlen-type (polygonal) chrysotile ("Pov"). (Micrograph by P.P.K. Smith, as reproduced in Cressey, 1979).

may result in part from the larger area that can be examined with x-rays. A great deal more work using both methods will clearly be required before the highly complex relationships involved in the serpentinization of pyroxenes are fully understood.

TEM results confirm petrographic observations that orthopyroxenes are usually more easily serpentinized than clinopyroxenes. Figure 41 shows an incompletely altered clinopyroxene that was studied by P. P. K. Smith (as reported in Cressey, 1979). Orthopyroxene lamellae parallel to (100) have been completely replaced by fine-grained serpentine, whereas the clinopyroxene has been only partially replaced, in part by Povlen-type chrysotile. The remnant clinopyroxene also contains (010) lamellae of amphibole. Such textures appear to be typical of incompletely serpentinized bastites.

Non-polymorphic Replacement of Pyroxene by Pyroxenoids

A polymorphic relationship exists between certain pyroxenes and pyroxenoids. For example, bustamite is a high-temperature polymorph of

200

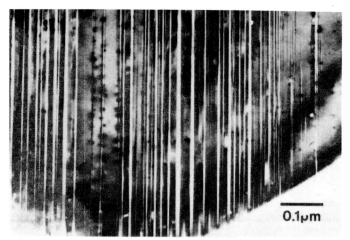

Figure 42. Low-resolution TEM image of johannsenite that has been partially replaced by lamellae of pyroxmangite and rhodonite. The pyroxene lamellae appear white, while the lamellae of pyroxenoid are gray. The planes of intergrowth (vertical) are parallel to (011) of the pyroxene. Dark spots along some of the lamellar interfaces result from structural strain and more rapid electron beam damage to the specimen in these regions. (Veblen, in preparation).

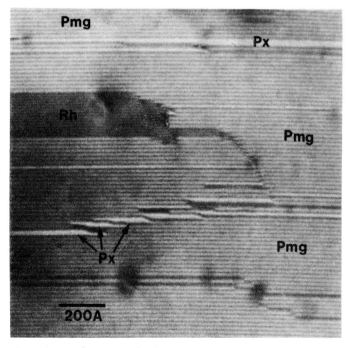

Figure 43. HRTEM image of an area of johannsenite that has been almost completely replaced by pyroxmangite ("Pmg") and rhodonite ("Rh"). The pyroxenoids in this area have some chain-offset disorder, as well as other defects. Narrow lamellae of remnant pyroxene are labeled "Px." Intergrowth takes place on planes parallel to (011) of the pyroxene (horizontal). (Veblen, in preparation).

johannsenite, and it is possible for one of these minerals to invert to the other without altering the chemical composition of the crystal. Pyroxene-pyroxenoid reactions in which there is a change in composition can also occur, however. Such an example of johannsenite from a skarn being replaced by pyroxmangite and rhodonite has recently been examined with HRTEM by Veblen (in preparation).

Pyroxenoids can be represented as members of a polysomatic series consisting of mixtures of (011) pyroxene slabs and slabs of wollastonite structure (Koto *et al.*, 1976; Thompson, 1978). Because of this structural relationship, it is to be expected that replacement reactions involving pyroxenes and pyroxenoids should be oriented reactions in which parts of the parent structure are inherited by the mineral that replaces it. This is, indeed, the case in the specimen studied. The details of the replacement are not simple, however.

As in many other examples of alteration reactions, several distinct reaction mechanisms appear to have operated in the pyroxene-pyroxenoid specimen studied by Veblen. In some places isolated slabs of pyroxenoid structure have reacted into the pyroxene, producing, in effect, disordered pyroxenoid having very long distances between some of the chain offsets, as predicted by Burnham (1966). Much more commonly, pyroxene appears to have been replaced by a bulk mechanism in which the reaction product is a pervasively disordered pyroxenoid having chains with offsets occurring every five and seven tetrahedra. This material is thus a completely disordered mixture of rhodonite and pyroxmangite structures. In still other parts of this specimen, pyroxene has been partially replaced by lamellae of rhodonite and pyroxmangite. Figure 42 is a low-resolution micrograph of such an area undergoing lamellar replacement of pyroxene by two ordered pyroxenoids. HRTEM images of the products of this type of replacement additionally reveal some chain-offset disorder of the pyroxenoids, as well as more complex intergrowth microstructures (Fig. 43).

ACKNOWLEDGEMENTS

We thank Janet Teshima, Donna Colletta, Kathleen Consagra, Connie Gilbert, Ellen Thurnav, Jenny Needham and Liz McCurdy for assistance during the preparation of this paper. Drs. M. A. Carpenter, R. H. McCallister, J. Tullis, R. A. Yund, M. Ross, J. S. Huebner, D. L. Kohlstedt, P. Robinson,

P. P. K. Smith, B. Cressey, P. E. Champness, and S. Iijima kindly supplied figures from their work. G. L. N. thanks M. Ross and J. S. Huebner for critical reviews of the section on exsolution. Partial support for this research was provided by the Microstructure Analysis Project, U. S. Geological Survey, NSF Earth Sciences Division Grants EAR77-00128 A01 and EAR7926375 to PRB and EAR7927094 to DRV, as well as an Arizona State University Faculty Grant in Aid to DRV.

REFERENCES

Aaronson, H. I., C. Laird and K. R. Kinsman (1970) Mechanisms of diffusional growth of precipitate crystals. In, *Phase transformations*, p. 313-396. Am. Soc. Metals, Metals Park, Cleveland.

Allen, S. M. and J. W. Cahn (1976) On tricritical points resulting from the intersection of lines of higher-order transitions with spinodals. Sci. Metall., 10, 451-454.

Bailey, J. C., P. E. Champness, A. C. Dunham, J. Esson, W. S. Fyfe, W. S. MacKenzie, E. P. Stumpfl and J. Zussman (1970) Mineralogy and petrology of Apollo 11 lunar samples. Proc. Apollo 11 Lunar Sci. Conf., 169-194.

Berner, R. A., E. L. Sjoberg, M. A. Velbel and M. D. Krom (1980) Dissolution of pyroxenes and amphiboles during weathering. Science, 207, 1205-1206.

Buerger, M. J. (1951) Crystallographic aspects of phase transformations. In R. Smoluchowski, J. E. Mayer and W. A. Weyl, Eds., *Phase Transformations in Solids*, p. 183-209. John Wiley, New York.

Boland, J. N. (1974) Lamellar structures in low-calcium orthopyroxenes. Contrib. Mineral. Petrol., 47, 215-222.

Boland, J. N. and R. A. Eggleton (1978) Orthopyroxene to talc transformation. Ninth Int. Congr. Electron Microscopy, 478-479.

_____ and M. Van Gijlswijk (1980) An electron microscopy study of multiple exsolution reactions in augite (abstr.). Int. Mineral. Assoc., Orleans, France.

Bollmann, W. and H. -U. Nissen (1968) A study of optimal phase boundaries: the case of exsolved alkali feldspars. Acta Crystallogr., A24, 546-557.

Bonnichsen, B. (1969) Metamorphic pyroxenes and amphiboles in the Biwabik Iron Formation, Dunkha River Area, Minnesota. Mineral. Soc. Am., Spec. Paper 2, 217-239.

Bown, M. G. and P. Gay (1959) The identification of oriented inclusions in pyroxene crystals. Am. Mineral., 44, 592-602.

Boyd, F. R. and G. M. Brown (1968) Electron-probe study of exsolution in pyroxenes. Carnegie Inst. Wash., Year Book, 66, 353-359.

_____, and J. F. Schairer (1964) The system $MgSiO_3-CaMgSi_2O_6$. J. Petrol., 5, 275-309.

Brady, J. B. and R. H. McCallister (1980, In press) Diffusion kinetics of homogenization and coarsening of pigeonite lamellae in subcalcic diopsides (abstr.). Geol. Soc. Am., Abstracts with Programs, 12.

Brown, G. E., C. T. Prewitt, J. J. Papike and S. Sueno (1972) A comparison of the structures of low and high pigeonite. J. Geophys. Res., 77, 5778-5789.

Brown, G. M. (1957) Pyroxenes from the early and middle stages of fractionation of the Skaergaard intrusion, east Greenland. Mineral. Mag., 31, 511-543.

_____ (1972) Pigeonite pyroxenes: A review. Geol. Soc. Am., Mem., 132, 523-534.

Brown, W. L., N. Morimoto and J. V. Smith (1961) A structural explanation of the polymorphism and transitions of $MgSiO_3$. J. Geol., 69, 609-616.

Burnham, C. W. (1966) Ferrosilite III: A triclinic pyroxenoid-type polymorph of ferrous metasilicate. Science, 154, 513-516.

_____ (1967) Ferrosilite. Carnegie Inst. Wash. Year Book, 65, 285-290.

Buseck, P. R. and S. Iijima (1974) High resolution electron microscopy of silicates. Am. Mineral., 59, 1-21.

_____, _____ (1975) High resolution electron microscopy of enstatite. II: geological application. Am. Mineral., 60, 771-784.

_____ and D. R. Veblen (1978) Trace elements, crystal defects, and high resolution electron microscopy. Geochim. Cosmochim. Acta, 42, 669-678.

Bystrom, A. (1943) Röntgenuntersuchung des Systems $MgO-Al_2O_3-SiO_2$. Ber. Dtsch. Keram Ges., 24, 2-12.

Cahn, J. W. (1961) On spinodal decomposition. Acta Met., 9, 795-801.

_____ (1962) Coherent fluctuations and nucleation in isotropic solids. Acta Met., 10, 907-913.

_____ (1966) The later stages of spinodal decomposition and the beginnings of particle coarsening. Acta Met., 14, 1685-1692.

_____ (1968) Spinodal decomposition. Trans. AIME, 242, 166-180.

Cameron, M., S. Sueno, C. T. Prewitt and J. J. Papike (1973) High temperature crystal chemistry of acmite, diopside, hedenbergite, jadeite, spodumene and ureyite. Am. Mineral., 58, 594-618.

Carpenter, M. A. (1978a) Nucleation of augite at antiphase boundaries in pigeonite. Phys. Chem. Minerals, 2, 237-251.

_____ (1978b) Kinetic control of ordering and exsolution in omphacite. Contrib. Mineral. Petrol., 67, 17-24.

_____ (1978c) Topotactic replacement of augite by omphacite in a blueschist rock from north-west Turkey. Mineral. Mag., 42, 435-438.

_____ (1979a) Contrasting properties and behavior of antiphase domains in pyroxenes. Phys. Chem. Minerals, 5, 119-131.

_____ (1979b) Experimental coarsening of antiphase domains in a silicate mineral. Science, 206, 681-683.

_____ (1979c) Omphacites from Greece, Turkey, and Guatemala: composition limits of cation ordering. Am. Mineral., 64, 102-108.

_____ (1980) Mechanisms of exsolution in sodic pyroxenes. Contrib. Mineral. Petrol., 71, 289-300.

Champness, P. E. (1973) Speculation on an order-disorder transformation in omphacite. Am. Mineral., 58, 540-542.

_____ and P. A. Copley (1976) The transformation of pigeonite to orthopyroxene. In H.-R. Wenk, Ed., Electron Microscopy in Mineralogy, p. 228-233. Springer-Verlag, New York.

_____ and G. W. Lorimer (1971a) An electron microscopic study of a lunar pyroxene. Contrib. Mineral. Petrol., 33, 171-183.

_____, _____ (1971b) Electron microscopic studies of some lunar minerals. In Electron Microscopy and Microanalysis, p. 324-327. Inst. Physics Conf. Ser. No. 10, Academic Press, London.

_____, _____ (1972) Electron microscopic studies of some lunar and terrestrial pyroxenes. In Proc. 5th Int. Mater. Sym., p. 1245-1255. Univ. Calif. Press, Berkeley.

_____, _____ (1973) Precipitation (exsolution) in an orthopyroxene. J. Mater. Sci., 8, 467-474.

_____, _____ (1974) A direct lattice-resolution study of precipitation (exsolution) in orthopyroxene. Phil. Mag., 30, 357-366.

_____, _____ (1976) Exsolution in silicates. In H.-R. Wenk, Ed., Electron Microscopy in Mineralogy, p. 174-204. Springer-Verlag, New York.

_____, A. C. Dunham, F. G. F. Gibb, H. N. Giles, W. S. MacKenzie, E. F. Stumpfl and J. Zussman (1971) Mineralogy and petrology of some Apollo 12 lunar samples. Proc. Lunar Sci. Conf., 2nd, 359-376.

Christian, J. W. (1965) The Theory of Transformations in Metals and Alloys. Pergamon Press, Oxford.

_____ (1975) The Theory of Transformations in Metals and Alloys, Part I. Equilibrium and General Kinetic Theory, 2nd ed., Pergamon Press, Oxford.

Christie, J. M., J. S. Lally, A. H. Heuer, R. M. Fisher, D. T. Griggs and S. V. Radcliffe (1971) Comparative electron petrography of Apollo 11, Apollo 12, and terrestrial rocks. Proc. Lunar Sci. Conf., 2nd, 69-90.

Clark, J. R., D. E. Appleman and J. J. Papike (1969) Crystal-chemical characterization of clinopyroxenes based on eight new structure refinements. Mineral. Soc. Am., Spec. Paper, 2, 31-50.

Cliff, G. and G. W. Lorimer (1975) The quantitative analysis of thin specimens. J. Microscopy, 103, 203-207.

Coe, R. S. (1970) The thermodynamic effect of shear stress on the ortho-clino inversion in enstatite, and other coherent phase transitions characterized by finite simple shear. Contrib. Mineral. Petrol., 26, 247-264.

_____ and S. H. Kirby (1975) The orthoenstatite to clinoenstatite transformation by shearing and reversion by annealing: mechanism and potential applications. Contrib. Mineral. Petrol., 52, 29-56.

_____ and Muller, W. F. (1973) Crystallographic orientation of clinoenstatite produced by deformation of orthoenstatite. Science, 180, 64-66.

Copley, P. A., P. E. Champness and G. W. Lorimer (1974) Electron petrography of exsolution textures in an iron-rich clinopyroxene. J. Petrol., 15, 41-57.

Cressey, B.A. (1979) Electron microscopy of serpentinite textures. Can. Mineral., 17, 741-756.

Dallwitz, W. B., D. H. Green and J. E. Thompson (1966) Clinoenstatite in a volcanic rock from the Cape Vogel area, Papua. J. Petr., 7, 375-403.

Davenport, E. S. and E. C. Bain (1930) Transformation of austenite at constant subcritical temperatures. Trans. AIME, 90, 117-154.

Deer, W. A., R. A. Howie and J. Zussman (1978) *Single-Chain Silicates, Vol 2A*, J. Wiley & Sons, New York.

Desnoyers, C. (1975) Exsolutions d'amphibole, de grenat et de spinelle dans les pyroxenes de roches ultrabasiques: peridotite et pyroxenolites. Bull. Soc. Franc. Mineral. Cristallogr., 98, 65-77.

Ditchek, B. and L. H. Schwartz (1980) Diffraction study of spinodal decomposition in Cu-10 w/o Ni-6 w/o Sn. Acta Met., 28, 807-822.

Dungan, M. A. (1979) Bastite pseudomorphs after orthopyroxene, clinopyroxene and tremolite. Can. Mineral., 17, 729-740.

Egerton, R. F. (1978) Microanalysis of light elements by electron energy loss spectrometry. Mat. Res. Bull., 13, 389-397.

Eggleton, R. A. (1975) Nontronite topotaxial after hedenbergite. Am. Mineral., 60, 1063-1068.

Finger, L. W. and Y. Ohashi (1976) The thermal expansion of diopside to 800°C and a refinement of the crystal structure at 700°C. Am. Mineral., 61, 303-310.

Fleet, M. E., C. T. Herzberg, G. M. Bancroft and L. P. Aldridge (1978) Omphacite studies, I. The P2/n → C2/c transformation. Am. Mineral., 63, 1100-1106.

Fletcher, R. C. and R. H. McCallister (1974) Spinodal decomposition as a possible mechanism in the exsolution of clinopyroxene. Carnegie Inst. Wash. Year Book, 73, 396-399.

Francis, D. M. (1976) Amphibole pyroxenite xenoliths: Cumulate or replacement phenomena from the upper mantle, Nunivak Island, Alaska. Contrib. Mineral. Petrol., 58, 51-61.

Ghose, S., G. Ng and L. S. Walter (1972) Clinopyroxenes from Apollo 12 and 14: exsolution, domain structure and cation order. Proc. Lunar Sci. Conf., 3rd, 507-531.

Gibbs, J. W. (1948) Collected works, Vol. 1. p. 105-115 and 252-258. Yale University Press, New Haven, Conn.

Goldschmidt, V. M. (1911) Die Kontaktmetamorphose im Kristianiagebiet. Vidensk. Skrifter. I. Mat. - Naturv, K. no. 11., 345-351.

Goldstein, J. I. (1979) Principles of thin film x-ray microanalysis. In J. J. Hren, J. I. Goldstein and D. C. Joy, Eds., *Introduction to Analytical Electron Microscopy*, p. 83-120. Plenum Press, New York.

Gordon, W. A., D. R. Peacor, P. E. Brown and E. J. Essene (1980, In Press) Exsolution relationships in a clinopyroxene of average composition $Ca_{43}Mn_{69}Mg_{82}Si_2O_6$ from Balmat, New York: X-ray diffraction and analytical electron microscopy. Am. Mineral., 65.

Grove, T. L. (1979) An experimental calibration of submicroscopic textures in lunar pyroxenes: a transmission electron microscope study. Lunar Planet. Sci., X, 467-469.

Grover, J. E. (1972) The stability of low-clinoenstatite in the system $Mg_2Si_2O_6$-$CaMgSi_2O_6$ (abstr.). Trans. Am. Geophys. Union, 53, 539.

Harlow, G. E., C. E. Nehru, M. Prinz, G. J. Taylor, and K. Keil (1979) Pyroxenes in Serra de Mage: cooling history in comparison with Moama and Moore County. Earth Planet. Sci. Lett., 43, 173-181.

Hess, H. H. (1941) Pyroxenes of common mafic magmas, part 2. Am. Mineral., 26, 573-594.

_____ (1952) Orthopyroxenes of the Bushveld type, ion substitution and changes in unit cell dimensions. Am. J. Sci. (Bowen Vol.), 173-188.

_____ and A. H. Phillips (1938) Orthopyroxenes of the Bushveld type. Am. Mineral., 23, 450-456.

_____, _____ (1940) Optical properties and chemical composition of magnesian orthopyroxenes. Am. Mineral., 25, 271-285.

Heuer, A. H. and G. L. Nord Jr. (1976) Polymorphic phase transitions in minerals. In H.-R. Wenk, Ed., *Electron Microscopy in Mineralogy*, p. 274-303. Springer-Verlag, New York.

Hilliard, J. E. (1970) Spinodal decomposition. In *Phase transformations*, p. 497-560. Am. Soc. Metals, Cleveland.

Huebner, J. S. and A. C. Turnock (1980) The melting relations at 1 bar of pyroxenes composed largely of Ca-, Mg-, and Fe-bearing components. Am. Mineral., 65, 225-271.

_____, M. Ross and N. Hickling (1975) Significance of exsolved pyroxenes from lunar breccia 77215, Proc. Lunar Sci. Conf. 6th., 529-546.

Iijima, S. and P. R. Buseck (1974) High resolution electron microscopy of polymorphism in Ca-poor pyroxene. In J. V. Sands and D. J. Goodchild, Eds., *Electron Microscopy 1974*, p. 500-501. Aust. Acad. Sci.

_____, _____ (1975) High resolution electron microscopy of enstatite I: twinning, polymorphism and polytypism. Am. Mineral., 60, 758-770.

_____, _____ (1976) High-resolution electron microscopy of unit cell twinning in enstatite. In H.-R Wenk, Ed., *Electron Microscopy in Mineralogy*, p. 319-323. Springer-Verlag, New York.

Ishii, T. (1973) Zoning, exsolution and phase equilibria in the Kintoki-san pyroxenes, Hakone volcano (abstr.). Geol. Soc. Japan 80th Ann. Meeting, 178.

_____ and H. Takeda (1974) Inversion, decomposition and exsolution phenomena of terrestrial and extraterrestrial pigeonites. Geol. Soc. Japan, Mem., 11, 19-36.

Ito, T. (1935) On the symmetry of the rhombic pyroxenes, Z. Krist., 90, 151-162.

_____ (1950) *X-ray Studies on Polymorphism.* Maruzen Co., Tokyo.

Jaffe, H.W., P. Robinson, R. J. Tracy and M. Ross (1975) Orientation of pigeonite lamellae in metamorphic augite: correlation with composition and calculated optimal phase boundaries. Am. Mineral., 60, 9-28.

Joy, D. C. (1979) The basic principles of electron energy loss spectroscopy. In J. J. Hren, J. I. Goldstein and D. C. Joy, *Introduction to Analytical Electron Microscopy*, p. 223-244. Plenum Press, New York.

Kohlstedt, D. L. and J. B. Vander Sande (1973) Transmission electron microscopy investigation of the defect microstructure of four natural orthopyroxenes. Contrib. Mineral. Petrol., 42, 169-180.

_____, _____ (1976) On the detailed structure of ledges in an augite-enstatite interface. In H.-R. Wenk, Ed., *Electron Microscopy in Mineralogy*, p. 234-237. Springer-Verlag, New York.

Koto, K., N. Morimoto and H. Narita (1976) Crystallographic relationships of the pyroxenes and pyroxenoids. J. Jap. Assoc. Mineral., Petrol., Econ. Geol., 71, 248-254.

Krstanovic, I. (1975) X-ray diffraction studies of orthoenstatite (abstr.). Acta Cryst., A31, S73.

Kuno, H. (1966) Review of pyroxene relations in terrestrial rocks in the light of recent experimental works. Mineral. Jour. (Japan), 5, 21-43.

Kushiro, I. (1972) Determination of liquidus relations in synthetic silicate systems with electron probe analysis: The system forsterite-diopside-silica at 1 atmosphere. Am. Mineral., 57, 1260-1271.

Lally, J. S., R. M. Fisher, J. M. Christie, D. T. Griggs, A. H. Heuer, G. L. Nord, and S. V. Radcliffe (1972) Electron petrography of Apollo 14 and 15 rocks. Proc. Lunar Sci. Conf., 3rd, 401-422.

_____, A. H. Heuer, G. L. Nord, Jr. and J. M. Christie (1975) Subsolidus reactions in lunar pyroxenes: an electron petrographic study. Contrib. Mineral. Petrol., 51, 263-282.

Laughlin, D. E. and J. W. Cahn (1975) Spinodal decomposition in age hardening of copper-titanium alloys. Acta Met., 23, 329-339.

Lorimer, G. W. and P. E. Champness (1973) Combined electron microscopy and analysis of an orthopyroxene. Am. Mineral., 58, 243-248.

_____ and G. Cliff (1976) Analytical electron microscopy of minerals. In H.-R. Wenk, Ed., Electron Microscopy in Mineralogy, p. 506-519. Springer-Verlag, New York.

_____ and R. B. Nicholson (1970) The nucleation of precipitates in aluminum alloys. In The Mechanism of Phase Transformations in Crystalline Solids, p. 36-42. Inst. of Metals, London.

Massalski, T. B. (1970) Massive transformations. In Phase Transformations, p. 433-486. Am. Soc. Metals, Cleveland.

Matsumoto, T. (1974) Possible structure types derived from Pbca-orthopyroxene. Mineral. Jour. (Japan), 7, 374-383.

_____ M. Tokonami and N. Morimoto (1975) The crystal structure of omphacite. Am. Mineral., 60, 634-641.

McCallister, R. H. (1978) The coarsening kinetics associated with exsolution in an iron-free clinopyroxene. Contrib. Mineral. Petrol., 65, 327-331.

_____ and R. A. Yund (1977) Coherent exsolution in Fe-free pyroxenes. Am. Mineral., 62, 721-726.

McConnell, J. D. C. (1975) Microstructures of minerals as petrogenetic indicators. Annu. Rev. Earth Planet. Sci., 3, 129-155.

McLaren, A. C. and M. A. Etheridge (1976) A transmission electron microscope study of naturally deformed orthopyroxene. I. Slip mechanisms. Contrib. Mineral. Petrol., 57, 163-177.

Morimoto, N. (1974) Crystal structure and fine texture of pyroxenes. Fortsch. Mineral., 52, 52-80.

_____ and M. Tokonami (1969a) Domain structure of pigeonite and clinoenstatite. Am. Mineral., 54, 725-740.

_____, _____ (1969b) Oriented exsolution of augite in pigeonite. Am. Mineral., 54, 1101-1117.

Müller, W. F. (1974) One-dimensional lattice imaging of a deformation-induced lamellar intergrowth of orthoenstatite and clinoenstatite [(Mg,Fe)SiO$_3$]. N. Jahrb. Mineral. Monatsh., 2, 83-88.

Nakajima, Y. and S. S. Hafner (1980) Exsolution in augite from the Skaergaard intrusion. Contrib. Mineral. Petrol., 72, 101-110.

_____ and P. H. Ribbe (1980) Alteration of pyroxenes from Hokkaido, Japan, to amphibole, clays, and other biopyriboles. N. Jahrb. Mineral. Monatsh., 6, 258-268.

Nakazawa, H. and S. S. Hafner (1977) Orientation relations of augite exsolution lamellae in pigeonite hosts. Am, Mineral., 62, 79-88.

Newbury, D. E. and R. L. Myklebust (1979) Monte Carlo electron trajectory simulation of beam spreading in thin foil targets. Ultramicroscopy, 3, 391-395.

Nicholson, R. B. (1970) Nucleation at imperfections. In Phase Transformations, p. 269-312. Am. Soc. Metals, Cleveland.

Nobugai, K. and N. Morimoto (1979) Formation mechanism of pigeonite lamallae in Skaergaard augite. Phys. Chem. Minerals, 4, 361-371.

_____, M. Tokonami and N. Morimoto (1978) A study of subsolidus relations of the Skaergaard pyroxenes by analytical electron microscopy. Contrib. Mineral. Petrol., 67, 111-117.

Nord, G. L., Jr. (1980a) Decomposition kinetics in clinopyroxenes (abstr.). Geol. Soc. Am., Abstracts with Programs, 12, 492.

_____ (1980b) The composition, structure and stability of Guinier-Preston zones in lunar and terrestrial orthopyroxene. Phys. Chem. Minerals, 6, 109-120.

_____, A. H. Heuer, J. S. Lally (1976) Pigeonite exsolution from augite. In H.-R. Wenk, Ed., *Electron Microscopy in Mineralogy*, p. 220-227. Springer-Verlag, New York.

_____, _____, _____ and J. M. Christie (1975) Substructures in Lunar clinopyroxenes as petrologic indicators. Lunar Sci. VI, 601-603.

_____, J. S. Huebner, and M. Ross (1977) Structure, composition and significance of G. P. zones in 76535 orthopyroxene. Lunar Sci. VIII, 732-733.

_____ and O. B. James (1978) Consortium breccia 73255: thermal and deformational history of bulk breccia and clasts, as determined by electron petrography. Proc. Lunar Planet. Sci. Conf., 9th, 821-839.

_____ and R. H. McCallister (1979) Kinetics and mechanism of decomposition in $Wo_{25}En_{31}Fs_{44}$ clinopyroxene (abstr.). Geol. Soc. Am., Abstracts with Programs, 11, 488.

_____ and J. J. McGee (1979) Thermal and mechanical history of granulated norite and pyroxene anorthosite clasts in breccia 73255. Proc. Lunar Planet. Sci. Conf., 10th, 817-832.

O'Keefe, M. A. (1973) n-beam lattice images IV. computed two-dimensional images. Acta Crystallogr., A29, 389-401.

_____, P. R. Buseck and S. Iijima (1978) Computed crystal structure images for high resolution electron microscopy. Nature, 274, 322-324 and cover photograph.

Papike, J. J., A. E. Bence, G. E. Brown, C. T. Prewitt, and C. H. Wu (1971) Apollo 12 clinopyroxenes: exsolution and epitaxy. Earth Planet. Sci. Lett., 10, 307-315.

_____ and M. Cameron (1976) Crystal chemistry of silicate minerals of geophysical interest. Rev. Geophysics Space Physics, 14, 37-80.

_____, C. T. Prewitt, S. Sueno and M. Cameron (1973) Pyroxenes: comparisons of real and ideal structural topologies. Z. Krist., 69, 254-273.

_____, M. Ross and J. R. Clark (1969) Crystal-chemical characterization of clinoamphiboles based on five new structure refinements. Mineral. Soc. Am., Spec. Pap., 2, 117-136.

Park, M., T. E. Mitchell and A. H. Heuer (1976) Coarsening in a spinodally decomposing system: TiO_2-SnO_2. In H.-R. Wenk, Ed., *Electron Microscopy in Mineralogy*, p. 203-205. Springer-Verlag, New York.

Peacor, D. R., E. J. Essene, P. E. Brown and G. A. Winter (1978) The crystal chemistry and petrogenesis of a magnesian rhodonite. Am. Mineral., 63, 1137-1142.

Phakey, P. P. and S. Ghose (1973) Direct observation of anti-phase domain structure in omphacite. Contrib. Mineral. Petrol., 39, 239-245.

Phase Transformations, A. S. M. Seminar (1970) American Society for Metals, Metals Park, Ohio.

Poldervaart, A. and H. H. Hess (1951) Pyroxenes in the crystallization of basaltic magmas. J. Geol., 59, 472-489.

Prewitt, C. T., G. E. Brown and J. J. Papike (1971) Apollo 12 clinopyroxenes: High temperature X-ray diffraction studies. Proc. Lunar Sci. Conf., 2nd, 59-68.

Provost, A. and Y. Bottinga (1974) Rates of solidification of Apollo 11 basalt and Hawaiian tholeiite. Earth Planet. Sci. Lett., 15, 325-337.

Radcliffe, S. V., A. H. Heuer, R. M. Fisher, J. M. Christie and D. T. Griggs (1970) High voltage (800kV) electron petrography of Type B rock from Apollo 11. Proc. Apollo 11 Lunar Sci. Conf., 731-748.

Rietmeijer, F. J. M. (1979) Pyroxenes from iron-rich igneous rocks in Rogaland, SW Norway. Geol. Ultraiectina, 21, (Ph. D. thesis State University of Utrecht), 341.

_____ and P. E. Champness (1980) Inverted pigeonites from Rogaland, SW Norway. Inst. Phys. Conf. Ser. No. 52, 105-108.

Robin, P.-Y. F. (1977) Angular relationships between host and lamellae and the use of the Mohr circle. Am. Mineral., 62, 127-131.

Robinson, P., H. W. Jaffe, M. Ross and C. Klein, Jr. (1971) Orientation of exsolution lamellae in clinopyroxenes and clinoamphiboles: consideration of optimal phase boundaries. Am. Mineral., 56, 909-939.

_____, M. Ross, G. L. Nord, Jr., J. R. Smyth, and H. W. Jaffe (1977) Exsolution lamellae in augite and pigeonite: fossil indicators of lattice parameters at high temperature and pressure. Am. Mineral., 62, 857-873.

Ross, M. and J. S. Huebner (1979) Temperature-composition relationships between naturally occurring augite, pigeonite and orthopyroxene at one bar pressure. Am. Mineral., 64, 1133-1155.

_____, _____ and E. Dowty (1973) Delineation of the one atmosphere augite-pigeonite miscibility gap for pyroxenes from lunar basalt 12021. Am. Mineral., 58, 619-635.

Sadanaga, R., E. Okamura and H. Takeda (1969) X-ray study of the phase transformations of enstatite. Mineral. J. (Japan), 6, 110-130.

Shimizu, N., M. P. Semet and C. J. Allegre (1978) Geochemical applications of quantitative ion-microprobe analysis. Geochim. Cosmochim. Acta, 42, 1321-1334.

Smith, J. V. (1959) The crystal structure of protoenstatite $MgSiO_3$. Acta Crystallogr., 12, 515-519.

_____ (1969a) Crystal structure and stability of the $MgSiO_3$ polymorphs: physical properties and phase relations of Mg, Fe pyroxenes. Mineral. Soc. Am., Spec. Paper 2, 3-29.

_____ (1969b) Magnesium pyroxene at high temperature: inversion in clinoenstatite. Nature, 222, 256-257.

Smith, P. P. K. (1977) An electron microscope study of amphibole lamellae in augite. Contrib. Mineral. Petrol., 59, 317-322.

_____ (1978) Exsolution and inversion in pigeonite from the Whin Sill, Northern England. Ninth Int. Congr. Electron Microscopy, 474-475.

Smyth, J. R. (1969) Orthopyroxene-high-low clinopyroxene inversions. Earth Planet. Sci. Lett., 6, 406-407.

_____ (1971) Protoenstatite: a crystal structure refinement at $1100°C$. Z. Krist., 134, 262-274.

_____ (1974a) Experimental study on the polymorphism of enstatite. Am. Mineral., 59, 345-352.

_____ (1974b) Low orthopyroxene from a lunar deep crustal rock: a new pyroxene polymorph of space group $P2_1ca$. Geophys. Res. Lett., 1, 27-30.

_____ (1974c) The high temperature crystal chemistry of clinohypersthene. Am. Mineral., 59, 1069-1082.

_____ and C. W. Burnham (1972) The crystal structures of high and low clinohypersthene. Earth Planet. Sci. Lett., 14, 183-189.

Steele, I. M. (1975) Mineralogy of lunar norite 78235; second lunar occurrence of $P2_1ca$ pyroxene from Apollo 17 soils. Am. Mineral., 60, 1086-1091.

_____, R. Hervig and I. Hutcheon (1980) Quantitative ion microprobe analysis of Mg, Fe silicates. In D. B. Wittry, Ed., Microbeam Analysis-1980, p. 151-153. San Francisco Press, San Francisco.

Takeda, H. (1973) Inverted pigeonites from a clast of rock 15459 and basaltic chondrites. Proc. Lunar Sci. Conf. 4th, 875-885.

Thompson, J. B., Jr. (1970) Geometrical possibilities for amphibole structures: model biopyriboles. Am. Mineral., 55, 292-293.

_____ (1978) Biopyriboles and polysomatic series. Am. Mineral., 63, 239-249.

Tullis, J. and R. A. Yund (1979) Calculation of coherent solvi for alkali feldspar, iron-free clinopyroxene, nepheline-kalsilite, and hematite-ilmenite. Am. Mineral., 64, 1063-1074.

Turner, F. J., H. C. Heard and D. T. Griggs (1960) Experimental deformation of enstatite and accompanying inversion to clinoenstatite. Rept. 21st Int. Geol. Congr. Copenhagen, 18, 399-408.

Veblen, D. R. (1980) Anthophyllite asbestos: microstructures, intergrown sheet silicates, and mechanisms of fiber formation. Am. Mineral., 65, 1075-1086.

_____ and P. R. Buseck (1977) Petrologic implications of hydrous biopyriboles intergrown with igneous pyroxene (abstr.). Trans. Am. Geophys. Union (EOS), 58, 1242.

_____, _____ (1979a) Chain width order and disorder in biopyriboles. Am. Mineral., 64, 687-700.

_____, _____ (1979b) Serpentine minerals: Intergrowths and new combination structures. Science, 206, 1398-1400.

_____, _____ (1980) Microstructures and reaction mechanisms in biopyriboles. Am. Mineral., 65, 599-623.

Walker, K. R. (1969) The Palisades Sill, New Jersey: A reinvestigation. Geol. Soc. Am., Spec. Pap., 111, 178.

Warren, B. E. and D. L. Modell (1930) Structure of enstatite $MgSiO_3$. Z. Krist., 75, 1-14.

Wicks, F. J. and A. G. Plant (1979) Electron-microprobe and x-ray microbeam studies of serpentine textures. Can. Mineral., 17, 785-830.

_____ and E. J. W. Whittaker (1977) Serpentine textures and serpentinization. Can. Mineral., 15, 459-488.

_____ and J. Zussman (1975) Microbeam x-ray diffraction patterns of the serpentine minerals. Can. Mineral., 13, 244-258.

Willaime, C. and W. L. Brown (1974) A coherent elastic model for the determination of the orientation of exsolution boundaries: application to feldspars. Acta Crystallogr., A30, 316-331.

Yamaguchi, Y., J. Akai and K. Tomita (1978) Clinoamphibole lamellae in diopside of garnet lherzolite from Alpe Arami, Bellinzona, Switzerland. Contrib. Mineral. Petrol., 66, 263-270.

Yokoyama, K., S. Banno and T. Matsumoto (1976) Compositional range of P2/n omphacite from the eclogitic rocks of central Shikoku, Japan. Mineral. Mag., 40, 773-779.

Yund, R.A. (1975) Microstructure, kinetics, and mechanisms of alkali feldspar exsolution. In P.H. Ribbe, ed., *Feldspar Mineralogy*, Mineral. Soc. Am., Short Course Notes, p. Y29-Y57.

_____ and R. H. McCallister (1970) Kinetics and mechanisms of exsolution. Chem. Geol., 6, 5-30.

Zaluzec, N. J. (1979) Quantitative x-ray microanalysis: Instrumental considerations and applications to materials science. In J. J. Hren, J. I. Goldstein and D. C. Joy, Eds., *Introduction to Analytical Electron Microscopy*, p. 121-167. Plenum, New York.

Chapter 5

PYROXENE PHASE EQUILIBRIA at LOW PRESSURE

J. Stephen Huebner

INTRODUCTION AND NOMENCLATURE

The basic purpose of this review is to synthesize the past half cen-
tury of studies of pyroxene phase relations at low pressure. The admir-
ably comprehensive bibliographies of Deer *et al.* (1978) will be updated,
but emphasis will be placed on systematization of knowledge, rather than
on exhaustive documentation. As will become apparent, the systematization
is most successful for common quadrilateral pyroxenes composed largely of
calcium-, magnesium-, and iron-bearing metasilicate components. Extension
of this generalized approach to pyroxenes containing large amounts of non-
quadrilateral (i.e., aluminum-, sodium-, or titanium-bearing) components
or to uncommon structural states such as protopyroxene is not as successful; these less common pyroxenes are treated as a series of vignettes.
The manganese-bearing pyroxenes are an exception -- recent investigations
of natural manganiferous pyroxenes suggest possible manganese-bearing
pyroxene quadrilaterals.

The comprehension of pyroxene phase relations is not helped by the
complicated and composition-dependent nomenclature of many petrologists
and mineralogists. The actual compositions of the observed pyroxenes
are unknown in much of the published experimental work, but the pyroxene
structural state is usually reliably reported. We now know that the
various pyroxene structural states are associated with definite limits
of composition and that reports of the structural state thus connote at
least composition limits. For these reasons I have adopted a simplified
pyroxene nomenclature based on structural state and similar to that
originally proposed for the Pyroxene Nomenclature Committee by Malcolm
Ross and used for quadrilateral pyroxenes by Ross and Huebner (1979) and
by Huebner and Turnock (1980), summarized in Figure 1. According to this
simplified system of nomenclature, pyroxene compositions are expressed in
terms of percent Wo, En, and Fs, where Wo is simply the atomic proportion
100 Ca/(Ca + Mg + Fe), En = 100 Mg/(Ca + Mg + Fe), and Fs = 100 Fe/(Ca +
Mg + Fe). Aluminum-, titanium-, and sodium-bearing components such as
"CaTs," "FeTs," or "MgTs" ($CaAl^{VI}Al^{IV}SiO_6$, etc.) are *not* subtracted before

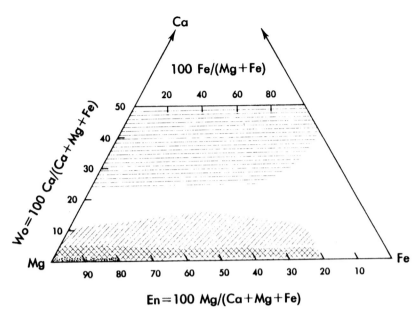

Ca

$100\ Fe/(Mg+Fe)$

$Wo=100\ Ca/(Ca+Mg+Fe)$

Mg Fe

$En=100\ Mg/(Ca+Mg+Fe)$

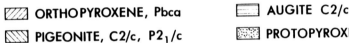

▨ ORTHOPYROXENE, Pbca   ▥ AUGITE C2/c

▧ PIGEONITE, C2/c, $P2_1/c$ ▩ PROTOPYROXENE Pbca

Figure 1. Simplified nomenclature for Ca-, Fe-, and Mg-pyroxenes, adopted for this summary. Compositions are given in Wo, En, and Fe/(Mg+Fe); the maximum composition range
known at low pressure is shown. All calcium-rich pyroxenes are augite. Where their
compositional ranges overlap, the calcium-poor pyroxenes are distinguished by their
structure (space group). Protopyroxene is not stable at low temperatures (see text);
pigeonite is not known to form in nature at temperatures below those of high grade metamorphic rocks.

calculation of Wo, En, and Fs; hence Wo, En, and Fs are comparable with
the [Ca], [Mg], and [Fe] used by Ross and Huebner (1979).

Augite (A) is a calcium-rich pyroxene, space group $C2/c$, with Wo $\geq$
25% (not known to exceed 50%). Values of atomic 100 Mg/(Mg + Fe) range
from 0 (as in hedenbergite, $CaFeSi_2O_6$, Hd) to 100 (as in diopside, CaMg
Si_2O_6, Di). Orthopyroxene (O) with symmetry $Pbca$ has Wo restricted to
<5%, but has a wide range of permissible values of 100 Mg/(Mg + Fe).
Pigeonite (P) has $C2/c$ space group at high temperature (isostructural
with augite) and space group $P2_1/c$ at low temperature (Prewitt et al.,
1971; Brown et al., 1972). The Wo value of pigeonite is <15% but can be
as small as 0% as in compositions termed clinoenstatite, clinohypersthene,
or clinoeulite by some investigators. It is not always possible to distinguish the monoclinic pyroxenes augite and pigeonite when examining

214

laboratory products by means of conventional techniques of optical microscopy and x-ray powder diffraction. In these cases the term clinopyroxene (Cpx) is used to describe the experimental product. Clinopyroxene is a particularly useful term for those apparently metastable single phases of composition Wo = 25% synthesized by Turnock *et al.* (1973). [Indeed, recent observations by transmission electron microscopy reveal that it is possible to synthesize at high pressure and quench single phase monoclinic pyroxene with Wo 15-25%, but such pyroxenes are not stable (Nord and McCallister, 1979 and oral communication).] Protopyroxene (Pr) with symmetry *Pbcn* is a polymorph of orthopyroxene and pigeonite that has low values of Wo (not known to exceed 2%) and an En value commonly exceeding 95%. Longhi and Boudreau (1980) used the term "clinoenstatite" for the monoclinic polymorph that forms from cooling protopyroxene, thereby reserving the term "pigeonite" for the monoclinic, Ca-poor volume that abuts the solidus. Other pyroxenes that have unique compositions (or structures) will be called by their specific and traditional mineralogical names.

ROCK-FORMING PYROXENES

Orthopyroxene, pigeonite, and augite, composed largely of quadrilateral components, are the pyroxenes commonly found in basalt, gabbro, and granulite. These rock-forming pyroxenes, which have been well studied by students of mineralogy and petrology, are present in rocks whose temperatures and pressures of formation have also been investigated. Because one of the goals of experimental petrology is to provide a basis by which geologists can interpret the conditions at which pyroxene-bearing rocks formed, it is appropriate first to review the results of natural pyroxene-forming processes. Conversely, it is the natural assemblages themselves that have provided many clues, without which the phase relations would not be so well understood today.

Assemblages of Naturally Occurring Quadrilateral Pyroxenes

Two diagrams will summarize adequately the phase relations of rock-forming pyroxenes that can be gleaned from studies of natural pyroxene-forming processes. More diagrams could of course be constructed, but to do so would imply an erroneously large degree of accuracy. Too little is yet known about the effects on the pyroxene phase relations of non-

215

quadrilateral components, fugacities of H_2O and O_2, pressure, and time necessary to achieve equilibrium, to be able to assign accurate values to the intensive parameters prevailing during natural pyroxene crystallization. Refinement of these parameters should be a primary goal of experimental petrologists in the years to come. In the meantime, precision (and accuracy) of estimates can be increased only by limiting the intensive and extensive parameters to narrow ranges close to those that have been used in the laboratory, such as those ranges characteristic of lunar mare basalts (low pressure) or iron formations (iron-rich, alkali- and alumina-poor compositions). For the generalized model of naturally occurring pyroxenes, one diagram each of the solidus and liquidus will suffice.

Subsolidus. Many studies support the contention that at temperatures well below the solidus, that is, at temperatures characteristic of many metamorphic rocks including pyroxene granulites, orthopyroxene (Wo < 2%) and augite (Wo > 40%) are the only two pyroxenes that occur. Where these pyroxenes coexist, they define a broad miscibility gap, apparent in quadrilateral (Ca-Mg-Fe) projection (Fig. 2). This gap has been shown by means of wet-chemical analyses of mechanically separated pyroxenes to persist over a range of Fe/(Mg+Fe) in granulites formed at moderate pressure (*e.g.*, Bartholomé, 1962; Subramaniam, 1962; Davidson; 1968; Wilson, 1976). Subsequent investigations, using the electron microprobe, analytical electron microscope, and x-ray precession camera to examine individual phases (crystals) of pyroxene refined the limits of orthopyroxene and augite miscibility but did not greatly change the overall understanding: Jaffe *et al.* (1975) analyzed granulite- and amphibolite-facies pyroxenes; Bohlen and Essene (1978) summarized data on pyroxenes from the Adirondack granulitic gabbro massif; Coleman (1978) and Nobugai *et al.* (1978) scrutinized Skaergaard pyroxenes; and Buchanan (1979) looked in detail at equilibrated pyroxenes from the Bushveld intrusion. Other investigators proved that the orthopyroxene-augite miscibility gap extends to the large Fe/(Mg+Fe) values characteristic of iron-formation bulk compositions (Bonnichsen, 1969; Simmons *et al.*, 1974; Immega and Klein, 1976) and Skaergaard ferrodiorites (Nwe and Copley, 1975) and is similar at the lower pressures characteristic of the lunar crust (Huebner *et al.*, 1975; McCallum and Mathez, 1975; Takeda and Miyamoto, 1977). The orthopyroxene-augite gap is also present in many meteorites (Duke and

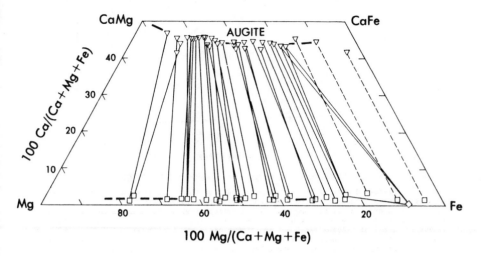

Figure 2. Generalized subsolidus diagram for quadrilateral pyroxenes based upon selected assemblages of coexisting terrestrial pyroxenes. Portrayed relations are applicable several hundred degrees below the solidus and at temperatures at which orthpyroxene + augite are stable (instead of pigeonite). Dashed tie lines indicate assemblages that could not exist unless stabilized by high pressure or a manganese component. The shapes of symbols are the same as those adopted by Huebner and Turnock (1980), will be used consistently for all original drawings in this chapter, and are as follows: squares represent orthopyroxene compositions, upright triangles represent pigeonite; inverted trianges represent augite; diamonds represent olivine; and a dot within a symbol indicates that a silica mineral is also present. The sources of data include both electron microbeam and bulk chemical analyses: Jaffe *et al.* (1975); Immega and Klein (1976); Bonnichsen (1969); Wilson (1976); Davidson (1968).

Silver, 1967), exemplified by the Moore County eucrite (Hess and Henderson, 1949; Stewart, Ross and Hickling, pers. comm., 1974, 1979; Harlow *et al.*, 1979). However, much of what is known about the temperatures and pressures of pyroxene crystallization in meteorites is inferred from the pyroxenes themselves; in the absence of independent evidence, the meteorites are not good indicators of low *vs* high temperature or pressure of pyroxene crystallization.

The range of Fe–Mg substitution in natural quadrilateral orthopyroxenes is not complete. Orthopyroxenes more iron rich than $Fe/(Mg+Fe)$ = 80% do not occur (unless stabilized by elevated pressure or $Mn_2Si_2O_6$ component); the characteristic assemblages for such bulk compositions are orthopyroxene + olivine (fayalite) $\pm$ quartz or olivine $\pm$ quartz (Smith, 1974). More calcium-rich compositions include hedenbergitic augite as an additional phase (Bonnichsen, 1969; Simmons *et al.*, 1974). The two fields, orthopyroxene + augite and fayalite + augite $\pm$ silica, are separated by a region of orthopyroxene + augite + fayalite $\pm$ silica (see Simmons *et al.*, 1974, Fig. 3). Quartz would be expected to accompany

217

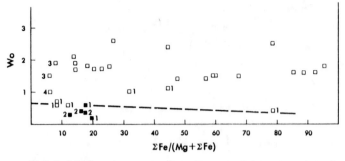

Unlabelled points are from Ross and Huebner (1979).

Figure 3. Distorted view of the calcium-poor part of the pyroxene quadrilateral projection at subsolidus conditions. The Wo component direction between 0% and 3% has been expanded to project clearly the microprobe analyses of orthopyroxenes that contain intergrown (exsolved) augite lamellae (open symbols). The dashed line suggests the maximum possible solubility limit of calcium (as % Wo) in orthopyroxene at "low temperature." The sources of analyses: (1) Ross & Huebner (1979), (2) Huebner *et al.* (1979), (3) unpublished; AMNH #G40403, Emali, Kenya, (4) unpublished; Opx "Kohlstedt," donated by A. Duba.

the fayalite for these pyroxene bulk compositions (see Huebner and Turnock, 1980, Fig. 10, 700-800°C); that it only rarely is reported to do so (Bonnichsen, 1969, sample #122; Morey *et al.*, 1972, p. 245) is probably because of inappropriate bulk compositions.

Recent studies give a clue to the minimum width (in terms of Wo) of the orthopyroxene-augite miscibility gap at low temperature (Fig. 3). The orthopyroxenes that coexist with augite have a surprisingly small calcium content. Ross and Huebner (1979) examined 30 orthopyroxenes spanning the Fe/(Mg+Fe) range 8-94%. Most had Wo < 2%, but only two orthopyroxenes (compositions $Wo_{0.6}En_{81.8}Fs_{17.6}$ and $Wo_{0.2}En_{80.4}Fs_{19.4}$) appeared devoid of exsolved augite when examined by x-ray precession methods. Other crystals with $Wo_{0.4}$ and $Wo_{0.6}$ contained augite. Huebner *et al.* (1979) reported analyses of three orthopyroxene crystals with Fe/(Mg+Fe) = 81-88% and Wo = 0.4%; in these crystals no exsolved augite (or any other phases) was detected by x-ray precession photography or transmission electron microscopy. When specifically sought, augite has always been found as an (exsolved) intergrowth in plutonic and metamorphic "orthopyroxenes" that contain more than 0.5% Wo. Therefore, under some conditions and for some pyroxene compositions, the orthopyroxene boundary of the miscibility gap at low temperature is apparently at least as calcium poor as Wo = 0.5%. Some analyses of "orthopyroxene" that report Wo > 0.5 may in fact be analyses of more than one phase, orthopyroxene plus a calcic phase (probably augite or G.P. zones such as those described by Nord, 1980 -- and see Buseck *et al.*, this volume).

The composition of augite that coexists with orthopyroxene in metamorphic rocks appears to be close to Wo_{45}. There is some scatter in the Wo contents of augites that coexist with orthopyroxenes of a given Fe/ (Mg+Fe) value (see Jaffe *et al.*, 1975; Immega and Klein, 1976). Despite this scatter it is possible to discern an arcuate trend: The Wo content of the augite decreases from Wo $\sim$ 47% to Wo $\sim$ 42% as Fe/(Mg+Fe) increases from 0 to $\sim$50%; with further iron enrichment, Wo increases to Wo $\sim$ 48%. Assuming a uniform Wo content of orthopyroxene, the orthopyroxene-augite tie lines appear shortest at Fe/(Mg+Fe) = 50%. The explanation of the change in tie line length is not known but it could be due to any (or all!) of several possibilities: (1) A warp in the augite solvus. (2) The shortest tie lines reflect the highest temperatures of equilibration. (3) Variations in the concentrations of minor elements such as aluminum may markedly affect tie line length. (4) The change in tie-line length reflects the crystal chemistry of the pyroxenes, perhaps the distribution of Fe, Mg, and Ca in the octahedral sites.

Pyroxenes that have compositions close to $FeSiO_3$ (orthoferrosilite) do not occur at low pressure in nature; the most iron-rich *quadrilateral* orthopyroxenes that occur have Fe/(Mg+Fe) $\sim$ 85%. Orthopyroxenes with Fe/(Mg+Fe) > 85% do occur, but they are characteristically slightly enriched in the non-quadrilateral manganese component, $Mn_2Si_2O_6$; sometimes zinc is also found (see Jaffe *et al.*, 1975, analyses of orthopyroxene Po-13, Sc-6, and Po-17 in Table 2b).

Quadrilateral pigeonite does not occur in metamorphic rocks. However, two reports of ferroan calcium-poor manganiferous clinopyroxene have been made: the uninverted manganiferous pigeonite ($Wo_{13.8}En_{10.7}$ $Fs_{75.5}$ with 13.0% MnO) from the Biwabik Iron Formation (Bonnichsen, 1969) and the clinoeulite ($Wo_{2.6-3.7}En_{19.5-19.9}Fs_{76.4-77.9}$ with 2.2% MnO) of Schreyer *et al.* (1978). In the very manganiferous pigeonite from the Biwabik Iron Formation, the closely associated manganese-poor pigeonites that were present at peak metamorphic temperatures "inverted" on cooling to orthopyroxene + augite ($\pm$ olivine $\pm$ silica), suggesting that Mn stabilizes pigeonite.

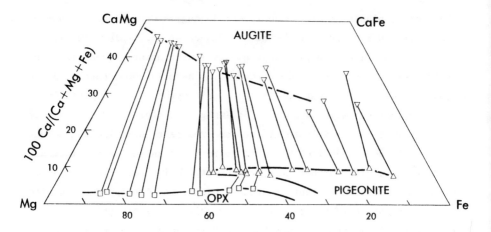

Figure 4. Generalized solidus diagram for quadrilateral pyroxenes based upon assemblages of coexisting pyroxenes that are inferred to have precipitated from a melt. Sources of data: Atkins (1969), Anderson & Wright (1972), Wheeler (1965), Nobugai et al. (1978), Buchanan (1979), Nakamura & Kushiro (1970), Philpotts (1966), and Bohlen & Essene (1978).

Solidus. Crystallization of basic magmas commonly results in orthopyroxene, pigeonite, and/or augite. The composition ranges of magmatic pyroxenes from volcanic and igneous rocks define a generalized "solidus" diagram for quadrilateral pyroxenes (Fig. 4). The "solidus" outlined in this manner is not a phase diagram because it is composed of small parts of the solidus of each of many compositionally unrelated, fractionating systems. However, there are sufficient data on natural magmatic assemblages of pyroxenes to suggest how the solidus varies with melt (or bulk) Ca, Fe, and Mg composition.

The orthopyroxene-augite miscibility gap on the solidus for basaltic magmas extends to compositions no more iron rich than $Fe/(Mg+Fe) \sim 45\%$. Representative values are 27% for the basic magma of the Bushveld intrusion (Atkins, 1969); 28-36% for the intrusions summarized by Campbell and Nolan (1974); 25% and 36% in Kilauea (Hawaii) basalts analyzed by Anderson and Wright (1972); and 40-45% for the Morin series of the Grenville province, Quebec (Philpotts, 1966). Orthopyroxene-augite assemblages that have $Fe/(Mg+Fe)$ values <76% are present in andesites (Kuno, 1966, his Fig. 5; Nakamura and Kushiro, 1970), and presumably crystallized at lower temperatures than the pyroxenes from the basalts. Compared with the sub-solidus, the "solidus" tie lines between the orthopyroxene and augite are short, extending from Wo $\sim$ 4% to Wo $\sim$ 35-45%, depending on the $Fe/(Mg+Fe)$ value.

At values of Fe/(Mg+Fe) greater than those of the orthopyroxene-augite field, pigeonite, not orthopyroxene, is the calcium-poor pyroxene that coexists with augite. The transition from orthopyroxene to pigeonite with fractionation is recorded by stratiform intrusions such as the Bushveld, Skaergaard, and Dufek, and the Palisades sill (Atkins, 1969; Brown and Vincent, 1963; Himmelberg and Ford, 1976; Walker *et al.*, 1973, respectively). Where the pigeonite became unstable on slow cooling and "inverted" (see discussion, next section), the presence of pigeonite at magmatic conditions is inferred using lines of evidence originally elucidated by Hess (1941, p. 580-582) and summarized by Brown (1972) and by Hess (1960).

Pigeonite is preserved (quenched) in more quickly cooled volcanic rocks, and it is in such volcanics that the pigeonite-augite assemblage must be sought. The two pyroxenes occur together in many volcanic rocks, but evidence that the pyroxenes precipitated from a melt while in equilibrium is elusive. For example, the coexisting pigeonite and augite in lunar basalt 12021, analyzed in such detail by Boyd and Smith (1971) and Ross *et al.* (1973), and in other lunar basalts, probably formed sequentially rather than simultaneously. Perhaps the best evidence for the equilibrium association of pigeonite and augite was found in the Weiselberg andesite studied by Nakamura and Kushiro (1970), who used the electron microprobe, and in the Hakone and Asio andesites studied by Kuno (1966, 1969), using optical microscopy and wet chemical methods. In these andesites, the pyroxene miscibility gap extends from Wo $\sim$ 10% to $\sim$ 38%. The calcium-rich and calcium-poor pyroxene trends found in individual lunar basalt samples suggest that in some cases the pigeonite-augite miscibility gap might be narrower (i.e., Wo = 15% and 35% in basalt 15499 in Bence and Papike, 1972; pigeonite having composition $Wo_{15}En_{55}Fs_{30}$ in basalt 12021, Ross *et al.*, 1973, p. 623), but in these cases there are the possibilities, respectively, that a fine two-pyroxene intergrowth was analyzed or that crystal-melt equilibrium was not achieved.

There is no conclusive evidence that in natural systems the pigeonite-augite miscibility gap closes with iron enrichment, a possibility raised originally by Hess (1941, p. 588) and suggested by some experiments at high pressure (Hensen, 1973, p. 528-529; Mori, 1978; Grover and Lindsley, 1972). The most suggestive evidence that the gap might close under some conditions in nature is the range of clinopyroxene compositions that

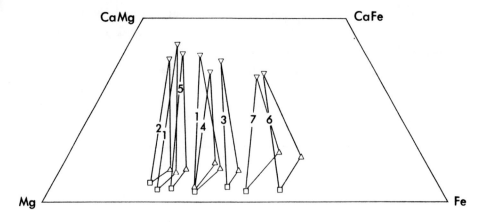

Figure 5. The range of naturally occurring three-pyroxene regions that coexist with silicate melt. Compositions of coexisting orthopyroxene, pigeonite, and augite are projected onto the pyroxene quadrilateral. Symbols are the same as before. Sources of data: (1) Bushveld; von Gruenwalt & Weber-Diefenbach (1977). (2) Jimberlana; Campbell & Borley (1974). (3) Weiselberg; Nakamura & Kushiro (1970). (4) Palisades; Walker *et al.* (1973). (5) Skaergaard (inferred); Brown & Vincent (1963). (6) Okubo-Yama; Kuno (1966). (7) Highashi-Yama; (cited by Kuno, 1966).

precipitated from fractionated mare basalt melts (see Bence and Papike, 1972; Ross *et al.*, 1973, their Fig. 3). However, such pyroxene compositions also can be explained by rapid and metastable crystallization (Papike *et al.*, 1976, p. 507) or by the possibility that finely intergrown pyroxenes were analyzed. (The iron-rich residual melts in the mare basalts sometimes precipitate the pyroxenoid pyroxferroite having composition Wo $\sim$ 15% (Boyd and Smith, 1971). This mineral does not appear miscible with pyroxene; when pyroxferroite occurs, the pyroxene and pyroxenoid form distinct crystals and have distinct compositions.)

The three-pyroxene region. The equilibrium of orthopyroxene, pigeonite, and augite, whether deduced from coexisting phenocrysts (Nakamura and Kushiro, 1970) or inferred from the orthopyroxene + augite to pigeonite + augite succession in fractionating systems such as the Bushveld, can occur at any Fe/(Mg+Fe) value within a wide range (Fig. 5). Hess (1941, p. 581-583) first recognized that the position of this orthopyroxene to pigeonite transition could serve as a geologic thermometer for magmatic crystallization. In magmatic systems, this transition is now recognized as a reaction involving *three* pyroxenes and melt:

orthopyroxene + melt $\rightarrow$ pigeonite + augite

(Huebner and Ross, 1972). The observed large range of Fe/(Mg+Fe) values for different occurrences of this magmatic three-pyroxene transition must represent a considerable range in temperature (Ross and Huebner, 1975).

The presence of a magmatic three-pyroxene field suggests the possibility that orthopyroxene + pigeonite can coexist without augite (see Virgo and Ross (1973) on the Mull andesite). An augite-absent region is necessary if the pigeonite field is to be as wide as is implied by the scatter in the Wo contents of magmatic pigeonite. For example, the data of Bence and Papike (1972) on the Apollo 15 basalts suggest that the pigeonite field might span a range of at least 5-10% Wo.

Forbidden zone. The pigeonite-augite miscibility gap in natural systems does not extend to the magnesium-free side of the quadrilateral but terminates at about Fe/(Mg+Fe) = 87%. (The most iron-rich pair known was reported by Wheeler (1965) and consists of "ferriferous pigeonite" (estimated composition $Wo_{10}En_{15}Fs_{75}$) and augite of $Wo_{36}En_8Fs_{56}$ composition.) More iron-rich pigeonites do not occur on the solidus; the augites are never richer in iron and approach the magnesium-free join only by changing their composition toward $CaFeSi_2O_6$, at approximately constant iron content. Hess (1941, p. 587) referred to the tie line of the limiting pyroxene *pair* as the two-pyroxene boundary, which for mafic magmas extrapolates to approximately $En_{40}Fs_{60}$ and to $Wo_{75}En_{25}$. Clearly, the position of this boundary is characteristic of the mafic magmas considered by Hess, rather than of pyroxenes themselves. Subsequently, Lindsley and Munoz (1969) coined the term "forbidden zone" to describe that part of the quadrilateral in which natural pyroxene compositions known at that time did not fall. Lindsley and Munoz used this term in reference to quadrilateral pyroxenes that crystallize from mafic magmas; we now know that possible exceptions can be caused by manganiferous bulk compositions and very high pressures.

Changes with decreasing temperature. Natural assemblages reveal that pyroxenes undergo two kinds of changes on cooling. The first kind of change is caused by intracrystalline exsolution, or intercrystalline exchange of Ca, Mg, and Fe, whereby the quadrilateral tie line between calcium-rich and calcium-poor pyroxene increases in length (and may rotate slightly). The second process is the "inversion" of pigeonite, whereby pigeonite transforms to orthopyroxene + augite. These processes

223

commonly operate together.

Natural pyroxene assemblages support the concept that tie-lines between pairs of pyroxenes become longer as temperature falls. According to this model or concept (Boyd and Schairer, 1964; Ross and Huebner, 1975; Lindsley and Dixon, 1976), based largely upon experimental data, the Wo contents of solidus (high temperature) and subsolidus (low temperature) orthopyroxenes are very similar; the lengthening of the tie line is accomplished largely by increasing the value of Wo in augite by 5-10% in the model (see Fig. 7 of Bohlen and Essene, 1978). This increase in tie-line length with decreasing temperature is most clearly shown in suites of pyroxene-bearing rocks that record both a high- and a low-temperature thermal history, such as the basic granulites of Broken Hill, New South Wales (Binns, 1962, his Fig. 5) and the anorthosites and mangerites of the Morin series, Quebec (Philpotts, 1966). Similarly, the increase in the pigeonite-augite tie line is shown by pyroxenes in the lunar basalts where $\Delta\beta$, the difference between the crystallographic β angles of the pigeonite and augite, has been equated with difference in Ca content (tie-line length) and cooling rate (Papike et $al.$, 1972; Takeda et $al.$, 1975). The slower the cooling rate, the better the chance for the pyroxenes to approach low-temperature equilibrium, and hence the longer the tie line. An increase in the pigeonite-augite tie-line length is also implied by the measurements of Schreyer et $al.$ (1978), who described a ferropigeonite that exsolved to clinoeulite (structurally a Ca-poor pigeonite) and ferroaugite.

That pigeonite is not found in slowly cooled rocks (plutons, granulites) is evidence that it is not stable at low temperature. Pigeonite is inferred to have crystallized in such rocks, but is now represented by bronzite (Philpotts and Gray, 1974) and by intergrowths of orthopyroxene + augite (see Hess, 1941, p. 580-581) or of orthopyroxene + augite + olivine $\pm$ quartz (see Bonnichsen, 1969). Hess (1960) concluded that orthopyroxene + augite intergrowths with a bulk CaO content of approximately 9.5% by weight originally crystallized as pigeonite that exsolved augite, then "inverted" isochemically to orthorhombic pyroxene; he termed such pyroxenes as "Stillwater type." Ishii and Takeda (1974) distinguished a second kind, Kintoki-san type, which they suggest is a decomposed rather than an inverted pigeonite. Huebner et $al.$ (1975) and Ishii and Takeda (1974) outlined possible paths by which pigeonite inverts or

decomposes; the initial pigeonite composition and the degree to which
equilibrium is maintained on cooling appear to influence the nature of
the pyroxene product. No matter what the path, the net result is an
increased length of tie line between augite- and calcium-poor pyroxene.
The increased separation in Ca-content takes place whether the transformation of pigeonite occurs at high temperature (close to the presumed
equilibrium temperature), as in pyroxenes from Kintoki-san (Ishii and
Takeda, 1974) or at low temperatures (below the presumed equilibrium
temperature), as in the lunar cataclastic norite 77215 (Huebner et $al.$,
1975) and the classic pyroxenes of the Stillwater Complex (Hess, 1960).
In many cases the distinction between high- and low-temperature pigeonite
transformation has not been made, but the lengthening of the tie line is
still apparent: the Bushveld (Boyd and Brown, 1969; Buchanan, 1979),
the Dufek (Himmelberg and Ford, 1976), and the Skaergaard pyroxenes
(Coleman, 1978; Nobugai et $al.$, 1978).

The mechanisms by which pigeonite might transform to orthopyroxene
$\pm$ augite are modeled after the metallurgical literature and reviewed
by Buseck et $al.$ (this volume). Champness and Copley (1976) concluded
that under some cases, the absence of a crystallographic orientation
relationship between reactant (high temperature) and product (low-temperature phase) may be caused by a "massive" transformation. Other mechanisms are theoretically possible and may help explain the diverse textures associated with pigeonite transformations (see Huebner et $al.$,
1975, Fig. 1). Possible mechanisms are discussed in the companion
chapter by Buseck et $al.$ (this volume).

There is evidence for two more transformations involving quadrilateral pyroxenes. "Ferrowollastonite" (now recognized to be the ferrobustamite of Rapaport and Burnham (1973), rather than the β ferrowollastonite of earlier investigators) occurred as a late-stage fractionation
product in the Skaergaard intrusion (Brown and Vincent, 1963), but subsequently transformed to augite. Lindsley et $al.$ (1969) proposed three
possible reactions whereby the pyroxenoid could transform to augite, but
they could not determine which reaction actually took place in the Skaergaard:

$$\text{ferrobustamite} \rightarrow \text{augite} \tag{1}$$

$$\text{ferrobustamite} \rightarrow \text{augite} + \text{fayalite} + \text{quartz} \tag{2}$$

$$\text{ferrobustamite} \rightarrow \text{augite} + \text{FeSiO}_3\text{-rich pyroxene} \tag{3}.$$

225

Nwe and Copley (1975) re-examined these Skaergaard pyroxenes using the electron microprobe and transmission electron microscope; they concluded that reactions (1) and (3) operated to produce, after subsequent exsolution of Ca-poor pigeonite, augites of composition Wo_{42} and Wo_{46}, respectively.

The second transformation is that of protopyroxene to form pigeonite or orthopyroxene on cooling. Protopyroxene is not found in nature, but its presence has been inferred from the diagnostic fractures and polysynthetic twin lamellae that occur in pigeonites from the magnesian andesites of Papua (Dallwitz *et al.*, 1966; Nakamura, 1971), the boninites of Japan (Shiraki *et al.*, 1980), and the Steinbach meteorite (Reid *et al.*, 1974). Dodd *et al.* (1975) suggest that a low-calcium orthopyroxene in the Shaw chondrite formed from protopyroxene on slow cooling; the evidence for slow cooling is the absence of structural stacking disorder. If correct, it is possible that other disordered Ca-poor orthopyroxenes in meteorites, such as those described by Pollack and Ruble (1964), were originally protopyroxene.

Rotation of tie lines. In addition to changes in length, the tie lines between calcium-poor pyroxene and augite are free to rotate with temperature as Fe and Mg reapportion themselves between the two pyroxene phases. The Fe-Mg partitioning between phases can be visualized by extrapolating the tie line from augite to the iron-free join of the quadrilateral (see Hess, 1941, p. 585 and Fig. 11). For a given bulk pyroxene composition, the point at which the tie line intersects the iron-free join may change with temperature, moving toward the Wo apex as temperature increases (Bartholomé, 1962). The corresponding exchange reaction for pyroxenes

$$Fe_{Ca-poor} + Mg_{Ca-rich} \rightleftarrows Fe_{Ca-rich} + Mg_{Ca-poor}$$

has been used as a geothermometer (see Bartholomé, 1962; Kretz, 1961; Philpotts, 1966; Davidson, 1968) in the sense that the equilibrium constant for the exchange reaction,

$$K_D = (\frac{X_{Fe}}{X_{Mg}})_{Ca-poor} (\frac{X_{Mg}}{X_{Fe}})_{Ca-rich}$$

varies as a function of temperature (and pressure). Values of $\leqslant 1.4$ are regarded as typically igneous, whereas values of $\geqslant 1.7$ correlate with

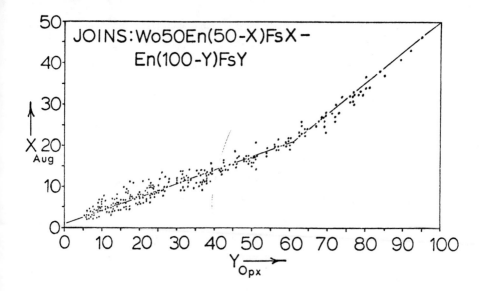

Figure 6. Fe-Mg partitioning between Ca-poor pyroxene and Ca-rich pyroxene. Data were normalized to exclude effects of variable Ca-content by extending Ca-Mg-Fe tie lines to the Wo = 0% and Wo = 50% joins. Points of tie-line intersection with these joins are plotted as X (in calcium-rich pyroxene, $Wo_{50}En_{50-x}Fs_x$) and Y (in calcium-poor pyroxene, $En_{100-y}Fs_y$). Coexisting pyroxenes from a variety of environments would yield trends that are indistinguishable from that of all pyroxenes together. The observed scattered cannot be attributed to different values of temperature and pressure and must be due to other causes. The break in slope is attributed to a change in degree of Fe-Mg partitioning when the orthopyroxene M(2) site has accepted as much Fe as it can tolerate, forcing further Fe to enter the M(1) site. (From unpublished work of Malcolm Ross, U.S. Geological Survey).

metamorphic temperatures. Binns (1962) noted that K_D was not simply a function of temperature and pressure (and explored the effect of composition).

It is important to distinguish true tie-line rotation from the apparent rotation calculated from the Fe/Mg ratios of coexisting pyroxenes. Bartholomé (1962, p. 7) first noted the correlation between the value of K_D and the Wo content of pyroxenes: Large values of K_D were associated with a wide miscibility gap. In order to isolate tie-line rotation from changes in Wo content, M. Ross (pers. comm., 1980) normalized the Wo contents of the Ca-poor and Ca-rich pyroxenes by extrapolating numerous quadrilateral tie lines to the En-Fs and Di-Hd joins. He found that when the points of intersection were plotted against each other (Fig. 6), the data could be expressed as two straight lines. The data base includes pyroxenes from both low- and high-temperature and low- and high-pressure environments. Pyroxenes from these different environments could not be distinguished on the basis of either the slopes or the positions of their trends. For this reason the observed scatter shown in Figure 6 cannot be due largely to changing temperature (or pressure). Other possible causes of the scatter are disequilibrium, analytical uncertainties, the assumption that $\Sigma Fe = Fe^{2+}$, and the presence of minor components, especially $Al_2Al_2O_6$, that might affect the activities of Fe and Mg pyroxene components. Clearly, the true tie-line rotation with change in temperature and pressure is not great, and its magnitude will be difficult to measure in natural assemblages.

The change in the partitioning constant with Fe content, shown so clearly in Figure 6, reflects the partitioning of iron in the M(2) and M(1) sites of orthopyroxene. As Fe is added to magnesian pyroxenes, the Fe preferentially enters the orthopyroxene. At Fs $\simeq$ 50 in the orthopyroxene, incremental additions of Fe are partitioned into the augite. Such behavior can be explained by reference to crystal chemistry. The augite M(2) site is filled or nearly filled with Ca, so Fe-Mg exchange must occur largely in the M(1) site in augite. Both orthopyroxene sites are available, but it is well known that at metamorphic temperatures Fe preferentially partitions into the M(2) site (Virgo and Hafner, 1969; Saxena and Ghose, 1971; Evans *et al.*, 1978). It would appear that as the pyroxene bulk composition becomes more iron rich, exchange of Fe for Mg in pyroxene pairs occurs first in $M(2)_{opx}$, then $M(1)_{aug}$. The site preference for Fe (over Mg) is $M(2)_{opx} > M(1)_{aug} > M(1)_{opx}$.

Minor- and trace-element partitioning. There have been surprisingly few attempts to evaluate or systematize the abundant available data on minor-element partitioning between the quadrilateral pyroxenes orthopyroxene and augite. For pigeonite, few data (for other than manganese) are available for occurrences that might represent equilibrium assemblages. Data in the literature are of two kinds: chemical analyses of coexisting pyroxenes and compilations of data. Some unevaluated data are summarized in Table 1 as values of D = (wt % oxide or element in the Ca-rich or monoclinic pyroxene)/(wt % in Ca-poor or orthorhombic pyroxene).

The analytical data are sufficient to generalize only about the augite-orthopyroxene pair. Divalent cations (Mn, Co, Zn, and perhaps Ni) favor the orthopyroxene whereas cations that have other valences (Na, Al,

Table 1. Minor and Trace Element Partitioning Data for Pyroxenes. Wt.% ratios

	$\frac{Aug}{D_{Opx}}$	$\frac{Aug}{D_{Pig}}$	$\frac{Pig}{D_{Opx}}$	Analyses or Compilation	Geology	Reference
Li	6.8			A6	charnockites, India	10
Na$_2$O	8.8			A9	Jimberlana, Australia	4
			0.89	A9	Bushveld, S. Africa	24
	2.7			A1	granulites, Australia	6
	4.2			A9	Grosse Pile ultramafics, Australia	18
	3.1			A6	amphibolites, New York (USA)	7
	10.5			A7	metamorphic rocks	13
	2.8			A3	iron formation, Canada	12
	6.6			A6	granulites, New South Wales	2
	8.3			A6	charnockites	10
	4.6			A3	gabbro sill	25
		2.6		A4	"	25
			1.8	A1	"	25
	10.4			A11	metamorphic rocks, New York	19
Al$_2$O$_3$	1.8				Jimberlana, Australia	4
			0.96	A9	Bushveld, S. Africa	24
	1.5			A12	granulites, Australia	6
	1.3			A11	Grosse Pile ultramafics, Australia	18
	1.3			A6	amphibolites, New York	7
	2.1			A13	metamorphic rocks	13
	1.6			A15	volcanics, Tonga	8
	2.9			A6	charnockite, Finland	22
	1.6			A9	various	15
	1.6			A6	mafic granulites	26
			0.5	A1	meteorite	20
		3.9		A2	meta iron formation	23
	2.1			A6	granulites, New South Wales	2
	1.4			A6	charnockites, India	10
	3.8			A3	gabbro sill	25
		1.7		A4	"	25
			3.7	A1	"	25
	1.7			A3	ultrabasic rocks, India	16
	1.6			A10	metamorphic rocks, New York	19
Sc	2.0			A15	basaltic andesites, Tonga	8
	3.9			A6	amphibolite, New York	7
	13			A1	charnockites, India	10
	4.9			A1	metamorphic rock, New York	19
TiO$_2$	2.0			A6	lunar norite	11
	1.9			A6	mafic granulites, Australia	26
	3.3			A10	Jimberlana, Australia	4
			1.2	A9	Bushveld, S. Africa	24
	0.97			A12	granulites, Australia	6
	3.0			A11	Grosse Pile ultramafics, Australia	18
	1.24			A4	amphibolites, New York	7
	2.26			A11	metamorphic rocks	13
	1.30			A15	volcanics, Tonga	8
	1.4			A5	charnockite, Finland	22
	1.3			A6	granulites, charnockites	15
	2.7			A3	ultramafic rocks	15
			1.0	A1	meteorite	20
	1.9			A3	gabbro, Norway	5
	1.2			A6	granulites	2
	3.2			A6	charnockites, India	10
	3.0			A3	gabbro sill	25
		1.7		A4	"	25
			3.2	A1	"	25

229

Table 1 (continued)

	Aug/D_{Opx}	Aug/D_{Pig}	Pig/D_{Opx}	Analyses or Compilation	Geology	Reference
TiO$_2$	2.6			A3	ultrabasic rocks, India	16
	1.41			A10	metamorphic rocks, New York	19
V	2.8			A6	amphibolites, New York	7
	1.9			A15	volcanics, Tonga	8
	2.4			A6	granulites & charnockites	15
	3.1			A3	ultramafic rocks	15
	2.0			A3	gabbro, Norway	5
	3.1			A6	charnockites, India	10
	3.3			A8	metamorphic rocks	19
Cr$_2$O$_3$		2.0		C3	lunar rocks	11
			0.61	A1	lunar poikiloblastic breccia	1
	2.6			C7	terrestrial rocks	11
	1.6				lunar rocks	11
	1.7			A8	Jimberlana, Australia	
	1.4			A11	Grosse Pile ultramafics, Australia	4
	3.9				metamorphic rocks	18
			1.0	A1	meteorite	13
	2.9$_5$			A3	ultrabasic rocks, India	20
	2.4			A6	amphibolites, New York	16
	1.7			A7	volcanics, Tonga	7
	2.1			A3	granulites	8
	2.0			A3	andesites	15
	1.7			A2	gabbro, Norway	15
	2.8			A5	charnockites, India	5
	2.0			A9	metamorphic rocks, New York	10 19
MnO	0.56			A1	lunar norite 77215	
		0.95		A1	Steinbach meteorite	11
	0.94			C	Mn/(Mg+Fe+Mn) <0.008	20
	0.73			C	Mn/(Mg+Fe+Mn) >0.008	14
		0.5		A2	meta iron formation	14
	0.48			A6	mafic granulites, Australia	23
	0.42			A11	metamorphic rocks, New York	26
	0.54			A3	gabbros, Norway	19
	0.47			A3	ultrabasic rocks, India	5
	0.60			A9	various	16
	0.47			A2	eclogites, Norway	
	0.42			A7	granulites, New South Wales	17
	0.49			A10	iron formation, Montana	2
	0.48			C	metamorphic rocks	12
	0.44			A6	amphibolites, New York	21
	0.70			A10	Jimberlana, Australia	7
			1.6	A9	Bushveld, S. Africa	4
	0.49			A12	granulites, Australia	24
	0.44			A26	meta iron formation, Canada	6
	0.68			A11	Grosse Pile ultramafics, Australia	3
	0.54			A13	metamorphic rocks	18
	0.69			A15	metamorphic rocks, Tonga	13
	0.46			A6	charnockite, Finland	8
	0.60			A9	various	22
		0.9		A1	meteorite	15
	0.47			A2	eclogites	20
	0.42			A7	granulites, New South Wales	17
	0.43			A6	charnockites, India	2
	0.69			A3	gabbro sill	10 25
		0.90		A4	"	25
			0.72	A1	"	25

230

Table 1 (continued)

	$\dfrac{Aug}{D_{Opx}}$	$\dfrac{Aug}{D_{Pig}}$	$\dfrac{Pig}{D_{Opx}}$	Analyses or Compilation	Geology	Reference
Co	0.62			A6	amphibolites, New York	7
	0.53			A15	volcanics, Tonga	8
	0.46			A9	various	15
	0.44			A2	eclogites	17
	0.82			A3	gabbro	5
	0.70			A6	charnockites, India	10
	0.43			A11	metamorphic rock, New York	19
Ni	0.71-1.39			A10	Parikkala, Finland	9
	0.96			A6	amphibolites, New York	7
	0.69			A15	metamorphic rocks	8
	0.58			A9	various	15
	0.60			A2	eclogites	17
	1.0			A3	gabbros	5
	1.0			A6	charnockites, India	10
	0.68			A10	metamorphic rocks, New York	19
	0.78			A10	Grosse Pile ultramafics, Australia	18
	2.2			A4	metamorphic rocks	13
Cu	0.72			A6	amphibolites, New York	7
	0.90			A14	metamorphic rocks, Tonga	8
	1.3			A9	various	15
	37			A2	eclogites	17
	1.1			A2	metamorphic rocks, New York	19
Zn	<0.54			A6	amphibolites, New York	7
	0.60			A3	metamorphic rocks	13
	0.32			A6	granulites & charnockites	15
	0.7			A3	ultramafic rocks	15
	0.30			A2	eclogites	17
Ga	0.75			A6	amphibolites, New York	7
	0.76			A5	charnockites, India	10
	1.0			A11	metamorphic rocks, New York	19
Sr	>1.3			A6	amphibolites, New York	7
	4.0			A1	metamorphic rock, New York	19
Y	>1.8			A6	amphibolites, New York	7
	6.8			A1	metamorphic rock, New York	19
Zr	2.05			A6	amphibolites, New York	7
	1.0			A1	charnockites, India	10
	1.0			A10	metamorphic rocks, New York	19
Mo	0.43			A2	charnockites, India	10
Ba	1.0			A6	amphibolites, New York	7
	1.0			A1	metamorphic rock, New York	19
Yb	>0.6			A6	amphibolites, New York	7

References for Table 1

1. Bence et al. (1973)
2. Binns (1962)
3. Butler (1969)
4. Campbell and Borley (1974)
5. Carstens (1958)
6. Davidson (1968)
7. Engel et al. (1964)
8. Ewart et al. (1973)
9. Hakli (1968)
10. Howie (1955)
11. Huebner et al. (1976)
12. Immega and Klein (1976)
13. Jaffe et al. (1975)
14. Lindh (1974)
15. Liotard and Dupuy (1980)
16. Mall (1973)
17. Matsui et al. (1966)
18. Moore (1971)
19. Philpotts (1966)
20. Reid et al. (1974)
21. Saxena (1968)
22. Saxena (1969)
23. Schreyer et al. (1978)
24. von Gruenewalt and Weber-Diefenbach (1977)
25. Walker et al. (1973)
26. Wilson (1976)

Sc, Ti, V, and Cr^{+3}) favor an augite host. Ideal thermodynamic behavior of most minor divalent cations would be explained by the fact that orthopyroxene has two sites available for substitution (M(1) *and* M(2)), whereas the M(2) of augite is almost completely filled with large calcium cations, leaving only M(1) available for the smaller M^{2+} cations. On this ideal basis we would expect the divalent elements to have values of D equal to 0.5. The unweighted means (with standard deviations and standard errors) of the divalent elements shown in Table 1 are D_{Mn} = 0.55 (0.13, 0.03); D_{Co} = 0.60 (0.15, 0.06); D_{Zn} = 0.48 (0.20, 0.10); and D_{Ni} = 0.98 (0.52, 0.18). Thus, manganese, cobalt, and zinc appear to behave ideally and, if substituting in small quantities, would not be expected to affect greatly the orthopyroxene-augite phase relations.

The substitutions of minor and trace elements that have a valence other than 2 must be considered in terms of coupled substitutions that preserve charge balance by exchanging $[Al^{3+}]^{IV}$ for $[Si^{4+}]^{IV}$, $[Al^{3+}]^{VI}$ for $[M^{2+}]^{VI}$, and/or $[Na^{+}]^{VI}$ for $[M^{2+}]^{VI}$ in conjunction with another substitution for a $[M^{2+}]^{VI}$. Most such substitutions dilute the quadrilateral components (change the activities) of augite to a greater extent than those of orthopyroxene. It is likely that such minor-element substitutions (i.e., $NaAlSi_2O_6$, $M^{+2}AlAlSiO_6$, or $M^{+2}TiAl_2O_6$) will significantly affect the orthopyroxene-augite phase relations, particularly with respect to Al and Ti, which commonly occur at a level of several percent. If the foregoing is correct, petrologists must consider the minor elements present when using data on simple laboratory systems to model naturally-occurring pyroxenes.

Various authors have summarized particular aspects of the minor and trace-element partitioning between pyroxenes. For example, Wilson (1976) examined Na, Mg, Al, Ti, and Fe in coexisting pyroxenes from two mafic granulite localities and found that Al could be used to distinguish the metamorphic conditions. Huebner *et al.* (1976) summarized the Cr contents of coexisting lunar and terrestrial orthopyroxene and augite pairs. Overall, lunar pyroxenes have greater Cr contents, but the lunar D_{Cr} values are similar to terrestrial values despite the presumed difference in oxidation state ($Cr^{3+} \gg Cr^{2+}$ in terrestrial rocks; $Cr^{3+} < Cr^{2+}$ in lunar rocks). Lindh (1974) summarized the Mn partitioning in 115 nonalkaline pyroxenes; his values of D for orthopyroxene + augite pairs are larger (0.94 at low Mn concentration; 0.73 at higher concentration) than any values shown in Table 1. Liotard and Dupuy (1980) found that the divalent elements Fe^{2+}, Mn, Co, Ni, and Zn favored orthopyroxene whereas cations that have other values of valence (Ti, V, Cr^{+3}, and Cu) favored coexisting augite. The authors suggest that Ni and Co partitioning between the pyroxenes may be temperature dependent, whereas Cr and V partitioning may be a sensitive indicator of pressure. Finally, Fleet (1974) examined the partitioning of Na, Mg, Al, Si, Ca, Ti, Mn, Fe^{2+}, and Fe^{3+} between coexisting augite and Ca-poor pyroxenes from igneous and metamorphic rocks and found that the Na, Mg, Ti, Fe^{2+}, and Mn values were useful in distinguishing petrologic groups and conditions of formation. Most significant, he examined the effect of subsolidus cooling rate on partitioning, concluding that the degree of Mg, Mn, and Fe^{2+} partitioning *decreases* with increasing cooling rate.

A review of element partitioning between pyroxene and other minerals would be so voluminous that it would best be undertaken with a specific problem in mind. A thorough consideration is beyond the scope of this chapter because so little of the data have been compiled and analyzed with regard to attainment of equilibrium. The element partitioning most germane to pyroxene-phase relations involves minerals that have bulk chemistry similar to pyroxenes, because these are the minerals that so often participate in reactions that bound the pyroxene stability fields. For example, data on pyroxenes coexisting with amphiboles have been given by Saxena (1968, 1969), Engel *et al.* (1964), Neumann (1976), Howie (1955), Immega and Klein (1976), and Bonnichsen (1969). Lists of references could also be compiled for micas, olivine, and oxides using the extensive bibliographies in Deer *et al.* (1978).

Element partitioning between pyroxene and melt is extremely important for understanding the partial melting relations of pyroxenes, their crystallization from real magmas, and the compositional evolution of the residual melts or crystals. In view of the fundamental importance of such information for understanding igneous processes, it is surprising that the literature does not contain sufficient analyses of *coexisting* pyroxenes and melts to glean the major (Ca, Fe, Mg, Si) -element partitioning, not to mention the minor-element partitioning. Most of what we know is derived from compositional trends (of pyroxene and bulk rocks) and from laboratory studies of the last decade. A possible exception is some of the data of Anderson and Wright (1972), and even those data do not indicate clearly that the analyzed pyroxenes were actually in equilibrium with the analyzed melt. With regard to the minor elements, the pyroxenes clearly contain far less Al_2O_3, TiO_2, and Na_2O than the melts from which they crystallize, at least for low-pressure basaltic melts. Current studies on mid-ocean ridge basalts and samples from Hawaiian flows and lava lakes may provide the necessary data against which the laboratory results can be tested.

Experimental Studies on Pure and Natural Systems

Knowledge of pyroxene phase relations in the subsolidus has increased measurably since the MSA-sponsored "Short Course on Chain Silicates" in 1966. At that time Boyd (1966, p. 12) pointed out the need to understand the inversion relations of Ca-poor pyroxenes, especially pigeonite, and to determine better the hedenbergite solidus. Boyd also noted (p. 9) that the quadrilateral phase relations were "still most imperfectly known." In the following sections we will see that the efforts of the past 14 years have addressed these points and resulted in sufficient experimental data to establish the phase topology within the quadrilateral.

Subsolidus. Experimental petrologists have focused their attentions on the iron-free and magnesium-free joins of the pyroxene quadrilateral. The resulting studies have modeled with simple systems the temperature-composition relations for the orthopyroxene-augite and the augite-fayalite ± silica miscibility gaps suggested by the natural assemblages described earlier. The remaining problems appear to be largely concerned with the achievement of equilibrium in the laboratory and with the effects of minor components upon the simple model systems. The phase relations at intermediate values of Fe/(Mg+Fe) are not known with great certainty, but the topology appears sufficiently well understood to explain most natural occurrences. Much work is still needed to finish determining the temperature-composition relations of even the simplest Ca-, Mg-, and Fe-bearing pyroxene systems.

Phase relations near the bounding joins. The phase relations of the $Mg_2Si_2O_6$-$CaMgSi_2O_6$ join at one bar pressure are summarized in Figure 7.

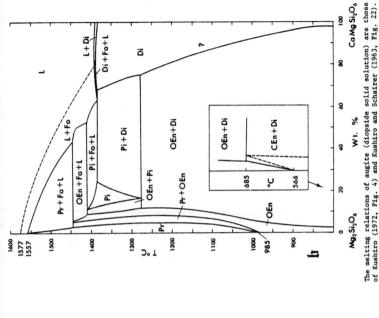

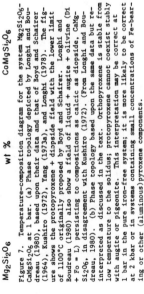

Figure 7. Temperature-composition diagram for the system Mg₂Si₂O₆-CaMgSi₂O₆ at 1 bar. (a) Phase topology depicted by Longhi and Boudreau (1980), based upon their data and that of Boyd and Schairer (1964), Kushiro (1972), Yang (1973a), and Longhi (1978). The figure shows the protopyroxene + diopside field with the lower limit of 1100°C originally proposed by Boyd and Schairer. Longhi and Boudreau (1980) also show a field of liquid + augite + olivine (Di + Fo = L) persisting to compositions as calcic as diopside, CaMgSi₂O₆, in disagreement with Kushiro (1972). (From Longhi and Boudreau, 1980). (b) Phase topology based upon the same data but reinterpreted as discussed in the text. Orthopyroxene is stable from low temperature to the solidus; protopyroxene cannot coexist stably with augite or pigeonite. This interpretation may be correct at 1 bar in the pure iron-free system; it is more likely to be correct at 2 kbar or in systems containing small concentrations of Fe-bearing or other (aluminous) pyroxene components.

The melting relations of augite (diopside solid solution) are those of Kushiro (1972, Fig. 4) and Kushiro and Schairer (1963, Fig. 22). The (?) in the augite field refers to the suggestion of Warner and Luth (1974) that there may be a compositional hiatus within the range of augite solid solution. The minimum temperature for pigeonite stability is placed at 1278 ± 10°C to agree with the inferred intersection of the limbs bounding the pigeonite field in the system anorthite–diopside–enstatite–silica (see microprobe analyses of Yang, 1973, Fig. 3), a reversed determination pigeonite ⇌ protoenstatite (?) + augite at 1288.5 ± 11.5°C (M. D. Feldman, personal communication, 1980), and an extrapolation at 1300°C and Fe/(Mg+Fe) = 0 of the results of Ross and Huebner (1979). The inset shows the stability of "low clinoenstatite" determined by Grover (1972).

235

(Note that compositions in this figure are given in weight percent of pyroxene end-member components $Mg_2Si_2O_6$ and $CaMgSi_2O_6$.) The orthopyroxene-augite solvus is taken from Boyd and Schairer (1964), whose study of this miscibility gap is still definitive. The augite limb of their proto-pyroxene-augite two-phase region also appears to have survived the test of time. The calcium-poor limb of this miscibility gap has been revised by the recognition of a field of iron-free pigeonite that separates two-phase regions of Ca-poor pyroxene + pigeonite and of pigeonite + augite (Yang and Foster, 1972; Kushiro, 1972). These investigators believed the Ca-poor pyroxene to be protopyroxene, Yang and Foster citing the ortho-rhombic morphology and the appearance of polysynthetic twinning in some quenched crystals, whereas Kushiro noted the low content of CaO (<0.8 wt %) and the characteristic cracks attributed to inverted protoenstatite. Yang (1973a) subsequently investigated Ca-poor pyroxenes in the system anorthite-diopside-enstatite-silica and described two "protoenstatites." The "protoenstatite" lacking the characteristic cracks but having CaO contents of 1.6-2.4% by weight was the only orthorhombic pyroxene found in runs for "extended periods" (p. 493). On the basis of an oral com-munication (1973) with Yang, Ross and Huebner (1979, p. 1149) proposed that this calcic orthorhombic pyroxene was actually orthopyroxene at higher temperatures than previously recognized for orthopyroxene sta-bility. Foster and Lin (1975) inserted an orthopyroxene field between those of protopyroxene and pigeonite, extending from the liquidus (1445°C) to temperatures at least as low as 1405°C. Recently, Longhi and Boudreau (1980) chose to extend the lower stability of orthopyroxene to 1375 $\pm$ 10°C, where it is terminated by decomposition to form protopyroxene + pigeonite. Their reason for choosing 1375°C is not clear; in view of the fact that orthopyroxene appears to persist to temperatures at least as low as 1245°C in the system anorthite-diopside-enstatite-silica (the temperature of the misidentified calcic "protoenstatites" of Yang, 1973a), the orthopyroxene field may persist stably to much lower temperatures (perhaps stabilized by the small amount of Al_2O_3 incorporated in the structure). Two lines of reasoning suggest that it is not inconceivable that at low pressure orthopyroxene is continuously stable from low tem-perature (<1000°C) to the solidus (Fig. 7b). (a) Protopyroxene does not appear to be stable in the presence of augite at pressures $\geqslant$2 kbars

236

(Warner and Luth, 1974; Warner, 1975; see also Mori and Green, 1975, their Fig. 4). (b) When heating natural magnesian orthopyroxenes *in the presence of augite*, Ross and Huebner (1979) did not detect evidence of protopyroxene (in a later section, we will see that they did detect protopyroxene when heating orthopyroxenes that have Wo < 1).

At high pressure for pure $MgSiO_3$ composition, there is a stability field for "low clinoenstatite" ($P2_1/c$ pigeonite) that lies at lower temperatures than orthopyroxene (Sclar *et al.*, 1964; Boyd and England, 1965). Extrapolated to 1 bar, the inversion lies at 550-625°C. Because pigeonite is not a low-temperature phase in nature, and because "clinoenstatite" was heretofore only known as an inversion product of protopyroxene, these results have been regarded with suspicion. Grover (1972) eliminated the possibility that the "low clinoenstatite" phase was produced only in nonisostatic environments by reversing its reaction to orthopyroxene in a hydrostatic environment (a hydrothermal solution of $MgCl_2$). At 1 bar the inversion reaction occurs at 566°C, in good agreement with the extrapolation of the results of Sclar *et al.* (1964) at high pressures. Grover (1972) also found that addition of $Ca_2Si_2O_6$ component to the pyroxenes raised the equilibrium markedly at 2 kbars pressure: from 575° at Wo = 0% to 685°C at Wo = 3.8%, the solubility limit at that pressure. The low field of pigeonite may be a stable or metastable continuation of the pigeonite field that lies at higher temperatures (see inset, Fig. 7b), but if stable it does not occur at the conditions of temperature, pressure, and/or bulk composition encountered in nature and in most laboratory investigations. (Use of the terms "high" and "low" in reference to pigeonite or clinoenstatite is ambiguous. They may refer to the high- and low-temperature structural states ($C2/c$ and $P2_1/c$, respectively), or they may refer to high- and low-temperature stability fields.)

Unlike the Ca-poor part of the system, the calcic portion of the system $Mg_2Si_2O_6$-$CaMgSi_2O_6$ does not show several discrete pyroxene phases that have narrow composition limits. However, Warner and Luth (1974, p. 104-105) noted that between 1100° and 1150°C at 2 kbar, there was a break in the temperature-composition slope of the augite-solubility limit (with respect to Ca-poor pyroxene) and peak splitting in the x-ray diffraction patterns of the augites. A transformation of the Ca-poor pyroxene, from orthopyroxene to protopyroxene, cannot account for the peak splitting.

The authors raised the possibilities of a miscibility gap or polymorphism to explain the observations but could provide no conclusive evidence for other than a single-phase region.

The temperature-composition diagram for the system $Mg_2Si_2O_6-Fe_2Si_2O_6$ was originally drawn by Bowen and Schairer (1935), who showed a narrow subsolidus inversion loop extending from 1145°C at $Mg_2Si_2O_6$ to 960° at Fe/(Mg+Fe) = ∿88%. Olivine + quartz was present at more iron-rich compositions. Orthopyroxene was not stable at temperatures above 1150°C and nowhere reached the solidus. Bowen and Schairer (1935) had no knowledge of protopyroxene and thus did not include it at $Mg_2Si_2O_6$ composition or show a protopyroxene field on the calcium-free join.

To appreciate fully the modified subsolidus phase diagram for the $Mg_2Si_2O_6-Fe_2Si_2O_6$ join (Fig. 8) it is necessary to understand the kinds of data and observations available to Bowen and Schairer. Data for the melting region were obtained for the pure system $MgO-"FeO"-SiO_2$. The quenched pyroxenes produced from experiments on the pure system were described as "clinopyroxenes"; we now know that they could have been either protopyroxene or pigeonite at temperature, but the records of observations do not permit a distinction. The six orthopyroxenes used to determine the subsolidus loop were natural orthopyroxenes that have compositions that do not strictly fall in the pure system. The addition of NaF flux to several samples further complicates the interpretation. Following the discussion of Ross and Huebner (1979, p. 1150-1151), it is probable that the extremely pure orthopyroxene from the Bishopville meteorite (and perhaps also the enstatite from Espedalen, Norway, Fe/(Mg+Fe) = 8%) transformed to protopyroxene on heating at 1145°C, then inverted to pigeonite on quenching. Two bronzites (Fe/(Mg+Fe) = 14%) were heated in flux at 1120°C; the monoclinic reaction product is probably fluoramphibole -- the CaO of one sample (1.55 wt %) is probably too great for the formation of protoenstatite at temperature, and the 1120°C temperature is too low for formation of pigeonite from orthopyroxene + augite with these Fe/(Mg+Fe) values (discussed later). Clinopyroxene was probably correctly identified in the products of two annealed iron-rich pyroxenes; the high CaO contents (1.5 and 1.05 wt %) suggest that augite may have been present as an exsolved phase, causing the observed transformation temperature to be less than in the calcium-free system. Thus it is reasonable to show a protopyroxene + orthopyroxene

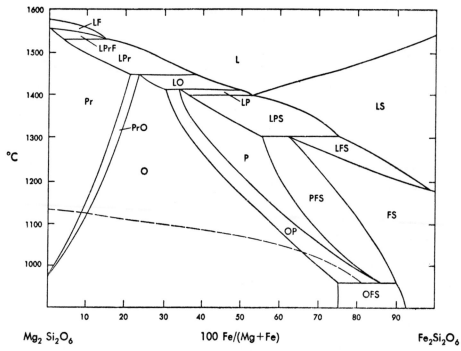

Figure 8. Temperature-composition relations from the pure system $Mg_2Si_2O_6-Fe_6Si_2O_6$. This figure is modified from that of Huebner and Turnock (1980) by the distinction of two orthorhombic pyroxenes, protopyroxene (Pr) and orthopyroxene. The dashed line represents the trend of the original orthopyroxene-clinopyroxene inversion of Bowen and Schairer (1935); reinterpretation of the nature of their run products and recent understandings of the effects of Ca-component on the system (see text) makes the earlier experiments consistent with this diagram.

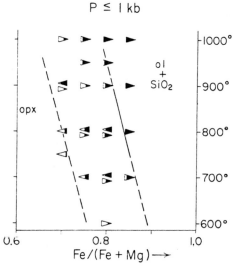

Figure 9. A part of the iron-rich end of the $Mg_2Si_2O_6-Fe_2Si_2O_6$ join showing the shift in composition of the three phase region orthopyroxene + olivine + quartz. The field boundaries are dashed. The axes are atomic proportion and °C. Triangles pointing to the right indicate that orthopyroxene decomposed to olivine + silica; triangles pointing to the left show that olivine + quartz reacted to produce orthopyroxene. The position of each symbol indicates the composition of a ferromagnesian reactant; the percentage of shading corresponds to the percentage of olivine + silica in the products. From Smith (1971).

239

field extending from 985°C at $Mg_2Si_2O_6$ composition to the solidus at Fe/(Mg+Fe) = 20-22% (Fig. 8). The trend of the original orthopyroxene-clinopyroxene loop of Bowen and Schairer (1935) crosses regions of protopyroxene, orthopyroxene and pigeonite in agreement with recent reinterpretations of their results.

At iron-rich compositions (Fe/(Mg+Fe) > 75%) orthopyroxene alone is not stable but is joined by olivine + quartz; ultimately olivine + quartz alone are stable (Fig. 8). The positions of the reactions bounding these fields at P < 1 kbar have been determined by Smith (1971) whose data are reproduced as Figure 9, and at higher temperatures by Nafziger and Muan (1967) and by Wood and Strens (1971). With increasing temperature, the respective field boundaries move toward more magnesian compositions; however, increasing pressure has a strong and opposite effect. At 950°C, the composition limit of orthopyroxene moves from Fe/(Mg+Fe) < 70% at < 1 kbar to 100% at 14 kbar (Smith, 1971). Clearly, pressure is an important variable and must be known before the composition limit of orthopyroxene (coexisting with olivine and quartz) can be used as a geothermometer.

In the foregoing discussion I have shown ample evidence that ortho-pyroxene is the *stable* calcium-poor metasilicate in many natural systems. However, the experimental evidence is not so compelling. We have al-ready discussed the stability of a "low" pigeonite of $Mg_2Si_2O_6$ compo-sition. In addition, it is difficult to synthesize orthopyroxene with Fe/(Mg+Fe) > 30% at low pressure. When attempting to synthesize Ca-free orthopyroxenes, Turnock *et al.* (1973) found that clinopyroxene (pigeonite) formed unless they used high pressures (10 kbar). Schwab and Schwerin (1975, p. 232) encountered similar difficulties. Orthopyroxenes formed at high pressure have cell dimensions that vary regularly over the Fe/ (Mg+Fe) range of 0-70% at standard temperature and pressure. Orthopy-roxene saturated with respect to augite has been produced by Simmons *et al.* (1974, p. 557) at 750-800°C and 2-4 kbar, and by Turnock and Lindsley (pers. comm., 1980) at 900° and 1000°C and P < 1 kbar, but in these cases the orthopyroxene is the product of a carefully "reversed" reaction. When crystallizing a high-energy reactant such as an oxide mix, clino-pyroxene grows and tends to persist if Fe/(Mg+Fe) > 30%.

The situation is further confused by experimental investigations of very magnesium-rich pyroxenes summarized recently by Ross and

240

Huebner (1979). Riecker and Rooney (1967) and Coe (1970) found that orthopyroxene transforms easily to pigeonite at low temperature in an environment in which shearing stress was present. Raleigh *et al.* (1971) concluded that under hydrostatic (isostatic) conditions, orthopyroxene was stable relative to pigeonite. Subsequent results seem to indicate the opposite conclusion. Grover (1972) reversed the reaction isostatically (and at P < 1 bar), finding that pigeonite is stable relative to orthopyroxene at temperatures below 556°C. Similarly, Coe and Kirby (1975) concluded that non-isostatic (shearing) stress favored formation of pigeonite by expanding its field and, at conditions where pigeonite is "stable," by increasing the rate at which orthopyroxene transforms to pigeonite. If pigeonite is truly stable at low temperature, it is puzzling that the mineral occurs so rarely in rocks that must have cooled slowly in a non-isostatic environment. The situation appears to be one in which the activation energy of a transformation is relatively large, relative to the free-energy difference between stable and metastable states.

A complete clinopyroxene solution series can be synthesized at 1 kbar along the $CaMgSi_2O_6$-$CaFeSi_2O_6$ join (Nolan, 1969; Rutstein and Yund, 1969). Cell dimensions vary regularly, giving no hint of the presence of more than one phase. Turnock (1962) first outlined the augite (hedenbergite)-ferrobustamite transition; he collected more data which were incorporated by Huebner and Turnock (1980), and subsequently further modified the join with recent subsolidus runs defining the augite solubility limit at 1000°C (Turnock and Lindsley, pers. comm., 1980), shown in Figure 10.

The magnesium-free join $Fe_2Si_2O_6$-$CaFeSi_2O_6$($-Ca_2Si_2O_6$) was originally investigated by Bowen *et al.* (1933) as a part of the system CaO-"FeO"-SiO_2. They emphasized the melting relations; the subsolidus was not closely defined. Ernst (1966) derived some data on the hedenbergite (Fe-augite) solubility limit from a study of the ferrotremolite stability limits, but the major revisions to the subsolidus are by Lindsley and Muñoz (1969) who carefully determined the temperature-composition relations of the augite + olivine + quartz and the ferrobustamite + olivine + quartz miscibility gaps (Fig. 11). That part of the join more calcic than Wo = 50% has been reinvestigated by Rutstein (1971) and Matsueda (1974). These calcic compositions are germane to the pyroxene quadrilateral

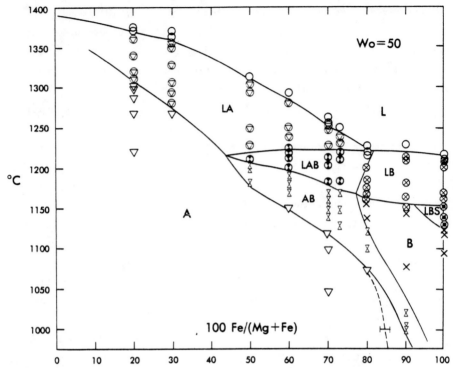

Figure 10. Temperature-composition section along the join CaMgSi₂O₆-CaFeSi₂O₆, showing the augite-ferrobustamite reaction interval. Placement of the field boundaries are self-explanatory except between LB and B, where the solidus is constrained by runs with compositions that do not lie in the plane of section, Wo = 50%. From Huebner and Turnock (1980) except for a revised augite solvus (dashed line and bracket) which is from Turnock and Lindsley (pers. comm.. 1980).

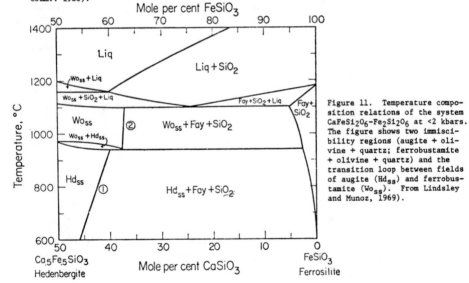

Figure 11. Temperature composition relations of the system CaFeSi₂O₆-Fe₂Si₂O₆ at <2 kbars. The figure shows two immiscibility regions (augite + olivine + quartz; ferrobustamite + olivine + quartz) and the transition loop between fields of augite (Hd$_{ss}$) and ferrobustamite (Wo$_{ss}$). From Lindsley and Munoz, 1969).

242

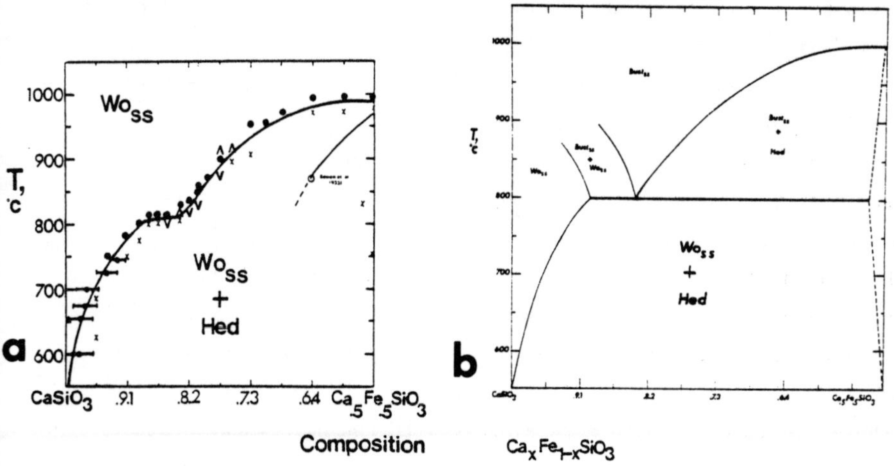
a
1000
900
800
700
600
T,
°C
Wo$_{ss}$
Wo$_{ss}$
+
Hed
CaSiO$_3$
.91
.82
.73
.64
Ca$_{.5}$Fe$_{.5}$SiO$_3$
Composition
b
Wo$_{ss}$
+
Hed
Ca$_x$Fe$_{1-x}$SiO$_3$

Figure 12. Subsolidus relations in the system $Ca_2Si_2O_6$-$CaFeSi_2O_6$. Although these compositions are not part of the pyroxene quadrilateral, tie lines between augite and ferrobustamite cross the Di-Hd join, terminating at Wo values exceeding 50%. (a) Data of Rutstein (1971) obtained at 1 kb and at f_{O_2} values defined by the FMQ buffer; complete reversals are indicated with V and Λ. (b) Preferred interpretation by Rutstein (1971) of the data in (a). Although the phase boundaries are schematic, this topology accounts for the inflected solvus shown in (a) and is compatible with Figure 11. (From Rutstein, 1971).

because many tie lines between augite and ferrobustamite (and between melt and metasilicate) cross the join $CaMgSi_2O_6$-$CaFeSi_2O_6$. Rutstein's interpretation of the subsolidus relations of the system $Ca_2Si_2O_6$-$CaFeSi_2O_6$ is reproduced in Figure 12 and shows a wide immiscibility field between hedenbergite ($CaFeSi_2O_6$) and wollastonite solid solution below 800°C; at higher temperatures hedenbergite is in equilibrium with a bustamite solid solution that approaches $CaFeSi_2O_6$ composition with increasing temperature. Shimazaki and Yamanaka (1973) on the basis of natural occurrences and Matsueda (1974) on the basis of natural occurrences plus heating experiments, propose a limited solution of "iron wollastonite" ($Ca_5FeSi_6O_{18}$ with the bustamite structure) at temperatures below 800°C. Shimazaki and Yamanaka (1973, their Fig. 2) suggested that ordinary wollastonite $(Ca_{.95}Fe_{.05})_2Si_2O_6$ can coexist with clinopyroxene of approximate composition $CaMg_{.15}Fe_{.85}Si_2O_6$, whereas the "iron wollastonite" ($Ca_5FeSi_6O_{18}$) coexists with clinopyroxene of approximate composition $CaMg_{.05}Fe_{.95}Si_2O_6$.

Phase relations within the quadrilateral. The principal features of the ternary subsolidus phase relations are suggested by relations found on the bounding joins: the Ca-poor pyroxene + augite miscibility volume, pigeonite and augite solvi, a surface at which the two volumes meet, and a region in which olivine + quartz is present instead of Ca-poor metasilicate.

The only successful investigation of the orthopyroxene + augite miscibility gap in the iron-bearing system at low pressure is that of Turnock and Lindsley (1981)[*] for pressure < 1 kbar in the pure system. At 900° and 1000°C, the most Fe-rich tie line is at $Fe/(Mg+Fe)$ ≃ 55% and 45%, respectively; the presence of pigeonite precludes an orthopyroxene + augite field at less magnesian compositions. At 900°C, orthopyroxene coexisting with augite increases from Wo = 3% to 4% as $Fe/(Mg+Fe)$ increases from 12% to 50%. Correspondingly, the augites range from Wo = 42% to 37%. It is interesting to note that the lengths of these tie lines compare more closely with the higher temperature solidus tie lines (Fig. 4) discussed earlier than with the subsolidus tie lines (Fig. 2). (A similar situation exists at 810°C and 15 kbar, conditions at which Lindsley *et al.* (1974) found tie lines that were much shorter than those of granulites.) Application of such laboratory data on pure systems to compositionally complex natural systems must be made with caution. It is possible that additional minor components such as $Al_2Al_2O_6$ may affect the pyroxene tie-line length by markedly decreasing the activity coefficient of the $Ca_2Si_2O_6$ component in augite.

The only comprehensive study of the pigeonite-augite solvus is that of Ross *et al.* (1973), who progressively homogenized submicroscopic intergrowths of lunar clinopyroxenes (pigeonite + augite). The crest of the solvus was not detected; instead, the grains homogenized or began to melt while there was still an appreciable difference in the Wo contents of the pigeonite and augite (Fig. 13). The solvus was measured over the range $Fe/(Mg+Fe)$ = 34% to 85% (Fig. 14) and can be described as forming a steep ridge whose crest is truncated by the solidus. Ross *et al.* determined the compositions of the intergrown pigeonite and augite by measuring the crystallographic β angles in x-ray precession photographs of the heated crystals. Although the accuracy of this method is not great, the precision is sufficient to show that whereas the solvus has a symmetric cross section at iron-rich compositions (Fig. 14c and d), it is asymmetric at more magnesium compositions with the pigeonite limb being steeper than the augite limb (Figs. 14a and b). The entire solvus is oriented so that with increasing $Fe/(Mg+Fe)$, corresponding parts of
_ _ _ _ _ _ _
[*]See Canadian Mineralogist 19, 255-267.

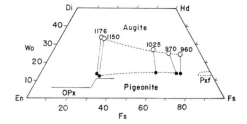

Figure 13. A section or slice through the pigeonite + augite miscibility gap at the solidus (temperatures of first melting) for exsolved pyroxenes from lunar basalt 12021. Compositions were determined by x-ray precession photography and microprobe analyses of zoned crystals and thus have a large uncertainty. (A simplified version of Ross *et al.*, 1973, Figure 10).

the solvus lie at progressively lower temperatures. The investigation did not reveal a "warp" in the augite limb, causing the tie lines to be shortest at intermediate Fe/(Mg+Fe), that might have been expected from the trends of coexisting pyroxenes in tholeiitic intrusions. (As we will see later, the augite minimum *can* be explained by the position of the solidus.)

Turnock and Lindsley (1981 - footnote p. 244) encountered the pigeonite-augite solvus in their subsolidus investigation of the pure system. Their solubility limits at 900 and 1000° lie inside the solvus of Ross *et al.* (1973), but it is not clear whether the difference can be attributed to failure to achieve equilibrium in heating the lunar pyroxenes, real differences in the natural and pure systems that are being compared, or analytical errors associated with the x-ray diffraction method used to determine composition. Schwab (1969) synthesized a series of compositions from $Mg_{1.5}Fe_{0.5}Si_2O_6$ to $CaMg_{0.5}Fe_{0.5}Si_2O_6$; although he found a field of "*Pigeonit-Diopsid Mischkristalle*," he did not find a corresponding field of pigeonite + diopside; rather, he drew an inferred solvus with a crest at lower temperature than that of the mixed pigeonite–diopside that formed.

The composition volume for pigeonite lies between the orthopyroxene + augite two-phase region and the solidus. At appropriate conditions it extends over a wide range of Fe/(Mg+Fe), but permissible values of Wo are always restricted. Ross and Huebner (1979) studied the lower temperature stability limit of pigeonite by heating natural orthopyroxene + augite "crystal" intergrowths and measuring the "modes" of the reactants and products in the reaction orthopyroxene + augite → pigeonite by measuring x-ray precession photograph spot intensities. Melt accompanied pigeonite in many of the magnesian pyroxene runs and probably acted as a catalyst; although not reversed, the data probably define equilibrium conditions

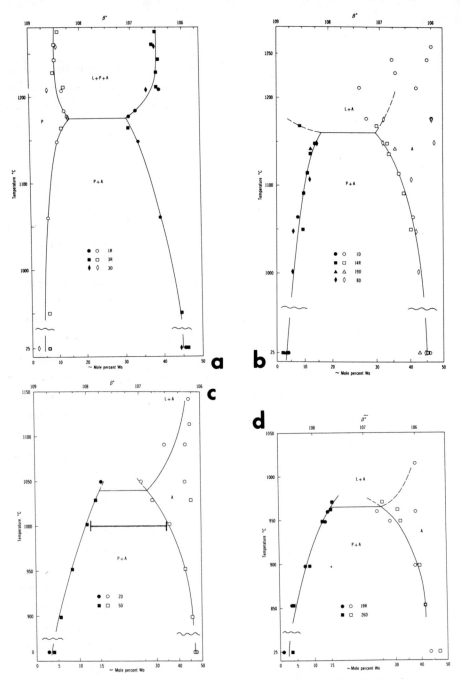

Figure 14. Temperature-composition sections intersecting the pigeonite-augite solvus at Fe/(Mg+Fe) ∿ 33% (Ca-rich pigeonite, a); ∿ 34% (Mg-rich augite, b); ∿ 60% (Augite, c); ∿ 85% (Ferro-augite, d). All pyroxene compositions are bulk compositions of separates made from lunar basalt 12021. The pigeonite-augite solvus estimated from data of Turnock and Lindsley (pers. comm., 1980) has been superimposed as a tie line on Figure 9C of Ross *et al.* (1973). (Note that (c) and (d) were interchanged as originally published but have now been corrected.) Orthopyroxene is probably the stable Ca-poor pyroxene at the lowest temperature of (a) and (b), but did not appear in the run products.

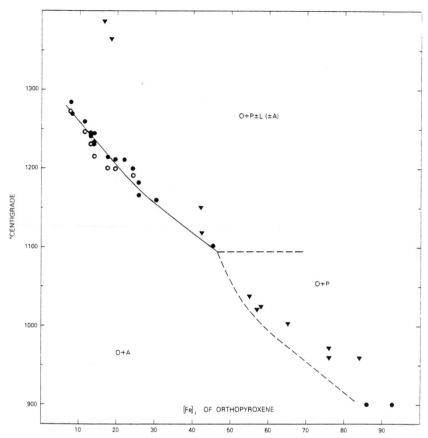

Figure 15. Temperatures at which pigeonite first appears in terrestrial orthopyroxenes heated by Ross and Huebner (1979). At Fe/(Mg+Fe) ≤45 (solid portion of the curve), the plane of section passes through the minimum temperature for pigeonite stability (pigeonite coexists with orthopyroxene + augite). At more iron-rich compositions, most data do not represent three-phase equilibrium; the plane of section can be said to lie at Wo contents less than that for the minimum stability of pigeonite (no augite is present). Closed circles indicate orthopyroxene in which augite was observed to coexist with pigeonite. Open circles indicate orthopyroxene in which augite was detected, but pigeonite was not observed. Triangles designate either orthopyroxenes in which augite was dissolved before the appearance of pigeonite or orthopyroxene that never contained augite. The triangle points in the direction in which the three-pyroxene equilibrium must lie. The solid line, which is thought to be close to equilibrium, is not reversed. The dashed lines are schematic. Note that [Fe]$_i$ is the value of 100 Fe/(Ca+Fe+Mg) for orthopyroxene thought to be in equilibrium with pigeonite and augite.

for Fe/(Mg+Fe) = 10-45% (Fig. 15). Extrapolation of the reaction trend to Fe/(Mg+Fe) = 0% gives 1300°C, in satisfactory agreement with the reversed temperature of 1278 ± 10°C for synthetic, slightly aluminous pyroxenes (Yang, 1973a). At compositions more iron-rich than Fe/(Mg+Fe) = 55%, Ross and Huebner (1979) found that for most crystals used, the exsolved augite dissolved into the host *before* pigeonite formed (Fig. 15); thus, these runs define the maximum possible temperature for the lower

247

stability limit of pigeonite. Better control of the more iron-rich part of the curve is provided by six additional determinations summarized in Table 2. All experimental data pertaining to the conditions for the reaction between pigeonite and orthopyroxene + augite are summarized in Figure 16. The results appear consistent despite the use of natural and synthetic materials in complex (dirty) and simple (pure) systems. Although with increasing iron content the temperature of reaction drops steeply through the subsolidus, high-grade metamorphic temperatures are not reached until $Fe/(Mg+Fe) > 60\%$. Thus, the rarity of pigeonite in metamorphic rocks is due only in part to slow cooling, causing pigeonite to decompose. In most cases temperatures were not high enough for pigeonite stability, even at the peak temperatures of metamorphism.

The pigeonite phase volume extends from the iron-free join, where it appears as a wedge pointing downwards (Fig. 7), continuously across the quadrilateral with decreasing temperature to about $Fe/(Mg+Fe) \sim 0.7$, where the one-phase region abuts regions containing olivine + quartz. Here, the pigeonite volume is terminated by fields of pigeonite + olivine + silica and pigeonite + augite + olivine + silica, depending on the Wo content. Huebner and Turnock (1980) show a possible topology for the disappearance of pigeonite: With decreasing temperature, $P \rightarrow O + A + F + S$ between temperatures of 800-900°C and 700-800°C (Fig. 17). Lindsley and Grover (1980) discovered that in the pure system this univariant assemblage is essentially independent of temperature and occurs at 825 ± 15°C over the pressure range <1 to 11 kbar.

The thermal stability of pigeonite is at a minimum where it coexists with both orthopyroxene and augite. Loss of *either* the augite *or* the calcium-poor pyroxene causes the temperature to rise in the iron-free system (Fig. 7). This relationship has been noted when heating natural iron-bearing pyroxene assemblages. Ross and Huebner (1975, 1979) found that when augite is not present, the reaction of orthopyroxene takes place at temperatures as much as 150°C above the minimum temperature (see Fig. 15). The effect on pigeonite stability of the loss of orthopyroxene has not been determined directly, but an increase in temperature can be inferred from the shape of the pigeonite-augite solvus (Fig. 14).

The subsolidus phase relations at $Fe/(Mg+Fe) > 70\%$ are complicated by the fact that even in the pure system, quaternary assemblages are

Table 2. Data for the Reaction Orthopyroxene + Augite ± Pigeonite

P,bar	T,°C	Reaction or Assemblage	Ref
0	1278	$Wo_4En_{96}Fs_0$ + $Wo_{35}En_{65}Fs_0$ + $Wo_7En_{93}Fs_0$ + Forsterite + L	7
0	1012	$Wo_3En_{50}Fs_{46}$ + $Wo_{43}En_{36}Fs_{21}$ → Pigeonite	1
0	1000	Aug + Cpx → Aug + Pig (bulk $Wo_{25}En_{38}Fs_{38}$)	2
1000	900	Opx + Aug → Opx + Aug + Pig (bulk $Wo_{10}En_{36}Fs_{54}$)	2
2000	850	Opx + Aug → Opx + Aug + Pig (bulk $Wo_{10}En_{22.5}Fs_{67.5}$)	2
2000	825	Opx + Aug → Opx + Aug + Pig (bulk $Wo_{10}En_{22.5}Fs_{67.5}$)	2
$\leq$2000	825$\pm$15	Opx + Aug + Pig + Olivine + Quartz	3
2000	800	Opx + Aug + Pig $\pm$ Olivine $\pm$ Quartz (bulk $Wo_{25}En_{19}Fs_{56}$)	4
1	1270	Opx + Aug → Pig Fe/(Mg+Fe) = 8%	5
1	1102	Opx + Aug → Pig Fe/(Mg+Fe) = 46%	5
$\leq$1000	900	Opx + Pig + Aug (Fe/(Mg+Fe) of Opx = 58%)	6
$\geq$1000	1000	Opx + Pig + Aug (Fe/(Mg+Fe) of Opx = 45%)	6

References for Table 2

1 Huebner and Nord (unpublished). Formation of pigeonite at orthopyroxene - augite
    interface; annealed pyroxene from Moore County eucrite.
2 Podpora and Lindsley (1979), pure system
3 Lindsley and Grover (1980), pure system
4 Simmons et al. (1974), pure system
5 Ross and Huebner (1979), natural orthopyroxene crystals
6 Turnock and Lindsley (personal communication, 1980), pure system
7 Yang (1973a), system anorthite-diopside-enstatite-silica

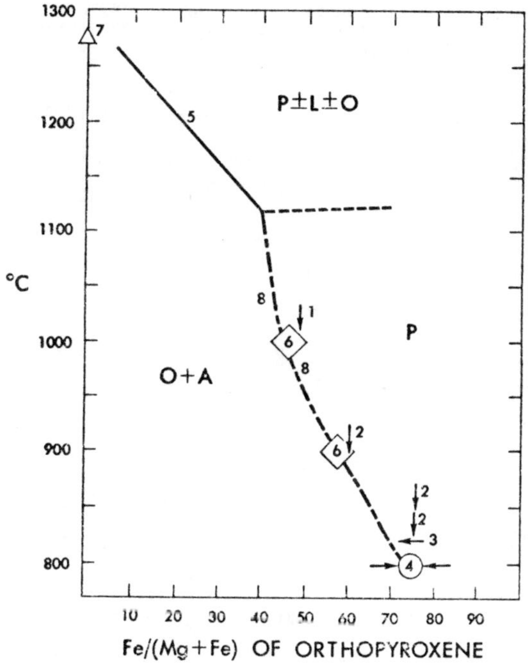

Figure 16. Lower thermal stability limit of pigeonite (the reaction Opx + Aug ⇄ Pig) at low pressure as a function of inferred orthopyroxene composition. Triangle, circle, and diamond represent reversed determinations; arrows represent unreversed data and indicate the direction in which the equilibrium must lie. The solid line from Ross and Huebner (1979) is believed to be close to equilibrium despite the fact that the experiments were not reversed. Most natural magnesian orthopyroxenes melt at about the same temperature at which pigeonite first forms. Numbers 1 to 7 are keyed to Table 2; 8 is from D. H. Lindsley (pers. comm.) at 1 kbar.

Fe/(Mg+Fe) OF ORTHOPYROXENE

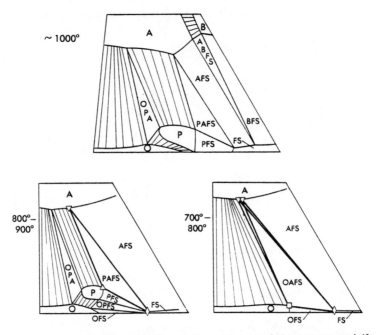

Figure 17. Subsolidus phase relations for the iron-rich part of the pyroxene quadrilateral. At 700°C-800°C the heavy tie lines indicate the natural orthopyroxene + augite + olivine + quartz assemblage of Bonnichsen (1969). Similarly at 800°C-900°C, the assemblage pigeonite + augite + olivine + silica is from Simmons *et al.* (1974). The two assemblages are related by the equilibrium O+A+F+S = P, which has been determined by Lindsley and Grover (1980) to lie at 825 ± 15°C in the pure system. The topology at ∿1000ᵇC is drawn to be consistent with both the natural assemblages and the melting relations of natural pyroxenes as determined by Huebner and Turnock (1980), from which these sections are reproduced.

present. One possible topology, that of Huebner and Turnock (1980) and shown in Figure 17, is consistent with the coexistence of augite with olivine + quartz (at low temperature) observed in natural assemblages (Bonnichsen, 1969; Simmons *et al.*, 1974) and with the melting relations of natural pyroxenes. This topology precludes the equilibrium association of pigeonite + ferrobustamite, an assemblage not known to occur in nature (but suggested by Hoover and Irvine, 1978, p. 780), but it must be realized that there is no direct experimental control for most of the illustrated phase relations.

Little is known about how additional components might affect the subsolidus phase relations of the Ca-, Mg-, and Fe-bearing quadrilateral pyroxenes. I have alluded to the idea that the orthopyroxene-augite tie-line length may depend upon the Al_2O_3 content of the pyroxenes, but that the orthopyroxene + augite ⇄ pigeonite reaction surprisingly may *not* be sensitive to minor-element contents. A systematic study is needed if

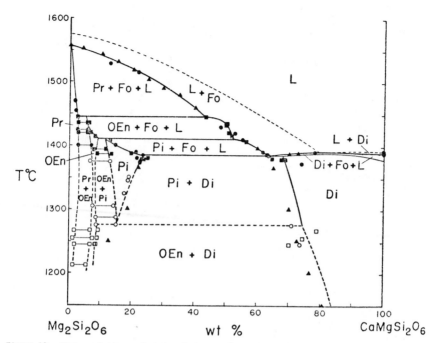

Figure 18. Phase relations of parts of the systems Mg_2SiO_4-$CaMgSi_2O_6$, Mg_2SiO_4-$CaMgSi_2O_6$-SiO_2, and $CaAl_2Si_2O_8$-$CaMgSi_2O_6$-$Mg_2Si_2O_6$-SiO_2 projected onto the enstatite-diopside join. Solid symbols are for aluminum-free systems; open symbols are for the aluminous system. Solid circles - Kushiro (1972). Solid triangles - Boyd & Schairer (1964). Solid squares - Longhi & Boudreau (1980). Open circles - Yang (1973a). Open squares - Longhi (1978 and unpublished). (From Longhi and Boudreau, 1980).

results obtained on simple laboratory systems are to be used to model the more complex natural systems, especially at high pressure. In lieu of such information, great uncertainties must be assigned to temperature and pressure estimates of natural processes. The best that can be done here is to discuss the isolated studies that have been performed.

Small amounts of Al_2O_3 do not appear to affect significantly the subsolidus pyroxene phase relations of iron-free pigeonite and augite at one bar pressure. In a study of the system $CaAl_2Si_2O_8$-$CaMgSi_2O_6$-$MgSiO_3$-SiO_2, Yang (1973a) found 0.1-0.5% Al_2O_3 by weight in coexisting subsolidus pyroxenes. His data for the mildly aluminous pyroxenes are consistent with data obtained in the aluminum-free systems $Mg_2Si_2O_6$-$CaMgSi_2O_6$ (Boyd and Schairer, 1964) and $CaMgSi_2O_6$-Mg_2SiO_4-SiO_2 (Kushiro, 1972; Longhi and Boudreau, 1980), shown in Figure 18.

Al_2O_3 appears to affect the temperature range over which iron-free orthopyroxene is stable by lowering the decomposition reaction orthopyroxene $\rightleftharpoons$ protopyroxene + pigeonite from 1375° (Fig. 7a) to lower temperatures (Fig. 18), and by increasing the stability of orthopyroxene

251

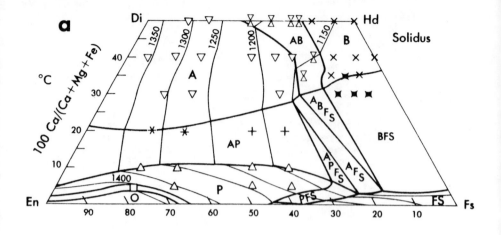

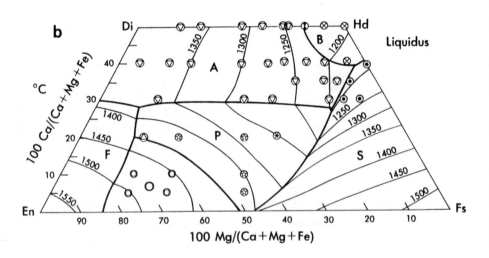

Figure 19. Solidus (a) and liquidus (b) at low oxygen fugacity values experimentally determined for pyroxene quadrilateral bulk compositions in the pure system. The pigeonite field extends almost all the way across the solidus surface, separating orthopyroxene and augite. Temperatures are higher than would be expected in most magmas that are precipitating pyroxene at low pressure. The liquidus shows fields of olivine (F, for forsterite or fayalite), orthopyroxene (O), pigeonite (P), augite (A), ferrobustamite (B), and silica (S). Because of incongruent melting and solid solution, the liquidus and solidus do not meet. From Huebner and Turnock (1980).

relative to that of protopyroxene (Anastasiou and Seifert, 1972). It is conceivable that the addition of $\leqslant 2\%$ by weight Al_2O_3 could eliminate entirely a stable field of protopyroxene + augite (Fig. 7b).

The melting interval. The melting relations of both pure and natural quadrilateral pyroxenes have been discussed recently by Huebner and Turnock (1980). For the pure system they blended new data on ternary (Ca-, Mg-, and Fe-bearing) compositions with earlier data on the bounding joins (Figs. 7, 8, and 11b). (They found it necessary to modify the $Mg_2Si_2O_6$-$Fe_2Si_2O_6$ join of Bowen and Schairer (1935) by bringing the orthopyroxene field to the solidus. Because they did not observe protoenstatite, Huebner and Turnock did not attempt to resolve the conflicting data of earlier investigators and distinguish the two orthorhombic pyroxenes in their diagrams.) The solidus (Fig. 19a) shows fields of orthopyroxene, pigeonite, augite, and bustamite; the pigeonite field extends continuously from the iron-free join to $Fe/(Mg+Fe) = 72\%$. Orthopyroxene and augite cannot coexist on the solidus; instead, there is a long pigeonite-augite miscibility gap that extends across most of the quadrilateral. The corresponding liquidus diagram for the pure system (Fig. 19b) has fields of olivine, orthopyroxene, pigeonite, augite, bustamite, and silica, thereby providing more detail about the nature of the pyroxene phases than did the earlier liquidus of Roedder (1965). The pigeonite + augite cotectic is prominent, crossing much of the quadrilateral and falling in temperature from 1380 to 1245°C. The prominence of this cotectic is emphasized by the stereographic projections of the liquidus (Fig. 20) which show that the olivine, orthopyroxene, and pigeonite surfaces are nearly co-planar and in contrasting orientation to the augite surface. The extensive solid solutions, plus the fact that for many bulk compositions the phase assemblages must be considered in a quaternary system, combine to prevent the liquidus and solidus from meeting. The resulting T-X and isothermal sections intersect three- and four-phase volumes (see Figs. 3 and 5 of Huebner and Turnock, 1980).

It is now possible to show, in a schematic sense, the position of protoenstatite fields on the solidus and liquidus (Fig. 21), using the new data of Longhi and Boudreau (1980) for the iron-free system, and drawing the L(Pr,O) cotectic to intersect the $Mg_2Si_2O_6$-$Fe_2Si_2O_6$ join as revised (Fig. 8). This proposed topology should be tested by experiment.

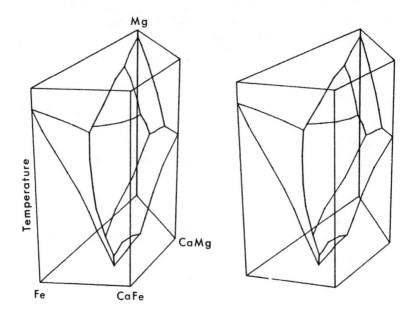

Figure 20. Liquidus surface of the metasilicate plane in the pure system CaO–MgO–FeO–SiO$_2$. Temperature is plotted along the vertical axis. This stereographic projection of the data in Figure 19b is viewed from the CaFeSi$_2$O$_6$ (hedenbergite) corner of the pyroxene quadrilateral. The surfaces of olivine, orthopyroxene, and pigeonite are nearly coplanar. Drawings prepared by using a BASIC program written by R.F. Sanford of the U.S. Geological Survey and a program for true perspective supplied with the Tektronic, Inc. "Plot 50" package of graphics programs (Revision A, September 1978).

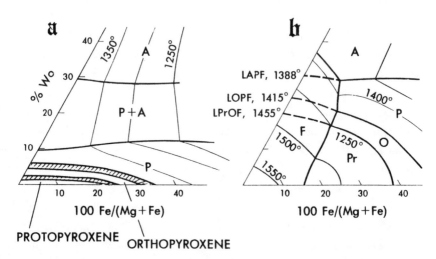

Figure 21. The Mg$_2$Si$_2$O$_6$-rich region of the quadrilateral showing inferred protopyroxene-bearing fields as discussed in the text. (a) Solidus; (b) Liquidus. Compared with Figure 19, Figure 21 incorporates some preliminary revisions in the placement of field boundaries near the iron-free join. Except for the addition of protopyroxene, the topology is unchanged.

The solidus for pure pyroxene quadrilateral compositions fails to model the solidus deduced from natural occurrences of pyroxenes (Figs. 4 and 5). In particular, the solidus for the pure system does not permit the equilibrium assemblage liquid + orthopyroxene + augite; relative to the natural solidus, the pigeonite + augite miscibility gap of the pure, synthetic solidus is surprisingly narrow; and the pure solidus temperatures are considerably higher than the few measured temperatures for magmas that precipiate pyroxene (Wright and Okamura, 1977; Ishii, 1975; Wright and Weiblen, 1967).

Minor-element partitioning between pyroxene and melt. An understanding of the minor-element distribution between crystalline pyroxenes and melt is fundamental to understanding the melting relations of natural pyroxenes and the crystallization of pyroxenes from real magmas which have additional components. Analyses of pyroxenes and the now-crystalline melts from which they precipitated are only sufficient to lead us to conclude that magmas precipitate pyroxenes that contain considerably less Al_2O_3 and TiO_2. For example, a basalt containing 15-20% Al_2O_3 and 3% TiO_2 might reasonably be expected to crystallize pyroxene containing 2-3% Al_2O_3 and 0.5-1.0% TiO_2. But the laboratory studies give clues to the true degree of partitioning of these and many other elements and contribute to the uncertainties as well. An understanding of the *kinetics* of crystal growth and element partitioning appears essential.

The distribution of an element between two phases may be considered in terms of a simple partitioning constant (D_i = (% i in phase A)/(% i in phase B)) or as a reaction constant (K_D) that accounts for all coupled substitutions. Use of K_D is the rigorous approach that can be related to thermodynamic potentials. Unfortunately, so little is known about pyroxene component activity-coefficients and melt speciation that it is misleading to write coupled substitutions for pyroxene-melt equilibria. The simple partitioning constants (D_i) will be used in the following discussions in the hope that the observed ranges of pyroxene compositions and melt compositions will be so small that they will not affect the results significantly.

The only published attempts to summarize the partition coefficients of minor elements between melt and pyroxenes are those of Irving (1978) and of Huebner and Turnock (1980, Table 5). (Nielsen and Drake (1979,

Table 1) compiled many pertinent analyses of coexisting pyroxenes and melts but made no attempt to summarize the minor-element partitioning values.) Huebner and Turnock found that in their partial melting experiments, using single crystals of natural pyroxenes, Al_2O_3 and TiO_2 were more strongly partitioned into the melt than when pyroxene crystals grew from basaltic melts in experiments by other investigators. Grove and Bence (1977) had found that at increased cooling rates, the (apparent) degree of Al_2O_3 and TiO_2 partitioning between crystals and melt decreased. Although the melt bulk compositions in the partial melting and the crystal growth experiments were different, Huebner and Turnock (p. 267) attributed the systematic difference to experimental disequilibrium and implied that at rapid crystal growth rates, resultant determinations of the Al_2O_3 and TiO_2 partitioning constants will yield anomalously small values. During growth of a pyroxene crystal, Al_2O_3 and TiO_2 are excluded from the crystalline phase. At sufficiently slow crystal growth rates, these components can diffuse away from the growth interface, resulting in a homogeneous melt. At rapid growth rates, there is not sufficient time for diffusive homogenization of the melt (see also discussion by Grove and Raudsepp, 1978). Melt immediately adjacent to the growing pyroxene crystal is enriched in Al_2O_3 and TiO_2, relative to melt farther from the crystal, and precipitates pyroxene that is correspondingly enriched in Al_2O_3 and TiO_2. The Al^{3+} and Ti^{4+} cations diffuse very slowly in (solid) pyroxene; most experimentalists do not make runs of sufficient duration for subsequent reequilibration of Al and Ti between melt and crystal. Analyses of the pyroxene and melt at some distance from the interface yield an erroneous (apparent) partition coefficient because the melt (glass) that was analyzed is not the same composition as the melt that actually precipitated the pyroxene. The partial melting process does not appear to offer the same difficulties, *provided* sufficient time is allowed for segregation of the melt into large bodies that can be analyzed (or avoided) with the microprobe beam. Migration and coalescence of melt blebs (by dissolution and reprecipitation) appears to be a faster process than solid-state equilibration. Values of D_i determined by partial melting experiments are thus likely to be at least as good as some, and superior to many, values determined in crystal growth experiments.

Huebner (unpublished data) has compiled 146 paired analyses of melt
and coexisting pyroxene. The partition coefficients have a wide range,
especially for aluminum and titanium; the observed ranges are D_{Al} = 2.4-
72, D_{Ti} = 1.1-29, D_{Cr} = 0.1-1.5, and D_{Mn} = 0.3-9.7. Some values of D_{Al}
and D_{Ti} are anomalously small and were not included in the compilation
because the small values can now be attributed to rapid crystal growth
(see Lindstrom, 1976; Ho Sun, 1973) rather than to equilibrium conditions.
Some of the very large values included in the compilation can be attributed to analytical uncertainties in analyzing small amounts of Al_2O_3
and TiO_2 in pyroxenes by standard major-element microprobe methods.
Preferred values should account for the results of the partial melting
experiments and for some of the results of crystal growth experiments.
Such favored values are 7-20 for Al_2O_3, 5-10 for TiO_2, 0.5 for Cr_2O_3,
and 1.5 for MnO. Clearly, melts that coexist with pyroxenes will be
significantly richer in Al_2O_3 and TiO_2 than the pyroxene.

Melting relations of natural pyroxenes. The melting relations of
natural pyroxenes with compositions close to those of the quadrilateral
have been reported by Huebner and Turnock (1980) and are reproduced as
Figures 22a and b. Compared with the pure system (Fig. 19), the natural
solidus and liquidus lie at temperatures that are 100-200°C lower. This
occurs because Al_2O_3 and TiO_2 are partitioned into the melt, diluting it
with respect to Ca-, Mg-, and Fe-pyroxene components, thereby lowering
the melting temperature. In the example shown (Fig. 22a) the solidus
temperatures have been lowered sufficiently to intersect the orthopyroxene + augite ⇄ pigeonite reaction at Fe/(Mg+Fe) ∿ 15% and 1225°C
(compare with Figs. 15 and 16), permitting liquid to coexist at equilibrium with both orthopyroxene and augite (an orthopyroxene-augite miscibility gap is present). The three-phase triangle orthopyroxene +
pigeonite + augite on the solidus shown appears to be relatively magnesian compared with most natural solidus assemblages (Fig. 5), but the
melt compositions were not as aluminous as most basalts. At plagioclase
saturation, the solidus temperatures should be depressed even more. The
greater the temperature depression, the greater is the Fe/(Mg+Fe) of the
three-pyroxene tie triangle on the solidus. If we assume that variations in minor-element concentrations and pressure do not affect the
temperature *vs* Fe/(Mg|Fe) relationship, the three-pyroxene reaction

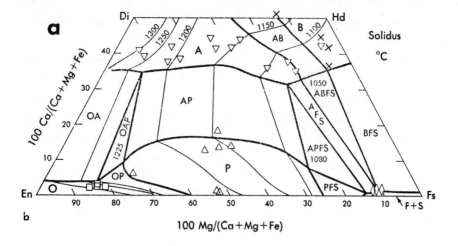

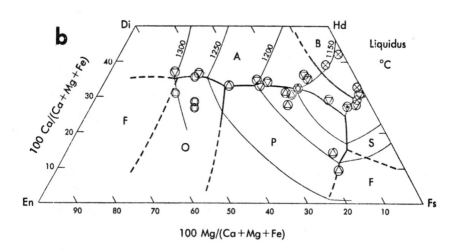

Figure 22. The liquidus and solidus determined experimentally using natural pyroxene crystals and reducing conditions. The solidus (a) intersects the orthopyroxene + augite ⇌ pigeonite reaction (Figs. 15 and 16) to reveal a field in which magnesian orthopyroxene and augite coexist. The solidus temperatures are 100-200°C lower than in the pure system, due to partitioning of Al_2O_3 and TiO_2 into the liquid phase on partial melting. The corresponding liquidus (b) shows orthpyroxene coexisting with augite. The compositions plotted on the liquidus are partial melt compositions, not bulk compositions, and the compositions plotted on the solidus are residual pyroxene compositions, not initial pyroxene (or bulk) compositions. (From Huebner and Turnock, 1980).

shown in Figure 16 can be used as a geothermometer to estimate temperatures of the three-pyroxene assemblages shown in Figure 5. As the
solidus temperature is depressed, the Fe/(Mg+Fe) of the three-pyroxene
triangle increases and the pigeonite-augite tie lines lengthen. Thus,
the solidus for impure quadrilateral pyroxenes *appears* to model fairly
adequately the compositions of pyroxenes crystallized from natural
magmas. The effectiveness of this model should now be tested against
measured temperatures of real magma.

Quaternary liquidus model. It is obvious in the preceding discussions that the melting relations of even the pure quadrilateral pyroxenes cannot be understood in terms of a ternary model system. In some
metasilicate bulk compositions, the solids (olivine, silica) and melt
present do not plot on the metasilicate plane. There are also "minor"
elements, present in natural systems, that tend to reside in the melt,
diluting the liquid with respect to SiO_2 and other components. Clearly,
SiO_2 must be considered as a compositional variable in the system.
Because pyroxenes are the focus of attention, it is convenient to use
a six-oxygen pyroxene formula unit in the following discussion. When
the activities of SiO_2 are successfully determined, conversion to a_{SiO_2}
will be relatively easy.

The phase relations of the pure system can be shown in a ternary
prism by plotting the number of Si cations per six oxygens on the prism
axis (Fig. 23). Each vertical prism face is a ternary system such as
$CaO-MgO-SiO_2$. The following discussion is limited to compositions more
siliceous than the olivine plane (at 1.5 Si cations per formula unit).
The bounding ternary joins (Figs. 24a-c) are constructed from the classic
quenching studies of the systems $CaO-MgO-SiO_2$ (Osborn and Muan, 1960;
Ricker and Osborn, 1954), $MgO-FeO-SiO_2$ (Bowen and Schairer, 1935), and
$CaO-FeO-SiO_2$ (Bowen *et al.*, 1933) augmented by the recent microprobe
data of Kushiro (1972), Yang (1973a), and Longhi and Boudreau (1980) on
the melting of iron-free pyroxenes (see Fig. 18), and the reinterpretation
of the $MgSiO_3$-$FeSiO_3$ join by Huebner and Turnock (1980). Much of the data
is summarized in Table 3a. Data on the phase relations within the interior of the prism are sparse; Huebner and Turnock (1980) determined
the liquidus of the metasilicate plane and Ricker (1952) gives data on
the joins $CaMgSiO_4$-$FeO-SiO_2$ and $MgSiO_3$-$CaSiO_3$ FeO, which penetrate the
region of interest.

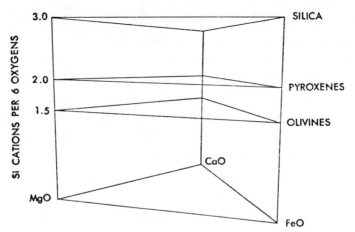

Figure 23. Sketch of prism used to show phase relations of the quaternary system CaO-MgO-FeO-SiO₂ after Huebner (1979, 1980). Adopting a six-oxygen pyroxene formula unit, the olivine plane ideally lies at 1.50 Si-cations per unit and the pyroxene plane at 2.00 cations.

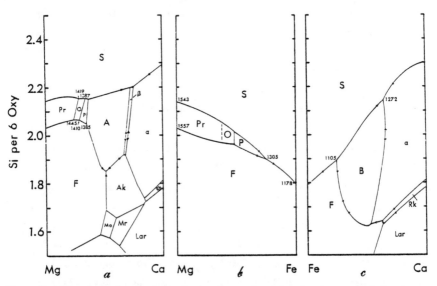

Figure 24. Flow charts (projections of univariant and pseudo-univariant curves) for the system CaO-MgO-FeO-SiO₂ (a) and the system having additional components exemplified by Al₂O₃ and TiO₂ (b). Sufficient additional components have the effect of lowering the solidus in temperature and restricting liquids that are in equilibrium with pigeonite to Fe/(Mg+Fe) values >45%.

Liquidus relations within the prism can be summarized by projecting the univariant liquidus curves onto the base of the prism (Fig. 25a). The part of this network that is accessible by metasilicate *bulk* compositions has been presented by Huebner and Turnock (1980, their Fig. 4c), but here it has been completed by considering *bulk* compositions that do not lie on the metasilicate plane. By distinguishing pyroxene phases,

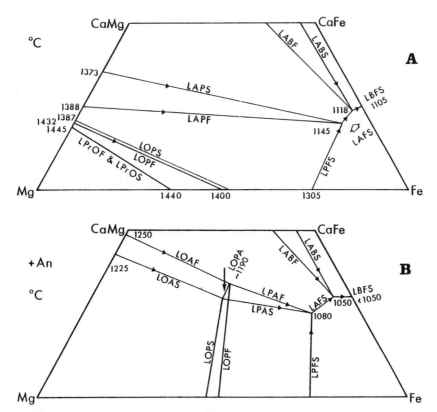

Figure 25. Flow charts (projections of univariant and pseudo-univariant curves) for the system CaO–MgO–FeO–SiO$_2$ (a), and (b) the system having additional components Al$_2$O$_3$ and TiO$_2$. Sufficient additional components effectively lower the solidus temperature and restrict liquids in equilibrium with pigeonite to Fe values >45%.

this diagram is superior to the earlier efforts of Ricker (1952) and Yoder *et al.* (1963). There are only two invariant points involving quadrilateral pyroxenes; both lie between the pyroxene and olivine planes near the Mg-free join. The minimum liquidus temperature is placed at the Mg-free join and 1105°C.

I have attempted to show the liquidus diagram for the pure system in stereographic projection in Figure 26. The calcium-poor pyroxene volumes occupy a thin wedge that separates the extensive regions of olivine and silica -- except at Fe$_2$Si$_2$O$_6$-rich compositions, where the olivine and silica volumes meet at a plane. Orthopyroxene and pigeonite appear stable over a surprisingly small range of Si concentration -- and presumably Si activity as well. However, olivine precipitation from melts that lie below (at lower

261

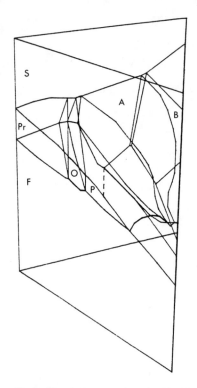
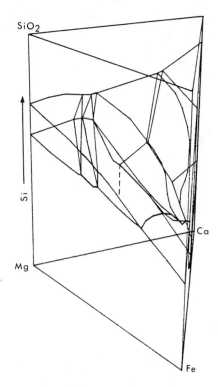

Figure 26. Attempt to show the liquidus diagram for pyroxenes in the pure system CaO-MgO-FeO-SiO$_2$, in stereographic projection. Pyroxenes are restricted to a small range of compositions that form a wedge separating the large volumes of silica and olivine. The pyroxene wedge is not present near Fe$_2$Si$_2$O$_6$-Fe$_2$SiO$_4$ compositions; in that region olivines and silica coexist. The boundaries separating the 3 pyroxene volumes (orthopyroxene, pigeonite, and augite) can be seen.

Si concentrations than) the pyroxene-bearing volumes drives the melt composition to the pyroxene saturation surface, explaining in part why pyroxene is such a common mineral in mafic rocks. As in the case of the metasilicate bulk compositions, the pigeonite volume separates the volumes of orthopyroxene and augite. A change in Si content is not sufficient to permit melt to be in equilibrium simultaneously with orthopyroxene and augite. The calcium-poor pyroxene volumes are entirely *above* the metasilicate plane at iron-free compositions and entirely *below* the metasilicate plane at magnesium-poor compositions. Thus, iron- and calcium-poor pyroxenes melt incongruently to olivine + liquid, whereas calcium- and magnesium-poor pyroxene compositions have silica on the liquidus.

The surfaces of the liquidus prism (Fig. 26) at which metasilicate is saturated with silica or olivine can be projected onto the prism base

262

and contoured in terms of temperature and the silica content (silicon cations per 6 oxygen formula unit) of the melt (Fig. 27). These contours are to a large extent a best guess; control is present only along the bounding faces of the prism and at the pyroxene quadrilateral plane. With this caveat, however, several interesting observations can be made. (1) At melt compositions with Fs = Fe/Ca+Mg+Fe) = 0.70-0.75, the saturation surfaces meet. Although pyroxene saturation does not occur at more Fe-rich compositions, the olivine-silica surface which does occur is contoured for completeness. (2) At Wo = Ca/(Ca+Mg+Fe) = 0.51-0.63, the surface of mutual olivine and metasilicate saturation terminates; at more calcic compositions, metasilicate coexists with akermanite or rankinite. (3) The difference in Si content between the olivine and silica saturation surfaces bounding the Ca-poor pyroxene volumes is no more than 0.15 cations per 6 oxygens, but this difference is as great as 0.50 cations where the Ca-rich metasilicates are bounded. (4) For any given value of Wo, En, and Fs, the temperature is approximately the same on the metasilicate-olivine and the metasilicate-silica surfaces, implying that isotherms pass steeply through the metasilicate volumes (nearly parallel to the Si axis).

Liquidus Diagram, System $CaO-MgO-FeO-SiO_2$-"OTHERS"

The simple quaternary system and the more complex system containing other components have different phase relations because components such as Al_2O_3, TiO_2, and K_2O (the "OTHERS" components) are concentrated in the residual melt during crystallization of olivine or pyroxene. We have already seen that, in the case of Al_2O_3 and TiO_2, this effect reduces the temperature of the pyroxene solidus relative to a system without Al_2O_3 and TiO_2, to expose a field of orthopyroxene + augite on the solidus. As the solidus temperature range is progressively reduced, the iron-rich limit of the orthopyroxene + augite field moves closer to the join $Fe_2Si_2O_6$-$CaFeSi_2O_6$.

Early attempts to deduce a topology for the quaternary system with "OTHERS" were only partly successful. Yoder *et al.* (1963, p. 87) presented a "flow sheet" for compositions on or near the quadrilateral. Their diagram was based upon a mixture of previously published determinations of the simple bounding ternary systems and new experiments in which they partially melted natural (and presumably aluminous) pyroxenes. In this manner, they deduced the existence of the two invariant points LAPFS

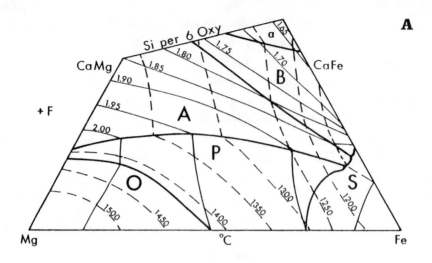

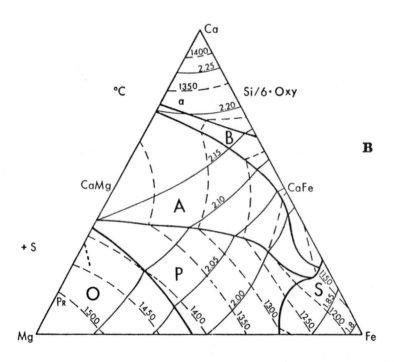

Figure 27. Liquidus surfaces bounding the pyroxene and pyroxenoid volumes: orthopyroxene (O), pigeonite (P), augite (A), and ferrobustamite (B) in the pure system CaO-MgO-FeO-SiO₂. (a) the olivine saturation surface; (b) the silica saturation surfaces (cristobalite and tridymite); dashed contours, °C; solid contours, Si cations. Note that pyroxene is not stable at Fe > ∿75%

and LABFS. The existence of these points, which appear in *both* the
simple quaternary (Fig. 25a) and the more complicated system containing
"OTHERS" (Fig. 25b), does not appear to depend on the presence or absence of "OTHERS" or anorthite. More recently Yang (1973a, p. 492)
published a "flow sheet" for pigeonite crystallization in the iron-free
system $Mg_2Si_2O_6$-$CaMgSi_2O_6$-$CaAl_2Si_2O_8$-SiO_2. As anorthite component (or
Al_2O_3) is added to the simple melt, the solidus temperatures are depressed until iron-free pigeonite ceases to be stable in the presence
of crystalline anorthite. In the presence of anorthite component in
the iron-bearing prism, there are two invariant points (LOPAF and LOPAS)
which do not appear in the simple ternary system CaO-MgO-SiO_2 or in the
pure quaternary prism.

The predicted "flow sheet" for the system at anorthite saturation
is shown in Figure 25b. In the iron-free system, orthorhombic pyroxene
and/or augite, but not pigeonite, is stable in the presence of anorthite
(Yang, 1973a), causing the appearance of two univariant or pseudo-univariant curves (LOASAn, LOAFAn) not found in the alumina-free system
(compare with Fig. 25a). These curves terminate at LOPASAn and LOPAFAn;
the calcium-poor pyroxene in equilibrium with more iron-rich liquids is
pigeonite, not orthopyroxene. The three curves LOPSAn, LOPAAn, and LOPFAn
bound a surface that separates the liquidus volumes of orthopyroxene and
pigeonite. Thus, with increasing concentration of "OTHERS", there appears
a liquidus surface that separates the primary volumes of orthopyroxene and
pigeonite and that moves through the prism to the position shown schematically in Figure 28. At this point, further additions of "OTHERS"
components causes the crystallization of "OTHERS"-bearing phases, principally plagioclase when Al_2O_3 is the major "OTHERS" component.

Sufficient data are available in the literature on quinary + quaternary phase relations to plot in the liquidus prism (Fig. 28) the compositions of melts that coexist with anorthite *and* pyroxene(s) with or
without olivine. Such a scheme corresponds to projecting from Al^{3+} and
any "OTHERS" ions that may be present. Data are summarized in Table 3b.
Interior elements in the liquidus prism for the "OTHERS"-bearing system
are positioned schematically in Figure 28. The diagram will become a
useful model for the crystallization of real magmas only if these elements
can be positioned quantitatively with respect to the prism coordinates.

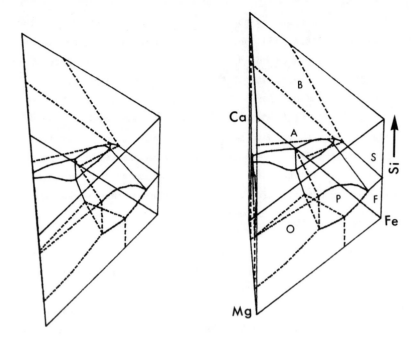

Figure 28. Attempt to portray the liquidus relations of the system CaO-MgO-FeO-SiO₂="Others"
at close to anorthite saturation. The liquidus volume of pigeonite is confined to Fe/(Mg+Fe)
> 50%; at more magnesian compositions there is a large liquidus volume of orthopyroxene.

Quantification of the model diagram could in principle be achieved through
calculations from fundamental thermodynamic data, but the necessary data
are not available. In practice, the best approach appears to be the in-
corporation of experimental phase-equilibrium data for which the compositions of coexisting crystals and melt have been measured or are otherwise known with reasonable certainty (Huebner, 1980). Such data include
the classic phase-equilibrium studies on simple (ternary or pseudoternary)
systems and more recent investigations using natural materials or their
synthetic analogues.

The effect of orthoclase, rather than anorthite, on the pyroxene
phase relations was explored by Hoover and Irvine (1978) who added
$Or_{56}Qz_{44}$ (wt %) glass to compositions in the plane Mg_2SiO_4-Fe_2SiO_4-
$CaMgSi_2O_6$-$CaFeSi_2O_6$. In both the Mg-free and the Fe-free bounding joins,
addition of the Or component ($KAlSi_3O_8$) suppressed the silica liquidus
field while favoring fields of mafic silicates. In the Fe-free part of
the system, protopyroxene and augite do not meet but are instead separated by a field of pigeonite over the temperature range 1345-1300°C.

266

Table 3. Summary of "Quenching" Experiments Used to Construct Liquidus Diagrams

a. Melt Compositions at Invariant and Piercing Points, System $CaO-MgO-FeO-SiO_2$

Assemb.	Ref.	T°C	Si/6·O	Wo	En	Fs
$LW_\beta FS$	1	1105	1.892	0.245	-	0.755
LPFS	2	1305	1.903	-	0.258	0.742
$LPxnS_cS_t$	2	1470	2.058	-	0.677	0.323
LOF	3	1557	2.03	-	1.00	-
LOS_c	3	1543	2.141	-	1.00	-
LFS_t	2	1178	1.799	-	-	1.00
LBS	1	1200	2.067	0.50	-	0.50
LAF	3	?	1.840	0.50	0.50	-
LAS_t	3	?	2.184	0.50	0.50	-
$LOPF^1$	13	1420	2.000	0.20	0.67	0.24
$LPAF^1$	13	1375	2.000	0.28	0.61	0.11
$LPAS^1$	13	1195	2.000	0.28	0.15	0.58
$LABS^1$	13	1170	2.000	0.40	0.02	0.58
LPAF	14	1385	2.047	0.355	0.645	-
LPAS	14	1373	2.153	0.370	0.630	-
LPrOS	15	1419	2.167	0.295	0.705	-
LOPS	15	1387	2.158	0.346	0.654	-

b. Melt Compositions, System $CaO-MgO-FeO-SiO_2-Al_2O_3$ at Anorthite Saturation

Assemb.	Ref.	T°C	Si/6·O	Wo	En	Fs	Al_2O_3,wt.%
LOF·An	4	1214	1.842	29.3	39.1	31.6	15.7
LOS·An	4	1181	2.017	32.9	35.9	31.2	15.5
LOAF·An	5	1242	1.930	48.1	51.9	-0-	15.4
LAF·An	6	1254	1.836	48.7	51.3	-0-	16.2
LOF·An	6	1263	1.868	39.0	61.0	-0-	18.7
LFS·An	7	1070+4	1.828	15.6	-0-	84.5	10.3
LFSpAn	7	1108+4	1.522	14.2	-0-	85.8	10.2
$L\alpha\beta An$	7	1186+5	1.579	71.2	-0-	28.8	13.01
$L_\beta Mel·An$	7	1125+5	1.547	61.1	-0-	38.9	11.21
LSpMel·An	7	1130+5	1.419	50.2	-0-	49.8	13.01
$L\alpha\beta·An$	8	1245+3	1.798	82.8	17.2	-0-	14.1
$LA\beta An$	8	1236+3	1.792	81.8	18.2	-0-	13.9
LFSp·An	4	1264‾	1.670	28.0	41.8	30.2	18.2
LFSp·An	9	1320	1.663	34.3	65.7	-0-	22.4
LOF·An	9	1260	1.829	35.3	64.7	-0-	20.2
LOS·An	9	1222	2.046	44.0	56.0	-0-	18.5
LAF·An	10	1270	1.742	54.0	46.0	-0-	15.9
LFSp·An	10	1317	1.614	44.2	55.8	-0-	20.7
LAF·An	11	1262+5	1.783	48.9	51.1	-0-	15.8
LAF·An	12	1272‾	1.723	54.7	45.3	-0-	15.9
LAF·An	12	1266	1.702	56.3	43.7	-0-	15.5
LAF·An	16	?	1.622	55.2	44.8	-0-	~ 18
LFSp·An	16	?	1.551	51.4	48.6	-0-	~ 21.5

^{1}Piercing point, metasilicate plane

Abbreviations for phases

L	melt
Pr	protopyroxene
O	orthopyroxene
P	pigeonite
A	augite
Pxn	pyroxene, polymorph unknown
B	ferrobustamite
S_c	cristobalite
S_t	tridymite
S	crystalline silica
An	anorthite
Mel	melilite
Sp	spinel
α	α-wollastonite
β	β-wollastonite

References for Table 3

1. Bowen, Schairer, & Posnjak (1933)
2. Bowen and Schairer (1935)
3. Ricker and Osborn (1954)
4. Lipin (1978)
5. Presnall et al. (1979)
6. Longhi (1978)
7. Schairer (1942)
8. Osborn (1942)
9. Andersen (1915)
10. Osborn and Tait (1952)
11. Hytonen and Schairer (1960)
12. Presnall et al. (1978)
13. Huebner and Turnock (1980)
14. Kushiro (1972)
15. Longhi and Boudreau (1980)
16. Segnit (1956)

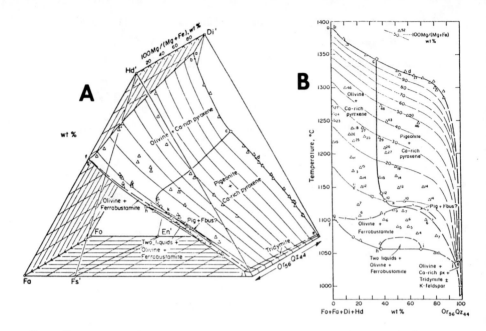

Figure 29. Augite and ferrobustamite saturation surfaces in part of the "Others"-bearing system Mg_2SiO_4-Fe_2SiO_4-$Ca_2Si_2O_6$-SiO_2-$KAlSi_3O_8$ (orthoclase, Or) at <1 bar pressure; iron-bearing compositions were run in equilibrium with iron. (a) Composition diagram; the saturation surfaces are contoured in terms of wt% $Fe/(Mg+Fe)$. (b) Saturation surfaces are plotted in terms of °C and wt proportion $(Or_{56}Qa_{44})/(Fo+Fa\_Hd+Di+Or_{56}Qz_{44})$ and contoured in terms of wt% $Mg/(Mg+Fe)$. The intersection of projected univariant curves near k is schematic. Both (a) and (b) are from Hoover and Irvine (1978).

It is surprising that the additional components do not stabilize orthopyroxene over protopyroxene (Fig. 7b) and depress the solidus to bring orthorhombic pyroxene and augite together on the liquidus (Figs. 22b, 25b).

In the quinary system, Hoover and Irvine (1978) located extensive olivine + augite and pigeonite + augite saturation surfaces (Fig. 29a,b) separated by a "cotectic" that corresponds to the pseudounivariant curve LPAF shown in Figure 25b. With decreasing temperature, $Fe/(Mg+Fe)$ of the melt increases (Fig. 29a) until at $Fe/(Mg+Fe)$ = 9% and 1115-1101°C, first pigeonite, then augite give way to a field of ferrobustamite + olivine (Fig. 29b). Fractional crystallization ceases at 1059°C. This fractionation sequence is similar to that inferred from Figures 22b and 25b, but with the silica activity held sufficiently low to prevent crystallization of silica (a strong probability, considering the extremely restricted silica liquidus field in the orthoclase-bearing system).

Table 4. Protopyroxene Composition Limits

Protopyroxene Composition	Assemblage	Reversed or Synthesis	T°C	P,bar	P_{H_2O}	Ref
20 wt% $CaMgSi_2O_6$	Proto + Aug	Syn	1365	1	0	1
2.43% CaO, 0.37% Al_2O_3	L+Proto+Olivine	Syn	1302	1	0	2
3.72 0.22	L+Proto+Tridy	Syn	1302	1	0	2
4.10 0.43	L+Proto+Aug+Oliv	Syn	1264	1	0	2
3.98 0.19	L+Proto+Aug+Tridy	Syn	1264	1	0	2
- 1.62	L+Proto+Oliv	Syn	1378	1	0	2
$Li_{.35}Sc_{.35}Mg_{1.3}Si_2O_6$		Syn	1350-650	1	0	3
1.75 wt% Al_2O_3	Proto+Oliv	Syn	1300	1	0	4
0.48%CaO, 3.69%FeO, 0.88%Al_2O_3	Proto+Opx+Oliv +Magnetite+Liq	Syn	1303	1	0	5
2% Al_2O_3	Proto+Opx	Rev	1200	1000	1000	6
$Ca_{.04}Mg_{1.96}Si_2O_6$	Proto+Opx+Aug+Gas	Syn	1240	1000	1000	7
<3% wt% $CaMgSi_2O_6$ in $Mg_2Si_2O_6$		Rev	1400-1460	1	0	8
0.81% CaO			1365	1	0	9
$Wo_{0.5}En_{86.1}Fs_{13.4}$ with 0.12% Al_2O_3		Syn	1151	1	<0.5	10

References for Table 4
1 Boyd and Schairer (1964)   6 Anastasiou and Seifert (1972)
2 Biggar and Clarke (1972)   7 Warner (1975)
3 Smyth and Ito (1977)   8 Longhi and Boudreau (1980)
4 Onuma and Arima (1975)   9 Kushiro (1972)
5 Clarke and Biggar (1972)   10 Huebner and Ross (unpublished)

Protopyroxene. The phase relations of the $MgSiO_3$ polymorphs including protopyroxene have been reviewed thoroughly by Smith (1969). Subsequent studies, although not all in agreement, have done much to eliminate
the enigma that has surrounded this mineral. Carefully reversed experiments reveal that protopyroxene is not a metastable phase but has a true
stability field relative to the other pure $MgSiO_3$ polymorphs, orthopyroxene
and pigeonite (Anastasiou and Seifert, 1972; Longhi and Boudreau, 1980).
Reversed experiments have established that protopyroxene forms limited
solid solutions (Table 4) with respect to $Al_2Al_2O_6$ (Anastasiou and Seifert, 1972) and $CaSiO_3$ (Kushiro, 1972; Longhi and Boudreau, 1980). Synthesis experiments prompted reports of somewhat greater solubilities, but
the reliability of such data is not known. Additional synthesis experiments demonstrate that the protopyroxene structure can be retained as the
$LiScSi_2O_6$ (Smyth and Ito, 1977) or the $Fe_2Si_2O_6$ (Clarke and Biggar, 1972;
Ross and Huebner, unpublished data) components are added to the pure

$Mg_2Si_2O_6$ composition (Table 4). There is a protopyroxene solid-solution series between $Mg_2Si_2O_6$ and $Li_{.35}Sc_{.35}Mg_{1.30}Si_2O_6$; compositions with 10-30% (mol) $LiScSi_2O_6$ component are quenchable to protopyroxene. The protopyroxene field is bounded at 1550-1385°C by a monoclinic phase whose composition range at the solidus includes $Li_{.20}Sc_{.20}Mg_{1.60}Si_2O_6$ (Ito and Steele, 1976); this phase has been termed "En-IV" and has space group $P2/a$ (Takéuchi et $al.$, 1977). At lower temperatures and with increasing $LiScSi_2O_6$ component, there are fields of orthopyroxene (at $Li_{.5}Sc_{.5}MgSi_2O_6$) and clinopyroxene (between $Li_{.6}Sc_{.6}Mg_{.8}Si_2O_6$ and $LiScSi_2O_6$) (Ito and Steele, 1976). Pure $LiScSi_2O_6$ has a structure similar to that of spodumene (Hawthorne and Grundy, 1977).

The orthopyroxene-protopyroxene transition temperature, determined by Atlas (1952) to lie at 985 $\pm$ 10°C, has been reaffirmed by Anastasiou and Seifert (1972), using reversed experiments. This temperature is much lower than that reported by Schwab and Jablonski (1973) for the formation of protopyroxene from "hochklinoenstatite" at 1250°C. Reversal of the orthopyroxene-protopyroxene transition must be done with care; on rapid cooling the protopyroxene usually inverts to pigeonite ("clinoenstatite") with a characteristic combination of lamellar twinning and cracks (Tilley et $al.$, 1964, p. 97).

Protopyroxene normally inverts to pigeonite on cooling. This transformation is essentially completed during quenching in dry systems; x-ray precession photographs of crystals quenched dry from 1100-1200°C reveal no more than a trace of protopyroxene, the remainder being twinned pigeonite (Ross and Huebner, unpublished data). Boyd and Schairer (1964, p. 299) noted that protopyroxene synthesized hydrothermally can be quenched and ground in a mortar but will invert to pigeonite within 12 months if stored in the laboratory. Addition of $LiScSi_2O_6$ component apparently stabilized protopyroxene — the $Li_{.3}Sc_{.3}Mg_{1.4}Si_2O_6$ composition has persisted at standard laboratory conditions for several years.

The classic irreversible path involving orthopyroxene, protoenstatite, and pigeonite is well illustrated with single-crystal x-ray-diffraction measurements (Smyth, 1974) and with powder-diffraction measurements obtained from synthetic $MgSiO_3$ (Ito, 1975) with a Gunier camera operated at high temperature by Howard Evans (unpublished data). The orthopyroxene cell parameters vary smoothly as a function of temperature until conversion to protopyroxene at 1192 $\pm$ 8°C in the example

270

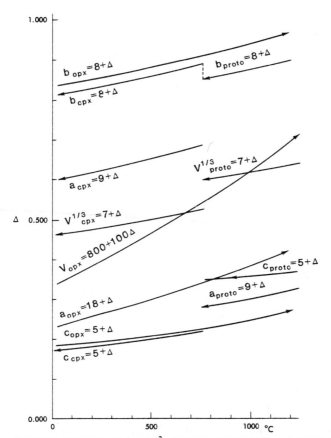

Figure 30. Unit cell dimensions in A and A³ of MgSiO₃ polymorphs as a function of temperature, obtained by H. T. Evans, Jr. (unpublished data). Orthopyroxene synthesized by Ito
(1975) was used as the starting material. The orthopyroxene transformed to protopyroxene at
1190°C, which on cooling transformed to pigeonite (clinoenstatite) at 700-800°C. The volume
decrease at the protopyroxene-pigeonite transition, 2.9%, is largely accounted for by the 2.4%
decrease in c.

illustrated in Figure 30. Protopyroxene has a larger molar volume than

other common magnesian pyroxenes (thus its instability at pressure).

Polymorphic transitions between protopyroxene and other pyroxenes involve

anisotropic changes in cell parameters, most of the volume change being

caused by the large change in c dimension. The protopyroxene to pigeonite

transformation on cooling to 800-600°C is responsible for the character-

istic (100) twinning and sets of cracks approximately perpendicular to c

(because protopyroxenes grown directly from a melt develop these charac-

teristic features on quenching, these features cannot be a relic of the

orthopyroxene to protopyroxene transition). On inverting to pigeonite

at 780°C, b increased 0.4%; a and c and V decreased by 0.8%, 2.4%, and

2.9%, respectively. Most of the accommodation is due to the change in

c, so it is probably not surprising that the cracks appear at a high angle to c.

MANGANIFEROUS PYROXENES

Pure manganese metasilicate, $Mn_2Si_2O_6$, has a pyroxenoid structure at pressures characteristic of the earth's crust (rhodonite or pyroxmanganite, depending upon the temperature and pressure; see Maresch and Mottana, 1976) and is thus beyond the arbitrarily set scope of this chapter. However, $Mn_2Si_2O_6$ does enter quadrilateral pyroxenes to form limited solid solutions (significantly changing their stability fields) and to form ordered pyroxenes (johannsenite, $CaMnSi_2O_6$; and kanoite, $MnMgSi_2O_6$). To this limited extent manganiferous pyroxenes will be discussed in the following paragraphs.

Most existing knowledge of the phase relations for manganiferous pyroxenes has been gleaned from analyses of pyroxenes and pyroxenoids in Mn-rich skarns and metasedimentary pods. Brown et $al.$ ((1980) have summarized the metasilicate assemblages in amphibolite- to granulite-facies rocks at estimated values of $600 \pm 50°C$ and 3-6 kbar (Fig. 31). We can infer from these diagrams that augite can accept substantial $Mn_2Si_2O_6$ component into solid solution but that orthopyroxene is less tolerant, just the opposite of the relationship found in coexisting orthopyroxenes and augites that contain only small amounts of manganese (Table 1). There also appears to be a complete pyroxene solution between $CaMgSi_2O_6$ (diopside) and $CaMnSi_2O_6$ (johannsenite), but not between $CaFeSi_2O_6$ (hedenbergite) and $FeMnSi_2O_6$ (which falls in the pyroxmangite field). Bulk compositions with $Mn/(Ca+Mg+Fe+Mn) > 50\%$ are invariably associated with pyroxenoids at the intermediate temperatures considered by Brown et $al.$ (1980).

The effects of changing temperature (and pressure) on the relationships shown in Figure 31 are only partly known. Gordon et $al.$ (1978, 1981)   infer that a pyroxene of composition $Ca_{.43}Mn_{.69}Mg_{.82}Si_2O_6$, formed at $625 \pm 25°C$ and 6.5 ± 9.5 kbar, exsolved two $C2/c$ clinopyroxenes (Fig. 32), corresponding to manganiferous "augite" (Wo_{34}) and manganiferous "pigeonite" (Wo_6). Although the Ca-poor phase now has space group $P2_1/c$, they inferred crystallization in $C2/c$ because of the presence of anti-phase domain boundaries. Using the transmission electron microscope, Gordon et $al.$ observed a $C2/c$ to $P2_1/c$ transition in the Ca-poor phase

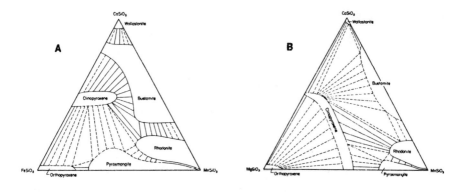

Figure 31. Composition limits of metasilicates with observed (——) and inferred (---) tie lines based upon assemblages in amphibolite granulite facies rocks, 3-6 kbars and 600 ± 50°C. (a) Ca-Fe-Mn pyroxenes and pyroxenoids. The orthopyroxene field is schematic; for these P-T conditions no FeSiO₃-rich metasilicate is stable. (b) Ca-Mg-Mn pyroxenes and pyroxenoids. From Brown *et al.* (in press).

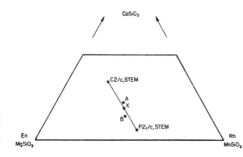

Figure 32. Plot of C2/c and P2₁/c clino-pyroxene compositions in the Ca-Mn-Mg quad-rilateral. The points "A" and "B" are aver-age compositions for the clinopyroxene deter-mined by microprobe: "A" corresponds to the average microprobe composition of the sample subsequently analyzed using analytical elec-tron microscopy (AEM) and X is the average AEM composition. The C2/c↔P2₁/c intergrowth was presumed to have been a single phase at 600°C and 3-6 kbars (see Fig. 31b); subse-quently it exsolved to Ca-rich and Ca-poor phases. Simplified from Figure 1 of Gordon *et al.* (1981).

at 330 ± 20°C, a transition that is very similar crystallographically to the "high"-"low" transition of common Ca-Mg-Fe pigeonite. From this measurement we can infer that the exsolution to form the Ca-poor $C2/c$ phase took place between the temperature of peak metamorphism, 625°C, and the $C2/c \rightarrow P2_1/c$ transition temperature, 330°C (neglecting the effect of pressure).

There is some evidence that pyroxenes may be prevalent at low tem-peratures and pyroxenoids at high temperature for Ca-Mg-Fe-Mn bulk compo-sitions. Lamb *et al.* (1972) experimentally determined that $CaMnSi_2O_6$ clinopyroxene transforms to the bustamite structure (high-temperature form of $CaMnSi_2O_6$) at temperatures *below* 400°C at 2 kbar; at 16 kbar, the transformation is at 530 ± 100°C. By means of synthesis experiments at one atm, Yuquan *et al.* (1977) found a much higher inversion temperature, 830°C, and found that the temperature rises to 900°C as CaTs component

273

(CaAlAlSiO$_6$) is substituted to its limit, 35 mol %. Ito (1972, his Fig.
1) found that with increasing temperature over the range 1200–1400°C at
1 atm, the pyroxmangite composition synthesis field increases relative
to that of pyroxenes. At 1400°C, Ito synthesized pyroxmangite at compositions where Huebner (unpublished data) synthesized orthopyroxene at
800°C. Studies of manganese-free compositions along the join Ca$_2$Si$_2$O$_6$-
Fe$_2$Si$_2$O$_6$ (Lindsley and Munoz, 1969, and Fig. 11 of this review; Shimazaki
and Bunno, 1978, their Fig. 1) also suggest that pyroxenoids are more
prevalent at high than low temperature. Thus, as temperatures are increased above 600°C, pyroxenoids conceivably may replace pyroxenes near
CaFeSi$_2$O$_6$ and CaMnSi$_2$O$_6$ compositions. At 900–1000°C, there will probably
be extensive regions of bustamite and pyroxmangite solid solution in the
ternaries shown in Figure 31; a small field of orthopyroxene is likely to
be the only remaining field of pyroxenes.

Manganjadeite (or manganacmite, NaMn^{3+}Si$_2$O$_6$) is a possible component
by which trivalent manganese could enter pyroxene. The significance of
this substitution in nature is in doubt, even for the violet diopsides
and omphacites termed violan (Mottana et $al.$, 1979). There appear to be
no experimental phase-equilibrium data on Mn^{3+}-bearing pyroxene.

FERRIC IRON IN PYROXENES

Although iron is almost totally ferrous in pyroxenes formed at low
oxygen fugacities, particularly on the moon, ^{VI}Fe$_2^{3+}$ ^{IV}Al$_2$O$_6$ is an important
component in terrestrial basalts (Papike and White, 1979). Most labora-
tory studies of the pyroxene phase relations have been conducted at low
f_{O_2} values to minimize the proportion of ferric iron; there has been no
systematic investigation of the effect of Fe^{3+} on the quadrilateral pyroxene relations. Because Fe^{3+} does not commonly substitute for ^{IV}Si,
one might have expected Fe^{3+} (or Fe$_2$O$_3$) to behave as an inert component
in the pure system CaO–MgO–FeO–SiO$_2$–O$_2$ (diluting the melt, or forming
iron-rich oxide plus a silica mineral in the subsolidus). However,
Huckenholz et $al.$ (1969) reported that at magmatic temperatures, CaMgSi$_2$O$_6$
can accept up to 7 wt % Fe$_2$O$_3$ (at 1292°C) and 33% CaFe$_2^{3+}$SiO$_6$ (at 1175°C)
in solid solution at 1 atm in air. There is abundant evidence to show
that, in the presence of alkali or alumina, at moderate to high f_{O_2}
values (above those of the assemblages iron-wüstite), ^{VI}Fe^{3+} is an important constituent in pyroxene.

Acmite $(NaFe^{+3}Si_2O_6)$ enters both $CaMgSi_2O_6$ and $CaFeSi_2O_6$ to form a complete ternary solid solution, at least at 700°C and 2000 kbar H_2O at moderate f_{O_2} (Nolan, 1969). Yagi (1966) has shown that at 1 atm in air, the progressive addition of $NaFeSi_2O_6$ to $CaMgSi_2O_6$ causes the liquidus temperature to fall ~200°C (at ~70-100 wt % acmite, hematite becomes the liquidus phase) and the solidus to fall continuously to 988°C, the incongruent melting point of pure acmite in air. Ohta *et al.* (1977) studied the system $CaFeSi_2O_6$-$NaFeSi_2O_6$ as a function of f_{O_2}; the addition of acmite component and the raising of the f_{O_2} both appear to lower the liquidus temperature. Two investigators have considered acmitic pyroxenes in model igneous systems. Nolan (1966) first investigated the effects of $NaAlSi_3O_8$ and $NaAlSiO_4$ components on the melting relations of acmite at 1 kbar H_2O pressure; then, by adding $CaMgSi_2O_6$, he was able to model the genesis of alkaline rocks. Bailey and Schairer (1966) considered acmite stability relations in the system Na_2O-Al_2O_3-Fe_2O_3-SiO_2, a model for the peralkaline residua system.

CHROMIUM-BEARING PYROXENES

Experimental studies of chromium partitioning between pyroxenes and silicate melts suggest that both Cr^{+2} and Cr^{+3} can substitute in pyroxene (Schreiber and Haskin, 1976; Huebner *et al.*, 1976). The components likely in these experiments were $Cr_2^{+2}Si_2O_6$ and $Cr_2^{+3}Al_2O_6$; the melts were too deficient in soda to evaluate the ureyite component $(NaCrSi_2O_6)$ and too rich in silica for tetrahedrally-coordinated Cr^{3+} to occur (if indeed it can occur) in the pyroxene as $^{VI}Cr_2^{3+IV}Cr_2^{3+}O_6$. The two most important factors noted were the redox state (which determines the proportion of Cr^{+2}) and the availability of Al^{3+} to permit entry of Cr^{3+} into pyroxene by means of the coupled substitution $^{VI}Cr_2^{+3IV}Al_2O_6$. At redox conditions thought to be characteristic of most terrestrial magmas, chromium is dominantly trivalent, and little or no Cr^{2+} is expected in the pyroxenes. At the more reducing conditions characteristic of basalts of the lunar surface, Cr^{2+} is an important component and both Cr^{2+} and Cr^{3+} are expected to reside in the pyroxenes.

Past investigators of the chromium contents of pyroxene in the subsolidus have adopted laboratory conditions at which all chromium is expected to be trivalent. Vredevoogd and Forbes (1975) determined the solubility of ureyite ($NaCr^{3+}Si_2O_6$) in diopside; their value of 20-25 mol % at 1150°C (in the presence of eskolaite, Cr_2O_3) confirms the earlier value (24%) of Ikeda and Yagi (1972). (Yoder and Kullerud (1970) found a continuous pyroxene solid-solution series, $CaMgSi_2O_6-NaCrSi_2O_6$ at 500-700°C and P_{H_2O} = 2 kbar, rather than the solvus anticipated from natural occurrences.) Vredevoogd and Forbes (1975) further proposed that the ureyite solubility in diopside would diminish with increasing pressure, attaining 0% at 45 kbar. Chromium that enters diopside at such high pressures probably does so as a component *other* than ureyite. Dickey *et al.* (1971) studied the substitution of $CaCr^{+3}AlSiO_6$ in diopside and found that 14 ± 4 wt % could be accommodated. Ikeda and Yagi (1977) found that the *maximum* solubility of $Ca^{VI}Cr^{3+IV}CrSiO_6$ in diopside at 1 atm is 6.4 wt % at 940°C.

ALUMINOUS PYROXENES

Aluminum can enter both the tetrahedral and the octahedral M(1) site of pyroxene by means of several different coupled substitutions. Recent experimental investigations have explored the effects of pyroxene components that are characterized by Ca in M(2) plus Al in tetrahedral coordination: $CaAlAlSiO_6$ (calcium-Tschermak's molecule, CaTs), $CaFe^{3+}AlSiO_6$ (fassaite), and $CaTiAl_2O_6$. In an attempt to organize the recent efforts, this section will consider joins involving $CaMgSi_2O_6$ and modifications to those joins. Investigations of several additional joins will then be mentioned.

The liquidus of the join $CaMgSi_2O_6-CaTiAl_2O_6$ has fields of clinopyroxene, forsterite, perovskite, and spinel (Gupta *et al.*, 1973; Onuma and Kimura, 1978; Akasaka and Onuma, 1980). The minimum temperature on the liquidus, about 1230°C, is associated with the perovskite-olivine field boundary and lies at 34 wt % $CaTiAl_2O_6$. The subsolidus relations on the join have not been determined. The clinopyroxene liquidus field extends toward the compositions $CaAl_2Si_2O_6$ (Onuma and Kimura, 1978), $CaFeAlSiO_6$ (Akasaka and Onuma, 1980), SiO_2 (Gupta *et al.*, 1973), and $CaMgSi_2O_7$ (Onuma and Yagi, 1971). The perovskite field does not extend more than a few percent toward any component other than akermanite. In the case of the $CaAl_2SiO_6$ component, the perovskite field passes into a field of Al- and Ti-rich clinopyroxene that has 16-20% Al_2O_3 and 8-13% TiO_2 (Yang, 1973b). This Ti-rich clinopyroxene occurs at 1250-1300°C and is distinct from the diopsidic clinopyroxene: There is a common liquidus boundary between the two clinopyroxenes (Onuma and Kimura, 1978). Onuma *et al.* (1979) have compared this Ti-rich pyroxene with similar Ti-rich pyroxenes in the Allende meteorite.

Several additional joins have been considered in recent years, but it is impossible to systematize the results because there are no phases or joins common to these investigations. Yoshikawa and Onuma (1975) and Yoshikawa (1977) have reported on the join $NaFeSi_2O_6-CaAl_2SiO_6$; approximately 6% by weight CaTs can dissolve in acmite; pyroxene does not appear on the liquidus. Yoshikawa (1977) added $CaMgSi_2O_6$ to the system; diopside-acmite solutions can contain up to 5-12 wt % $CaAl_2SiO_6$ (depending upon composition) at 1050°C; the solubility limit increases markedly with pressure. The system $CaFeAlSiO_6-CaTiAl_2O_6$ was investigated by Onuma and Akasaka (1979); a subsolidus clinopyroxene field extends from 0% to about 19% $CaTiAl_2O_6$.

FUTURE WORK

Existing knowledge of pyroxene phase relations is derived from studies of *both* natural pyroxene systems and their synthetic analogues. Results of the first kind of study are used to formulate the questions asked in the second kind, and vice versa. There appear to be two major problem areas in pyroxene phase equilibria; one would best be resolved by laboratory phase equilibrium research, the other by analysis of natural pyroxene-bearing assemblages. In addition, many smaller questions might be resolved by laboratory experiments carefully designed to demonstrate attainment of equilibrium.

The first major problem area is the effect of minor elements or components on subsolidus pyroxene assemblages that are potential geothermometers and geobarometers. I have alluded to the fact that orthopyroxene-augite tie lines determined experimentally at ≤1 kbar and 15 kbar appear anomalously short when compared with tie lines measured in granulite-facies metamorphic rocks. Assuming that equilibrium was achieved in the laboratory, the most likely cause of the anomaly is the Al_2O_3- and TiO_2-bearing components that occur in natural pyroxenes but were not considered in the laboratory studies. Similarly, the orthopyroxene + augite = pigeonite reaction should be reinvestigated.

277

Manganese appears to stabilize pigeonite; perhaps other minor elements will significantly affect the pigeonite stability field.

The second major area concerns the melting relations of pyroxenes and their crystallization from magmas. Recent laboratory studies have resulted in complicated model phase diagrams that have not yet been tested by comparison with natural assemblages. Surprisingly, data on natural assemblages of pyroxenes and melt (glass) is sparse. Before investing significant effort in further laboratory studies, it would appear wise to examine pertinent volcanic rocks to determine how good the existing models for tie-line length and orientation, minor-element partitioning, and temperature really are.

Several topical laboratory investigations, suggested by this review, might form the basis of graduate studies research. The stability of protopyroxene in the simple system $CaO-MgO-SiO_2$ is now fairly well known, but data are inadequate to apply the results to naturally-occurring systems. How does Al_2O_3 affect the stability of protopyroxene and orthopyroxene? Under what conditions is there a protopyroxene solid solution from $Mg_2Si_2O_6$ toward $Fe_2Si_2O_6$? Is the liquidus phase always protopyroxene or is it sometimes a "new" pyroxene? A second topical study concerns the manganiferous pyroxene and pyroxenoid. Assemblages containing these phases have potential as geothermometers because the solubility limits appear to change with temperature. This potential could be evaluated in the laboratory. Finally, the possible existence of a "low" pigeonite or clinoenstatite at temperatures below the orthopyroxene field should be examined again, particularly to determine if the observed phenomena occur at the more iron-rich compositions of most natural pyroxenes.

ACKNOWLEDGEMENTS

Preparation of a review chapter presents an author with the opportunity to exhibit his bias and the products of his musing. Many of the attempts to synthesize pyroxene phase relations have not had the benefit of exhaustive review and may not endure with time. Some ideas may bring discredit not only to their perpetrator, but to those who let me do it. Perhaps apologies should be made in advance to those I acknowledge. Mary Woodruff procured many obscure references, Pat Dick labored over versions of the manuscript, typing to give real-time turn around, and Shirely Brown drafted many of the illustrations. Richard Sanford made operational the program for the stereographic projections -- and appeared to enjoy the task immensely! Philip Brown, Donald Lindsley, Donald Peacor and Allen C. Turnock provided preprints of papers and abstracts "in press," Malcolm Ross and Gordon L. Nord, Jr., kindly reviewed the typescript. Finally, the National Aeronautic & Space Administration shared in supporting a decade of U.S.G.S. research on pyroxenes.

ADDENDUM FROM THE PYROXENE SHORT COURSE LECTURE (November 15, 1981)

Recent experiments in the pure system at 900°C and 1 kbar (Turnock and Lindsley, 1981)
.nd at 810°C and 15 kbar (Lindsley *et al.*, 1974), summarized in Figure 33, (p. 288), help de-
'ine the orthopyroxene-augite miscibility gap. The experimental miscibility gap is much nar-
'ower than is revealed in the natural pyroxene subsolidus; in fact, the experimental augite
.olvus at 800-900°C and 1 to 15 kbar approximates the miscibility gap at the presumed solidus
.f gabbroic rocks (Fig. 34, p. 288) more closely than at the subsolidus. Experimental data,
.articularly those of Turnock and Lindsley (1981) at 1 kbar, cannot be reinterpreted to permit
. wider miscibility gap. Some other explanation must be sought.

The problem can be seen in a different perspective by using a T-X section at Fe/(Mg +
'e) = 0.3 (Fig. 35, p. 288). The shaded regions indicate compositions of natural coexisting
.ugite and calcium-poor pyroxenes. Metamorphic pyroxene pairs do not represent temperatures
.uch in excess of 800°C (the upper limit of the granulite facies). Most pyroxenes that
.ormed at higher temperatures re-equilibrated on slow cooling so that the final temperature re-
.orded is 800°C or lower. Compositions of solidus pyroxenes in Figures 4, 5, and 34 were ob-
.ained by analyzing exsolved pyroxenes by bulk analysis methods, assuming that the composi-
.ions represent those of the single-phase precursors to the present pyroxene assemblages, or
.y averaging analyses of individual exsolved phases, weighting each composition according to
.ts relative abundance. The results indicate that orthopyroxene, pigeonite, and augite have
.ompositions of approximately 3, 9, and 39 percent Wo, respectively.

Laboratory data for the range 0 to 15 kbars are also summarized on Figure 35. If ex-
.erimental investigations of pyroxene phase relations accurately model the assemblages in
.ature, we should be able to draw solvi that pass through the compositions of pyroxenes that
.ccur both in nature and in the laboratory. We can sketch a reasonable orthopyroxene solvus,
.ut its change with temperature is slight; we cannot chemically analyze orthopyroxenes with
.ufficient accuracy to make the orthopyroxene solvus a good geothermometer. The augite
.olvus appears to show a greater change in position with change in temperature, but its use
.s a geothermometer is limited by uncertainties in the position of solvus. We are forced to
.raw very strangely curved solvi or to decide that plutonic pyroxenes crystallized at low
.emperatures, on the order of 900°C. Neither alternative is acceptable.

The laboratory tielines are invariably shorter than tie lines in nature, and it is the
.ugite end of the tie line that appears to be most discrepant. Laboratory temperatures ap-
.ear reasonable on the basis of direct observation of igneous systems and numerous experi-
.entally-calibrated geothermometers. The effect of increasing pressure tends to be small.
.inor-element composition is a possible explanation in those cases where the natural augite
.ontains appreciable Al or Ti^{4+}: if the minor-element component were to decrease the
.hermodynamic activity of $CaSiO_3$, the tie lines plotted in Figures 33 and 34 would be longer
.han those found in laboratory experiments on pure pyroxenes, and a correction to the Wo con-
.ent of augite to be plotted on the quadrilateral could be made by removing calcium as
.aAl^{VI}Al^{IV}SiO$_6$, CaTi^{VI}Al$^{IV}_2$O$_6$, etc. (Of course, we should develop a rational basis for re-
.oving Ca, rather than Mg or Fe, with the minor elements.) However, this scheme will not be
.pplicable to relatively pure augites.

Another possibility is that we fail to appreciate the extent of post-crystallization
.agmatic (or near magmatic) re-equilibration. It is conceivable that very sub-calcic augites,
.recipitated at relatively high solidus temperatures (p. 257-259), could exchange magnesium
.nd iron for calcium shortly after precipitation, perhaps while the pyroxene was in chemical
.ommunication with the melt. Such an exchange would be promoted by a gently sloping augite
.olvus ($\partial T/\partial Wo$ is small at high temperatures): the first few degrees of cooling could bring
.bout a marked change in the Wo content of the augite. The petrologist, using textural
.riteria to define the grains of pyroxene precipitated with the magma, would obtain an anom-
.lously long tie line from chemical analysis of the grains.

One of the most significant contributions to pyroxene mineralogy that could be made would
.e a resolution of this discrepancy between what I call the natural and experimental pyroxene
.olvi. This research area poses two difficult questions. First, does the addition of minor
.omponents (Al_2O_3, TiO_2, H_2O, etc.) to the pure pyroxene system $CaO-MgO-FeO-SiO_2$ significantly
.ffect the subsolidus phase relations, perhaps by displacing the augite solvus? Secondly, is
.t possible that igneous augite-pigeonite and augite-orthopyroxene pairs have re-equilibrated
.o that they record a temperature no greater than 1000°-1100°C, even when higher temperatures
.re sought by the most advanced mineralogical techniques?

279

REFERENCES

Akasaka, M. and K. Onuma (1980) The join $CaMgSi_2O_6$-$CaFeAlSiO_6$-$CaTiAl_2O_6$ and its bearing on the Ti-rich pyroxenes. Contrib. Mineral. Petrol., 71, 301-312.

Anastasiou, P. and F. Seifert (1972) Solid solubility of Al_2O_3 in enstatite at high temperatures and 1-5 kb water pressure. Contrib. Mineral. Petrol, 34, 272-287.

Andersen, O. (1915) The system anorthite-forsterite-silica. Am. J. Sci., 39, 407-454.

Anderson, A. T. and T. L. Wright (1972) Phenocrysts and glass inclusions and their bearing on oxidation and mixing of basaltic magmas, Kilauea volcano, Hawaii. Am. Mineral., 57, 188-216.

Atkins, F. B. (1969) Pyroxenes of the Bushveld intrusion, South Africa. J. Petrol., 10, 222-249.

Atlas, L. (1952) The polymorphism of $MgSiO_3$ and solid-state equilibria in the system $MgSiO_3$-$CaMgSi_2O_6$. J. Geol., 60, 125-147.

Bailey, D. K. and J. F. Schairer (1966) The system Na_2O-Al_2O_3-Fe_2O_3-SiO_2 at 1 atmosphere, and the petrogenesis of alkaline rocks. J. Petrol., 7, 114-170.

Bartholome, P. (1962) Iron-magnesium ratio in associated pyroxenes and olivines. Geol. Soc. Am. Bull., Buddington Vol., 1-20.

Bence, A. E. and J. J. Papike (1972) Pyroxenes as recorders of lunar basalt petrogenesis: Chemical trends due to crystal-liquid interaction. Proc. 3rd Lunar Sci. Conf., 431-469.

Bence, A. E., J. J. Papike, S. Sueno, and J. W. Delano (1973) Pyroxene poikiloblastic rocks from the lunar highlands. Proc. 4th Lunar Sci. Conf., 597-611.

Biggar, G. M. and D. B. Clarke (1972) Protoenstatite solid solution in the system CaO-MgO-Al_2O_3-SiO_2. Lithos, 5, 125-129.

Binns, R. A. (1962) Metamorphic pyroxenes from the Broken Hill district, New South Wales. Mineral. Mag., 33, 320-338.

Bohlen, S. R. and E. J. Essene (1978) Igneous pyroxenes from metamorphosed anorthosite massifs. Contrib. Mineral. Petrol., 65, 433-442.

Bonnichsen, B. (1969) Metamorphic pyroxenes and amphiboles in the Biwabik Iron formation, Dunka River area, Minnesota. Mineral. Soc. Am. Spec. Paper, 2, 217-239.

Bowen, N. L. and J. F. Schairer (1935) The system, MgO-FeO-SiO_2. Am. J. Sci., 29, 151-217.

Bowen, N. L., J. F. Schairer, and E. Posnjak (1933) The system CaO-FeO-SiO_2. Am. J. Sci., 26, 193-284.

Boyd, F. R. (1966) Phase relations in the pyroxene quadrilateral. Short Course Lecture Notes. *Chain Silicates*. Am. Geol. Inst., Washington, D. C.

Boyd, F. R. and G. M. Brown (1969) Electron-probe study of pyroxene exsolution. Mineral. Soc. Am. Spec. Paper, 2, 211-216.

Boyd, F. R. and J. L. England (1965) The rhombic enstatite-clinoenstatite inversion. Carnegie Inst. Washington Year Book, 64, 117-120.

Boyd, F. R. and J. F. Schairer (1964) The system $MgSiO_3$-$CaMgSi_2O_6$. J. Petrol., 5, 275-309.

Boyd, F. R. and D. Smith (1971) Compositional zoning in pyroxenes from Lunar Rock 12021, Oceanus Procellarum. J. Petrol., 12, 439-64.

Brown, G. E., C. T. Prewitt, Jr., J. J. Papike, and S. Sueno (1972) A comparison of the structures of low and high pigeonite. J. Geophys. Res., 77, 5778-5789.

Brown, G. M. (1972) Pigeonitic pyroxenes: A review. Geol. Soc. Am. Mem., 132, 523-534.

Brown, G. M. and E. A. Vincent (1963) Pyroxenes from the late stages of fractionation of the Skaergaard Intrusion, East Greenland. J. Petrol., 4, 175-197.

Brown, P. E., E. J. Essene and D. R. Peacor (1980) Phase relations inferred from field data for Mn pyroxenes and pyroxenoids. Contrib. Mineral. Petrol., 74, 417.

Buchanan, D. L. (1979) A combined transmission electron microscope and electron microprobe study of Bushveld pyroxenes from the Bethal Area. J. Petrol., 20, 327-54.

Butler, P. (1969) Mineral compositions and equilibria in the metamorphosed iron formation of the Gagnon region, Quebec, Canada. J. Petrol., 10, 56-101.

Campbell, I. H. and G. D. Borley (1974) The geochemistry of pyroxenes from the lower layered series of the Jimberlana Intrusion, western Australia. Contrib. Mineral. Petrol., 47, 281-297.

Campbell, I. H. and J. Nolan (1974) Factors affecting the stability field of Ca-poor pyroxene and the origin of the Ca-poor minimum in Ca-rich pyroxenes from tholeiitic intrusions. Contrib. Mineral. Petrol., 48, 205-219.

Carstens, H. (1958) Note on the distribution of some minor elements in co-existing ortho- and clinopyroxene. Norsk. Geol. Tidsskr., 38, 257-260.

Champness, P. E. and P. A. Copley (1976) The transformation of a pigeonite to ortho-pyroxene. In *Electron Microscopy in Mineralogy*. pp. 228-233. H.-R. Wenk, ed. New York: Springer-Verlag.

Clarke, D. B. and G. M. Biggar (1972) Calcium-poor pyroxene in the system $CaO-MgO-Al_2O_3-SiO_2-Fe-O_2$. Lithos, 5, 203-216.

Coe, R. S. (1970) The thermodynamic effect of shear stress on the ortho-clino inversion in enstatite and other coherent phase transitions characterized by a finite simple shear. Contrib. Mineral. Petrol., 2, 247-264.

Coe, R. S. and S. H. Kirby (1975) The orthoenstatite to clinoenstatite transformation by shearing and reversion by annealing: Mechanism and potential applications. Contrib. Mineral. Petrol., 52, 29-55.

Coleman, L. C. (1978) Solidus and subsolidus compositional relationships of some coexisting Skaergaard pyroxenes. Contrib. Mineral. Petrol., 66, 221-227.

Dallwitz, W. B., D. H. Green and J. E. Thompson (1966) Clinoenstatite in a volcanic rock from the Cape Vogel area, Papua. J. Petrol., 7, 375-403.

Davidson, L. R. (1968) Variation in ferrous iron-magnesium distribution coefficients of metamorphic pyroxenes from Quairading, Western Australia. Contrib. Mineral. Petrol., 19, 239-259.

Deer, W. A., R. A. Howie, and J. Zussman (1978) *Rock-forming Minerals, Volume 2A: Single Chain Silicates*. New York: John Wiley & Sons, 668 p.

Dickey, J. S., Jr., H. S. Yoder, Jr., and J. F. Schairer (1971) Chromium in silicate-oxide systems. Carnegie Inst. Washington Yearbook, 70, 118-122.

Dodd, R. T., J. E. Grover, and G. E. Brown (1975) Pyroxenes in the Shaw (L-7) chondrite. Geochim. Cosmochim. Acta, 39, 1585-1594.

Duke, M. B. and L. T. Silver (1967) Petrology of eucrites, howardites and mesosiderites. Geochim. Cosmochim. Acta, 31, 1637-1665.

Engel, A. E. J., C. G. Engel, and R. G. Havens (1964) Mineralogy of amphibolite interlayers in the gneiss complex, northwest Adirondack Mountains, New York. J. Geol., 72, 131-156.

Ernst, W. G. (1966) Synthesis and stability relations of ferrotremolite. Am. J. Sci., 264, 37-65.

Evans, H. T., Jr., J. S. Huebner, and J. A. Konnert (1978) The crystal structure and thermal history of orthopyroxene from lunar anorthosite 15415. Earth Planet. Sci. Lett., 37, 476-484.

Ewart, A., W. B. Bryan and J. B. Gill (1973) Mineralogy and geochemistry of the younger volcanic islands of Tonga, S. W. Pacific. J. Petrol., 14, 429-65.

Fleet, M. E. (1974) Partition of major and minor elements and equilibration in coexisting pyroxenes. Contrib. Mineral. Petrol., 44, 259-274.

Foster, W. R. and H. C. Lin (1975) New data on the forsterite-diopside-silica system. Trans. Am. Geophys. Union, 56, 470.

Gordon, W. A., D. R. Peacor and P. E. Brown (1978) Exsolution relationships in a pyroxene of composition $(Mn_{.89}Mg_{.82}Ca_{.28}Fe_{.01})Si_2O_6$. Geol. Soc. Am. Abstr., 10, 409.

Gordon, W. A., D. R. Peacor, P. E. Brown, E. J. Essene, and L. F. Allard (1981) Exsolution relationships in a clinopyroxene of average composition $Ca_{.43}Mn_{.69}Mg_{.82}Si_2O_6$ from Balmat, New York: X-ray diffraction and analytical electron microscopy. Am. Mineral., 66, 127-141.

Grove, T. L. and A. E. Bence (1977) Experimental study of pyroxene-liquid interaction in quartz-normative basalt 15597. Proc. 8th Lunar Sci. Conf., 1549-1579.

Grove, T. L. and M. Raudsepp (1978) Effects of kinetics on the crystallization of quartz normative basalt 15597: An experimental study. Proc. 9th Lunar Planet. Sci. Conf., 585-599.

Grover, J. (1972) The stability of low-clinoenstatite in the system $Mg_2Si_2O_6$-$CaMgSi_2O_6$. Trans. Am. Geophys. Union, 53, 539.

Grover, J. E. and D. H. Lindsley (1972) Ca-Mg-Fe pyroxenes: Subsolidus phase relations in iron-rich portions of the pyroxene quadrilateral. Geol. Soc. Am. Abstr. Programs, 4, 521-522.

Gruenewaldt, G. von and K. Weber-Diefenbach (1977) Coexisting Ca-poor pyroxenes in the Main Zone of the Bushveld Complex. Contrib. Mineral. Petrol., 65, 11-18.

Gupta, A. K., K. Onuma, K. Yagi, and E. G. Lidiak (1973) Effect of silica concentration on the diopsidic pyroxenes in the system diopside-$CaTiAl_2O_6$-SiO_2. Contrib. Mineral. Petrol., 41, 333-344.

Hakli, T. A. (1968) An attempt to apply the Makaopuhi nickel fractionation data to the temperature determination of a basic intrusive. Geochim. Cosmochim. Acta, 32, 449-460.

Harlow, G. E., C. E. Nehru, M. Prinz, G. J. Taylor, and K. Keil (1979) Pyroxenes in Serra de Mage: Cooling history in comparison with Moama and Moore County. Earth Planet. Sci. Lett., 43, 173-81.

Hawthorne, F. C. and H. D. Grundy (1977) Refinement of the crystal structure of $LiScSi_2O_6$ and structural variations in alkali pyroxenes. Canadian Mineral., 15, 50-58.

Hensen, B. J. (1973) Pyroxenes and garnets as geothermometers and barometers. Carnegie Inst. Washington Year Book 72, 527-534.

Hess, H. H. (1941) Pyroxenes of common mafic magmas. Part 2. Am. Mineral. 26, 573-594.

Hess, H. H. (1960) Stillwater Igneous Complex, Montana, a quantitative mineralogical study. Geol. Soc. Am. Mem., 80, 1-230.

Hess, H. H., and E. P. Henderson (1949) The Moore County meteorite: A further study with comment on its primordial environment. Am. Mineral. 36, 494-507.

Himmelberg, G. R. and A. B. Ford (1976) Pyroxenes of the Dufek intrusion, Antarctica. J. Petrol., 17, 219-243.

Ho Sun, C.-O. (1973) *Experimental study of plagioclase/liquid and clinopyroxene/liquid distribution coefficients for Sr and Eu in oceanic ridge basalt system.* Master's Thesis, Lamont-Doherty Geological Observatory, Columbia University.

Hoover, J. D. and T. N. Irvine (1978) Liquidus relations and Mg-Fe partitioning on part of the system Mg_2SiO_4-Fe_2SiO_4-$CaMgSi_2O_6$-$CaFeSi_2O_6$-$KAlSi_3O_8$-SiO_2. Carnegie Inst. Washington Year Book 77, 774-784.

Howie, R. A. (1955) The geochemistry of the Charnockite Series of Madras, India. Trans. Roy. Soc. Edinburgh, 62, 625-768.

Huckenholz, H. G., J. F. Schairer, and H. S. Yoder, Jr. (1969) Synthesis and stability of ferri-diopside. Mineral. Soc. Am. Spec. Paper, 2, 163-177.

Huebner, J. S. (1979) The system CaO-MgO-FeO-SiO_2-"others": Possible model for the crystallization of basaltic magmas. Geol. Soc. Am. Abstr. Programs, 11, 447.

Huebner, J. S. (1980) Refinement of a model for basalt crystallization. Geol. Soc. Am. Abstr. Programs, 12, 452.

Huebner, J. S., B. R. Lipin, and L. B. Wiggins (1976) Partitioning of chromium between silicate crystals and melts. Proc. 7th Lunar Sci. Conf., 1195-1220.

Huebner, J. S. and M. Ross (1972) Phase relations of lunar and terrestrial pyroxenes at one atmosphere. Lunar Sci. III, 410-412. The Lunar Science Institute: Houston, Texas.

Huebner, J. S., M. Ross, and N. Hickling (1975) Significance of exsolved pyroxenes from lunar breccia 77215. Proc. 6th Lunar Sci. Conf., 529-546.

Huebner, J. S., A. Duba, and L. B. Wiggins (1979) Electrical conductivity of pyroxene which contains trivalent cations: laboratory measurements and the lunar temperature profile. J. Geophys. Res., 84, 4652-4656.

Huebner, J. S. and A. C. Turnock (1980) The melting relations at 1 bar of pyroxenes composed largely of Ca-, Mg-, and Fe-bearing components. Am. Mineral., 65, 225-71.

Hytonen, K. and J. F. Schairer (1960) The system enstatite-anorthite-diopside. Carnegie Inst. Washington Year Book, 59, 71-72.

Ikeda, K. and K. Yagi (1972) Synthesis of kosmochlor and phase equilibria in the join $CaMgSi_2O_6$-$NaCrSi_2O_6$. Contrib. Mineral. Petrol., 36, 63-72.

Ikeda, K. and K. Yagi (1977) Experimental study on the phase equilibria in the join $CaMgSi_2O_6$-$CaCrCrSiO_6$ with special reference to the blue diopside. Contrib. Mineral. Petrol., 61, 91-106.

Immega, I. P. and C. Klein, Jr. (1976) Mineralogy and petrology of some metamorphic Precambrian iron formations in southwestern Montana. Am. Mineral., 61, 117-1144.

Irving, A. J. (1978) A review of experimental studies of crystal/liquid trace element partitioning. Geochim. Cosmochim. Acta, 42, 743-770.

Ishii, T. and H. Takeda (1974) Inversion, decomposition and exsolution phenomena of terrestrial and extraterrestrial pigeonites. Mem. Geol. Soc. Japan, 11, 19-36.

Ishii, T. (1975) The relations between temperature and composition of pigeonite in some lavas and their application to geothermometry. Mineral. J. (Japan), 8, 48-57.

Ito, J. (1972) Rhodonite-pyroxmangite peritectic along the join $MnSiO_3$-$MgSiO_3$ in air. Am. Mineral., 57, 865-876.

Ito, J. (1975) High temperature solvent growth of orthoenstatite, $MgSiO_3$, in air. Geophys. Res. Lett., 2, 533-536.

Ito, J. and I. M. Steele (1976) Experimental studies of Li^+ and Sc^{3+} coupled substitution in the Mg-silicates: olivine, clinopyroxene, orthopyroxene, protoenstatite and a new high temperature phase with c = 27 A. Geol. Soc. Am. Abstr. Programs, 8, 937-938.

Jaffe, H. W., P. Robinson, R. J. Tracy and M. Ross (1975) Orientation of pigeonite exsolution lamellae in metamorphic augite: Correlation with composition and calculated optimal phase boundaries. Am. Mineral., 60, 9-28.

Kretz, R. (1961) Some applications of thermodynamics to coexisting minerals of variable composition. Examples: orthopyroxene-clinopyroxene and orthopyroxene-garnet. J. Geol., 69, 361-387.

Kuno, H. (1966) Review of pyroxene relations in terrestrial rocks in the light of recent experimental works. Mineral. J. (Japan), 5, 21-43.

Kuno, H. (1969) Pigeonite-bearing andesite and associated dacite from Asio, Japan. Am. J. Sci., Schairer Vol., 267-A, 257-268.

Kushiro, I. (1972) Determination of liquidus relations in synthetic silicate systems with electron probe analysis: The system forsterite-diopside-silica at 1 atmosphere. Am. Mineral., 57, 1260-1271.

Kushiro, I. and J. F. Schairer (1963) New data on the system diopside-forsterite-silica. Carnegie Inst. Washington Year Book, 62, 95-103.

Lamb, C. L., D. H. Lindsley, and J. E. Grover (1972) Johannsenite-bustamite: Inversion and stability range. Geol. Soc. Am. Abstr. Programs, 4, 571-572.

Lindh, A. (1974) Manganese distribution between coexisting pryoxenes. N. Jahrb. Mineral. Monatsh., 8, 335-345.

Lindsley, D. H., G. M. Brown, and I. D. Muir (1969) Conditions of the ferrowollastonite-ferrohedenbergite inversion in the Skaergaard intrusion, East Greenland. Mineral. Soc. Am. Spec. Paper, 2, 193-201.

Lindsley, D. H. and S. A. Dixon (1976) Diopside-enstatite equilibria at $850°$ to $1400°C$, 5 to 35 kb. Am. J. Sci., 276, 1285-1301.

Lindsley, D. H. and J. E. Grover (1980) Fe-rich pigeonite: a geobarometer. Geol. Soc. Am. Abstr. Programs, 12, 472.

Lindsley, D. H., H. E. King, Jr., and A. C. Turnock (1974) Compositions of synthetic augite and hypersthene coexisting at $810°C$: Application to pyroxenes from lunar highlands rocks. Geophys. Res. Letters, 1, 134-136.

Lindsley, D. H. and J. L. Munoz (1969) Subsolidus relations along the join hedenbergite-ferrosilite. Am. J. Sci., Schairer Volume 267-A, 295-324.

Lindstrom, D. J. (1976) *Experimental study of the partitioning of the transition metals between clinopyroxene and coexisting silicate liquids.* Ph. D. Thesis. University of Oregon, Eugene, Oregon.

Liotard, J. M. and C. Dupuy (1980) Partage des elements de transition entre clino-pyroxene et orthopyroxene-variations avec la nature des roches. Chem. Geol., 28, 307-319.

Lipin, B. R. (1978) The system $Mg_2SiO_4-Fe_2SiO_4-CaAl_2Si_2O_8-SiO_2$ and the origin of Fra Mauro basalts. Am. Mineral., 63, 350-364.

Longhi, J. (1978) Pyroxene stability and the composition of the lunar magma ocean. Proc. 9th Lunar Planet. Sci. Conf., 285-306.

Longhi, J. and A. E. Boudreau (1980) The orthoenstatite liquidus field in the system forsterite-diopside-silica at one atmosphere. Am. Mineral., 65, 563-573.

Mall, A. P. (1973) Distribution of elements in coexisting ferromagnesian minerals from ultrabasics of Kondapalle and Gangineni, Andhra Pradesh, India. N. Jahrb. Mineral. Monatsh., 323-336.

Maresch, W. V. and A. Mottana (1976) The pyroxmangite-rhodonite transformation for the $MnSiO_3$ composition. Contrib. Mineral. Petrol., 55, 69-79.

Matsueda, H. (1974) Immiscibility gap in the system $CaSiO_3-CaFeSi_2O_6$ at low temperatures. Mineral. J. (Japan), 7, 327-343.

Matsui, Y., S. Banno, and I. Hernes (1966) Distribution of some elements among minerals of Norwegian eclogites. Norsk. Geol. Tidsskr., 46, 364-368.

McCallum, I. S. and E. A. Mathez (1975) Petrology of noritic cumulates and a partial melting model for the genesis of Fra Mauro basalts. Proc. 6th Lunar Sci. Conf., 395-414.

Moore, A. C. (1971) The mineralogy of the Gosse pile ultramafic intrusion, central Australia. II. Pyroxenes. J. Geol. Soc. Australia, 18, 243-258.

Morey, G. B., J. J. Papike, R. W. Smith, and P. W. Weiblen (1972) Observations on the contact metamorphism of the Biwabik iron-formation, East Mesabi District. Geol. Soc. Amer. Mem., 135, 225-264.

Mori, T. (1978) Experimental study of pyroxene equilibria in the system of CaO-MgO-FeO-SiO_2. J. Petrol., 19, 45-65.

Mori, T., and D. H. Green (1975) Pyroxenes in the system $Mg_2Si_2O_6-CaMgSi_2O_6$ at high pressure. Earth Planet. Sci. Letters, 26, 277-286.

Mottana, A., G. Rossi, A. Kracher, and G. Kurat (1979) Violan revisited: Mn-bearing omphacite and diopside. Tschermaks Min. Petr. Mitt., 26, 187-201.

Nafziger, R. H. and A. Muan (1967) Equilibrium phase compositions and thermodynamic properties of olivines and pyroxenes in the system MgO-"FeO"-SiO_2. Am. Mineral., 52, 1364-1385.

Nakamura, Y. (1971) Equilibrium relations in Mg-rich part of the pyroxene quadri-lateral. Mineral. J. (Japan), 6, 264-276.

Nakamura, Y. and I. Kushiro (1970) Equilibrium relations of hypersthene, pigeonite and augite in crystallizing magmas: Microprobe study of a pigeonite andesite from Weiselberg, Germany. Am. Mineral., 55, 1999-2015.

Neumann, E.-R. (1976) Compositional relations among pyroxenes, amphiboles and other mafic phases in the Oslo Region plutonic rocks. Lithos, 9, 85-109.

Nielsen, R. L. and M. J. Drake (1979) Pyroxene-melt equilibria. Geochim. Cosmochim. Acta, 43, 1259-1272.

Nobugai, K., M. Tokonami, and N. Morimoto (1978) A study of subsolidus relations of the Skaergaard pyroxenes by analytical electron microscopy. Contrib. Mineral. Petrol., 67, 111-117.

Nolan, J. (1966) Melting-relations in the system $NaAlSi_3O_8-NaAlSiO_4-NaFeSi_2O_6-CaMgSi_2O_6-H_2O$, and their bearing on the genesis of alkaline undersaturated rocks. Quarterly J. Geol. Soc. London, 122, 119-157.

Nolan, J. (1969) Physical properties of synthetic and natural pyroxenes in the system diopside-hedenbergite-acmite. Mineral. Mag., 37, 216-229.

Nord, G. L., Jr. (1980) The composition, structure, and stability of Guinier-Preston zones in lunar and terrestrial orthopyroxene. Phys. Chem. Minerals, 6, 109-128.

Nord, G. L., Jr., and R. H. McCallister (1979) $C2/c \rightarrow P2_1/c$ transition and exsolution microstructure in quenched Wo_{17-32} natural and synthetic clinopyroxenes. Am. Crystallogr. Assoc. Program and Abstracts Ser. 2, vol. 6, 63.

Nwe, Y. Y. and P. A. Copley (1975) Chemistry, subsolidus relations and electron petrography of pyroxenes from the late ferrodiorites of the Skaergaard intrusion, East Greenland. Contrib. Mineral. Petrol., 53, 37-54.

Ohta, K., K. Onuma, and K. Yagi (1977) The system $NaFe^{3+}Si_2O_6-CaFe^{2+}Si_2O_6$ at low oxygen fugacity. J. Fac. Sci. Hokkaido Univ., 17, 487-504.

Onuma, K. and M. Akasaka (1979) A reconnaissance of the system $CaFe^{3+}AlSiO_6-CaTiAl_2O_6$ at 1 atm. J. Fac. Sci. Hokkaido Univ., 19, 29-35.

Onuma, K. and M. Arima (1975) The join $MgSiO_3-MgAl_2SiO_6$ and the solubility of Al_2O_3 in enstatite at atmospheric pressure. J. Japan Assoc. Mineral. Petrol. Econ. Geol., 70, 53-60.

Onuma, K. and M. Kimura (1978) Study of the system $CaMgSi_2O_6-CaFe^{3+}AlSiO_6-CaAl_2SiO_6-CaTiAl_2O_6$: II. The join $CaMgSi_2O_6-CaAl_2SiO_6-CaTiAl_2O_6$ and its bearing on Ca-Al-rich inclusions in carbonaceous chondrite. J. Fac. Sci. Hokkaido Univ., 18, 215-236.

Onuma, K., M. Kimura and K. Yagi (1979) Significance of the system $CaMgSi_2O_6-CaAl_2SiO_6-CaTiAl_2O_6$ to Ca-Al-rich inclusions in carbonaceous chondrites. Mem. National Inst. Polar Res. Spec. Issue no. 12. *Proc. of the Third Symposium on Antarctic Meteorites.* T. Nagata, ed., National Inst. Polar Res. Tokyo, 134-143.

Onuma, K. and K. Yagi (1971) The join $CaMgSi_2O_6-Ca_2MgSi_2O_7-CaTiAl_2O_6$ in the system $CaO-MgO-TiO_2-SiO_2$ and its bearing on the titanpyroxenes. Mineral. Mag., 38, 471-480.

Osborn, E. F. (1942) The system $CaSiO_2$-diopside-anorthite. Am. J. Sci., 240, 751-788.

Osborn, E. F. and A. Muan (1960) Phase equilibrium diagrams of oxide systems, plate 2, published by Am. Ceramic Soc. and Edward Orton, Jr., Ceramic Foundation.

Osborn, E. F. and D. B. Tait (1952) The system diopside-forsterite-anorthite. Am. J. Sci., Bowen vol., 413-433.

Papike, J. J., A. E. Bence, and M. A. Ward (1972) Subsolidus relations of pyroxenes from Apollo 15 basalts. In *The Apollo 15 Lunar Samples.* pp. 144-148, J. W. Chamberlain and C. Watkins, eds., The Lunar Sci. Inst., Houston.

Papike, J. J., F. N. Hodges, A. E. Bence, M. Cameron, and J. M. Rhodes (1976) Mare basalts: Crystal chemistry, mineralogy and petrology. Rev. Geophys. Space Phys., 14, 475-540.

Papike, J. J. and C. White (1979) Pyroxenes from planetary basalts: Characterization of "other" than quadrilateral components. Geophys. Res. Lett. 6, 913-916.

Philpotts, A. R. (1966) Origin of the anorthosite-mangerite rocks in southern Quebec, J. Petrol., 7, 1-64.

Philpotts, A. R. and N. H. Gray (1974) Inverted clinobronzite in eastern Connecticut diabase. Am. Mineral., 59, 374-377.

Podpora, C. and D. H. Lindsley (1979) Fe-rich pigeonites: Minimum temperatures of stability in the Ca-Mg-Fe quadrilateral. Trans. Am. Geophys. Union, 60, 420-421.

Pollack, S. S. and W. D. Ruble (1964) X-ray identification of ordered and disordered ortho-enstatite. Am. Mineral., 49, 983-992.

Presnall, D. C., S. A. Dixon, J. R. Dixon, T. H. O'Donnell, N. L. Brenner, R. L. Schrock, and D. W. Dycus (1978) Liquidus phase relations on the join diopside-forsterite-anorthite from 1 atm. to 20 kbar: their bearing on the generation and crystallization of basaltic magma. Contrib. Mineral. Petrol., 66, 203-220.

Presnall, D. C., J. R. Dixon, T. H. O'Donnell, and S. A. Dixon (1979) Generation of mid-ocean ridge tholeiites. J. Petrol., 20, 3-35.

Prewitt, C. T., G. E. Brown and J. J. Papike (1971) Apollo 12 clinopyroxenes: High temperature X-ray diffraction studies. Proc. 2nd Lunar Sci. Conf., 59-68.

Raleigh, C. B., S. H. Kirby, N. L. Carter and H. G. Ave Lallemant (1971) Slip and the clinoenstatite transformation as competing rate processes in enstatite. J. Geophys. Res., 76, 4011-4022.

Rapoport, P. A., and C. W. Burnham (1973) Ferrobustamite: The crystal structures of two Ca, Fe bustamite-type pyroxenoids. Z. Kristallogr., 138, 419-438.

Reid, A. M., R. J. Williams, and H. Takeda (1974) Coexisting bronzite and clino-bronzite and the thermal evolution of the Steinbach meteorite. Earth Planet. Sci. Letters, 22, 67-74.

Ricker, R. W. (1952) Phase equilibria in the Quaternary system. Ph. D. Thesis, Pennsylvania State University, University Park, Pennsylvania.

Ricker, R. W. and E. F. Osborn (1954) Additional phase equilibrium data for the system $CaO-MgO-SiO_2$. J. Am. Ceram. Soc., 37, 133-139.

Riecker, R. E. and T. P. Rooney (1967) Deformation and polymorphism of enstatite under shear stress. Geol. Soc. Am. Bull., 78, 1045-1054.

Roedder, E. (1965) A laboratory reconnaissance of the liquidus surface in the pyroxene system En-Di-Hd-Fs ($MgSiO_3$-$CaMgSi_2O_6$-$CaFeSi_2O_6$-$FeSiO_3$) Am. Mineral., 50, 696-703.

Ross, M. and J. S. Huebner (1975) A pyroxene geothermometer based on composition-temperature relationships of naturally occurring orthopyroxene, pigeonite, and augite. Internat. Conf. Geothermometry and Geobarometry. The Pennsylvania State University, University Park, PA.

Ross, M. and J. S. Huebner (1979) Temperature-composition relationships between naturally occurring augite, pigeonite, and orthopyroxene at one bar pressure. Am. Mineral., 64, 1133-1155.

Ross, M., J. S. Huebner, and E. Dowty (1973) Delineation of the one atmosphere augite-pigeonite miscibility gap for pyroxenes from Lunar Basalt 12021. Am. Mineral., 58, 619-635.

Rutstein, M. S. (1971) Re-examination of the wollastonite-hedenbergite ($CaSiO_3$-$CaFeSi_2O_6$) equilibria. Am. Mineral., 56, 2040-2052.

Rutstein, M. and R. A. Yund (1969) Unit-cell parameters of synthetic diopside-hedenbergite solid solutions. Am. Mineral., 54, 238-245.

Saxena, S. K. (1968) Crystal-chemical aspects of distribution of elements among certain coexisting rock-forming silicates. N. Jahrb. Mineral. Abh., 108, 292-323.

Saxena, S. K. (1969) Distribution of elements in coexisting minerals and the problem of chemical disequilibrium in metamorphosed basic rocks. Contrib. Mineral. and Petrol., 20, 177-197.

Saxena, S. K. and S. Ghose (1971) Mg^{2+}-Fe^{2+} order-disorder and the thermodynamics of the orthopyroxene crystalline solution. Am. Mineral., 56, 532-559.

Schairer, J. F. (1942) The system CaO-FeO-Al_2O_3-SiO_2: 1, results of quenching experiments on five joins. Am. Ceramic Soc., 25, 241-274.

Schreiber, H. D. and L. A. Haskin (1976) Chromium in basalts: Experimental determination of redox states and partitioning among synthetic silicate phases. Proc. 7th Lunar Sci. Conf., 1221-1259.

Schreyer, W., D. Stepto, K. Abraham, and W. F. Muller (1978) Clinoeulite (magnesian clinoferrosilite) in a eulysite of a metamorphosed iron formation in the Vredeford structure, South Africa. Contrib. Mineral. Petrol., 65, 351-61.

Schwab, R. G. (1969) Die Phasenbeziehungen im System $CaMgSi_2O_6$-$CaFeSi_2O_6$-$MgSiO_3$-$FeSiO_3$. Fortschr. Mineral., 46, 188-273.

Schwab, R. G. and K. H. Jablonski (1973) Der Polymorphismus der Pigeonite. Fortschr. Mineral., 50, 223-263.

Schwab, R. G. and M. Schwerin (1975) Polymorphie und Entmischungsreaktionen der Pyroxene im System Enstatit ($MgSiO_3$)-Diopsid ($CaMgSi_2O_6$). N. Jb. Miner. Abh., 124, 223-245.

Sclar, C. B., L. C. Carrison and C. M. Schwartz (1964) High-pressure stability field of clinoenstatite and the orthoenstatite-clinoenstatite transition. Trans. Am. Geophys. Union, 45, 121.

Segnit, E. R. (1956) The section $CaSiO_3$-$MgSiO_3$-Al_2O_3. Mineral. Mag. 31, 255-264.

Shimazaki, H. and M. Bunno (1978) Subsolidus skarn equilibria in the system Ca-SiO_3-$CaMgSi_2O_6$-$CaFeSi_2O_6$-$CaMnSi_2O_6$. Canadian Mineral., 16, 539-545.

Shimazaki, H. and T. Yamanaka (1973) Iron-wollastonite from skarns and its stability relation in the $CaSiO_3$-$CaFeSi_2O_6$ join. Geochem. J., 7, 67-79.

Shiraki, K., N. Kuroda, H. Urano, and S. Maruyama (1980) Clinoenstatite in boninites from the Bonin Islands, Japan. Nature, 285, 31-32.

Simmons, E. C., D. H. Lindsley and J. J. Papike (1974) Phase relations and crystallization sequence in a contact-metamorphosed rock from the Gunflint Iron Formation, Minnesota. J. Petrol., 15, 539-65.

Smith, D. (1971) Stability of the assemblage iron-rich orthopyroxene-olivine-quartz. Am. J. Science, 271, 370-382.

Smith, D. (1974) Pyroxene-olivine-quartz assemblages in rocks associated with the Nain anorthosite massif, Labrador. J. Petrol., 15, 58-78.

Smith, J. V. (1969) Crystal structure and stability of the $MgSiO_3$ polymorphs; physical properties and phase relations of Mg,Fe pyroxenes. Mineral. Soc. Am. Spec. Pap., 2, 3-29.

Smyth, J. R. (1974) Experimental study on the polymorphism of enstatite. Am. Mineral., 59, 345-352.

Smyth, J. R. and J. Ito (1977) The synthesis and crystal structure of a magnesium-lithium-scandium protopyroxene. Am. Mineral., 62, 1252-1257.

Subramaniam, A. P. (1962) Pyroxenes and garnets from charnockites and associated granulites. Geol. Soc. Am. Bull., Buddington Vol., 21-36.

Takeda, H. and M. Miyamoto (1977) Inverted pigeonites from lunar breccia 76255 and pyroxene-crystallization trends in lunar and achondritic crusts. Proc. 8th Lunar Sci. Conf., 2617-2626.

Takeda, H., M. Miyamoto, T. Ishii, and G. E. Lofgren (1975) Relative cooling rates of mare basalts at the Apollo 12 and 15 sites as estimated from pyroxene exsolution data. Proc. 6th Lunar Sci. Conf. 987-996.

Takeuchi, Y., Y. Kudoh, and J. Ito (1977) High-temperature derivative structure of pyroxene. Proc. Japan Acad., 53, 60-63.

Tilley, C. E., H. S. Yoder, Jr., and J. F. Schairer (1964) New relations on melting of basalts. Carnegie Inst. Washington Year Book, 63, 92-97.

Turnock, A. C. (1962) Preliminary results on melting relations of synthetic pyroxene on the diopside-hedenbergite join. Carnegie Inst. Washington Year Book, 61, 82.

Turnock, A. C., D. H. Lindsley, and J. E. Grover (1973) Synthesis and unit-cell parameters of Ca-Mg-Fe pyroxenes. Am. Mineral., 58, 50-59.

Virgo, D. and S. S. Hafner (1969) Fe^{2+}, Mg order-disorder in heated orthopyroxenes. Mineral. Soc. Am. Spec. Pap., 2, 67-81.

Virgo, D. and M. Ross (1973) Pyroxenes from Mull andesites. Carnegie Inst. Washington Year Book, 72, 535-540.

Vredevoogd, J. J. and W. C. Forbes (1975) The system diopside-ureyite at 20 Kb. Contrib. Mineral. Petrol., 52, 147-156.

Walker, K. R., N. G. Ware, and J. F. Lovering (1973) Compositional variations in the pyroxenes of the differentiated Palisades Sill, New Jersey, Geol. Soc. Am. Bull., 84, 89-110.

Warner, R. D. (1975) New experimental data for the system $CaO-MgO-SiO_2-H_2O$ and a synthesis of inferred phase relations. Geochim. Cosmochim. Acta, 39, 1413-1421.

Warner, R. D., and W. C. Luth (1974) The diopside-orthoenstatite two-phase region in the system $CaMgSi_2O_6-Mg_2Si_2O_6$. Am. Mineral., 59, 98-109.

Wheeler, E. P., 2nd (1965) Fayalite olivine in Northern Newfoundland-Labrador, Canadian Mineral., 8, 339-346.

Wilson, A. F. (1976) Aluminium in coexisting pyroxenes as a sensitive indicator of changes in metamorphic grade within the mafic granulite terrane of the Fraser Range, Western Australia. Contrib. Mineral. Petrol., 56, 255-277.

Wood, B. J. and R. G. J. Strens (1971) The orthopyroxene geobarometer. Earth Planet. Sci. Lett., 11, 1-6.

Wright, T. L. and R. T. Okamura (1977) Cooling and crystallization of tholeiitic basalt, 1965 Makaopuhi Lava Lake, Hawaii. Geol. Surv. Prof. Pap., 1004, 1-78.

Wright, T. L. and P. W. Weiblen (1967) Mineral composition and paragenesis in tholeiitic basalt from Makaopuhi lava lake, Hawaii. Geol. Soc. Am. Abstr. Programs, 242.

Yagi, K. (1966) The system acmite-diopside and its bearing on the stability relations of natural pyroxenes of the acmite-hedenbergite-diopside series. Am. Mineral., 51, 976-1000.

Yang, H.-Y. (1973a) Crystallization of iron-free pigeonite in the system anorthite-diopside-enstatite-silica at atmospheric pressure. Am. J. Sci. 273, 488-497.

Yang, H.-Y. L973b) Synthesis of an Al- and Ti-rich clinopyroxene in the system $CaMgSi_2O_6-CaAl_2SiO_6-CaAlTiAl_2O_6$. Trans. Am. Geophys. Union, 54, 478.

Yang, H.-Y. and W. R. Foster (1972) Stability of iron-free pigeonite at atmospheric pressure. Am. Mineral., 57, 1232-1241.

Yoder, H. S., Jr., and G. Kullerud (1970) Kosmochlor and the chromiteplagioclase association. Carnegie Inst. Washington Year Book, 69, 155-157.

Yoder, H. S., Jr., C. E. Tilley, and J. F. Schairer (1963) Pyroxenes and associated minerals in the crust and mantle. Carnegie Inst. Washington Year Book, 62, 84-95.

Yoshikawa, K. (1977) Phase relations and the nature of clinopyroxene solid solutions in the system $NaFe^{3+}Si_2O_6$-$CaMgSi_2O_6$-$CaAl_2SiO_6$. J. Fac. Sci., Hokkaido Univ., 17, 451-485.

Yoshikawa, K. and K. Onuma (1975) The join $NaFeSi_2O_6$-$CaAl_2SiO_6$ at 1 atmospheric and high pressure: Part I. Phase relations at 1 atmospheric pressure in air. J. Jpn. Assoc. Mineral. Petrol. Econ. Geol., 70, 335-346.

Yuquan, S., Y. Danian, and G. Jingxiong (1977) Experimental studies of $CaMnSi_2O_6$-$CaAlSiAlO_6$ system. Scientia Geologia Sinica, 4, 343-354.

ADDENDUM: Figures 33, 34 and 35 are referred to on p. 279.

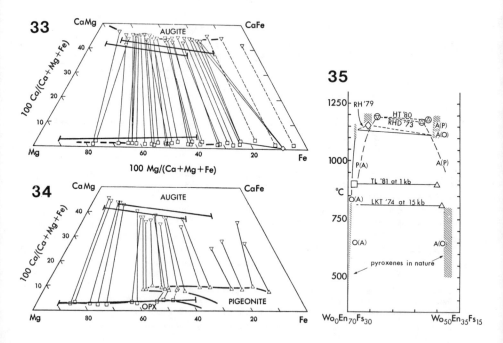

Figure 33. Generalized subsolidus diagram for coexisting quadrilateral pyroxenes at temperatures several hundred degrees below the solidus. This figure is identical to Figure 2 but for the addition of three solvi determined by laboratory experiment using pure quadrilateral pyroxene compositions and shown as straight lines. From top to bottom, these solvi are augite coexisting with orthopyroxene at 810°C and 15 kbar (Lindsley *et al.*, 1974); augite coexisting with orthopyroxene at 900°C and 1 kbar (Turnock and Lindsley, 1981); and orthopyroxene coexisting with augite, also at 900°C and 1 kbar (Turnock and Lindsley, 1974).

Figure 34. Generalized solidus diagram for coexisting quadrilateral pyroxenes. This figure is identical to Figure 4 but for the addition of the same experimentally determined solvi described in Figure 33.

Figure 35. Generalized temperature-composition section across the pyroxene quadrilateral at Fe/(Mg + Fe) = 0.30. Shaded areas represent compositions of coexisting orthopyroxene, augite, and pigeonite in nature; temperatures are best guesses based on available geothermometry (granulite facies <800°C; basaltic magma crystallization at 1150-1250°). The miscibility gaps implied by the natural pyroxenes are wider than the gaps suggested by laboratory data: HT '80 (Huebner and Turnock, 1980); RHD '73 (Ross *et al.*, 1973), including the dashed pigeonite and augite solvi, P(A) and A(P); RH '79 (Ross and Huebner, 1979); TL '80 (Turnock and Lindsley, 1981); and LKT '74 (Lindsley *et al.*, 1974).

Chapter 6

PHASE EQUILIBRIA OF PYROXENES at PRESSURES > 1 ATMOSPHERE

Donald H. Lindsley

INTRODUCTION

This chapter presents a summary of phase-equilibrium data on Ca-Mg-Fe- ("quadrilateral") and some Mn-bearing pyroxenes for pressures greater than 1 atm. The following chapter treats experiments that include mono- and trivalent cations--especially Al. The literature through early 1977 has been exhaustively surveyed by Deer *et al.* (1978); accordingly, this review makes no attempt to duplicate that comprehensive coverage. Instead, the emphasis here is on more recent work, together with critical summaries where older studies appear to in conflict. The notation, "DHZ, p. xx" directs the reader to the more extensive references in Deer, Howie, and Zussman as appropriate.

The approach is to consider the effect of increasing pressure (usu- ally load or total pressure) on (1) the phase relations of the end-member pyroxenes enstatite (En), ferrosilite (Fs), diopside (Di), and hedenbergite (Hd); (2) the phase relations of the binaries that bound the quadrilateral; and (3) phase relations within the quadrilateral itself. Some studies in- volving H_2O or CO_2 or both are also included.

PHASE RELATIONS OF END-MEMBER PYROXENES

Enstatite

The complex behavior of enstatite at 1 atm becomes simpler with in- creasing pressure, as shown in Figure 1. The celebrated incongruent melting of enstatite dies out in the first few kilobars, although the exact pressure (shown as 2 kbar in Fig. 1) is not well known. Kushiro *et al.* (1968) have shown that the incongurent melting persists to higher pressures under water-saturated conditions, with a concomitant depression of both liquidus and solidus. Eggler (1973) demonstrated that the mole fraction of H_2O in H_2O-CO_2 mixtures must be at least 0.6 to produce in- congruent melting at 20 kbar.

There is disagreement in the literature over the position of the OEn- PEn boundary (DHZ, p. 54), but the results of Chen and Presnall (1975) strongly support the boundary shown in Figure 1. The OEn-CEn boundary

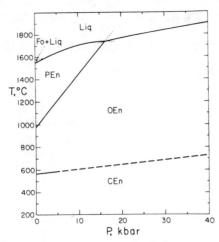

Figure 1. Phase relations for En (Mg₂Si₂O₆). Fo (forsterite) + Liq (liquid) field and high-P melting from Boyd and England (1965); PEn (protoenstatite) field from Chen and Presnall (1975); OEn-CEn (orthoenstatite-clinoenstatite) boundary from Grover (1972; low P) and from Boyd *et al.* (1964; high P).

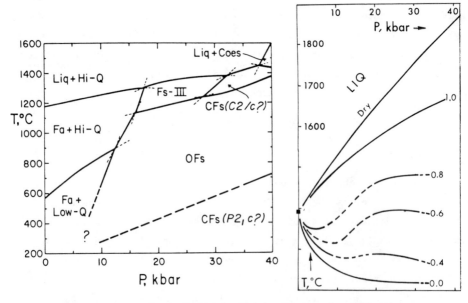

Figure 2. Phase relations for Fs (Fe₂Si₂O₆). High-low Qtz curve from Cohen and Klement (1967); OFs-CFs (orthoferrosilite-clinoferrosilite) boundary at low T from Akimoto *et al.* (1965); Fa + Qtz - OFs boundaries from Bohlen *et al.* (1980a); high-T relations from Lindsley (1965).

Figure 3. Melting relations of Di (CaMgSi₂O₆), both dry and in the presence of H₂O and CO₂. Dry melting from Williams and Kennedy (1969); vapor-saturated melting from Rosenhauer and Eggler (1975). Curves are labelled with values of CO₂/(CO₂ + H₂O).

must remain controversial, because most of it is based on piston-cylinder experiments that involved nonhydrostatic pressure, and it seems clear that shearing stress at least enhances the rate of formation of CEn and possibly enlarges its stability field (DHZ, p. 32-33; 52-53) as well. The conversion of OEn to CEn in an unquestionably hydrostatic environment (in molten $MgCl_2$ within sealed silica-glass tubes; Grover, 1972) is here taken as proof that CEn is stable at low temperatures. The dashed boundary from 5 to 40 kbar in Figure 1 should be viewed as an upper limit because of the possible stabilizing effect of shearing stress on CEn.

Ferrosilite

Pyroxene of $FeSiO_3$ composition is not stable relative to fayalite + SiO_2 at low pressures, as shown in Figure 2. Earlier experiments on the reaction Fa + Q = Fs have been largely supplanted by the very careful study of Bohlen *et al.* (1980a). The OFs-CFs ($P2_1/c$) boundary is extrapolated from data obtained at 40-70 kbar in a tetrahedral press by Akimoto *et al.* (1965); it is more plausible than the nearly isothermal boundary at $\sim$800° shown by Lindsley (1965) which was later shown (Lindsley and Munoz, 1969a) to have been affected by shearing stress in the piston-cylinder apparatus. Shearing stresses are presumed to have been smaller in the tetrahedral press, but they probably were still present, and, like the case for CEn, the boundary in Figure 2 should be considered as an upper limit for $P2_1/c$ CFs. (Afficionados of the bizarre may note with wry amusement this rejection of a bracketed curve for one that is extrapolated from a set of synthesis points plus a single reversal!)

The fields of Fs-III and CFs ($C2/c$) shown at temperatures above the OFs field deserve some comment. Arguing mainly by analogy with protoenstatite, Lindsley (1965) tentatively interpreted those two phases as inversion products of a hypothetical protoferrosilite. Later studies on a variety of Ca-Fe pyroxenes (Lindsley, 1967) showed the general trend that, at the solidus, a pyroxenoid is replaced by Cpx with increasing pressure. Hence, the fields of Fs-III and CFs ($C2/c$) are now interpreted as being stable.

Diopside

The melting of Di as a function of pressure is shown in Figure 3. The dry melting curve is based on differential thermal analysis (DTA; Williams and Kennedy, 1969) and is slightly steeper than the melting curve based on the quenching method (Boyd and England, 1963), although the discrepancy probably mainly reflects different corrections for friction in piston-cylinder apparatus. The still lower melting temperatures reported by Yoder (1952) are very similar to those in the presence of pure CO_2 (Rosenhauer and Eggler, 1975) and may reflect solution of the argon pressure medium in the melt. The effects of water, of CO_2, and of H_2O-CO_2 mixtures on the melting of Di are also shown in Figure 3 (see Yoder, 1965; Eggler, 1973; Rosenhauer and Eggler, 1975). The (relatively minor) incongruent melting reported for Di at 1 atm (see preceding chapter) has not been identified positively at high pressures, although Rosenhauer and Eggler may have detected it using DTA techniques (1975, p. 477).

Hedenbergite

Like En and Fs, Hd shows simpler phase relations with increasing pressure. The inversion of Hd to ferrobustamite (B) with increasing temperature has ended by ∿13 kbar. Phase-rule considerations do not require the Hd-B boundary and the melting curves to coincide at a point, but they appear to do so within experimental uncertainty (Fig. 4). It is possible that the B + Liq field terminates in a singular point a fraction of a kilobar below the pressure of the Hd + B + Liq point. Solid solution with $CaMnSi_2O_6$ extends the stability of the bustamite phase to higher pressures. For example, pure $CaMnSi_2O_6$ (bustamite) remains stable relative to the Cpx form (johannsenite) at pressures up to 25 ± 2 kbar at 950°C (Lamb *et al.*, 1972). The redox stability of Hd at 2 kbar has been determined by Gustafson (1974).

PHASE RELATIONS OF BINARY PYROXENES

Enstatite-Ferrosilite

Most experimental studies on the En-Fs join at pressures greater than 1 atm have concentrated on Fe-rich compositions at subsolidus temperatures (DHZ, p. 77-82). Increasing pressure and $MgSiO_3$ content both stabilize Fe-rich pyroxene whereas increasing temperature destabilizes it relative to olivine + quartz; their combined effects are shown in Figure 5. $MnSiO_3$ also stabilizes Fe-rich pyroxene (Fig. 6), the effect per mole percent $MnSiO_3$ being about 0.4 of that for $MgSiO_3$ (Bohlen *et al.*, 1980b; Bohlen and Boettcher, 1980). It is hoped that these results will put an end to

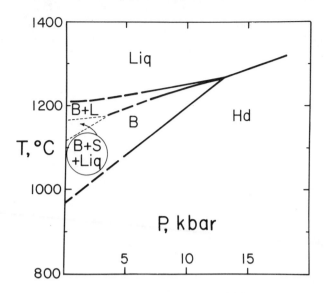

Figure 4. Phase relations of Hd ($CaFeSi_2O_6$). After Lindsley (1967): B = ferro-bustamite; S = SiO_2-phase; Liq (or L) = liquid.

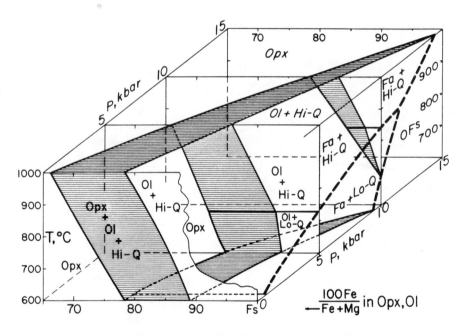

Figure 5. Block diagram illustrating the effects of T, P, and Fe/(Fe + Mg) on the stability of Fe-rich Opx (orthopyroxene) relative to fayalitic olivine (Ol;Fa) + Qtz. The stable SiO_2 phase on the front face (0 kbar isobar) is tridymite above ~848°C. Curves for pure Fs are from Bohlen et al. (1980a); relations in the Mg-bearing portion are adapted from Smith (1971) and from Bohlen and Boettcher (1980). The ruled surfaces show the volume in which Opx + Ol + Q (high or low) are stable.

293

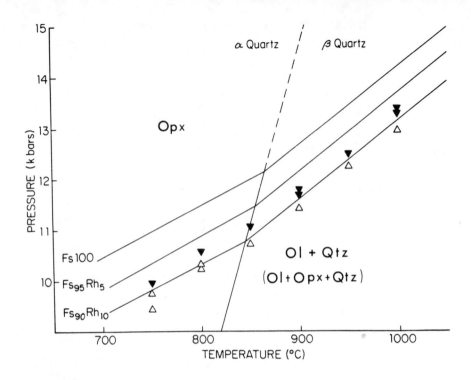

Figure 6. Effect of 5% and of 10% MnSiO$_3$(Rh) on the stability of Fs. Opx of composition Fs$_{100}$, Fs$_{95}$Rh$_5$, and Fs$_{90}$Rh$_{10}$ is stable at pressures above the labelled lines. At lower pressures the assemblages Ol + Qtz or Ol + Opz + Qtz are stable, depending on bulk composition and pressure. After Bohlen *et al*. (1980b). Data points are shown for Fs$_{90}$Rh$_{10}$.

the widespread, but pernicious, convention of combining Fe^{2+} and Mn for calculation of the Fs component of pyroxenes: Mn follows Fe geochemically, but their effects on pyroxene stability are quite different, and they should be kept separate!

Figure 5 shows boundaries for the reaction Fe-rich Opx = Fe-depleted Opx + Fe-rich Ol + Q; this reaction should be stable approximately between the fayalite-magnetite-quartz (FMQ) and iron-quartz-fayalite (IQF) oxygen buffers. Conditions *slightly* more oxidizing than FMQ should oxidize the Fe-rich olivine by the reaction $3Fe_2SiO_4 + O_2 = 2Fe_3O_4 + 3SiO_2$, but should otherwise have negligible effect on the phase boundaries. Accordingly, one might expect a redox phase diagram qualitatively like the isobaric sections in Figure 5, but with Mt replacing Ol. Fonarev *et al*. (1976) show a 1 kbar T-X diagram for NNO buffer (which is very slightly more oxidizing than FMQ) in which the Opx *vs* Opx + Mt + Q boundary has

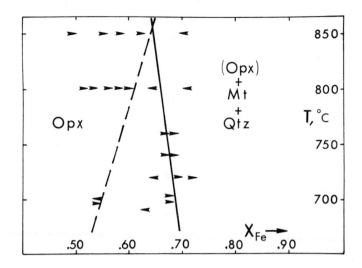

Figure 7. Phase diagram (P_{H_2O} = 1 kbar; fO_2 of the NNO buffer) for Fe-rich Opx co-existing with Mt (magnetite) + Qtz. Data and dashed curve from Fonarev *et al.* (1976); solid curve, boundary considered to be more plausible (see text). Symbols: ▶ , Opx + Mt + Qtz → Fe-enriched Opx + O_2; ◀ , Opx + O_2 → Fe-depleted Opx + Mt + Qtz.

a *positive* slope--that is, the Opx *gains* $FeSiO_3$ with increasing temper-ature (Fig. 7). The boundary shown by Fonarev *et al.* (1976) actually violates several of their data points (some of which are mutually incom-patible), and in this reviewer's opinion, a phase boundary with a nega-tive slope (solid line in Fig. 7) would be in better accord both with their data and with theoretical calculations of the redox reaction. Note, however, that the revised boundary also violates some data.

Several workers have determined the distribution of Fe and Mg be-tween Opx and olivine. Matsui and Nishizawa (1974) reported exchange experiments (unreversed) at 800-1300°C, 30 kbar, and developed a thermo-dynamic solution model for Ca-free Opx. See their paper or DHZ, p. 74-81, for additional references.

Diopside-Enstatite

Phase relations on the Di-En join have been studied very extensive-ly at high pressures (DHZ, p. 86-89). A thermodynamic solution model (Lindsley *et al.*, 1981; see also the chapter by Grover in this volume) has been fit to the reversed data, and has been used to calculate the isobaric, subsolidus T-X diagrams shown in Figure 8a-e. The symbols on the figures show the experimental data (brackets and half-brackets) used

295

as input to the model. The agreement between the calculated phase boundaries and the original data indicates the success of the model. Because
the model is compatible with virtually all the reversed experimental data
in the range 1 atm to 40 kbar, the diagrams in Figure 8 are preferred to
earlier diagrams that were usually fit by eye by the original experimen-
ters to their data. As noted by recent workers, the width of the En_{ss}-Di_{ss}
field increases with increasing pressure, and this effect is enhanced at
high temperatures.

The solution model assumes that Pig has the same $C2/c$ structure as
Di at high temperatures (see chapter by Cameron in this volume). Pig,
as a discrete phase that can coexist with Di_{ss}, is not found above ∿21
kbar; above that pressure there is an extensive, continuous field of Cpx
at high temperatures (Fig. 8d,e).

Kushiro's (1969) study at 20 kbar indicated a large Pig + Di_{ss} field
extending to the solidus, but a re-interpretation of Kushiro's x-ray data
and some additional experiments convinced Mori and Green (1976) that any
two-Cpx field at 20 kb must be much smaller than that shown by Kushiro.
The solution model predicts a minimal two-Cpx field at 20 kbar; Kushiro's
melting curves have been re-interpreted accordingly and are given as light
lines in Figure 8c.

Melting relations of the Di-En join at P_{H_2O} = 20 kbar were determined
by Kushiro (1969).

Diopside-Hedenbergite

Very little work at high pressures is available for the Di-Hd join.
The complications caused by ferrobustamite at low pressure should die out
with pressure, and above 13 kbar the join may well show a simple melting
loop between Cpx and liquid.

Hedenbergite-Ferrosilite

Subsolidus phase equilibria on the Hd-Fs join at high pressures are
summarized in Figure 9. Fe-rich pyroxenes are not stable below approximately 11-12 kbar, the exact pressure depending on the temperature.

Subsolidus Cpx-Opx relations on the Hd-Fs join above 11-12 kbar are
topologically similar to those for Di-En. Between 11.5 and approximately
18 kbar, Hd_{ss} + OFs_{ss} react with increased temperature to form pigeonite,

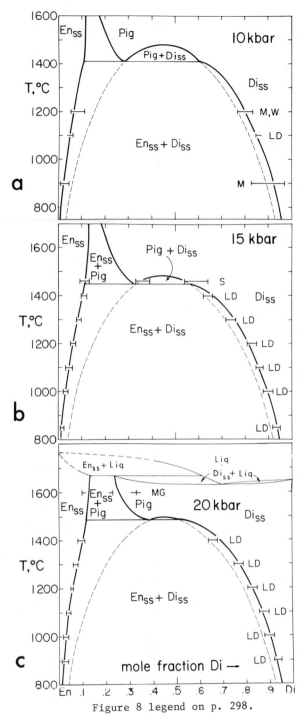

Figure 8 legend on p. 298.

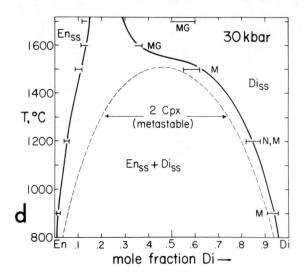

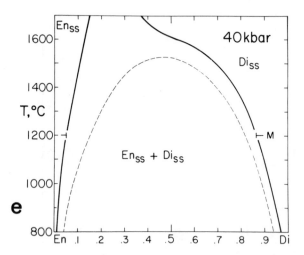

Figure 8. Phase diagrams calculated for the join Di-En at (a) 10 kbar; (b) 15 kbar; (c) 20 kbar; (d) 30 kbar; and (e) 40 kbar, using the thermodynamic solution model of Lindsley *et al.* (1981). Calculations are for En_{SS}-Pig-Di_{SS} equilibria only, and do not include a field for PEn at 10 kbar, nor melting relations. Melting curves at 20 kbar (light lines above 1600°C) are adapted from Kushiro (1969). Data from the literature used in generating the model are shown by the following symbols: ⊢⊣ full bracket; ⊣ or ⊢, half-bracket (vertical line shows direction from which equilibrium was approached); +, homogenization experiment. Sources of data: LB, Longhi and Boudreau (1980) (1 atm; not shown); M, Mori and Green (1975); W, Warner and Luth (1973); LD, Lindsley and Dixon (1976); S, Schweitzer (1977); MG, Mori and Green (1976); N, Nehru and Wyllie (1974). Dashed lines within En_{SS} + Di_{SS} fields show metastable 2-Cpx fields.

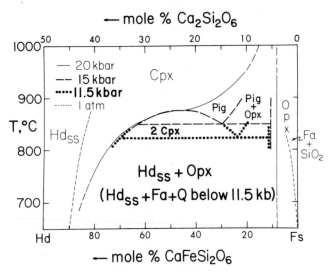

Figure 9. Subsolidus phase relations for the Hd-Fs join at several pressures. Data at 1 atm and at 20 kbar from Lindsley and Munoz (1969b); 11.5 and 15 kbar, new data. Note that Mg-free Pig, as a distinct phase that can coexist with Hd_{ss}, is stable only above 11.5 kbar to somewhat below 20 kbar.

thereby generating a three-pyroxene line (isothermal for a given pressure) and, above it, a small Hd_{ss} + Pig field. By 20 kbar the Hd_{ss} + Pig field has disappeared, and, instead, a broad field of Cpx (with an inflected solvus) coexists with OFs_{ss}. Not shown in Figure 9 is a series of pyroxenoids that are found just below the solidus, ranging from ferrobustamite (3-repeat chain) near Hd to Fs-III (9-repeat) near Fs. Of passing interest is pyroxferroite (7-repeat; $CaFe_6Si_7O_{21}$): It crystallized at the near-vacuum of the lunar surface and has been reported in an intergrowth synthesized at 1 atm (Ried *et al.*, 1979), but is not stable below 10 kbar (Lindsley and Burnham, 1970). Melting relations on the Hd-Fs join up to 15 kbar were outlined by Lindsley (1967; 1981 → Am.Min.66, 1175)

PHASE RELATIONS IN THE DI-EN-HD-FS QUADRILATERAL

Relatively little has been published on the phase relations of "quadrilateral" pyroxenes at pressures above 1 atm, and the work that has been done is mainly subsolidus. Reasons for the scarcity of data include (1) problems with simple containers, (2) difficulty in attaining-- and demonstrating--equilibrium, (3) difficulty in analyzing run products, and (4) a need first to understand the bounding binaries. Early work includes that by Brown (1968); Turnock (1970); Simmons *et al.* (1974);

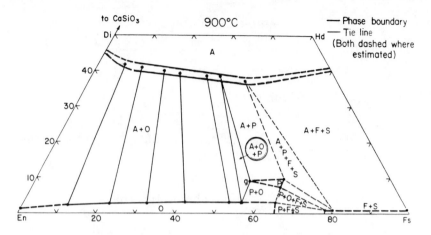

Figure 10. Phase relations in the pyroxene quadrilateral at 900° and $P_{H_2O} = P_{tot} =$ 1 kbar (Turnock and Lindsley, 1981). A, augite; O, orthopyroxene; P, pigeonite; F + S, equimolar mixture of Fe-rich olivine + quartz. The band at the base of the augite field shows the experimental uncertainty in determining the compositions of augite that coexist with Ca-poor phases. Compositions in mole percent.

Smith (1972); Lindsley *et al.* (1974a,b). Much of that work was of pre-liminary nature, and most has been supplanted by the results shown here, which include (a) an isothermal, isobaric section at 900°C and 1 kbar; (b) four isotherms at 15 kbar; and (c) a 1200°C isotherm for 30 kbar.

Phase relations at 900°C, 1 kbar

Turnock and Lindsley (1981) have determined phase relations in the quadrilateral at 900°C and 1 kbar. The pressure medium was water, which was used to enhance reaction rates. They attempted to bracket both Ca-contents and Fe/Mg ratios in coexisting pyroxenes. The combination of incomplete reaction and difficulties in analyzing fine-grained, zoned crystals by electron microprobe precluded a close determination of the composition of Aug coexisting with Opx or Pig; accordingly, the lower limit of the Aug field is shown as a band in Figure 10. The Aug + Opx field in Figure 10 is generally consonant with pyroxenes from granulites (e.g., Davidson, 1968): The two-pyroxene field is narrower and the K_D(Fe-Mg) closer to one for the 900° experiments than for the granulitic pyroxenes, which probably formed at temperatures 150-200° lower. The dashed lines near Fe/(Fe + Mg) $\sim$ 0.70 that bound phase fields containing olivine + quartz (F + S) are the boundaries of the "forbidden zone" at 900°C and 1 kbar.

The Fe-poor portions of Figure 10 are difficult to reconcile with pyroxene compositions from igneous rocks. For example, the most Mg-rich (and, therefore, highest temperature) augites from the Skaergaard intrusion (Brown and Vincent, 1963) appear to have higher Wo contents than do the 900° augites! At least part of this discrepancy probably reflects the relatively high Al_2O_3 contents (up to 3.22 wt. %) of the early Skaergaard augites. Much of the "excess" $CaSiO_3$ may be present as a $CaAlSiAlO_6$ component. But the earliest pyroxenes were also the most likely candidates for re-equilibration during cooling, and the present bulk compositions of Mg-rich Aug and Opx in the Skaergaard may well reflect temperatures below those of original crystallization.

Stability of Fe-rich pigeonite

Figures 8a-c and 9 illustrate that pigeonite (Pig) is a high-temperature phase at low to moderate pressures. With decreasing temperature it decomposes ("inverts") to Opx + Aug, the decomposition temperature being raised by increased pressure or Mg content. The isothermal decomposition lines on the Fe-free and Mg-free joins become three-pyroxene triangles within the quadrilateral because the X_{Fe} = Fe/(Fe + Mg) ratios of the coexisting pyroxenes are different: X_{Fe} is greatest in Pig, least in Aug. For a given pressure (and ignoring effects of polymorphism or instability of Fe-rich pyroxenes), the three-pyroxene field in the quadrilateral is a volume in T-X space, decreasing in temperature from Mg-rich to Fe-rich compositions, and showing a triangular cross section at either constant temperature or constant bulk X_{Fe}. The trace of the Pig corner in T-X_{Fe} space gives the minimum stability of Pig at a given pressure. The variation of the decomposition temperature of pigeonite with X_{Fe} is shown in the previous chapter of this volume (Huebner).

The observation that increasing Fe content lowers the stability temperature of Pig does not apply to very Fe-rich Pigs (X_{Fe} > 0.70). This is because Fe-rich pyroxenes require increased pressure to become stable, and the tendency of Fe-enrichment to stabilize Pig is just balanced by the destabilizing effect of pressure. Thus, the temperature of the five-phase assemblage Aug + Pig + Opx + Ol + Qtz remains virtually constant at 825 $\pm$ 15° from 0.70 < X_{Fe} < 1.0, and the assemblage can serve as a barometer (Fig. 11).

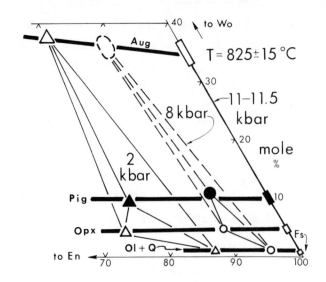

Figure 11. Polythermal, polybaric sections showing the pressures at which three pyroxenes (A + P + O; Aug + Pig + Opx) coexist with olivine (Ol) and Qtz at 825 ± 15°C (Lindsley and Grover, unpublished data). The heavy lines (horizontal and nearly horizontal) show the compositional trends of the three pyroxenes and of olivine + quartz with increasing pressure. Lighter lines are tie lines. Symbols: triangles, compositions at 2 kbar (data from Podpora and Lindsley, 1979); circles, at 8 kbar; rectangles, at 11-11.5 kbar. Dashed tie lines and data point for Aug at 8 kbar indicate composition from x-ray rather than electron microprobe analysis. Ol compositions are projected from SiO_2, so their Ca:Mg:Fe ratios are shown.

Isothermal sections at 15 kbar

Figure 12a-d shows phase equilibria in the quadrilateral for 810°, 905°, 990°, and 1200°C, all at 15 kbar. In Figure 12a-c, most tie lines have been bracketed with respect both to Ca-contents and Fe-Mg distribution. Figure 12d is based on exsolution experiments only, and thus, shows *minimum* widths of the two-pyroxene fields. With the exception of the 810° section, all the augites (coexisting with low-Ca pyroxene) seem to have Ca contents that are too low when compared to compositions for igneous augites. Analogy with the Di-En join suggests that the relatively high pressure of the experiments should have resulted in *higher* Ca contents in the augites. As argued above, the discrepancy probably reflects the effects of Al in the natural pyroxenes. It seems clear that experiments are needed on pyroxenes that contain Al and other minor constituents.

Figure 12a-d illustrates the behavior of Pig at 15 kbar. At 810°, not even Mg-free Pig is stable (it becomes stable at 850 ± 10°C; see Fig. 9). At 905°, Pigs with $X_{Fe} > 0.76$ are stable; the Pig field widens

302

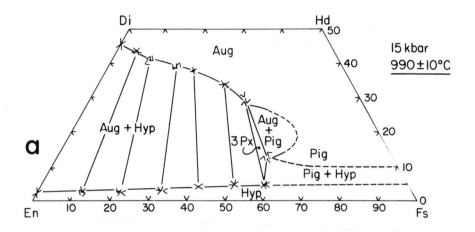

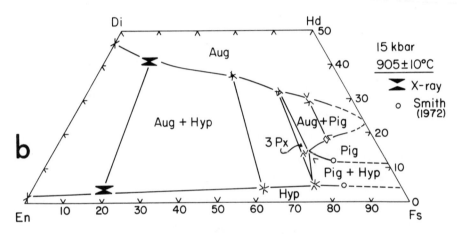

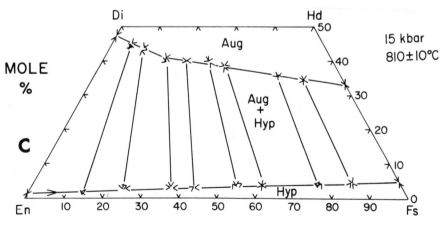

Legend on p. 304.

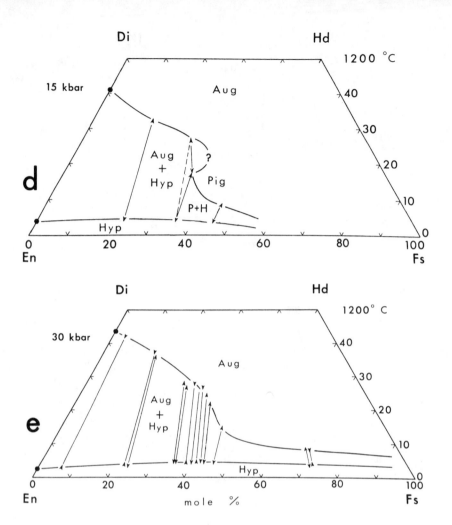

Figure 12. Isobaric, isothermal phase diagrams for the quadrilateral. (a) 810°C; (b) 905°C; (c) 990°C; (d) 1200°C, all at 15 kbar. (e) 1200°, 30 kbar. (d) and (e) adapted from Mori (1978). (a) From Lindsley *et al.* (1974a) plus new data. (b) New data; circles showing Pig + Hyp (= Opx) from Smith (1972) give Wo content only but not X_{Fe}. Similarly for tie line at X_{Fe} = 0.20 labelled "x-ray." (c) Lindsley *et al.* (1974b) plus new data. Arrowheads point away from composition(s) of starting materials (either single-phase Cpx or mechanical mixtures of high- and low-Ca pyroxene) and toward the equilibrium value; tightly clustered arrowheads, therefore, reflect a close approach to equilibrium.

to $X_{Fe} \approx 0.63$ at 990° and to $X_{Fe} \approx 0.40$ at 1200°. The broad, continuous field of high-to-low-Ca Cpx above 905°C is noteworthy; it is similar to the Cpx fields at 20 kbar for Hd-Fs (Fig. 9) and above 20 kbar for Di-En (Fig. 8c-e). The decomposition temperature of Pig increases with pressure by approximately 6-11°C/kbar.

Section at 30 kbar, 1200°C

Figure 12e shows Opx-Cpx relations at 30 kbar and 1200°C. The Aug-Hyp field is slightly wider at 30 than at 15 kb, in keeping with the pressure effect observed for Di-En. Likewise, there is no discrete Aug + Pig field, but only a continuous Cpx solution ranging from high-to-low Ca contents; compare with Figure 8d,e and Figure 9.

CONCLUSIONS

The phase relations shown here for quadrilateral pyroxenes should be very useful in developing thermodynamic solution models for pure pyroxenes which is a first step in modelling real (natural) pyroxenes. Phase equilibria on pure pyroxenes do not seem to be directly applicable to natural igneous pyroxenes, especially as regards the Ca content of augites that coexist with Opx or Pig. It is clear that experiments are needed on Ca-Mg-Fe pyroxenes with added Al and other minor constituents. Some such experiments are reported in the next chapter, but many of those studies were more concerned with the solubility of Al than with its effect on the Wo, En, and Fs components of the pyroxenes.

ACKNOWLEDGEMENTS

Steven R. Bohlen generously released data on ferrosilite stability prior to publication. Clara Podpora assisted greatly in the search for references. John Grover and Charles Prewitt provided helpful discussions. The comments of David Andersen and Paula Davidson helped clarify the presentation. Research supported by National Science Foundation, Earth Sciences Section, Grant EAR7622129. I am grateful to them all.

Akimoto, S., Katsura, T., Syono, Y., Fujisawa, H., and Komada, E. (1965) Polymorphic transition of pyroxenes FeSiO₃ and CoSiO₃ at high pressures and temperatures. J. Geophys. Res., 70, 5269-5278.

Bohlen, S. and Boettcher, A. L. (1980) The effect of magnesium on orthopyroxene-olivine-quartz stability: Orthopyroxene geobarometry. EOS, 61, 393.

_____, Essene, E. J., and Boettcher, A. L. (1980a) Reinvestigation and application of olivine-quartz-orthopyroxene barometry. Earth Planet. Sci. Letters, 47, 11-20.

_____, Boettcher, A. L., Dollase, W. A., and Essene, E. J. (1980b) The effect of manganese on olivine-quartz-orthopyroxene stability. Earth Planet. Sci. Letters, 47, 1-10.

Boyd, F. R. and England, J. L. (1963) Effect of pressure on the melting of diopside, CaMgSi₂O₆, and albite NaAlSi₃O₈ in the range up to 50 kilobars. J. Geophys. Res., 68, 311-323.

_____ and _____ (1965) The rhombic enstatite-clinoenstatite inversion. Carnegie Inst. Washington Year Book, 64, 117-120.

_____, _____, and Davis, B. T. C. (1964) Effects of pressure on the melting and polymorphism of enstatite, MgSiO₃. J. Geophys. Res., 69, 2101-2109.

Brown, G. M. (1968) Experimental studies on inversion relations in natural pigeonitic pyroxenes. Carnegie Inst. Washington Year Book, 66, 347-353.

_____ and Vincent, E. A. (1963) Pyroxenes from the late stages of fractionation of the Skaergaard Intrusion, East Greenland. J. Petrol., 4, 175-197.

Chen, C.-H. and Presnall, D. C. (1975) The system Mg₂SiO₄-SiO₂ at pressures up to 25 kilobars. Am. Mineral., 60, 398-406.

Cohen, L. H. and Klement, Jr., W. (1967) High-low quartz inversion: Determination to 35 kilobars. J. Geophys. Res., 72, 4245-4251.

Davidson, L. R. (1968) Variation in ferrous iron-magnesium distribution coefficients of metamorphic pyroxenes from Quairading, Western Australia. Contrib. Mineral. Petrol., 19, 239-259.

Deer, W. A., Howie, R. A., and Zussman, J. (1978) Rock-Forming Minerals, Vol. 2A, Second Ed., Single-Chain Silicates. John Wiley and Sons, Inc., New York, 668 p.

Eggler, D. H. (1973) Role of CO₂ in melting processes in the mantle. Carnegie Inst. Washington Year Book, 72, 457-467.

Fonarev, V. I., Korol'kov, G. Ya., and Dokina, T. N. (1976) Stability of the ortho-pyroxene + magnetite + quartz association under hydrothermal conditions. Geo-chem. Int., 13, 134-146.

Grover, J. E. (1972) The stability of low-clinoenstatite in the system Mg₂Si₂O₆-CaMgSi₂O₆. EOS, 53, 539.

Gustafson, W. I. (1974) The stability of andradite, hedenbergite, and related minerals in the system Ca-Fe-Si-O-H. J. Petrol., 15, 455-496.

Kushiro, I. (1969) The system forsterite-diopside-silica with and without water at high pressures. Am. J. Sci., 267-A (Schairer Vol.), 269-294.

_____, Yoder, Jr., H. S., and Nishikawa, M. (1968) Effect of water on the melting of enstatite. Geol. Soc. Am. Bull., 79, 1685-1692.

Lamb, C. L., Lindsley, D. H., and Grover, J. E. (1972) Johannsenite-bustamite: Inversion and stability range. Geol. Soc. Am., Abstr. with Progr., 4, 571-572.

Lindsley, D. H. (1965) Ferrosilite. Carnegie Inst. Washington Year Book, 64, 148-149.

_____ (1967) The join hedenbergite-ferrosilite at high pressures and temperatures. Carnegie Inst. Washington Year Book, 65, 230-234.

_____ and Burnham, C. W. (1970) Pyroxferroite: Stability and x-ray crystallography of synthetic Ca₀.₁₅Fe₀.₈₅SiO₃ pyroxenoid. Science, 168, 364-367.

_____ and Dixon, S. A. (1976) Diopside-enstatite equilibria at 850°-1400°C, 5-35 kbar. Am. J. Sci., 276, 1285-1301.

_____ and Munoz, J. L. (1969a) Ortho-clino inversion in ferrosilite. Carnegie Inst. Washington Year Book, 67, 86-88.

_____ and _____ (1969b) Subsolidus relations along the join hedenbergite-ferrosilite. Am. J. Sci., 267-A (Schairer Vol.), 295-324.

_____, Grover, J. E., and Davidson, P. M. (1981) The thermodynamics of the Mg₂Si₂O₆-CaMgSi₂O₆ join: A review and a new model. In, R.C. Newton *et al.*, eds. *Advances in Physical Geochemistry*, Vol. 1, Springer-Verlag, 149-175.

_____, King, Jr., H. E., and Turnock, A. C. (1974a) Composition of synthetic augite and hypersthene coexisting at 810°C: Application to pyroxenes from lunar highlands rocks. Geophys. Res. Letters, 1, 134-136.

_____, _____, and _____ (1974b) Phase relations in the pyroxene quadrilateral at 980°C and 15 kbar. Geol. Soc. Am., Abstr. with Progr., 1974 Annual Meetings, 6, 846.

Longhi, J. and Boudreau, A. E. (1980) The orthoenstatite liquidus field in the system forsterite-diopside-silica at one atmosphere. Am. Mineral., 65, 563-573.

Matsui, Y. and Nishizawa, O. (1974) Iron(II)-magnesium exchange equilibrium between olivine and calcium-free pyroxene over a temperature range 800 degrees C to 1300 degrees C. Soc. franc. Mineral. Cristallogr., Bull., 97, 122-130.

Mori, T. (1978) Experimental study of pyroxene equilibria in the system of CaO-MgO-FeO-SiO₂. J. Petrol., 19, 45-65.

_____ and Green, D. H. (1975) Pyroxenes in the system Mg₂Si₂O₆-CaMgSi₂O₆. Earth Planet. Sci. Letters, 26, 277-286.

_____ and _____ (1976) Subsolidus equilibria between pyroxenes in the CaO-MgO-SiO₂ system at high pressures and temperatures. Am. Mineral., 61, 616-625.

Nehru, C. E. and Wyllie, P. J. (1974) Electron microprobe measurement of pyroxenes coexisting with H₂O undersaturated liquid in the join CaMgSi₂O₆-Mg₂Si₂O₆-H₂O at 30 kilobars with applications to geothermometry. Contrib. Mineral. Petrol., 48, 221-228.

Podpora, C. and Lindsley, D. H. (1979) Fe-rich pigeonites: Minimum temperatures of stability in the Ca-Mg-Fe quadrilateral. EOS, 60, 420-421.

Ried, H., Schroepfer, L., and Korekawa, M. (1979) Synthesis of pyroxferroite and Fe-rhodonite at low pressure. Z. Kristallogr., 149, 121-123.

Rosenhauer, M. and Eggler, D. H. (1975) Solution of H₂O and CO₂ in diopside melt. Carnegie Inst. Washington Year Book, 74, 474-479.

Schweitzer, E. (1977) *The reaction pigeonite = diopside + enstatite at 15 kbar.* M.S. thesis, State Univ. of New York, Stony Brook, NY. [Am. Min. 67, 54 (1982)]

Simmons, E. C., Lindsley, D. H., and Papike, J. J. (1974) Phase relations and crystallization sequence in a contact-metamorphosed rock from the Gunflint Iron Formation, Minnesota. J. Petrol., 51, 539-565.

Smith, D. (1971) Stability of the assemblage iron-rich orthopyroxene-olivine-quartz. Am. J. Sci., 271, 370-382.

_____ (1972) Stability of iron-rich pyroxene in the system CaSiO₃-FeSiO₃-MgSiO₃. Am. Mineral., 57, 1413-1428.

Turnock, A. C. (1970) A pyroxene solvus section. Canadian Mineral., 10, 744-747.

_____ and Lindsley, D. H. (1981) Experimental determination of pyroxene solvi for P ≤ 1 kb, 900 and 1000°C. Canadian Mineral., 19, 255-267.

Warner, R. D. and Luth, W. C. (1974) The diopside-orthoenstatite two-phase region in the system CaMgSi₂O₆-Mg₂Si₂O₆. Am. Mineral., 59, 98-109.

Williams, D. W. and Kennedy, G. C. (1969) Melting curve of diopside to 50 kilobars. J. Geophys. Res., 74, 4359-4366.

Yoder, Jr., H. S. (1952) Change of melting point of diopside with pressure. J. Geol., 60, 364-374.

_____ (1965) Diopside-anorthite-water at five and ten kilobars and its bearing on explosive volcanism. Carnegie Inst. Washington Year Book, 64, 82-89.

Chapter 7
PHASE EQUILIBRIA at HIGH PRESSURE of PYROXENES CONTAINING MONOVALENT and TRIVALENT IONS

Tibor Gasparik & Donald H. Lindsley

INTRODUCTION

Phase equilibrium studies on aluminous pyroxenes are hampered by extreme difficulty in achieving--and demonstrating--equilibrium. This difficulty stems mainly from the reluctance of Al to diffuse in silicates and oxides, especially in tschermakitic pyroxenes where charge-balance requires coupled diffusion of octahedral and tetrahedral Al. Many (although not all) early studies on aluminous pyroxenes were based on crystallization of glass and thus were synthesis experiments, with no demonstration that equilibrium was achieved. Wood (1974) has shown that in at least some cases, glasses initially crystallize to pyroxenes that have excess Al_2O_3; with increased run duration, the excess Al_2O_3 "exsolves" from the pyroxene to form garnet, spinel, or other aluminous phases. Thus experiments using glass are necessarily suspect, unless they are combined with rate studies, as done by Wood. Although much interesting work involving pyroxenes has been done for complex systems at high pressures, we concentrate here on the relatively simple systems where the achievement (and, preferably, the demonstration!) of equilibrium is more probable. Many references to more complex systems can be found in Deer, Howie and Zussman (1978).

We first consider pyroxenes formed at high pressure in the system $CaO-Al_2O_3-SiO_2$ and then in $MgO-Al_2O_3-SiO_2$ (silica-undersaturated) and in that same system under silica-saturated conditions. Next comes the $CaO-MgO-Al_2O_3-SiO_2$ system, also divided into silica-undersaturated and -saturated cases. We conclude with sections on jadeitic pyroxenes and on pyroxenes containing Fe^{3+}, Cr^{3+}, and other elements.

The pyroxenes and most phases considered here are complex solid solutions. To simplify the writing of reactions, throughout this chapter we use abbreviations of the names of mineral end members *to stand for chemical components*. Thus the reaction En + MgTs = Py should be read as "$Mg_2Si_2O_6$ (component in pyroxene) plus $MgAlSiAlO_6$ (Mg-Tschermak component in pyroxene) yield $Mg_3Al_2Si_3O_{12}$ (pyrope component in garnet)." Abbreviations used in the chapter are listed in alphabetical order in Table 1.

Table 1

Abbreviations, names of components and their chemical formulas

Abbreviation	Name	Formula	Solid solution
Ab	Albite	$NaAlSi_3O_8$	Plagioclase
An	Anorthite	$CaAl_2Si_2O_8$	Plagioclase
CaEs	Ca–Eskola pyroxene	$Ca_{0.5}AlSi_2O_6$	Pyroxene
CaTs	Ca–Tschermak pyroxene	$CaAlSiAlO_6$	Pyroxene
Cd	Cordierite	$Mg_2Al_4Si_5O_{18}$	Cordierite
Cor	Corundum	Al_2O_3	
Cpx		$(Na,Ca,Mg)(Mg,Al)$ $(Si,Al)_2O_6$	Clinopyroxene
Di	Diopside	$CaMgSi_2O_6$	Pyroxene
En	Enstatite	$Mg_2Si_2O_6$	Pyroxene
Fo	Forsterite	Mg_2SiO_4	Olivine
Ga		$(Ca,Mg)_3Al_2Si_3O_{12}$	Garnet
Ge	Gehlenite	$Ca_2Al_2SiO_7$	Melilite
Gr	Grossular	$Ca_3Al_2Si_3O_{12}$	Garnet
Jd	Jadeite	$NaAlSi_2O_6$	Pyroxene
Ky	Kyanite	Al_2SiO_5	
L	Liquid		
MgTs	Mg–Tschermak pyroxene	$MgAlSiAlO_6$	Pyroxene
Ne	Nepheline	$NaAlSiO_4$	
Opx		$(Ca,Mg)(Mg,Al)$ $(Si,Al)_2O_6$	Orthopyroxene
Pl		$NaAlSi_3O_8-CaAl_2Si_2O_8$	Plagioclase
Px		$(Na,Ca,Mg)(Mg,Al)$ $(Si,Al)_2O_6$	Pyroxene
Py	Pyrope	$Mg_3Al_2Si_3O_{12}$	Garnet
Q	Quartz	SiO_2	
Sa	Sapphirine	$Mg_2Al_4SiO_{10}$	Sapphirine
Sil	Sillimanite	Al_2SiO_5	
Sp	Spinel	$MgAl_2O_4$	Spinel

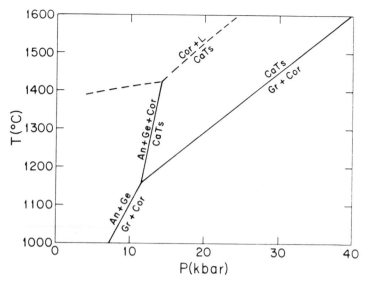

Figure 1. Stability field of calcium Tschermak pyroxene (CaTs). After Hays (1966).

CLINOPYROXENES IN THE SYSTEM $CaO-Al_2O_3-SiO_2$

Clinopyroxene in the system is a binary solid solution of two end members: calcium Tschermak pyroxene--$CaAl_2SiO_6$ (CaTs) and $Ca_{0.5}AlSi_2O_6$. Calcium Tschermak pyroxene was first synthesized by Clark *et al.* (1962), and Hays (1966) determined its stability field (Fig. 1). CaTs coexisting with anorthite or silica-rich liquid can dissolve excess silica balanced by vacancies in the M2 site. Such "non-stoichiometric" pyroxenes were first reported by Eskola (1921). The composition of clinopyroxenes with vacancies can be extrapolated to the end member $Ca_{0.5}AlSi_2O_6$. Following Khanukova *et al.* (1976a) we tentatively use for this member the name calcium Eskola pyroxene (CaEs).

Clinopyroxenes on the joint CaTs-CaEs were synthesized by Gasparik and Lindsley (1980) (Fig. 10). At 1450°C and 31.5 kbar (5 hours), synthetic anorthite converted to a metastable assemblage Px+Ky+Ga+Cor+L. The pyroxene composition determined by the electron microprobe varied from $CaTs_{74}CaEs_{26}$ to $CaTs_{59}CaEs_{41}$ with the average of 44 analyses $CaTs_{69}CaEs_{31}$ (mole %). This may be compared with Wood's (1976a) extrapolated value of $CaTs_{50}CaEs_{50}$.

The following exchange reactions are important in the system $CaO-Al_2O_3-SiO_2$:

$$An + Ge + Cor = 3CaTs \qquad (1)$$

311

$$2An + 2Ge = Gr + 3CaTs \qquad (2)$$
$$3CaTs = Gr + 2Cor \qquad (3a)$$
$$9An = 6CaEs + 2Gr + 4Cor \qquad (3b)$$
$$An = CaTs + Q \qquad (4a)$$
$$An + 2Q = 2CaEs \qquad (4b)$$
$$3CaTs + 2Q = Gr + 2Ky \qquad (5a)$$
$$6CaEs = Gr + 2Ky + 7Q \qquad (5b)$$
$$3An = 2CaTs + 2CaEs \qquad (6)$$
$$2CaEs = CaTs + 3Q \qquad (7)$$
$$7CaTs + 4CaEs = 3Gr + 6Ky \qquad (8)$$

The reaction curves (1), (2) and (3a) were reversed by Hays (1966) for the CaTs bulk composition. The reactions are univariant for that bulk composition, but, however, are divariant in the system with excess silica. The equations for the curves (1) and (3a) are respectively:

$$P(bar) = 12500 + 9.9(T°C - 1250); \; P(bar) = 17500 + 63.8(T°C - 1250)$$

The combined reactions (3), (4) and (5) represent the truly univariant equilibria in the system.

The reactions (4) and (5) were determined by Hariya and Kennedy (1968). However, the stability field of the assemblage CaTs+Q is almost completely outside the stability field of CaTs determined by Hays (1966). Moreoever, the assemblage Px+Q is metastable with respect to melting relations in the system (Wood and Henderson, 1978). Our experiments suggest that the invariant point of the five-phase assemblage An+Ga+Ky+Q+L lies below 1420°C and 30.5 kbar. This is in perfect agreement with Wood (1978) who estimated the invariant point to be between 1400 and 1430°C at 30.2 kbar (applying -7% pressure correction). Hence, it is impossible to reconcile the results of Hariya and Kennedy with the experiments of Hays by assuming that the extended stability field of CaTs pyroxene is caused by excess silica in the Px solid solution. Instead, the experiments of Hariya and Kennedy can be at least partially explained as a consequence of impurities in natural minerals used as the starting material. A detailed discussion of the complex phase relations in the pressure range 29-33 kbar and temperatures of 1350 to 1450°C is given by Wood (1978).

From the experimental study of mixing properties of clinopyroxenes in the system Di-CaTs-CaEs, discussed later, we have obtained the mixing

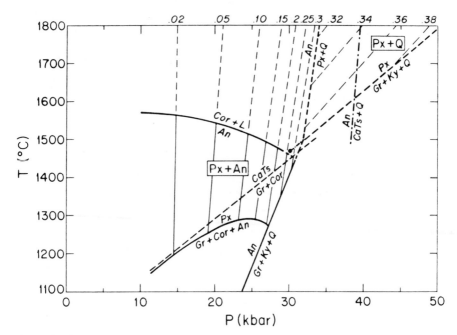

Figure 2. P-T phase relations for the system $CaO-Al_2O_3-SiO_2$ saturated with silica. Thick lines -- univariant equilibria; thin lines -- isopleths of CaEs ($Ca_{0.5}AlSi_2O_6$) component in the clinopyroxene solid solution (in mole fractions of CaEs). Solid lines -- stable; dashed lines -- metastable; dash-dot lines -- doubly metastable equilibria. The solid dot on the metastable extension of the univariant curve (5) is the doubly metastable singular point for the equilibrium (8).

properties along the binary CaTs-CaEs. The symmetric solution defines the mixing properties in terms of one Margules parameter (J/mol):

$$RTln\gamma_{CaTs} = -19590(X_{CaEs})^2; \quad RTln\gamma_{CaEs} = -19590(X_{CaTs})^2; \quad a_{CaTs} = X_{CaTs}\gamma_{CaTs}$$

The thermodynamic parameters for the reactions (4a) and (4b) are respectively (J,bar,K):

$$\Delta G°_{4a} = 44530 + 6.30T - 1.456P; \quad \Delta G°_{4b} = 16550 + 57.32T - 2.280P$$

Using these values and parameters for the reaction (3a) from Hays (1966) and for the anorthite breakdown reaction from Goldsmith (1980), we have calculated the phase relations for the system $CaO-Al_2O_3-SiO_2$ with excess silica (Fig. 2). For the anorthite breakdown reaction

$$3An = Gr + 2Ky + Q \tag{9}$$

we have used the equation: $P(bar) = 900 + 20.7T°C$, obtained by applying a pressure correction of -10.8% at $1150°C$ to all reversals of Goldsmith (1980).

Wood (1977) calculated the equation for reaction (4a) using Hays' data combined with the thermodynamic data of Robie and Waldbaum (1968). Our data indicate a smaller slope, 4.3 bar/°C compared to 11.3 bar/°C. At 1500°C, the position of the metastable curve is at about 3 kbar higher pressure than given by the Wood's equation.

ORTHOPYROXENES IN THE SYSTEM $MgO-Al_2O_3-SiO_2$ UNDERSATURATED WITH SILICA

Orthopyroxene in this system can be considered a binary solid solution with end members enstatite $(Mg_2Si_2O_6)$ and Mg-Tschermak pyroxene $(MgAlSiAlO_6)$. A number of experimental studies have investigated the phase relations in the system mainly because of its simplicity. There are three important reactions in the silica-undersaturated portion of the system:

$$2En + Sp = Fo + Py \qquad (10a)$$

$$2MgTs + Fo = Sp + Py \qquad (10b)$$

$$En + Sp = Fo + MgTs \qquad (11)$$

$$En + MgTs = Py \qquad (12)$$

Reaction (10) is univariant, reactions (11) and (12) are divariant; the assemblage Opx+Fo+Sp approximates spinel peridotite and the assemblage Opx+Fo+Py approximates garnet peridotite in nature.

Reaction (12) is the basis of the garnet-pyroxene geothermometer and geobarometer of Wood and Banno (1973). Their model assumes the orthopyroxene solid solution to be ideal with the exception of the volume for which an excess volume term was derived from volume measurements of Skinner and Boyd (1964). Wood and Banno calculated thermodynamic parameters for reaction (12) from the experimental data of Boyd and England (1964). Despite its simplicity, the model appears to satisfy the experimental data reasonably well and has been repeatedly applied in later experimental studies.

MacGregor (1964, 1974) experimentally determined the position of reaction (10) in the temperature range 1000-1600°C, using mostly crystalline starting material seeded with 10% of products. MacGregor (1974) also investigated the exchange equilibria (11) and (12) in both the spinel peridotite field and the garnet peridotite field, using glass as the starting material; hence, all his runs were synthesis experiments. (However, runs in the garnet field were seeded with garnet.) The orthopyroxene compositions were then obtained by electron microprobe. As MacGregor himself pointed out, his data were inconsistent with three reversals of Anastasiou and Seifert (1972) at ∿3 kbar and 900, 1000 and 1100°C, which show lower solubility of Al_2O_3 in orthopyroxene than MacGregor's data at given conditions. Their reversals were performed in a hydrothermal apparatus with crystalline starting material composed either of aluminous enstatite or the assemblage orthoenstatite + cordierite + forsterite or spinel. The alumina content of orthopyroxenes in their study was determined by x-ray diffraction and by phase-disappearance methods.

On the basis of MacGregor's data from the garnet peridotite field, Wood (1974) calculated a petrogenetic grid for alumina solubility in orthopyroxene using the ideal solution model of Wood and Banno (1973). Obata (1976) extended the thermodynamic modelling of Wood (1974) to the spinel peridotite field using the same model and the position of the univariant curve (10) as determined by MacGregor (1964, 1974). Obata's model, based on the data from the garnet peridotite field, is inconsistent with MacGregor's data in the spinel peridotite field, as the calculated isopleths in the spinel peridotite field are much less pressure sensitive than MacGregor's data suggest. Obata also showed that the curve for the univariant reaction (10) should flatten with decreasing temperature. Calculated parameters for equilibria (11) and (12), as obtained by several investigators, are shown in Table 2.

Fujii (1976) determined the solubility of alumina in orthopyroxene coexisting with forsterite and spinel at 15-25 kbar and 1150-1400°C. The starting material was a mixture of glass and crystals of orthopyroxene and spinel made by crystallizing the composition 4 moles $MgSiO_3$ + 1 mole $MgAl_2O_4$ at 1580°C. The composition of orthopyroxene in the starting material is not given so that the direction of the reaction is unknown. The analyses of run products were made by electron microprobe. Fujii used the ideal solution model of Wood and Banno (1973) to calculate isopleths for the alumina solubility in orthopyroxene; these are almost pressure insensitive and thus are also contrary to MacGregor's data. The agreement with the reversed data of Anastasiou and Seifert (1972) is reasonably good, but otherwise there is no *a priori* reason to favor Fujii's results over MacGregor's, since neither set of experiments was reversed.

Table 2

Thermodynamic parameters for the exchange reactions (11) and (12)

	$\Delta H°$ (J)	$\Delta S°$ (J/K)	$\Delta V°$ (J/bar)
Reaction (11)			
Obata (1976)	43510	16.95	
Fujii (1976)	43510	18.00	
Danckwerth & Newton (1978)	37660	15.06	
Dixon & Presnall (1980)	31560	10.59	0.151
This work	40250	15.72	0.196
Reaction (12)			
Wood & Banno (1973)	−17600	−11.25	
Wood (1974)	−29340	−16.28	
Fujii & Takahashi (1976)	−31850	−18.40	
This work	−25740	−14.93	−1.050

Danckwerth and Newton (1978) determined the position of the univariant curve (10) in the temperature range 900–1100°C by hydrothermal reversals. They confirmed the strong curvature of the equilibrium below 1200°C predicted theoretically by Obata (1976) and Fujii (1976). Three reversals of the equilibrium alumina content of orthopyroxene coexisting with spinel and forsterite were made at points adjacent to the univariant curve at 20 kbar and 950, 1000 and 1080°C. The alumina contents are higher than those predicted by MacGregor (1974) but close to the data of Fujii (1976).

The alumina-solubility isopleths for orthopyroxene calculated by Danckwerth and Newton are slightly discordant with the data of Anastasiou and Seifert (1972). Although the discrepancy is not serious, we have obtained a better fit of isopleths to the data by refining an excess volume term from the experiments, rather than using the volume term derived from the unit-cell-volume measurements. An asymmetric solution model with only one non-zero Margules parameter has been used for the binary giving the result (J/bar):

$$RT\ln\gamma_{En} = -0.272P(2X_{MgTs}^2 - 2X_{MgTs}^3);$$

$$RT\ln\gamma_{MgTs} = -0.272P(2X_{En}^3 - X_{En}^2);$$

$$a_{En} = X_{En}\gamma_{En}$$

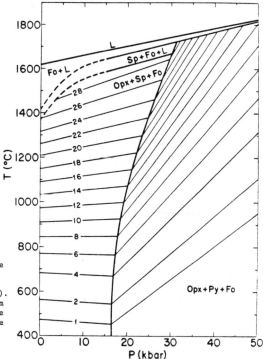

Figure 3. P-T phase diagram for the system MgO-Al₂O₃-SiO₂ undersaturated with silica. Thin lines are isopleths in mol % MgTs in the orthopyroxene solid solution calculated from the data of Anastasiou and Seifert (1972), Fujii (1976) and Danckwerth and Newton (1978). The flat univariant curve for the equilibrium (10) separates the field of spinel peridotite at lower pressures from the garnet peridotite field. Melting relations after MacGregor (1974).

Using these expressions and the reversals of MacGregor (1964, 1974) and Danckwerth and Newton (1978) for the univariant curve (10), we have calculated the parameters for reaction (12) and isopleths of the alumina solubility in Opx in the spinel- and garnet-peridotite fields (Fig. 3).

ORTHOPYROXENES IN THE SYSTEM $MgO-Al_2O_3-SiO_2$ SATURATED WITH SILICA

Exchange reactions in the system involve Opx, Py, Q, Al_2SiO_5 polymorphs, cordierite ($Mg_2Al_4Si_5O_{18}$) and sapphirine ($\approx Mg_2Al_4SiO_{10}$). Orthopyroxene is probably a binary solid solution En-MgTs. It has not been verified yet whether the end member $Mg_{0.5}AlSi_2O_6$, by analogy with the CaO-bearing system, is also present. Four univariant curves have been experimentally determined:

$$Cd = Sa + 4Q \tag{13}$$

$$Sa + 3Q = En + 2Sil \tag{14}$$

$$7En + 2Sa + 2Sil = 6Py \tag{15}$$

$$3En + 2Sil = 2Py + 2Q \tag{16}$$

317

The univariant reaction (13) was reversed by Newton (1972) between 1300 and 1400°C. At 1200°C only a half bracket was obtained due to the sluggishness of the reaction. Newton (1972) also reversed reaction (14); however, close brackets were obtained only at higher temperatures.

Better constraints on reaction (14) were obtained by Chatterjee and Schreyer (1972). The reaction was reversed between 1080 and 1400°C. The reaction curve has a pronounced curvature due to systematic compositional and structural variations in the coexisting solid phases. The univariant curves (13) and (14) intersect in an invariant point near 950°C and 8 kbar (Newton *et al.*, 1974).

The univariant curve (16) was reversed between 1000 and 1400°C by Hensen and Essene (1971). However, the authors pointed out that the curve was probably metastable inasmuch as reaction (15), which is the low pressure stability limit of pyrope, had been located by Boyd and England (1959) on the high pressure side of the reaction curve (16) as determined by Hensen and Essene. Hensen and Essene also suggested the existence of an invariant point at the intersection of curves (14), (15) and (16).

Hensen (1972), in a series of reversal experiments between 1050–1400°C and 12–22 kbar, determined that the reaction curves (14), (15) and (16) do not intersect and therefore do not form an invariant point. The univariant curve (16) was found to be stable and located at higher pressures than given by Hensen and Essene (1971), on the high pressure side of the pyrope forming reaction (15). The locations of curves (13)–(16) are given in Figure 4.

Arima and Onuma (1977) determined the solubility of alumina in orthopyroxene in the range of 1100–1500°C and 10–25 kbar, controlled by the divariant equilibria:

$$2MgTs = Sa + Q \tag{17}$$
$$6MgTs = 2Sa + En + 2Sil \tag{18}$$
$$En + MgTs = Py \tag{19}$$

Synthetic crystalline mixtures were used as starting materials, prepared by crystallizing oxide mixtures at 1300°C and 1 atm for six days. However, the composition of crystalline phases in starting materials was not given so that the direction of reactions is unknown. The orthopyroxene composition was determined by the x-ray diffraction. The alumina

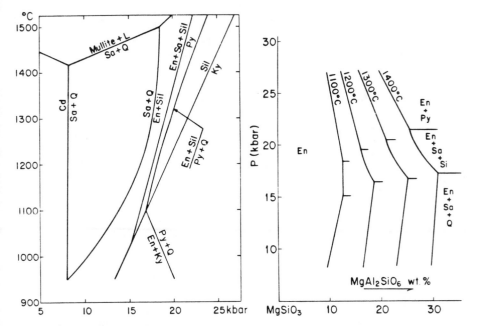

Figure 4. P-T phase relations in the system MgO-Al$_2$O$_3$-SiO$_2$ saturated with silica. The diagram is based on the experimental data of Newton (1972), Chatterjee and Schreyer (1972), Newton *et al.* (1974), Hensen and Essene (1971), Hensen (1972) and Richardson (1968).

Figure 5. Isothermal P-X sections in the join MgSiO$_3$-MgAl$_2$SiO$_6$ showing the solubility of alumina in orthopyroxene at various P-T conditions in the system MgO-Al$_2$O$_3$-SiO$_2$ saturated with silica. After Arima and Onuma (1977).

content in Opx coexisting with Sa+Q was found to increase with increasing temperature and pressure, while that in Opx coexisting with Sa+Sil or with Py decreases with increasing pressure and decreasing temperature (Fig. 5).

PYROXENES IN THE SYSTEM CaO-MgO-Al$_2$O$_3$-SiO$_2$ UNDERSATURATED WITH SILICA

The composition of pyroxenes in the system can be expressed in terms of four end members: Di, En, CaTs and MgTs. The univariant equilibria in the system involve two pyroxenes, usually ortho- and clinopyroxene. Only at very limited P-T-X conditions can Ca-rich clinopyroxene coexist with pigeonite. Three univariant reactions are important for the system:

$$An + 2Fo = Cpx + Opx + Sp \tag{20}$$

$$An + Cpx + Opx + Sp = Ga \tag{21}$$

$$Cpx + Opx + Sp = Ga + Fo \tag{22}$$

Univariant curve (20), experimentally located by Kushiro and Yoder (1965, 1966), Herzberg (1972, 1976a), and Presnall (1976), is the boundary

319

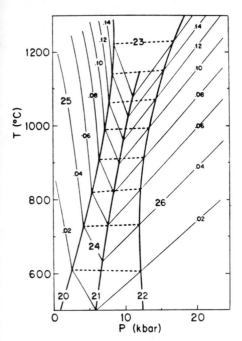

between plagioclase lherzolite and spinel lherzolite or between olivine gabbro and spinel gabbro. Reaction (21), experimentally determined by Kushiro and Yoder (1965, 1966) and Herzberg (1976b), marks the first appearance of garnet with pressure. Univariant reaction (22), determined by MacGregor (1965), Kushiro and Yoder (1965, 1966), and O'Hara *et al.* (1971), is the transition from spinel lherzolite to garnet lherzolite. Jenkins and Newton (1979) hydrothermally reversed the univariant reaction (22) at 900°C (15.0 ± 0.5 kbar) and 1000°C (16.0 ± 0.5 kbar).

Figure 6. P-T phase relations for the system CaO-MgO-Al$_2$O$_3$-SiO$_2$ undersaturated with silica showing isopleths of the alumina solubility in orthopyroxene in mole fractions of MgTs. Large numbers refer to the corresponding univariant and divariant equilibria. After Obata (1976).

Obata (1976) extended the thermodynamic modelling from the MgO-Al$_2$O$_3$-SiO$_2$ system to the system containing CaO assuming that thermodynamic parameters for the divariant equilibrium (11) do not change with addition of CaO. The experimental data of Kushiro and Yoder (1966) were used to calculate the univariant curves (20) and (22). The Ca/(Ca+Mg) ratio of coexisting ortho- and clinopyroxenes was estimated from data of Mori and Green (1975) assuming that the ratios are independent of the alumina content of pyroxenes. Alumina solubility isopleths were then calculated for orthopyroxene in plagioclase peridotite, spinel peridotite and garnet websterite or garnet peridotite (Fig. 6).

Akella (1976) determined the compositions of coexisting orthopyroxene, clinopyroxene and garnet at 1100-1500°C and 26-44 kbar. The assemblages were synthesized using glass of two bulk compositions on the join pyrope-diopside (Py$_{75}$Di$_{25}$; Py$_{70}$Di$_{30}$, wt %). Five reversals with natural mineral mixtures gave compositions in good agreement with the synthesis runs. The compositions of run products were obtained by electron microprobe. A friction correction of ±1 kbar was applied to the nominal pressures, depending on the direction of the piston motion. The experiments showed that the geothermometer of Boyd and Nixon (1973) based on the Ca/(Ca+Mg) ratios of coexisting Cpx and Opx is slightly pressure-dependent; the ratio decreases with increasing pressure. The form of isopleths for the solubility of Al in Opx is very similar to isopleths in the CaO-free system. The presence of a small amount of TiO_2 does not show any marked effect on the solubility of alumina in pyroxenes.

Presnall (1976) experimentally determined the melting relations for spinel lherzolite and reversed the alumina content of orthopyroxene on the spinel lherzolite solidus (11 kbar, 1350°C, 8.4 wt % Al_2O_3). This point, combined with data of Akella (1976) and the position of univariant curve (22), suggests that the alumina isopleths for orthopyroxene in the spinel lherzolite field have slightly negative slopes in contrast to the mainly positive slopes in the CaO-free system (Fig. 3).

Fujii (1977) experimentally determined the compositions of ortho- and clinopyroxene coexisting with forsterite and spinel at 16 kbar and 1100-1375°C, using both glass and crystalline mixtures of synthetic clinoenstatite, diopside and spinel as the starting material, so that the pyroxene compositions were considered reversed by homogenization and "exsolution" runs. Variable amounts of water were used as flux (7 wt % at 1100°C to less than 1% at 1375°C). Run time ranged from six hours at 1375°C to four days at 1100°C. The analyses of run products were done by electron microprobe. In the run products of homogenization experiments, the pyroxene grains were not completely equilibrated, as the compositions of the cores of large grains were close to the composition of the starting material. However, the analyses of small grains were similar to those of the rims of large grains. The compositions of ortho- and clinopyroxenes coexisting with forsterite and spinel were reversed at 1200, 1300 and 1375°C (16 kbar). At 1100°C only the pyroxenes from the homogenization experiments were suitable for analysis; the "exsolution" runs produced crystals smaller than five microns even after four days. The experiments showed that the width of the compositional gap between coexisting ortho- and clinopyroxene decreases with the addition of alumina, if compared with the data of Lindsley and Dixon (1976), the decrease being more pronounced at higher temperatures.

321

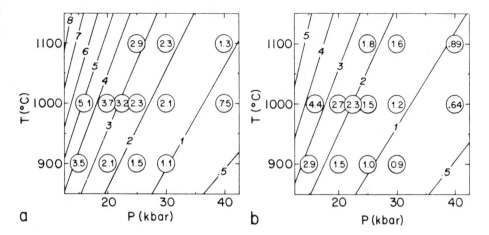

Figure 7. P-T phase relations for the system CaO-MgO-Al₂O₃-SiO₂ undersaturated with silica showing the alumina solubility isopleths (wt % Al₂O₃) in orthopyroxene (a) and clinopyroxene (b) coexisting with garnet in the garnet peridotite or garnet websterite field. After Perkins and Newton (1981).

While the Ca/(Ca+Mg) ratio of orthopyroxene does not change significantly with the addition of Al_2O_3, the same ratio in the coexisting clinopyroxene increases with the addition of Al_2O_3.

O'Hara and Howells (1979) experimentally determined the solubility of alumina in orthopyroxene coexisting with clinopyroxene, pyrope-rich garnet and forsterite at 20, 25, 30 and 35 kbar and between 1100 and 1600°C in long-lasting experiments using starting materials of different physical state. The solubility of alumina in orthopyroxene was found to be much lower than in previous studies by other investigators.

Perkins and Newton (1980, 1981) reversed the compositions of coexisting orthopyroxene, clinopyroxene and garnet in hydrothermal experiments at 15-40 kbar and 900-1100°C (Fig. 7). The garnet at equilibrium with Opx and Cpx contains 86 ± 1 wt % Py at all P-T conditions studied. The Ca/(Ca+Mg) ratios in Opx in equilibrium with Cpx and Ga do not vary measurably from values determined by Lindsley and Dixon (1976) for the Al-free composi- tions; however, the same ratios in Cpx are significantly greater than those in the Al-free system. The alumina content is slightly greater in Opx than in coexisting Cpx.

Dixon and Presnall (1980) reported 17 equilibrium experiments at 10-18 kbar and 1100-1405°C to determine mineral compositions for coexisting Opx+Cpx+Fo+Sp. Equilibrium has been demonstrated at four P-T conditions. The isopleths of wt % Al_2O_3 in orthopyroxene have negative slopes in P-T space ($-17°$/kbar for 6 wt % Al_2O_3) in contrast to slightly positive slopes in the CaO-free system. Using the ideal mixing model, they calculated the parameters for reaction (11) to be $\Delta H° = 43090J$, $\Delta S° = 9.83J/K$ and $\Delta V° = -4.05 \, cm^3$. The parameters differ from those calculated for the same reaction in the system $MgO-Al_2O_3-SiO_2$ using the data from Fujii (1976) and Danckwerth and Newton (1978): $\Delta H° = 31560J$, $\Delta S° = 10.59J/deg$, $\Delta V° = 1.51 \, cm^3$. They suggested that the differences between the two sets of values indicate non-ideal mixing of components in the orthopyroxene solid solution in at least one or possibly both of the systems. They concluded that addition of CaO to orthopyroxene apparently introduces an excess volume of mixing which is large enough to allow the reaction to be used as a geobarometer.

Herzberg and Chapman (1976) and Herzberg (1978) experimentally studied the mixing properties of clinopyroxenes in the system at 1 kbar to 30 kbar and 1100-1500°C using the divariant exchange equilibria:

$$Di + Sp = CaTs + Fo \qquad (23)$$
$$Di + Sp + An = 2CaTs + En \qquad (24)$$
$$An + Fo = CaTs + En \qquad (25)$$
$$Ga = Di + CaTs \qquad (26)$$

Reaction (23) defines the equilibrium of the assemblage Px+Fo+Sp (spinel lherzolite), reaction (24) is the equilibrium of the assemblage Px+An+Sp (spinel gabbro), reaction (25) is the equilibrium of the assemblage Px+An+Fo (olivine gabbro and plagioclase lherzolite), and reaction (26) defines the equilibrium of the assemblage Px+Ga (garnet pyroxenite and garnet lherzolite).

Gel mixtures previously crystallized at 1 atm and 1200°C for 28 days were used as the starting material. All runs were synthesis experiments and approach to equilibrium was not demonstrated. The pyroxene crystals in the experimental charges were very small (3-4 microns) and unsuitable for microprobe analysis. Therefore, the analyses were performed by x-ray diffraction using $22\bar{1}$ and 310 reflections. Although the runs produced both ortho- and clinopyroxene, only the compositions of clinopyroxenes could be measured and the composition of orthopyroxenes was estimated from Obata (1976) and Mori and Green (1975). Most of the experiments pertain to the equilibrium (23) where at all pressures and temperatures

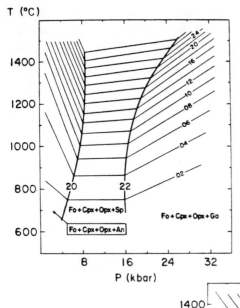

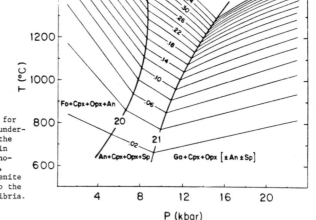

Figure 8. P-T phase relations for the system CaO-MgO-Al₂O₃-SiO₂ undersaturated with silica showing the alumina solubility isopleths (in mole fractions of CaTs) in clinopyroxene in the lherzolite facies. Large numbers refer to the corresponding univariant equilibria. After Herzberg (1978).

Figure 9. P-T phase relations for the system CaO-MgO-Al₂O₃-SiO₂ undersaturated with silica showing the alumina solubility isopleths (in mole fractions of CaTs) in clinopyroxene in the olivine gabbro, spinel gabbro and garnet pyroxenite facies. Large numbers refer to the corresponding univariant equilibria. After Herzberg (1978).

the clinopyroxene compositions have the molar ratio CaTs/En close to one. Therefore, Herzberg treated the ternary Di-En-CaTs as a pseudobinary for which he calculated the excess parameters: (J,K):

$$W_{G,CaTs(Di)} = 33890 - 20T; \quad W_{G,Di(CaTs)} = 0$$

Using the excess parameters and the parameters for reactions (20) through (26), Herzberg calculated the clinopyroxene alumina solubility grid for the lherzolite facies and for the olivine gabbro, spinel gabbro and garnet pyroxenite facies (Figs. 8, 9). His isopleths in the garnet field may be compared with those of Perkins and Newton (1981) in Fig. 7b. Herzberg also calculated the activity-composition relations of the pseudo-binary Di-CaTs using four activity models: complete order, complete disorder, charge-balance I (random mixing in the M1 and M2 site), and charge-balance II (random mixing in the M2 and the tetrahedral site). Only the completely disordered model gave activities smaller than the mole fractions, which is the result obtained by Wood (1977) using the disordered model and $\Delta G°$ for reaction (4a) based on Hays' data. However, this model would imply a local charge imbalance in the clinopyroxene structure that is an energetically unfavorable condition.

Gasparik (1980a) used data of Herzberg (1978) for the thermodynamic modelling of mixing properties on the binary En-CaTs, using the mixing properties on the binaries Di-CaTs and Di-En (Lindsley *et al.*, 1981). The best fit was found for the charge-balance I model and a symmetric binary En-CaTs.

CLINOPYROXENES IN THE SYSTEM $CaO-MgO-Al_2O_3-SiO_2$ SATURATED WITH SILICA

The composition of clinopyroxenes in the system can be described in terms of four end members: Di, En, CaTs, CaEs. However, it is possible to restrict the pyroxene composition to the ternary Di-CaTs-CaEs if the ratio CaO/MgO in the system is equal to one. Khanukhova *et al.* (1976a) investigated the system Di-CaTs-SiO_2 at 35 kbar and 1200°C. They synthesized a single clinopyroxene phase from the composition $Di_{75}CaEs_{25}$ using a gel mixture. Wood and Henderson (1978) reported solid-state-detector analyses and unit cell parameters for "non-stoichiometric" tschermakitic pyroxenes having as much as 20 mol % of CaEs component. Gasparik and Lindsley (1980) reversed the composition of clinopyroxene coexisting with An+Q at 1400°C and 29 kbar (Fig. 10). The average of 58 microprobe analyses gives the composition $Di_{18.4}CaTs_{55.6}CaEs_{26.1}$.

Wood (1976b, 1979) reported an experimental study of the mixing properties of pyroxenes in the system Di-CaTs-SiO_2 using the exchange reaction (4a), the equilibrium for which can be expressed as $RT\mathit{l}na_{CaTs}^{Cpx}$ = $-\Delta G°$, where $\Delta G°$ was estimated using the data of Hays (1966) and the thermodynamic data of Robie and Waldbaum (1968). The excess volume term was included using the unit cell volume measurements of Newton *et al.* (1977).

325

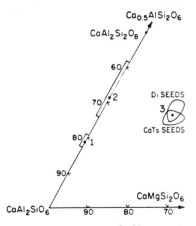

Figure 10. Compositions of clinopyroxenes in the system CaO-MgO-Al$_2$O$_3$-SiO$_2$ saturated with silica. Run 1: 1440°C, 29.7 kbar, 4 h., Px+An+Ga+Cor+L; run 2: 1450°C, 31.5 kbar, 5 h., Px+Ky+Ga+Cor+L; run 3: 1400°C, 28.8 kbar, 5 h., Px+An+Q+L.After Gasparik and Lindsley (1980).

Both glass and crystalline starting material were used. Wood (1974) observed that synthesis runs with glass tend to produce pyroxenes that are initially oversaturated in alumina; with increased run duration, the alumina content decreases toward the equilibrium value. Therefore, synthesis runs were used for the bracket from the high alumina side and crystalline material for the low-alumina bracket. The experiments were conducted in the pressure range 10-25 kbar and temperatures 900-1300°C. Because of the small size of the product pyroxene crystals (1-5 microns), the analyses were perfomed using the EMMA IV electron microscope-microanalyser with a Kevex solid state detector. The clinopyroxene compositions that showed significant non-stoichiometry at higher P and T were extrapolated to the Di-CaTs binary.

The activities of CaTs obtained correspond approximately to mole fractions which would suggest that the Di-CaTs solid solution is ideal (with the exception of the volume). However, the calorimetric measurements of Newton et al. (1977) give large positive excess enthalpies. In order to explain the discrepancy, Wood assumed that Al and Si are completely disordered in the tetrahedral site, so that the activity of CaTs is given by:

$$a_{CaTs} = (X_{CaTs})^2(2-X_{CaTs})\gamma_{CaTs}$$

This produces activities significantly smaller than mole fractions and the desired non-ideality of the Di-CaTs solid solution.

Wood's pioneering study failed to provide a definitive solution to the problem for several reasons. The small size of the pyroxene crystals prevented the use of the electron microprobe. The quality of analyses obtained by the solid state detector was relatively poor, and thus they are inadequate to provide the excess parameters, which are reflected in compositional changes much smaller than Wood's given uncertainties. Most of the experiments were done at and below 20 kbar where the variations in

326

the pyroxene composition are quite insensitive to P-T conditions.
Finally, the clinopyroxenes coexisting with An+Q contain a significant
proportion of the CaEs component which can not be ignored; instead, the
system is better treated as a ternary.

Gasparik and Lindsley (1980) outlined an approach which was used
for the experimental determination of mixing properties of clinopyroxenes
in the system Di-CaTs-CaEs using the exchange equilibria (4a), (4b), (6)
and (7). The main emphasis was placed on eliminating the problem of fine
grain size of the pyroxene crystals so that high-quality analyses could
be obtained by electron microprobe. This was achieved in two ways. First,
a flux, in this case PbO, was used in starting compositions. The flux
greatly increases the kinetics of exchange and the approach to equilibrium. Secondly, the seeds of two end members, Di and CaTs, were added
to starting mixtures which were made from reagent oxides and crystalline
compounds (MgO, SiO_2, PbO, $CaSiO_3$, $CaAl_2O_4$). The seeds were already large
enough to be probed easily. By reaction of both Di and CaTs seeds with
An+Q+L produced during the run, the equilibrium composition of pyroxene
was reversed in terms of CaTs and Di, but not CaEs. Runs produced zoned
pyroxenes from which only the outermost rims were analyzed. Typically,
20-50 high-quality analyses were obtained from each run. The average
composition from each kind of seed was calculated separately; these
values were then averaged to give the starting point used in the least-
squares solution modelling of the data. The asymmetric solution model
of Currie and Curtis (1976) was used in the least-squares calculations.
The binary Di-CaTs was treated as asymmetric, the remaining two binaries
as symmetric. The excess enthalpy term calculated from the calorimetric
measurements of Newton *et al.* (1977) was used as an additional constraint.
A pressure correction of -10% was applied to the nominal pressures to
bring the experimental results to an agreement with the experimental
determination of reaction (9) by Goldsmith (1980). Preliminary results
for the binary Di-CaTs were given by Gasparik (1980a). Since then additional data have been obtained and a new refinement gives the parameters (J,bar):

$$W_{G,CaTs(Di)} = 41240 - 30.90T; \quad W_{G,Di(CaTs)} = -12.25T - 0.167P$$

where $W_{G,CaTs(Di)}$ is the Margules parameter for CaTs in infinite dilution

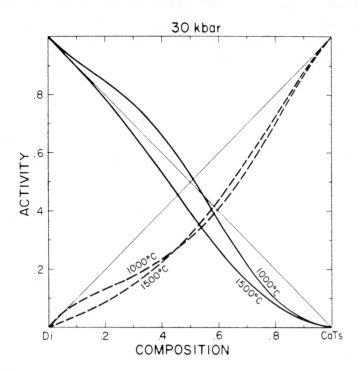

Figure 11. Activity-composition relations for the binary Di-CaTs at 1000 and 1500°C and 30 kbar.

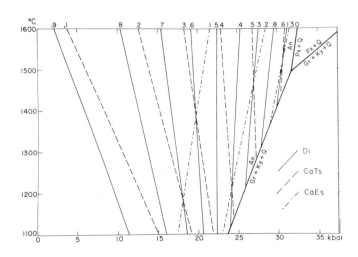

Figure 12. P-T phase relations for the system CaO-MgO-Al₂O₃-SiO₂ saturated with silica showing the isopleths of all three components (in mole fractions) in the clinopyroxene coexisting with anorthite and quartz. The high temperature part of the diagram is metastable with respect to the melting relations in the system, omitted from the diagram.

in the clinopyroxene solid solution, etc. The activities were defined assuming complete ordering. The activity-composition relations for the binary Di-CaTs at 1000 and 1500° and 30 kbar are shown in Figure 11 and the calculated isopleths for all three pyroxene components in the clino-pyroxene coexisting with An+Q are given in Figure 12. The isopleths intersect at relatively large angles which means that the clinopyroxene coexisting with anorthite and quartz can be used as both geothermometer and geobarometer.

Some preliminary experiments have been made on the solubility of Al-components in Cpx in the $CaO-MgO-FeO-Al_2O_3-SiO_2$ system (Wood, 1976b, 1979).

JADEITIC CLINOPYROXENES

Jadeite is an important component of most natural clinopyroxenes and particularly omphacites. However, very little is known about the thermodynamic properties of jadeitic clinopyroxenes. The main reason is that addition of Na_2O to the already complex system $CaO-MgO-Al_2O_3-SiO_2$ results in quaternary pyroxene solid solutions with end members Di-Jd-CaTs-CaEs in silica-saturated systems and Di-En-Jd-CaTs in silica undersaturated systems. Most of the experimental studies so far do not go beyond a binary treatment of pyroxene solid solutions.

The important subsolidus equilibria controlling the solubility of aluminum in jadeitic pyroxenes are univariant equilibria in the system $Na_2O-Al_2O_3-SiO_2$:

$$Ab = Jd + Q \qquad (27)$$

$$Ne + Ab = 2Jd \qquad (28)$$

Reaction (27) was experimentally determined by Birch and LeComte (1960), Newton and Smith (1967), Newton and Kennedy (1968), and most recently by Holland (1980). A detailed discussion of older experimental studies on reactions (27) and (28) is given by Bell and Roseboom (1969) in their experimental study of melting relations in the system $Na_2O-Al_2O_3-SiO_2$ (Fig. 13).

Holland (1980) determined reaction (27) by reversals in the temper-ature range 600-1200°C The equilibrium can be described by the equation: $P = 0.35 + 0.0265T(°C) \pm 0.50$ kbar. The starting material was synthetic

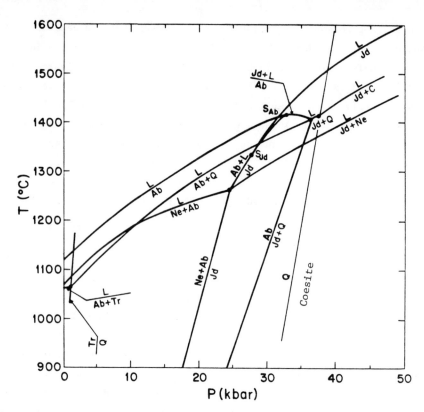

Figure 13. P-T phase relations for the system NaAlSiO₄-SiO₂. After Bell and Roseboom (1969). Modified to reconcile the diagram with the experimental data of Holland (1980) for reaction (27).

high albite, natural vein quartz and natural jadeite. The charge was moistened by breathing on the powder before sealing in Pt capsules. Complete reaction usually occurred in less than three hours at 1200°C and 24 hours at 800°C. Runs at and above 1100°C were not considered to be reversed as the albite broke down to Jd+Q during the approach to the final P-T conditions. Results are in excellent agreement with Johannes *et al.* (1971) and with the gas apparatus determinations by Birch and Le Comte (1960). The lack of curvature of the reaction implies that high albite does not undergo ordering down to 600°C at least within the duration time of the experiments.

Reaction (28) was experimentally determined by Robertson *et al.* (1957) in a gas apparatus. They reversed reaction (28) in the range 600-1100°C using crystalline starting materials. However, because they could not obtain close brackets, the position of the curve was further

estimated using glass of jadeite composition. The equilibrium is expressed by the equation: $P(bar) = 981 + 18.1T(°C)$. Two reversals at
700 and 800°C by Newton and Kennedy (1968) are in reasonable agreement
with the previous study.

In the system Di-Jd, invariant equilibria (27) and (28) become divariant. Effect of Di on equilibrium (28) was experimentally studied by
Bell and Kalb (1969) using the phase-disappearance technique and crystalline starting material. The reaction

$$\text{Omphacite}_1 = \text{Omphacite}_2 + \text{Pl} + \text{Ne} \qquad (29)$$

was reversed at 1150°C and 1225°C for bulk compositions $Jd_{50}Di_{50}$ and
$Jd_{70}Di_{30}$ (wt %).

Kushiro (1965a, 1969) studied the effect of Di on equilibrium (27)
in the system Di-Ab by the phase-disappearance technique. The reaction

$$\text{Omphacite}_1 + Q = \text{Omphacite}_2 + \text{Pl} + Q \qquad (30)$$

was located at 1150, 1250 and 1350°C using crystalline starting material
and glass, both giving nearly the same results.

The phase disappearance technique can provide useful constraints on
thermodynamic properties of solid solutions only if there is only one
compositional variable in the system studied which is then fixed by the
bulk composition. Moreover, it is assumed that the run products are
homogeneous. However, in reactions (29) and (30) not only omphacite but
also plagioclase and nepheline are solid solutions (Hamilton and MacKenzie,
1965; Mao, 1971). Moreover, aluminous pyroxenes, particularly when crystalline starting material is used, tend to produce zonation. Therefore,
determination of pyroxene composition by this technique is less precise
unless it is accompanied by direct analyses. Most recently, Holland (1979)
reversed reaction (30) at 600°C and 8, 10, 11, 12 and 13 kbar. Assuming
that plagioclase is pure albite and defining the activity of jadeite a_{Jd}
$= (X_{Jd})^2 \cdot \gamma_{Jd}$, he expressed γ_{Jd} in terms of a regular solution parameter
$\omega = 24$ kJ/mol.

Ganguly (1973) derived the activity-composition relations for natural
omphacite solid solutions considering mixing properties on all binaries to
be ideal with the exception of the join Di-Jd which he treated as a symmetric regular solution with only one parameter W_{12}. Ganguly first attempted to use the experimental data of Kushiro (1969) and Bell and Kalb

(1969) to estimate W_{12}, but he came to a conclusion that the Join Di-Jd is ideal at temperatures above 1100°C within the experimental error. Rationalizing that the presence of an ordered $P2$ structure in intermediate Di-Jd pyroxene indicates that the solid solution departs significantly from ideality at low temperatures, he used natural samples to estimate W_{12} and obtained $W_{12} \simeq -5860(\pm5860) + 4.85(\pm4.85)T°C$ J/mol.

Currie and Curtis (1976) formulated a multicomponent sub-regular solution model and applied it to natural omphacites. They derived the mixing properties on the binary Di-Jd using reversals by Kushiro (1969) for reaction (30) and by Bell and Kalb (1969) for reaction (29), unit cell volume measurements for the Di-Jd join by Kushiro (1969) and thermodynamic parameters from Robie and Waldbaum (1968). Assuming that Ab and Ne are pure end members, they obtained

$$W_{G,Di(Jd)} = -33420 + 41.8T - 0.11P; \quad W_{G,Jd(Di)} = -53760 + 0.475P \text{ (J,bar,K).}$$

We have updated the results of Currie and Curtis using data of Kushiro (1969) and Holland (1979) for reaction (30) and data of Holland (1980) for reaction (27) in the form (J,bar,K): $\Delta G° = -11940 + 45.96 \ T - 1.734P$. The excess volume terms were again calculated from the unit cell volume measurements of Kushiro (1969). Plagioclase in reaction (30) was considered to be pure albite. The least-squares solution modelling gives the following parameters for the binary Di-Jd (J,bar,K):

$$W_{G,Di(Jd)} = 30910 - 18.78T - 0.146P; \quad W_{G,Jd(Di)} = 35060 - 44.72T + 0.469P$$

$$RT\mathit{ln}\gamma_{Jd} = W_{G,Di(Jd)}(2X_{Di}^2 - 2X_{Di}^3) + W_{G,Jd(Di)}(2X_{Di}^3 - X_{Di}^2); \quad a_{Jd} = X_{Jd}\gamma_{Jd}$$

Activity-composition relations for the binary Di-Jd at 600, 1000 and 1400°C and 15 kbar are given in Figure 14. Figure 15 shows the isopleths of solubility of jadeite in omphacite coexisting with plagioclase and quartz.

The system Jd-An was experimentally studied by Mao (1971) who determined the phase relations for two bulk compositions $Jd_{95}An_5$ and $Jd_{65}An_{35}$ (wt %). At 1300°C and 40 kbar, he synthesized a single clinopyroxene phase from bulk compositions on the join Jd-An having 5, 15, 25 and 35 wt % An. Clinopyroxenes with excess silica in the system $Jd-CaTs-SiO_2$ were synthesized also by Khanukhova *et al.* (1976b) at 35 kbar and 1200°C.

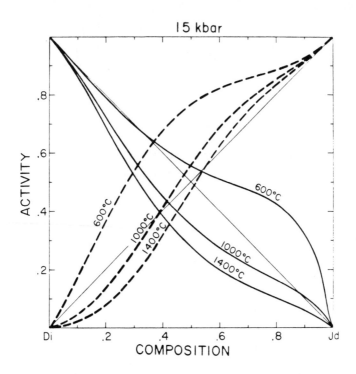

Figure 14. Activity-composition relations for the binary Di-Jd at 600, 1000 and 1400°C and 15 kbar.

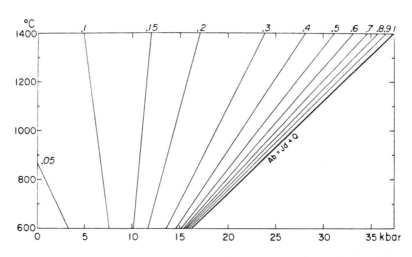

Figure 15. P-T phase diagram for the system Di-Jd-SiO$_2$ showing the isopleths in mole fractions of jadeite in omphacite. Based on data of Kushiro (1969) and Holland (1979; 1980).

Gasparik (1980b) in an experimental study of pyroxenes in the system Jd-CaTs-CaEs gave a microprobe analysis of clinopyroxene coexisting with plagioclase at 1200°C and 31.5 kbar. He used equilibria with plagioclase

$$Ab + CaTs = Jd + An \qquad (31)$$
$$2Ab + An = 2Jd + 2CaEs \qquad (32)$$
$$3An = 2CaTs + 2CaEs \qquad (6)$$

for experimental determination of mixing properties of clinopyroxenes in the ternary Jd-CaTs-CaEs, and equilibrium (3a) for the Ca-rich portion of the binary Jd-CaTs.

System $Na_2O-MgO-Al_2O_3-SiO_2$

The phase equilibria in this system are very important for petrogenesis of alkali basalts. Kushiro (1965b) in an experimental study of the system $Fo-Ne-SiO_2$ suggested two univariant reactions:

$$2Fo + Ab = Ne + 2En \qquad (33)$$
$$Ne + En = Jd + Fo \qquad (34)$$

The exact position of curves is presently unknown. Kushiro in preliminary runs with glass located reaction (33) between 9 and 11 kbar at 1100°C. Reaction (34) was located between 22 kbar and 1100°C (first appearance of jadeitic pyroxene) and 24 kbar and 1250°C also by synthesis from glass. At 1100°C and above 25 kbar glass of composition 18 wt % Fo, 62% Ne and 20% SiO_2 crystallized into jadeitic pyroxene + nepheline, from which composition of pyroxene should be $Jd_{62}En_{38}$ (mol %). This suggests that jadeitic clinopyroxenes in the system exhibit an extensive solid solution with En. However, the solid solution is not complete, as melting experiments at high pressures with the same bulk composition give both clino- and orthopyroxene (Kushiro, 1965c). Equilibrium (34) is univariant only if both pyroxenes are present. The equilibrium is divariant in the assemblage Opx+Ne+Fo at lower pressures and the assemblage Cpx+Ne+Fo at higher pressures.

Pyroxenes containing Fe^{3+}, Cr and Ti

Most studies on pyroxenes containing Fe^{3+}, Cr^{3+} and Ti^{4+} have been made at 1 atm (see Deer, Howie and Zussman, 1978, for references). We present here a review of some representative studies at high pressures.

Gilbert (1969) determined the incongruent melting temperature of acmite $NaFe^{3+}Si_2O_6$ to hematite + liquid as $T(°C). = 988 + 20.87P(kbar) - 0.155P^2$ up to 45 kbar. Bailey (1969) melted acmite in the presence of H_2O at 2 and 5 kbar. The temperature of (incongruent) melting was depressed to 865° (2 kbar) and 855° (5 kbar) for MH buffer and to 780° (2 and 4 kbar) for FMQ buffer. (The lower temperature for the more reducing buffer probably reflects increased Fe^{2+}/Fe^{3+} in the melt.) Flower (1974) synthesized titan-acmites from gels at 1000 bar P_{H_2O} and at the f_{O_2} of the $MnO-Mn_3O_4$ buffer. Some critical relations suggested by the syntheses were checked by experiments that used mechanical mixtures of crystalline starting material. The maximum solubilities in acmite of $NaTiFeSiO_6$ and of $NaTiAlSiO_6$ occur at the wet solidi: 35 mol %, 770°C and 18 mol %, 740°C, respectively. Using mainly oxide mixes of $NaFeSi_3O_8$-$NaAlSi_3O_8$ compositions as starting materials, Popp and Gilbert (1972) attempted to locate the boundary separating the Cpx + Q field from that of Cpx + Ab + Q at 4 kbar. At that pressure, the maximum solubility of Jd component in acmite decreases from 5-6 mol % at 500°C to 4-5 mol % at 600°C.

Abs-Wurmbach and Neuhaus (1976) reacted pure natural jadeite and synthetic ureyite ($NaCrSi_2O_6$) with 5% NaF as flux at 800°C and pressures from 1 atm to 20 kbar. The solid solution is complete above 18 kbar (which is their value for the breakdown pressure of pure jadeite). Vredevoogd and Forbes (1975) studied phase relations in the join diopside-ureyite ($NaCrSi_2O_6$) at 1 atm and at 1, 5, and 20 kbar. They reported that the solubility of ureyite in diopside decreases with increasing pressure. However, their experiments all involved synthesis from gels, so it is unclear whether these results represent equilibrium. Ikeda and Ohashi (1974) synthesized diopside-$CaCrSiCrO_6$ solid solutions at 15 kbar, and Akella and Boyd (1973a,b) have investigated the effects of Ti on the partitioning of Al between Opx, Cpx, and garnet.

ACKNOWLEDGEMENTS

We are grateful to James Dixon, Robert Newton, Dexter Perkins III, and Dean Presnall for permission to use data prior to publication. This work was supported in part by the National Science Foundation, Geochemistry Section, Grant #EAR76 22129.

Abs-Wurmbach, I. and A. Neuhaus (1976) The system $NaAlSi_2O_6$ (jadeite) - $NaCrSi_2O_6$ (cosmochlore) in the pressure range 1 bar to 25 kbars at 800 degrees C. Neues Jahrb. Mineral. Abh., 127, 213-241.

Akella, J. (1976) Garnet pyroxene equilibria in the system $CaSiO_3$-$MgSiO_3$-Al_2O_3 and in a natural mineral mixture. Am. Mineral., 61, 589-598.

_____ and F. R. Boyd (1973a) Partitioning of Ti and Al between coexisting silicates, oxides, and liquids. Proc. 4th Lunar Sci. Conf. (Supp. 4, Geochim. Cosmochim. Acta), 1, 1049-1059.

_____ and F. R. Boyd (1973b) Effect of pressure on the composition of co-existing pyroxenes and garnet in the system $CaSiO_3$-$MgSiO_3$-$FeSiO_3$-$CaAlTi_2O_6$. Carnegie Inst. Washington Year Book, 72, 523-526.

Anastasiou, P. and F. Seifert (1972) Solid solubility of Al_2O_3 in enstatite at high temperatures and 1-5 kb water pressure. Contrib. Mineral. Petrol., 34, 272-287.

Arima, M. and K. Onuma (1977) The solubility of alumina in enstatite and the phase equilibria in the join $MgSiO_3$-$MgAl_2SiO_6$ at 10-25 kbar. Contrib. Mineral. Petrol., 61, 251-265.

Bailey, D. K. (1969) The stability of acmite in the presence of H_2O. Am. J. Sci., Schairer Vol., 267-A, 1-16.

Bell, P. M. and J. Kalb (1969) Stability of omphacite in the absence of excess silica. Carnegie Inst. Washington Year Book, 67, 97-98.

_____ and E. H. Roseboom, Jr. (1969) Melting relationships of jadeite and albite to 45 kilobars with comments on melting diagrams of binary systems at high pressures. Mineral. Soc. Amer. Spec. Paper, 2, 151-161.

Birch, F. and P. LeComte (1960) Temperature pressure plane for albite composition. Am. J. Sci., 258, 209-217.

Boyd, F. R. and J. L. England (1959) Pyrope. Carnegie Inst. Washington Year Book, 58, 83-87.

_____ and J. L. England (1964) The system enstatite-pyrope. Carnegie Inst. Washington Year Book, 63, 157-161.

_____ and P. H. Nixon (1973) Structure of the upper mantle beneath Lesotho. Carnegie Inst. Washington Year Book, 72, 431-445.

Chatterjee, N. D. and W. Schreyer (1972) The reaction enstatite$_{ss}$ +sillimanite= sapphirine$_{ss}$+quartz in the system MgO-Al_2O_3-SiO_2. Contrib. Mineral. Petrol., 36, 49-62.

Clark, S. P., Jr., J. F. Schairer and J. de Neufville (1962) Phase relations in the system $CaMgSi_2O_6$-$CaAl_2SiO_6$-SiO_2 at low and high pressure. Carnegie Inst. Washington Year Book, 61, 59-68.

Currie, K. L. and L. W. Curtis (1976) An application of multicomponent solution theory to jadeitic pyroxenes. J. Geol., 84, 179-194.

Danckwerth, P. A. and R. C. Newton (1978) Experimental determination of the spinel peridotite to garnet peridotite reaction in the system MgO-Al_2O_3-SiO_2 in the range $900°C$-$1100°C$ and Al_2O_3 isopleths of enstatite in the spinel field. Contrib. Mineral. Petrol., 66, 189-201.

Deer, W. A., R. A. Howie and J. Zussman (1978) *Rock-forming minerals: Single-chain silicates*, Vol. 2A, 2nd ed. New York: John Wiley and Sons, 668 p.

Dixon, J. R. and D. C. Presnall (1980) Al_2O_3 content of enstatite: a spinel lherzolite barometer. Geol. Soc. Am. Abstr. Progr., 12, 414.

Eskola, P. (1921) On the eclogites of Norway. Vidensk.-selskapets Skr., Kristiania, I. Matematisk-Naturvidenskapelig Kl., 8, 1-118.

Flower, M. F. J. (1974) Phase relations of titan-acmite in the system Na_2O-Fe_2O_3-Al_2O_3-TiO_2-SiO_2 at 1000 bars total water pressure. Am. Mineral., 59, 536-548.

Fujii, T. (1976) Solubility of Al_2O_3 in enstatite coexisting with forsterite and spinel. Carnegie Inst. Washington Year Book, 75, 566-571.

_____ (1977) Pyroxene equilibria in spinel lherzolite. Carnegie Inst. Washington Year Book, 76, 569-572.

_____ and E. Takahashi (1976) On the solubility of alumina in orthopyroxene coexisting with olivine and spinel in the system MgO-Al_2O_3-SiO_2. Mineral. J. (Japan), 8, 122-128.

Ganguly, J. (1973) Activity-composition relation of jadeite in omphacite pyroxene: theoretical deductions. Earth Planet. Sci. Letters, 19, 145-153.

Gasparik, T. (1980a) Mixing properties of clinopyroxenes in the system Di-En-CaTs. Geol. Soc. Am. Abstr. Progr., 12, 432.

_____ (1980b) Role of aluminum in pyroxene equilibria. Geol. Soc. Am. Abstr. Progr., 12, 432.

_____ and D. H. Lindsley (1980) Experimental study of pyroxenes in the system $CaMgSi_2O_6-CaAl_2SiO_6-Ca_{0.5}AlSi_2O_6$. EOS, 61, 402-403.

Gilbert, M. C. (1969) High-pressure stability of acmite. Am. J. Sci., Schairer Vol., 267-A, 145-159.

Goldsmith, J. R. (1980) The melting and breakdown reactions of anorthite at high pressures and temperatures. Am. Mineral., 65, 272-284.

Hamilton, D. L. and W. S. Mackenzie (1965) Phase equilibrium studies in the system $NaAlSiO_4-KAlSiO_4-SiO_2-H_2O$. Mineral. Mag., 34, 214-231.

Hariya, Y. and G. C. Kennedy (1968) Equilibrium study of anorthite under high pressure and high temperature. Am. J. Sci., 266, 193-203.

Hays, J. F. (1966) Lime-alumina-silica. Carnegie Inst. Washington Year Book, 65, 234-239.

Hensen, B. J. (1972) Phase relations involving pyrope, enstatite$_{ss}$, and sapphirine$_{ss}$ in the system $MgO-Al_2O_3-SiO_2$. Carnegie Inst. Washington Year Book, 71, 421-427.

_____ and E. J. Essene (1971) Stability of pyrope-quartz in the system $MgO-Al_2O_3-SiO_2$. Contrib. Mineral. Petrol., 30, 72-83.

Herzberg, C. T. (1972) Stability fields of plagioclase- and spinel-lherzolite. In Progress in Experimental Petrology, D. 2, 145-148. Natural Environment Research Council Publications, London.

_____ (1976a) The plagioclase spinel-lherzolite facies boundary; its bearing on corona structure formation and tectonic history of the Norweigian caledonides. In Progress in Experimental Petrology, D. 3, 233-235. Natural Environment Research Council Publications, London.

_____ (1976b) The lowest pressure pyralspite garnet-forming reaction in $CaO-MgO-Al_2O_3-SiO_2$; the Seiland Ariegite subfacies boundary in simple spinel-iherzolites. In Progress in Experimental Petrology, D. 3, 235-237. Natural Environment Research Council Publications, London.

_____ (1978) Pyroxene geothermometry and geobarometry: experimental and thermodynamic evaluation of some subsolidus phase relations involving pyroxenes in the system $CaO-MgO-Al_2O_3-SiO_2$. Geochim. Cosmochim. Acta, 42, 945-957.

_____ and N. A. Chapman (1976) Clinopyroxene geothermometry of spinel-iherzolites. Am. Mineral., 61, 626-637.

Holland, T. J. B. (1979) Reversed hydrothermal determination of jadeite-diopside activities. EOS, 60, 405.

_____ (1980) The reaction albite = jadeite + quartz determined experimentally in the range 600-1200°C. Am. Mineral., 65, 129-134.

Ikeda, K. and H. Ohashi (1974) Crystal field spectra of diopside-kosmochlor solid solutions formed at 15 kb pressure. J. Japan. Min. Pet. Econ. Geol. 69, 103-109.

Jenkins, D. M. and R. C. Newton (1979) Experimental determination of the spinel peridotite to garnet peridotite inversion at 900°C and 1000°C in the system $CaO-MgO-Al_2O_3-SiO_2$, and at 900°C with natural garnet and olivine. Contrib. Mineral. Petrol., 68, 407-419.

Johannes, W., P. M. Bell, A. L. Boettcher, D. W. Chipman, J. F. Hays, H. K. Mao, R. C. Newton and F. Seifert (1971) An interlaboratory comparison of piston-cylinder pressure calibration using the albite-breakdown reaction. Contrib. Mineral. Petrol., 32, 24-38.

Khanukhova, L. T., V. A. Zharikov, R. A. Ishbulatov and Yu. A. Litvin (1976a) Excess silica in solid solutions of high-pressure clinopyroxenes as shown by experimental study of the system $CaMgSi_2O_6-CaAl_2SiO_6-SiO_2$ at 35 kilobars and 1200°C. Akad. Sci. USSR, Dokl., 229, 170-172.

_____, V. A. Zharikov. R. A. Ishbulatov and Yu. A. Litvin (1976b) Pyroxene solid solutions in the system $NaAlSi_2O_6-CaAl_2SiO_6-SiO_2$ at 35 kilobars and 1200°C. Acad. Sci. USSR, Dokl., 231, 140-142.

Kushiro, I. (1962a) Clinopyroxene solid solutions at high pressures. Carnegie Inst. Washington Year Book, 64, 112-117.

_____ (1965b) Coexistence of nepheline and enstatite at high pressures. Carnegie Inst. Washington Year Book, 64, 109-112.

_____ (1965c) The liquidus relations in the systems forsterite-$CaAl_2SiO_6$-silica and forsterite-nepheline-silica at high pressures. Carnegie Inst. Washington Year Book, 64, 103-109.

_____ (1969) Clinopyroxene solid solutions formed by reactions between diopside and plagioclase at high pressures. Mineral. Soc. Am. Spec. Pap., 2, 179-191.

_____ and H. S. Yoder, Jr. (1965) The reactions between forsterite and anorthite at high pressures. Carnegie Inst. Washington Year Book, 64, 89-94.

_____ and H. S. Yoder, Jr. (1966) Anorthite-forsterite and anorthite-enstatite reactions and their bearing on the basalt-eclogite transformation. J. Petrol., 7, 337-362.

Lindsley, D. H. and S. A. Dixon (1976) Diopside-enstatite equilibria at 850° to $1400^\circ C$, 5 to 35 kb. Am. J. Sci., 276, 1285-1301.

_____, J. E. Grover and P. M. Davidson (1981) The thermodynamics of the $Mg_2Si_2O_6$-$CaMgSi_2O_6$ join: A review and a new model. In *Advances in Physical Geochemistry*, vol. 1, S. K. Saxena, ed., Springer-Verlag, in press.

MacGregor, I. D. (1964) The reaction 4 enstatite + spinel = forsterite + pyrope. Carnegie Inst. Washington Year Book, 63, 157.

_____ (1965) Stability fields of spinel and garnet peridotites in the synthetic system $MgO-CaO-Al_2O_3-SiO_2$. Carnegie Inst. Washington Year Book, 64, 126-134.

_____ (1974) The system $MgO-Al_2O_3-SiO_2$: solubility of Al_2O_3 in enstatite for spinel and garnet peridotite compositions. Am. Mineral., 59, 110-119.

Mao, H. K. (1971) The system jadeite ($NaAlSi_2O_6$)-anorthite ($CaAl_2Si_2O_8$) at high pressures. Carnegie Inst. Washington Year Book, 69, 163-168.

Mori, T. and D. H. Green (1975) Pyroxenes in the system $Mg_2Si_2O_6$-$CaMgSi_2O_6$ at high pressure. Earth Planet. Sci. Lett., 26, 277-286.

Newton, R. C. (1972) An experimental determination of the high pressure stability limits of magnesian cordierite under wet and dry conditions. J. Geol., 80, 398-420.

_____, T. V. Charlu and O. J. Kleppa (1974) A calorimetric investigation of the stability of anhydrous magnesium cordierite with application to granulite facies metamorphism. Contrib. Mineral. Petrol., 44, 295-311.

_____, T. V. Charlu and O. J. Kleppa (1977) Thermochemistry of high pressure garnets and clinopyroxenes in the system $CaO-MgO-Al_2O_3-SiO_2$. Geochim. Cosmochim. Acta, 41, 369-377.

_____ and G. C. Kennedy (1968) Jadeite, analcite, nepheline, and albite at high temperatures and pressures. Am. J. Sci., 266, 728-735.

_____ and J. V. Smith (1967) Investigations concerning the breakdown of albite at depth in the earth. J. Geol., 75, 268-286.

Obata, M. (1976) The solubility of Al_2O_3 in orthopyroxene in spinel and plagioclase peridotites and spinel pyroxenite. Am. Mineral., 61, 804-816.

O'Hara, M. J. and S. Howells (1979) The enstatite-pyrope geobarometer. In *Progress in Experimental Petrology*, D. 4, 175-179. Natural Environment Research Council Publications, London.

O'Hara, M. J., S. W. Richardson and G. Wilson (1971) Garnet-peridotite stability and occurence in crust and mantle. Contrib. Mineral. Petrol., 32, 48-68.

Perkins III, D. and R. C. Newton (1980) Equilibrium compositions of orthopyroxene (Opx), clinopyroxene (Cpx), and garnet (GAR) in the $CaO-MgO-Al_2O_3-SiO_2$ system. EOS, 61, 402.

_____ and R. C. Newton (1980) The composition of coexisting pyroxenes and garnet in the system $CaO-MgO-Al_2O_3-SiO_2$ at 900°-$1100^\circ C$ and high pressures. Contrib. Mineral. Petrol., 75, 291-300.

Popp, R. K. and M. C. Gilbert (1972) Stability of acmite-jadeite pyroxenes at low pressure. Am. Mineral., 57, 1210-1231.

Presnall, D. C. (1976) Alumina content of enstatite as a geobarometer for plagioclase and spinel lherzolites. Am. Mineral., 61, 582-588.

Richardson, S. W., P. M. Bell, and M. C. Gilbert (1968) Kyanite-sillimanite
equilibrium between 700°C and 1500°C Am. J. Sci., 266, 513-541.

Robertson, E. C., A. F. Birch and G. J. F. MacDonald (1957) Experimental determination of jadeite stability relations to 25,000 bars. Am. J. Sci., 255,
115-137.

Robie, R. A. and D. R. Waldbaum (1968) Thermodynamic properties of minerals and
related substances at 298.15°K (25.0°C) and 1 atm (1.013 bars) pressure
and at higher temperatures. U. S. Geol. Surv. Bull., 1259, 256 p.

Skinner, B. J. and F. R. Boyd (1964) Aluminous enstatites. Carnegie Inst. Washington
Year Book, 63, 163-165.

Vredevoogd, J. J. and W. C. Forbes (1975) The system diopside-ureyite at 20 kb.
Contrib. Mineral. Petrol., 52, 147-156.

Wood, B. J. (1974) The solubility of alumina in orthopyroxene coexisting with
garnet. Contrib. Mineral. Petrol., 46, 1-15.

_____ (1976a) On the stoichiometry of clinopyroxenes in the system CaO-MgO-
Al_2O_3-SiO_2. Carnegie Inst. Washington Year Book, 75, 741-742.

_____ (1976b) Mixing properties of tschermakitic clinopyroxenes. Am. Mineral.,
61, 599-602.

_____ (1977) Experimental determination of the mixing properties of solid
solutions with particular reference to garnet and clinopyroxene solutions. In
Thermodynamics in geology, D. G. Fraser, ed., Dordrecht: Reidel, 11-27.

_____ (1978) Reactions involving anorthite and $CaAl_2SiO_6$ pyroxene at high
pressures and temperatures. Am. J. Sci., 278, 930-942.

_____ (1979) Activity-composition relationships in $Ca(Mg,Fe)Si_2O_6$-$CaAl_2SiO_6$
clinopyroxene solid solutions. Am. J. Sci., 279, 854-875.

_____ and S. Banno (1973) Garnet-orthopyroxene and orthopyroxene-clinopyroxene
relationships in simple and complex systems. Contrib. Mineral. Petrol., 42,
109-127.

_____ and C. M. B. Henderson (1978) Compositions and unit-cell parameters
of synthetic non-stoichiometric tschermakitic clinopyroxenes. Am. Mineral.,
63, 66-72.

Chapter 8

THERMODYNAMICS of PYROXENES

John E. Grover

INTRODUCTION

Geological Uses for Thermodynamics

Geologists use thermodynamics to study the chemical behavior of
pyroxenes and other minerals. Several kinds of important information
can be obtained using this approach:

Phase equilibria. Phase diagrams can be constructed using values
for the thermodynamic constants of end-member pyroxene components,
together with expressions describing the mixing properties of solutions
among those components. When a thermodynamic model is successful in
reproducing the phase equilibria observed experimentally at selected
pressures and temperatures, it is reasonable to use the model as a
basis for deriving thermochemical data.

Interpolation and extrapolation of thermochemical data. Phase
diagrams fashioned from chemical constants and thermodynamic theory
are useful in representing the behavior of systems at temperatures,
pressures or compositions that are not accessible to experimentation.
They may also be used to predict the phase relations in a system that
is to be studied in the laboratory.

Geothermometers and geobarometers. The equations used to generate
phase equilibria can often be rewritten in a form that gives temperature
(or pressure) as an explicit function of mineral composition, pressure
(or temperature), and several empirically determined thermochemical
constants. Expressions of this kind are important to geologists interested in fixing the temperature or pressure of equilibration for
natural, pyroxene-bearing assemblages.

Activities of pyroxene components. The mathematical expressions
used to represent chemical mixing in a pyroxene solid solution as a
function of composition, temperature and pressure can be recast in a
form that expresses the activity or activity coefficient for any chemical component. An expression for the activity of a component in pyroxene can be used either to monitor or to predict values for the activity

of that component in a pyroxene-bearing rock: when the activity is fixed by the composition of a pyroxene solution, the composition (or even the presence) of other minerals in the rock may be prescribed. Conversely, component activities defined by the composition of the *system* may control the specific identity of the pyroxene present. The magnitude or the extent of variation of component activities can affect the phase chemistry, element distribution and crystallization history of a natural rock system.

Sources of Thermodynamic Data

The thermochemical data necessary for these applications can be obtained in several ways; it is desirable if information from one source can be corroborated independently with data from another:

Calorimetry. The heats of solution for pure pyroxene phases and their component oxides can be measured directly at a variety of temperatures, depending upon the calorimetric solvent used (see Kleppa and Newton, 1975, and Navrotsky, 1977, for reviews). The enthalpies of formation for these phases and the enthalpies of chemical reactions involving them can be computed from these calorimetric data. Heat capacity measurements, taken both at cryogenic temperatures and at room temperature and above, can be used with values of enthalpy to determine standard free energies of formation or of reaction. Reactions among the pyroxene polymorphs and the "excess" energies (or enthalpies) associated with cation ordering on crystallographic sites within pyroxenes can also be studied using calorimetry.

Phase-equilibrium measurements. Thermochemical constants can be extracted from the measured slopes of univariant reaction curves (including those representing polymorphic inversions) or from an analysis of the mixing properties of solution phases coexisting in a given system. Expressions for the molar free energy of solution and the several component activities can be deduced from a solution model appropriate to the system if the compositions of minerals coexisting at a fixed temperature and pressure can be determined. Mössbauer, optical and x-ray spectroscopic measurements can be used to determine the site occupancies of certain atomic species within pyroxene phases (see, for example, Virgo and Hafner, 1969, or Aldridge *et al.*, 1978; Goldman and Rossman,

1979; and Ghose, 1965). The standard free energies for intracrystalline exchange as well as (model-dependent) values for component activities can be computed from these data.

Measurements of oxidation-reduction equilibria. Activity - composition relations and the free energies of reaction for chemical equilibria among phases containing elements of variable valence can be measured at high- and low-temperatures using either gas-mixing techniques or electrolytic cells with solid electrolytes (see, for example, Williams, 1971; Muan, 1967, and Sato, 1971, discuss procedures for extracting thermochemical data).

Calculations of thermochemical quantities. Thermochemical constants can be computed by means of (1) a statistical--mechanical analysis of spectroscopic determinations of the spacings among molecular energy levels; or (2) a third-law analysis of entropy or heat capacity data, usually obtained cryogenically. A statistical--mechanical approach to the calculation of thermochemical quantities will necessarily consider the effects of cation partitioning (site ordering), if any (see Kerrick and Darken, 1975, and Barrer and Klinowski, 1972; 1977; 1979). For a good review of statistical-mechanical theory as it relates to thermodynamics, with some examples of applications, see chapters 11 - 14 in Denbigh (1964).

Notes on Coverage

In this chapter I will review generally some of the problems inherent in using thermodynamics to treat pyroxene solutions, describe several important thermodynamic solution models applicable to the pyroxenes and show how, in a few recent cases, these models have been applied to published phase-equilibrium data. I have used the system $Mg_2Si_2O_6$(En) - $CaMgSi_2O_6$(Di) throughout as a model for pyroxene phase relations in general, and for those in the QUAD in particular. There are several reasons for this: (1) phase equilibria in this system have been investigated quite thoroughly (see reviews by Lindsley and Dixon, 1976, and Holland *et al.*, 1979); (2) many of the problems inherent in modelling pyroxene solutions thermodynamically are present and obvious in this system (or in closely related ones); (3) the phase relations in this system are generally analogous topologically to those in other

343

TABLE 1. SELECTED SOURCES FOR THERMODYNAMIC SOLUTION THEORY
(references and comments)

General review of thermodynamics

Darken & Gurry (1953) See esp. Chapters 10 (solutions); 11 (phase relations); 12 (heterogeneous equilibria); and 13 (free energy -- composition diagrams).

Denbigh (1964, or later editions) A sound, formal introduction to thermodynamics, with many worked problems. See esp. Chapters 6 through 10.

Prigogine & Defay (1954) An advanced treatment of chemical thermodynamics; carefully, thoroughly and quite clearly done. See esp. Chapters 12 ("condensed phases"); 13 and 14 (multi-component systems; phase changes); 15 (thermodynamic stability); 16 (critical phenomena and phase separation); 19 (configurational properties; order and disorder); 20 (solutions); and 24-26 (excess properties and configurational energy).

Swalin (1972) A short but remarkably thorough introduction to the thermodynamics of solids; examples are from metallurgy, but the principles are applicable to geology. See esp. Chapter 5 (phase transformations and chemical reactions); 7 (ideal and non-ideal solutions; ordering theory); 8 and 9 (equilibrium among phases of variable composition; binary phase diagrams and thermochemical calculations).

Thermodynamic theory, with applications to the phase relations,
stability and petrology of pyroxenes and related minerals

Carmichael, Turner & Verhoogen (1974) See esp. pp. 77-82, 98-102 (T,P effects on mineral equilibria; geothermometers and barometers); 180-217 (thermodynamic properties of solids).

Ernst (1976) See esp. Chapters 1 (thermodynamics, reviewed); 3 (methods for treating thermochemical data); and pp. 238-249 (element distributions among coexisting minerals).

Fraser (1977) See esp. Chapters 2 (mixing properties of clinopyroxene solutions) and 3 (thermochemistry of aluminous pyroxene solutions).

Greenwood (1977) See esp. Chapters 3 (entropy, activity); 9 (chemical equilibria in non-ideal and complex solutions); 13 and 14 (derivation of thermochemical data from phase equilibrium experiments and experimental uncertainty).

Mueller & Saxena (1977) See Chapters 1-3 for a concise review of thermodynamic and kinetic equations. Many applications of thermodynamics are given in subsequent chapters.

Saxena (1973) Macroscopic and microscopic (order-disorder) solution theory are presented systematically; see esp. pp. 1-32, 37-49, 59-71 and 79-89 for general relations. Pyroxenes are treated specifically on pp. 52-54; 90, *et seq.*

Turner & Verhoogen (1960) Principles of chemical equilibrium are reviewed succinctly in Chapter 2.

Wood & Fraser (1976) See esp. Chapter 3 (multicomponent solutions, pp. 81-114, with the summary and problems, pp. 118-124; Chapters 4 (geothermometry and barometry, pp. 127-139; 149-151) and 7 (estimating and using thermochemical data).

Treatment of Margules Equations and other models for
thermodynamic non-ideality

Blencoe (1977) Presentation of several important mixing models and of algorithms for their solution.

Broecker & Oversby (1971) See esp. Chapter 11, pp. 243-254.

Grover (1977) A review of the properties of the Margules equations, with lists (a) of papers from the geologic literature using that approach, and (b) of selected references for other diverse solution models.

King(1969) A comprehensive treatment of multicomponent solution theory.

Thompson (1967) The principal source in the geologic literature for the equations, justification and development of binary Margules theory.

Wisniak & Tamir (1978) The 35-page Introduction gives an excellent review of the thermodynamic solution models used most commonly in treating non-ideal behavior.

parts of the QUAD where x_{CaSiO_3} varies at a fixed value (more or less) for $(\frac{Fe}{Mg+Fe})$; and (4) many workers have used this system as the "join of departure" for attempts to treat the thermodynamics of QUAD pyroxenes. (This includes geologists - like Wood and Banno, 1973 - who have sought principally to use solution models of the pyroxenes as geothermometers for natural materials.)

I have tried in this chapter to deal chiefly with concepts and not to present a review of the literature (which is massive on this subject). For an exhaustive review of published data on pyroxenes and numerous examples of the application of thermodynamics to pyroxene problems, the reader should consult Deer, Howie and Zussman (1978; 'DHZ'). Other books or papers, notably those listed in Table 1, may profitably be reviewed for additional information on pyroxene thermodynamics or varying approaches to the thermodynamics of solid solutions.

A glossary of the notation used in this chapter is given below. Appendix A presents a brief review of the properties of the Gibbs free energy function for multicomponent solutions, showing how, in particular, the quantities μ_i, $\bar{G}_T$, γ_i, and $\bar{G}_{EX}$ are interrelated.

ABBREVIATIONS AND DEFINITIONS

Phases and Solutions

Opx -- Solid-solution series of orthorhombic (*Pbca*) pyroxenes, extending from $Mg_2Si_2O_6$ to approximately $Mg_{1.85}Ca_{0.15}Si_2O_6$; Thermodynamic expressions for more Ca-rich Opx do not have physical significance.

Cpx -- Solid-solution series of monoclinic (*C2/c*) pyroxenes, extending from $Mg_2Si_2O_6$ to $CaMgSi_2O_6$. Ca-poor phases in this series have the $P2_1/c$ space group at room temperature.

En$_{ss}$ -- Enstatite solid solution: a phase belonging to the Opx solid-solution series.

Di$_{ss}$ -- Diopside solid solution: a phase (usually relatively Ca-rich) belonging to the Cpx solid-solution series.

Pig -- Pigeonite: A Ca-poor phase belonging to the Cpx solid solution series. $P2_1/c$ space group at low temperature.

Wo$_{ss}$ -- Wollastonite solid solution: a $CaSiO_3$ phase containing some $MgSiO_3$.

345

CEn — clinoenstatite; low-CEn, $P2_1/c$, at low temperatures; high-CEn, *probably* $C2/c$, at high temperatures.

OEn — orthoenstatite.

ϕ — used as a superscript or subscript to mean "in a phase" or "of phase (ϕ)."

Components

En,Di,Wo — subscripts meaning the *components* $Mg_2Si_2O_6$, $CaMgSi_2O_6$, and $Ca_2Si_2O_6$.

A,B; 1,2,3 — indicate general components when used as subscripts.

Macroscopic Properties

x_i^ϕ — mole fraction of component i in phase ϕ, where i = A,B,1,2,3, En, Di, or Wo. Subscript omitted where not necessary.

γ_i^ϕ — rational activity coefficient of component i in phase ϕ. Subscript omitted where not necessary.

a_i^ϕ — activity of component i in phase ϕ; $a_i^\phi = \gamma_i^\phi x_i^\phi$. Subscript omitted where not necessary.

μ_i^ϕ — chemical potential of component i in phase ϕ.

$\mu_i^{o,\phi}$ — standard state chemical potential for component i in phase ϕ.

$\mu_i^{Ex,\phi}$ — the excess chemical potential for component i in phase ϕ;
$$\mu_i^{Ex} = RT ln \gamma_i.$$

$\Delta\bar{U}_A^o$, $\Delta\bar{U}_B^o$ — difference in standard-state internal energy per mole between Cpx and Opx series. Subscripts A and B refer to end-member reactions: (A) $(Mg_2Si_2O_6)^{Opx} \rightleftarrows (Mg_2Si_2O_6)^{Cpx}$; (B) $(CaMgSi_2O_6)^{Opx} \rightleftarrows (CaMgSi_2O_6)^{Cpx}$. Likewise for $\Delta\bar{H}_A^o$, $\Delta\bar{H}_B^o$ (entropy); $\Delta\bar{V}_A^o$, $\Delta\bar{V}_B^o$ (volume); $\Delta\bar{G}_A^o$, $\Delta\bar{G}_B^o$ (Gibbs free energy), all per mole.

$\bar{G}_{f,\phi}^o[T,P]$ — the molar free energy of formation for phase ϕ at T and P. ϕ subscript may be omitted where not necessary and T, P unspecified denote 298 K and 1 bar. This may also be given as $\Delta\bar{G}_{f,\phi}^o$. The corresponding enthalpy of formation is $\Delta\bar{H}_{f,\phi}^o$.

$\bar{S}_T^o$, $\bar{S}^o$ — standard entropy of a given phase at temperature T.

$\bar{H}^o - \bar{H}_{T_o}^o$ — enthalpy function, relative to standard temperature T_o.

$-(\bar{G}^o - \bar{H}_{T_o}^o)/T$ — Gibbs energy function, relative to standard temperature T_o. Unless otherwise noted, $T_o = 298°K$.

$\bar{C}p$ -- molar heat capacity at constant pressure.

$\bar{G}_T$ -- total Gibbs free energy per mole of a phase or a solution.

$\bar{G}_{ID}$ -- ideal Gibbs free energy of a solution (per mole).

$\bar{G}_{ID}^M$ -- that portion of the ideal Gibbs free energy of solution due to mechanical mixing; $\bar{G}_{ID}^M = \Sigma_i X_i \mu_i^o$.

$\bar{G}_{ID}^S$ -- that portion of the ideal Gibbs free energy of solution due to chemical mixing; $\bar{G}_{ID}^S = RT\Sigma_i X_i \ln X_i$ and $\bar{G}_{ID} - \bar{G}_{ID}^M + \bar{G}_{ID}^S$.

$\bar{G}_{EX}$ -- the non-ideal or "excess" portion of the molar Gibbs free energy of solution; $\bar{G}_T = \bar{G}_{ID} + \bar{G}_{EX}$ and $\bar{G}_{EX}^\phi = RT\Sigma_i X_i^\phi \ln\gamma_i^\phi$.

W_Z^ϕ -- symmetric Margules parameter giving the excess thermodynamic quantity $\bar{Z}_{EX}$ for phase ϕ, Z can be $\bar{U}$, $\bar{H}$, $\bar{S}$, $\bar{V}$, or $\bar{G}$.

W_{Z1}^ϕ, W_{Z2}^ϕ -- asymmetric Margules parameters for binary phase ϕ.

$W_{Z,ij}^\phi$ -- the Margules parameter for the binary joins of a ternary solution ϕ. Here i and j can be 1,2; 2,1; 1,3; 3,2 *etc.* but i $\neq$ j. The subscription of additional running indices, i, j, k, ℓ, m, and so on, can be used to accommodate systems having additional components. Other similar "fitting constants" --*e.g.* A_{ij}--are employed when convenient. A true ternary constant has the form W_{ijk}.

Properties Related to Intracrystalline Ordering

$\bar{G}_c$ -- the molar Gibbs free energy of mixing arising from the configural properties of atomic species on sub-lattice sites within a given phase; $\bar{G}_c = \bar{H}_c - T\bar{S}_c$, where $\bar{S}_c$ and $\bar{H}_c$ are:

$\bar{S}_c$ -- molar configurational entropy of mixing, and

$\bar{H}_c$ -- molar configurational enthalpy of mixing.

$\bar{H}_c^{ID}$ -- the ideal molar configurational enthalpy of mixing; that portion of $\bar{H}_c$ due to *ideal* partitioning of atomic species among the several sub-lattice sites within the phase. If $\bar{H}_c^{ID} \neq 0$, the phase must be regarded as *macroscopically* non-ideal.

$\bar{H}_c^N$ -- the non-ideal portion of the molar configurational enthalpy of mixing. This results from non-ideal partitioning of atomic species among the several sub-lattice sites within a given phase: In general, if $\bar{H}_c \equiv (\bar{H}_c^{ID} + \bar{H}_c^N) \neq 0$, the phase is *macroscopically* non-ideal.

M1 site -- an octahedral site, coordinated to six oxygens, in both Opx and Cpx structures. See Chapter 2 by Cameron and Papike.

M2 site -- a larger, less-regular site, coordinated to six, seven, or eight oxygens (depending on composition) in both Opx and Cpx structures.

α, β -- superscripts indicating specific (and distinct) crystallographic sites (such as M1 or M2).

X_j^k -- site mole fraction of species j on site k.

γ_j^k -- activity coefficient for species j on site k.

$\left.\begin{array}{l} e.g., \\ j = A,B,En,Wo \\ k = M1, M2 \end{array}\right.$

$\Delta \bar{G}_E^O$ -- a molar Gibbs free energy of intracrystalline exchange; the standard free energy for a site exchange reaction such as

$$A^\alpha + B^\beta \rightleftharpoons B^\alpha + A^\beta.$$

$\Delta \bar{G}_E^O$ is a function of temperature and pressure.

K_E -- the distribution coefficient for an intracrystalline exchange reaction such as that given above. Thus,

$$K_E = (X_B^\alpha X_A^\beta)/(X_A^\alpha X_B^\beta).$$

For *ideal* partitioning of species A and B between sites α and β, $\Delta \bar{G}_E^O = -RT \ln K_E$.

s -- a parameter measuring the *extent* of ordering of species A and B between sites α and β. Specifically, $s = (X_A^\beta - X_A^\alpha)$, so that $-1 \leq s \leq 1$.

Miscellaneous Variables and Constants

ε -- a multiplicity factor used to correct expressions for configurational entropy to a desired molecular basis.

L -- Avogadro's number ($6.02209 \times 10^{23} \text{mole}^{-1}$).

Z -- coordination number.

R -- the gas constant, taken as $8.3143 \text{ J·K}^{-1}\text{mole}^{-1}$, except where otherwise noted.

PROBLEMS IN MODELLING PYROXENE SOLUTIONS

Although the utility of a thermodynamic approach in representing phase relations among pyroxenes (and other minerals) is clear, measurement of values for the thermochemical constants and calibration of mixing models for the pertinent components are often difficult tasks, requiring both careful experiments on well-characterized materials (usually synthetic pyroxene phases) and the thoughtful formulation of mixing models. We do not yet have enough information to devise accurate, predictive solution models for any but the simplest pyroxene systems. This is due partially to the practical difficulties involved in: (1) synthesizing chemically homogeneous pyroxenes in large quantities (or as large single crystals); (2) reversing experiments designed to bracket reactions among pyroxenes and other minerals; and (3) determining the "correct" composition for any chemically complex product when both the bracketing experiments themselves and the multiple microprobe analyses of individual crystallites from those experiments give rise to statistical variations in composition.

Attempts to design rigorous and practical solution models for the pyroxenes are complicated further by polymorphism and cation ordering, widespread among both QUAD and non-QUAD compositions (see the chapter in this volume by Cameron and Papike), and by our growing awareness that certain obvious and stoichiometrically simple pyroxene compositions do not represent the best choice necessarily for end-members in the systems we want to treat thermodynamically.

The Problem of Cation Ordering

Pyroxene structures contain certain cation sites that are geometrically similar but crystallographically distinct in terms of their position, volume, molecular environment and therefore their energetic behavior as well. The octahedral M1 and M2 sites of the QUAD pyroxenes are classic examples (see Chapter 2 or DHZ, 1978, pp. 63-67, 165-166, 221-223, 424-427). In phases of variable composition, cations such as Ca^{2+}, Mg^{2+} or Fe^{2+} will be partitioned preferentially among these sites; and the extent of partitioning may vary as a function of T, P, f_{O_2} or bulk composition. In orthopyroxene, for example, Fe^{2+} tends to occupy M2 preferentially.

It has been apparent since the pioneering work of Mueller (1962a,b) that cation ordering--or intracrystalline partitioning as it is sometimes called--can have a pronounced influence on the thermodynamic properties of the pyroxenes. Ordering invariably gives rise to configurational contributions to the enthalpy, entropy, and heat capacity of mixing for a phase (see $e.g.$, Prigogine and Defay, 1954, pp. 293-305), and this affects--if only conceptually--the interpretation of thermochemical measurements on that material and the form of the mixing model applicable to solutions that it enters.

349

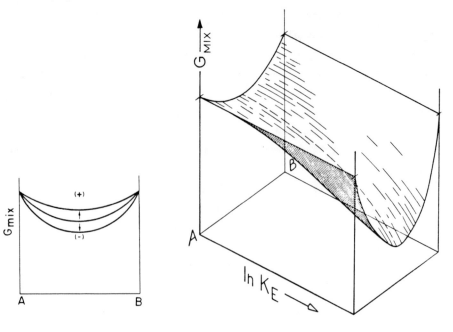

Figure 1. Schematic free energy-composition diagrams illustrating the effect of ideal intra-crystalline partitioning upon the free energy of mixing in a binary solid solution, A-B. (a) A free energy-composition section for a value of $\ln K_E = 0$. The free energy of mixing for ideal solution behavior is shown in relation to positive (+) and negative (-) deviations from ideality. If $\ln K_E$ is greater than zero, the solution A-B tends toward compound formation (negative deviation from ideality). (b) A perspective view showing, schematically, the effect of ideal intracrystalline partitioning on the free energy of mixing in the binary solid solution A-B. When $\ln K_E = 0$ (no partitioning between sites), the ideal free energy of mixing is obtained. As $\ln K_E$ increases (partitioning between sites becomes pronounced), the total free energy of mixing is depressed. (From Grover, 1974).

In general, the configurational enthalpy, $\bar{H}_c$, becomes increasingly negative, and the configurational entropy, $\bar{S}_c$ progressively less posi-tive as ordering increases. At a fixed temperature, the configurational contribution to the molar free energy of solution,

$$\bar{G}_c = \bar{H}_c - T\bar{S}_c,$$

thus becomes more negative with increased order, tending toward the for-mation of a stable compound of intermediate composition. This is shown schematically in Figure 1, where the free energy of solution for a binary system A-B is plotted against a measure of order, $\ln K_E$. If α and β are the two sites on which ordering takes place, K_E is given in terms of the mole fractions of species on sites as

$$K_E = \frac{X_A^\beta X_B^\alpha}{X_B^\beta X_A^\alpha} . \tag{1a}$$

350

Because the mole fractions of species on a given site must sum to one ($X_A^\alpha + X_B^\alpha = 1$), and because the bulk composition is related to site occupancies by the equations

$$\frac{X_A^\alpha + X_A^\beta}{2} = X_A \; ; \tag{2a}$$

and

$$\frac{X_B^\alpha + X_B^\beta}{2} = X_B \; , \tag{2b}$$

K_E can also be written in terms of bulk composition and Thompson's (1969) order parameter, s, as

$$K_E = \frac{(X_A + \frac{S}{2})(1 - X_A + \frac{S}{2})}{(1 - X_A - \frac{S}{2})(X_A - \frac{S}{2})} \; , \tag{1b}$$

where

$$s \equiv (X_A^\beta - X_A^\alpha) \; . \tag{3}$$

It is clear from equations (1) that as s increases (and component A is partitioned preferentially into site β), K_E also increases.

One form for the configurational enthalpy, appropriate to the case of *ideal* ordering (Grover, 1974), is

$$\bar{H}_c = \frac{s\Delta\bar{G}_E^O}{2} \; , \tag{4}$$

where $\Delta\bar{G}_E^O(T,P)$ is the standard free energy of exchange for a site-interchange reaction,

$$A^\alpha + B^\beta \rightleftarrows B^\alpha + A^\beta \; . \tag{5a}$$

We also have

$$\Delta\bar{G}_E^O = (\mu_A^{\beta o} + \mu_B^{\alpha o} - \mu_B^{\beta o} - \mu_A^{\alpha o}) \; , \tag{5b}$$

or, at internal equilibrium,

$$\Delta\bar{G}_E^O = -RT \ln K_E \tag{5c}$$

(see Grover and Orville, 1969, pp. 207-210). Several forms have been proposed for the configurational enthalpy in systems having *non-ideal* ordering (see especially Thompson, 1969, 1970). These allow for terms in addition to (4), and introduce flexibility in treating both negative and *positive* departures from ideality due to ordering. These will be discussed in a later section.

In the ideal case, when $\Delta\bar{G}_E^O$ is positive, reaction (5A) proceeds to the left; $X_A^\alpha > X_A^\beta$, so that $s < o$ (equation 3) and $\bar{H}_c$ is negative. When $\Delta\bar{G}_E^O$ is negative, $s > o$ and $\bar{H}_c$ is, again, negative. Thus, for ideal ordering (as defined by Grover, 1974; or Thompson, 1970), the configurational enthalpy is *always* negative. According to this model, $\Delta\bar{G}_E^O$ must be infinite in the theoretical limit where ordering is complete. Let us consider a bulk composition $X_A = \frac{1}{2}$, for example, so that $(X_A^\alpha + X_A^\beta)/2 = \frac{1}{2}$. If we specify arbitrarily that no A enters site β, then by equation (3),

$$s = -X_A^\alpha = 2X_A = -1 \ .$$

Substituting values for s and X_A in equation (1b), we find

$$K_E = 0 \ .$$

It follows from equation (5c) that $\Delta\bar{G}_E^O$ is infinite (positive), and we conclude from equation (4) that the ordered, intermediate compound of composition $A_1^\alpha B_1^\beta$ is infinitely stable ($\bar{H}_c \to \infty$ as $s \to 1$).

Does this mean that an existing intermediate compound with two distinct cation sites--like stoichiometric diopside, $Ca^{M2}Mg^{M1}Si_2O_6$--must have an infinite exchange energy, $\Delta\bar{G}_E^O \downarrow$ and an infinite configurational enthalpy, $\bar{H}_c$? If Ca never enters M1 in diopside, $\Delta\bar{G}_E^O$ is indeed infinite (formally) and $\bar{H}_c$ is simply undefined. In this case, solution models involving ordering are not applicable to diopside. If, however, M1 in diopside contains *very* small amounts of Ca (a concept many crystallographers consider to be anathema!) $\Delta\bar{G}_E^O$ is finite (but very large). It is not likely that $\bar{H}_c$ can be immeasurably large in this or any case; it represents part of the total enthalpy of formation for a compound and must, in principle, be measurable calorimetrically. If $X_{Ca}^{M1} \neq 0$ for diopside, it is probable that the magnitude of $\bar{H}_c$ is strongly influenced by non-ideal configurational terms, and cannot be computed using equation (4). Representative thermodynamic treatments of non-ideal ordering

----

[1] The corresponding expression for K_E is:

$$K_E(Di) = \frac{X_{Ca}^{M1} X_{Mg}^{M2}}{X_{Mg}^{M1} X_{Ca}^{M2}} = 0 \ ,$$

where A^β denotes Ca in M1, B^α is Mg in M2 and so forth.

are given by Thompson (1969, 1970), Saxena and Ghose (1970), Navrotsky and Loucks (1977), and Sack (1980).

The decrease in molar configurational entropy with increasing order may be ascribed quite simply to an increase in the number of $X \ln X$ terms in the expression

$$\bar{S}_c = -\varepsilon R \underset{jk}{\Sigma\Sigma} X_j^k \ln X_j^k \; ; \qquad (6)$$

here j and k index cation species and distinguishable sites, respectively, and ε is the multiplicity of positions, dictated by the number of molar units selected for comparison. (Conventionally, for two species ordered on two distinguishable sites in one mole of pyroxene, $(A,B)^{\alpha}(B,A)^{\beta}Si_2O_6$, $\varepsilon = 1$. For a pyroxene with two species filling a single site *in a mole of comparable size*, $\varepsilon = 2$.) Table 2 gives some typical values for $\bar{S}_c$ as a function of increasing order. The quantities listed are those defined and discussed above; $\bar{S}_{EX}$ is the excess entropy, given by

$$\bar{S}_{EX} = \bar{S}_{c(s)} - \bar{S}_{c(s=o)}.$$

The amount and, to some extent, the direction of variation in the configurational heat capacity with increasing order depend chiefly on the ordering model chosen; $\bar{C}_{p(config)}$ is positive, however, for standard formulations.

Only when the occupancy of crystallographically distinct sites in a mineral is constrained unambiguously by bulk composition can the thermodynamic effects of ordering be discounted categorically. (Clearly, the M1 and M2 sites in each of the stoichiometric pyroxene end-members $Mg_2Si_2O_6$ and $Fe_2Si_2O_6$ can contain only one species.) For complex pyroxene compositions, including some like $CaMgSi_2O_6$, taken classically to be end-members, there is at least the theoretical plausibility that ordering takes place. Without evidence to the contrary, the thermodynamic treatment given the calorimetric data and solution behavior for such phases should probably include an analysis of ordering.

The Problem of Components

The "correct" components for a system are those that are most convenient, and support the simplest thermodynamic interpretation of observed phase equilibria while conforming with crystal-chemical requirements. In equivocal cases, components should probably be chosen only after the system

Table 2

Values for $\bar{S}_c$ and $\bar{S}_{EX}$ $[= \bar{S}_{c(s)} - \bar{S}_{c(s=o)}]$, based on fabricated site occupancies*.

Site Occupancies				Order Parameters		Entropies	
X_A^α	X_A^β	X_B^α	X_B^β	$s = (X_A^\beta - X_A^\alpha)$	lnK_E	$\bar{S}_c/R$	$\bar{S}_{EX}/R$
0.4	0.4	0.6	0.6	0	0	1.3460	0
0.3	0.5	0.7	0.5	0.2	0.847	1.3040	-0.042
0.2	0.6	0.8	0.4	0.4	1.792	1.1734	-0.173
0.1	0.7	0.9	0.3	0.6	3.045	0.9359	-0.410

*Consistent with the material-balance equations:

$$X_A^\alpha + X_A^\beta = 2X_A; \quad X_B^\alpha + X_B^\beta = 2X_B; \quad X_A^\alpha + X_B^\alpha = 1; \quad X_A^\beta + X_B^\beta = 1.$$

The bulk composition is $X_A = 0.4$, $X_B = 0.6$. $\quad K_E = \dfrac{X_A^\beta \, X_B^\alpha}{X_B^\beta \, X_A^\alpha}$.

has been studied experimentally and analyzed thermodynamically in terms of each of the several compositions thought to be feasible as end-member components.

Some recent studies illustrate the simplification in apparent mixing properties of a system that can result when unconventional chemical components are chosen. Ganguly and Ghose (1979) found that the solution behavior of aluminous orthopyroxene is more nearly ideal (with activity coefficients more nearly equal to 1.0) when treated in terms of components $Mg_2Si_2O_6$-$Mg_{3/2}AlSi_{3/2}O_6$, rather than the more usual pair, $Mg_2Si_2O_6$-$MgAl(Al,Si)O_6$.

In measuring compositions and cell parameters for clinopyroxenes synthesized at 1300° to 1450°C and 25 to 32.3 kbar in the system $CaMgSi_2O_6$(Di)-$CaAl_2SiO_6$(CaTs)-SiO_2, Wood and Henderson (1978) detected vacancies in the M1 and M2 octahedral cation sites. They concluded that compositions such as $Ca_{3/4}^{M2}(Mg_{1/4}Al_{2/3})^{M1}Si_2O_6$ (approximate stoichiometry) or $(Ca,Mg)_{1/2}$ $AlSi_2O_6$ might represent suitable end-member "molecules" for aluminous clinopyroxene containing excess silica. Gasparik and Lindsley (1980) confirmed the existence of a component $Ca_{1/2}AlSi_2O_6$ in the system

$CaMgSi_2O_6-CaAl_2SiO_6-(CaAl_2Si_2O_8)-SiO_2$, and found that the mixing properties of tschermakitic pyroxene solutions could be treated most simply
using Di, CaTs and $Ca_{1/2}AlSi_2O_6$ as components. (Perhaps a suitable
acronym for $(Ca,Mg)_{1/2}AlSi_2O_6$ is BEDS: "balanced electrostatically,
defective structurally.") It is possible that other unconventional com-
positions--$Ca_{3/2}AlSi_{3/2}O_6$, with SiO_2 and CaTs, for example--will prove
useful in simplifying the apparent thermodynamic behavior of aluminous
clinopyroxene.

In some instances, selection of the components that seem appropriate
for crystal-chemical reasons does not lead to a simple thermodynamic model.
The system $Mg_2Si_2O_6$(En)-$CaMgSi_2O_6$(Di) is an example, where the possibility of Ca-Mg partitioning between M1 and M2 in $CaMgSi_2O_6$, discussed
earlier, has a profound influence on the theoretical treatment given
the phase equilibria. If Ca can, in fact, occupy M1 as well as M2, the
system must be extended to include the composition $Ca_2Si_2O_6$(Wo); this
insures that calcic clinopyroxenes (Cpx) are not arbitrarily limited to
the 1:1 Ca to Mg ratio imposed by the stoichiometry of the diopside component. (If M1 can accommodate Ca, compositions such as $Ca_{1.0}^{M2}(Ca_{.005}$
$Mg_{.995})^{M1}Si_2O_6$ should be accessible to phases in the system.)

The theoretical consequences of limiting this system to the components Di and En have been discussed by Lindsley and Grover (1978) and
by Lindsley *et al.* (1981). Briefly, if no Ca can enter M1, the free
energy curve for Cpx in this system must end with a vertical tangent
precisely at the $CaMgSi_2O_6$ composition. The curve is concave-upward
at compositions close to Di, owing to the strong influence of the ideal
mixing ($XlnX$) terms on the form of the first and second derivatives of
the free energy function.[2] Because of this geometric property, *pure*
diopside should never be stable: An assemblage consisting of Mg-enriched

----

[2]The expression for the second derivative, computed from equations A-6
and A-7 is:

$$\frac{d^2\bar{G}_T}{dx_{Di}^2} = \frac{RT}{X_{Di}X_{En}} + \frac{d^2\bar{G}_{EX}}{dx_{Di}^2} .$$

X_{Di} and X_{En} are never negative and $\bar{G}_{EX}$ and its derivatives are directly
proportional to powers of composition, $X_{Di}X_{En}$. Thus $\bar{G}_{EX} \to o$ as $X_{Di} \to 1$
or $X_{En} \to 0$, and $d^2\bar{G}_T/dx_{Di}^2$ is invariably positive for compositions, X_{Di},
close to 1.

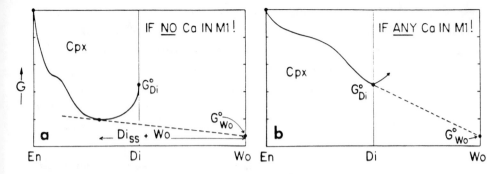

Figure 2. Schematic G-X diagrams illustrating two possible forms for the Cpx G-curve near CaMgSi2O6. (a) The G-curve ends with a vertical tangent precisely at CaMgSi2O6; pure diopside is always metastable with respect to wollastonite (Wo) and a Ca-depleted Di$_{ss}$ (here the amount of Ca-depletion is greatly exaggerated). (b) The G-curve passes through CaMgSi2O6 with a sharp upward inflection because small amounts of Ca enter the M1 site. Pure diopside can be stable. (From Lindsley et al., 1981).

Cpx plus wollastonite will have a lower total free energy than diopside alone (see Fig. 2a; the dashed line represents the free energy of mixing for the assemblage Mg-enriched cpx + wollastonite). If diopside coexisting with wollastonite is not, if fact, Ca-deficient, we must conclude that the free-energy curve for the Di-En system cannot properly end at composition CaMgSi$_2$O$_6$.

Kushiro (1964) and Davidson et al. (1979) have measured small excess amounts of Ca in diopside; it is possible--although by no means certain--that this excess Ca is located in M1. If so, the free-energy curve for Cpx must extend through CaMgSi$_2$O$_6$, as shown in Figure 2b. Thermodynamic ordering models for the system En-Wo capable of generating free-energy curves like that shown in Figure 2b have been developed by Navrotsky and Loucks (1977; convergent, non-ideal ordering) and by Davidson et al. (in preparation; non-convergent, non-ideal ordering). It is now clear that models that do not consider the effects of ordering cannot generate plausible free-energy curves for the system Mg$_2$Si$_2$O$_6$-Ca$_2$Si$_2$O$_6$(Wo).

We have considerable freedom in selecting chemical components: The compositions that we wish to treat must be linear combinations of the components chosen. We have more difficulty in deciding which components are justified on crystal-chemical grounds and in formulating expressions for the activities of those components in terms of the measurable compositions of phases that interest us.

The Problem of Polymorphism

Polymorphism is widespread among pyroxene compositions (see Chapter 2 or DHZ, 1978, pp. 32-34, 51-55, 164-165, 424-427). When the polymorphic forms of a phase are interrelated by means of a first-order structural inversion, each polymorph must have a unique thermodynamic equation of state. The reason for this is that there can be no single, *continuous* free-energy function having the general form (A-6) (or more particularly A-7b), that correctly pertains to a solution whose endmembers have different crystal structures: A first-order inversion, occurring either at a specific composition or over a range of compositions intermediate between the end-members, would necessarily generate an abrupt jump in energy, and eliminate the smooth functional continuity of the curve. This discontinuity in energy is especially evident at end-member compositions, where the reaction between two polymorphs, say $\phi 1$ and $\phi 2$, is

$$\phi 1 \rightleftarrows \phi 2 .$$

The free energy for such a reaction measures the difference in energy between polymorphs:

$$\Delta \bar{G}_\phi^o = (\mu_\phi^{o,2} - \mu_\phi^{o,1}) . \qquad (7a)$$

More generally,

$$\mu_\phi^{o,2}(T,P) - \mu_\phi^{o,1}(T,P) = \Delta \bar{U}_\phi^o + P\Delta \bar{V}_\phi^o - T\Delta \bar{S}_\phi^o . \qquad (7b)$$

(Not only is the inversion discontinuous energetically, but the magnitude of the discontinuity varies with temperature and pressure.) An understanding of the structural transformations that occur at end-member compositions is critical to the formulation of solution models because the *standard state chemical potentials* (μ_ϕ^o) used to "anchor" the ends of the thermodynamic mixing curves are defined at those compositions.

As an example, we will now consider the effects of polymorphism on free-energy-composition relations in the binary system $Mg_2Si_2O_6$(En)-$CaMgSi_2O_6$(Di). (I hope that this discussion will also serve readers in reviewing both the geometric properties of free energy curves and the ways by which those curves can be generated mathematically; see also the first two sections of Appendix A.)

Figure 3 is a schematic free-energy ($\bar{G}_T$)-composition (X_{Di}) diagram for the system $Mg_2Si_2O_6$(En)-$CaMgSi_2O_6$(Di) at a temperature and pressure

357

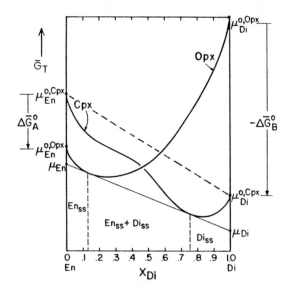

Figure 3. A schematic representation of the free energy-composition relations for the system Di-En at a temperature and pressure where En_{ss} (space group *Pbca*) and Di_{ss} (*C2/c*) coexist. Because these phases have different crystal structures, the free energy curves for the Opx and Cpx solutions are separate and independent. The standard state (end-member) chemical potentials for each solution are also independent. For En the orthorhombic form is stable at the pressure and temperature given. Thus, reaction (*A*) proceeds to the left because

$$\Delta \bar{G}_A^o = \mu_{En}^{o,Cpx} - \mu_{En}^{o,Opx} > 0.$$

For Di composition, the monoclinic form is stable at the prescribed conditions. Reaction (*B*), therefore, has

$$\Delta \bar{G}_B^o = \mu_{Di}^{o,Cpx} - \mu_{Di}^{o,Opx} << 0$$

and proceeds strongly to the right. At equilibrium, $\mu_{En}^{Opx} = \mu_{En}^{Cpx}$ and $\mu_{Di}^{Opx} = \mu_{Di}^{Cpx}$, as shown by the mutual tangent to both free-energy curves. (From Lindsley *et al.*, 1981).

where Di_{ss} and En_{ss} coexist (Lindsley *et al.*, 1981). Pure diopside (X_{Di} = 1) occurs only in the monoclinic form (space group *C2/c*) but enstatite (X_{Di} = 0) has both monoclinic (*C2/c*) and orthorhombic (*Pbca*) polymorphs. Each of the several polymorphs has a measurable standard free energy of formation and--presumably--calculable formation energies for the temperature and pressure represented here; these are to be taken as the standard state chemical potentials appropriate to the present system at T and P. Specifically,

$$\bar{G}_f(C2/c \ Di)_{T,P} \equiv \mu_{Di}^{o,Cpx}; \ \bar{G}_f(Pbca \ En)_{T,P} \equiv \mu_{En}^{o,Opx}; \ \bar{G}_f(C2/c \ En)_{T,P} \equiv \mu_{En}^{o,Cpx}$$

$\mu_{Di}^{o,Opx}$ is a fictive (calorimetrically unmeasurable) standard state potential for a hypothetical Di having space group *Pbca*; $\mu_{Di}^{o,Opx}$ cannot be measured because orthorhombic diopside is too unstable relative to the familiar monoclinic form to be synthesized. That is, $\Delta\bar{G}_B^o \ll 0$. Although, for these conditions, $C2/c$ En is unstable relative to $Pbca$ En ($C2/c$ En has the higher free energy), there are temperatures and pressures where the monoclinic form is stable as Fe-free pigeonite (= high clinoenstatite). The polymorphic inversion between these forms can be characterized by a free energy of reaction similar to (7):

$$\Delta\bar{G}_A^o = \mu_{En}^{o,Cpx} - \mu_{En}^{o,Opx} \qquad (8a)$$

and

$$\Delta\bar{G}_A^o = \Delta\bar{U}_A^o + P\Delta\bar{V}_A^o - T\Delta\bar{S}_A^o \qquad (8b)$$

where, for example, $\Delta\bar{V}_A^o = \bar{V}_{high\ CEn}^o - \bar{V}_{OEn}^o$ (molar volume).

If $C2/c$ En were stable at this temperature and pressure, it would be a simple matter to represent the observed solution behavior in this system with an analytic expression for the total molar free energy. Specifically, for $C2/c$ pyroxenes, curve "Cpx" is given by an equation of state,

$$\bar{G}_T(Cpx) = \bar{G}_{ID}^M(Cpx) + \bar{G}_{ID}^S(Cpx) + \bar{G}_{EX}(Cpx) , \qquad (9)$$

where the molar energy of mechanical mixing

$$\bar{G}_{ID}^M(Cpx) = \mu_{En}^{o,Cpx}X_{En} + \mu_{Di}^{o,Cpx}X_{Di}$$

$$= \mu_{En}^{o,Cpx} + (\mu_{Di}^{o,Cpx} - \mu_{En}^{o,Cpx})X_{Di}. \qquad (10a)$$

$\bar{G}_{ID}^M$ is represented by the straight line $\mu_{Di}^{o,Cpx} - \mu_{En}^{o,Cpx}$. The ideal molar energy of solution,

$$\bar{G}_{ID}^S(Cpx) = RT[(1-X_{Di})ln(1-X_{Di}) + X_{Di}lnX_{Di}] ; \qquad (10b)$$

and the non-ideal or excess molar energy of solution is

$$\bar{G}_{EX}(Cpx) = RT[(1-X_{Di})ln\gamma_{En}^{Cpx} + X_{Di}ln\gamma_{Di}^{Cpx}] \qquad (10c)$$

(for notation and further discussion, see Appendix A and below).

The orthorhombic polymorph is the more stable form of En for these conditions. It is formally incorrect to write a *single* expression for the total free energy of this system based (or "anchored") on the standard chemical potentials for the stable (but structurally distinct) end-member phases, $\mu_{En}^{o,Opx}$ and $\mu_{Di}^{o,Cpx}$. In order to treat the observed solution behavior of phases En_{ss} and Di_{ss} in the system Di-En it is necessary instead to use both equation (9) and a similar, but independent formulation for the Opx structure, anchored at $\mu_{En}^{o,Opx}$ for $X_{Di} = 0$ and at $\mu_{Di}^{o,Opx}$ for $X_{Di} = 1$. The resulting equation of state, $\bar{G}_T(Opx)$, delineates the minimum free-energy surface in the low-Ca portion of the system, consistent with the observed occurrence of orthorhombic En_{ss} in that region. The equilibrium compositions for the coexisting phases $En_{ss} + Di_{ss}$ are defined by a tie line, tangent simultaneously to both curves $\bar{G}_T(Cpx)$ and $\bar{G}_T(Opx)$, and ending at μ_{En} and μ_{Di}, the chemical potentials of the respective components in those coexisting phases.

THERMOCHEMICAL CONSTANTS

The Need for Thermochemical Data

As ordering, polymorphism and uncertainty regarding the choice of components appropriate for a system make modelling of pyroxene solutions challenging, so also do they impose formidable demands on the few calorimetrists currently working to provide thermochemical data on pyroxenes. Ideally, there would be abundant calorimetric data available to us, including not only standard enthalpies and free energies of formation (at several temperatures) for all pertinent end-member species (with the means for correcting these values to any temperature or pressure of interest); but also mixing and melting enthalpies, entropies and volumes for solutions intermediate among the end-member components; thermochemical reaction constants $(\Delta\bar{U}^o, \Delta\bar{V}^o, \Delta\bar{S}^o)$ for the polymorphic inversions among phases at each composition; and measurements of the configurational properties--$\bar{H}_{config}$, $\bar{S}_{config}$, and $\bar{C}_p(config)$--for pyroxenes having varying degrees of crystallographic order. Such data could be correlated and compared with analogous values obtained from the analysis of phase equilibria, and would thereby serve as a basis for the correction and regeneration of mixing models. In fact, we are at a very early stage in the development of a self-consistent system of solution models with thermochemical constants. Despite important progress in the techniques of high-pressure and high-temperature calorimetry and an acceleration in the publication of thermochemical data for pyroxenes and related phases (see reviews by Kleppa and Newton, 1975; and Navrotsky, 1977), many of the calorimetric data needed are simply unmeasured. Table 3 presents published thermochemical data not otherwise listed by Robie *et al.* (1978) or Ko *et al.* (1978) for pyroxenes. Owing to their

probable internal consistency and the care with which the synthetic
starting materials were prepared and analyzed, the data published by
Newton and coworkers are likely to be most trustworthy and usable.

Robie *et al.* (1978) tabulate polythermal data for the pyroxenes
$CaAl_2SiO_6$, $CaMgSi_2O_6$, $LiAlSi_2O_6$ (α and β forms), $Mg_2Si_2O_6$ (clinoenstatite) and $NaAlSi_2O_6$ (jadeite). Ko *et al.* (1978) present thermochemical data for synthetic acmite, $NaFeSi_2O_6$, including low-temperature heat capacities, measured incrementally in the range 5 to 308
Kelvin, and both low- and high-temperature data for $\bar{S}^o$, $(\bar{H}^o - \bar{H}^o_{T_o})$ and
$-(\bar{G}^o - \bar{H}^o_{T_o})/T$ (the high-temperature enthalpy increments were determined
between 298 and 1201 Kelvins using copper-block drop calorimetry).

Some significant progress has been made in devising computer-based
banks of self-consistent thermochemical data (Helgeson *et al.*, 1978;
Haas and Fisher, 1976). The data used include both calorimetric measurements and values extracted from phase-equilibrium experiments. As
such, the data for pyroxenes are necessarily incomplete, and may be
subject to uncertainties in either the quality of the experimental
brackets or the effect on the calorimetric measurements of physico-
chemical irregularities (ordering, twinning or polymorphism) in the
material studied.

Many geochemists needing to use pyroxene solution models to obtain
answers to pressing and specific geologic questions about temperature
or pressure have chosen to ignore or, for lack of information, have been
forced to approximate or improvise values for the thermochemical properties of end-member phases. In some few cases, theoretical solution
models have included the parameters associated with reactions among
end-member phases and their polymorphs ($\mu_A^{o,\phi1}$, $\mu_A^{o,\phi2}$, $\bar{U}_A^o$, $\bar{V}_A^o$, $\bar{S}_A^o$, for
example). In these cases best values for the "constants" and the
models' adjustable parameters have been determined simultaneously by
using a least-squares approach to fit compositional data from phase-
equilibrium experiments. Such constants, obtained using a variety of
models to treat two-phase data for the system Di-En, are given in Table
4.* It is clear that the values extracted for $\Delta\bar{U}_A^o$, $\Delta\bar{U}_B^o$, and so on are
strongly model-dependent.
- - - - - - - - - - - -
* See pages 396-397.

Table 3. Comparative table of thermochemical measurements for pyroxenes

Compound	Standard entropy, $\bar{S}^\circ_T$ (J/mole·K)	Enthalpy of solution, $\langle\bar{H}_{soln}(T)\rangle$ (kJ/mole)	Enthalpy of formation, $\Delta\bar{H}^\circ_f(T)$ (kJ/mole)	Free energy of formation, $\Delta\bar{G}^\circ_f(T)$ (kJ/mole)	T (Kelvin)	Ref.	Comment*
$Mg_2Si_2O_6$							
	135.72		-71.12	-70.798	298	RHF	Clinoenstatite; $\Delta\bar{G}^\circ_f$ based on ref. 15?
	136.80				298	7	Converted from $\bar{S}^\circ_T$ using data from RHF
	133.64				298	11	Computed; empirical estimator
			-67.95		298	12	$\Delta\bar{H}_{soln}$ measured by oxide melt calorimetry
			-72.72		298	15	HF calorimetry; see also RW
				-66.15	298	3	Nesbitt, 1973 pers. comm.
				-0.62	298	3	Estimated from regression equation, $\Delta\bar{G}^\circ_f = aEXP(bx) + c$
				-73.31	298	13	Ref. [5], Table 1, converted to oxide basis
				-84.86	298	13	Ref. [7], Table 1, converted to oxide basis
				-63.572	298	16	Computed from phase equilibria, here converted to oxide basis; See also Chernosky, 1976
		69.96	-72.63		965	12	
		73.47	-73.72		970	1	Melt calorimetry using $2PbO \cdot B_2O_3$ flux
		71.96			970	9	Melt calorimetry on materials synthesized at high T and P
	385.86				1000	RHF	For comparison
	385.68				1000	5	Computed; see p. 14 of ref.
	383.88				1000	6	Converted from $\bar{S}^\circ_T$ (1000) using oxide data from ref. RHF
		93.64	-68.20		1173	12	
	469.54				1400	RHF	For comparison
	469.61				1400	5	Computed; see p. 14 of ref. 5

Aluminous Orthopyroxenes

Compound	Standard entropy, $\bar{S}^\circ_T$ (J/mole·K)	Enthalpy of solution, $\langle\bar{H}_{soln}(T)\rangle$ (kJ/mole)	Enthalpy of formation, $\Delta\bar{H}^\circ_f(T)$ (kJ/mole)	Free energy of formation, $\Delta\bar{G}^\circ_f(T)$ (kJ/mole)	T (Kelvin)	Ref.	Comment*
$(MgSiO_3)_{1.8}(Al_2O_3)_{.2}$							
		64.77	-58.66		970	1	Melt calorimetry using $2PbO \cdot B_2O_3$ flux
$(MgSiO_3)_{1.92}(Al_2O_3)_{.08}$							
	389.20				1000	5	Computed; see p. 14 of ref.
$(MgSiO_3)_{1.84}(Al_2O_3)_{.16}$							
	475.14				1400	5	Computed; see p. 14
$(MgSiO_3)_{1.76}(Al_2O_3)_{.24}$							
	475.97				1400	5	Computed; see p. 14

$CaMgSi_2O_6$ (Di) - $Mg_2Si_2O_6$ (En) Clinopyroxene Series

Di	En						
0	100	71.96					
22	78	64.89					
30	70	63.47					
40	60	64.60			970	9	Melt calorimetry, using $2PbO \cdot B_2O_3$ flux
50	50	65.94					with samples synthesized at 1600°-1700°C
60	40	67.91					and 30 kbar. (+) denotes sample synthe-
70	30 (+)	72.38					sized in air at 1358°C for 24 hr.
70	30	71.38					
80	20 (+)	77.78					
80	20	76.15					
90	10 (+)	81.67					
90	10	80.42					
100	0	85.27					

$CaMgSi_2O_6$**

Compound	Standard entropy, $\bar{S}^\circ_T$ (J/mole·K)	Enthalpy of solution, $\langle\bar{H}_{soln}(T)\rangle$ (kJ/mole)	Enthalpy of formation, $\Delta\bar{H}^\circ_f(T)$ (kJ/mole)	Free energy of formation, $\Delta\bar{G}^\circ_f(T)$ (kJ/mole)	T (Kelvin)	Ref.	Comment*
	143.09		-152.781	-151.295	298	RHF	
	141.79				298	7	Converted from $\bar{S}^\circ_T$ using oxide data from RHF
	140.83				298	11	Computed; empirical estimator
			-140.67		298	2	From Nesbitt and Helgeson, pers. comm.
			-140.58		298	8	Corrected from 986 K using heat contents given by RW
			-145.18		298	8	Helgeson, pers. comm.
			-143.05		298	2	
				-139.363	298	4	Nesbitt, 1973, pers. comm., converted to oxide basis
			-146.89		298	13	Ref. [5], Table 1, converted to oxide basis
			-146.89		298	13	Ref. [7], Table 1, converted to oxide basis
					713	8	Measured calorimetrically in $2PbO \cdot B_2O_3$ flux; natural material
		87.65					
		87.11			713	8	As above; synthetic material
		<87.45>			713	8	Average, <all>
		85.27			970	9	Average of 6 measurements using $2PbO \cdot B_2O_3$ flux, material synthesized at 1600-1700°C
		85.90			970	10	Average value for natural Di, both with and without heat-treatment
		(85.90)	-146.40		970	2	$\bar{H}_{soln}$ (970) from ref. 10
		(87.86)	-143.51		986	8	$\bar{H}_{soln}$ taken as $\Delta\bar{H}_{react}$ for $(CaMgSi_2O_6)_{xl} \pm (CaMgSi_2O_6)_{glass}$

Aluminous Clinopyroxene**

CaAl$_2$SiO$_6$ (CaTs)

	$S°$	$\Delta H°_f$	$\Delta G°_f$	T	Ref	Notes
	156.0	-54.191	-61.767	298	RHF	
	134.89			298	11	Computed; empirical estimator
	140.54			298	14	
		-73.85		298	2	Material synthesized from glass at 1350°C, 20 kbar; lead borate melt used as flux
		-71.73	-75.52	298	RW	Converted to oxide basis using oxide data in ref. RW
			-94.89	298	3(a)	Estimated by regression technique,
			-92.06	298	3(b)	$\Delta G°_f$ = aEXP(bx) + c; (a) data from ref. RW; (b) data from other sources, converted to oxide basis using $\Delta G°_f$ for oxides from ref. RHF
	48.33		-76.27	970	2	Material synthesized from glass at 1250°C, 18 kbar; otherwise, as above

CaTs	Di		T	Ref	Notes
100	0	(48.33)	970	10	From ref. 2, above
70	30	56.48			Material in this series synthesized from
50	50	63.09			glass at 1250°C and 18 kbar. Melt
35	65	66.82	970	10	calorimetry using 2PbO·B$_2$O$_3$ flux
20	80	72.51			
10	90	78.07			
0	100	85.90	970	10	See ref. 11, p. 374

CaFeSi$_2$O$_6$

$S°$	$\Delta H°_f$	$\Delta G°_f$	T	Ref	Notes
172.56		-112.97	298	7	Estimated from thermochemical systematics; $S°_T$ here recalculated from formation entropy using $S°_{298}$ for constituent oxides from ref. RHF
		-107.95	1350	8	Estimated from the reaction clinopyroxene ⇌ pyroxenoid (pxd) and the approximations $\Delta G°_{f,pxd}$ = $\Delta G°_{f,CaSiO_3}$ + $\Delta G°_{f,FeSiO_3}$ - 2RTln2 and $\Delta G°_{f,cpx}$ = $\Delta G°_{f,pxd}$ + $P\Delta V°$

Fe$_2$Si$_2$O$_6$

$S°$	$\Delta H°_f$	$\Delta G°_f$	T	Ref	Notes
189.97		-29.62	298	7	Computed from high-P equilibria of Akimoto et al., 1963, J. Geophys. Res. 70, 5269-5278. $S°_T$ recalculated as above from oxide data in ref. RHF
		-22.59	298	6	
		-0.48	298	3	Estimated by regression technique from diverse thermochemical data; recalculated to oxide basis
		-31.50	298	3	From data of Karpov and Kashik, 1968, Geochem. Internat. 5, 706-713.
		-37.78	298	13	Table 1, ref. [6], converted to oxide basis
		-177.44	298	13	Table 1, ref. [8], converted to oxide basis Value is certainly too negative (cf. $\Delta G°_f$(298) for Mg$_2$Si$_2$O$_6$)
472.90			1000	6	Estimated from thermochemical systematics; recalculated from entropy of formation using $S°_{1000}$ data for oxides in ref. RHF
		-9.20	1353	4	Data from ref. [12] of 4
		-10.88	1423	6,4	Computed from phase-equilibrium data of Schwerdtfeger and Muan, 1966, Trans. A.I.M.E. 236, 201-211.
		-15.06	1427		Using data from Darken and Gurry, 1945,
		-12.55	1477	4	J. Am. Chem. Soc. 67, 1398-1412, for
		-15.06	1523		reactants wüstite + SiO$_2$
		-12.6	1427		Using data from ref. [9]
		-11.7	1477	4	"
		-13.4	1523		"
		-7.53	1523	6,4	Computed from phase-equilibrium data of Nafziger and Muan, 1967, Am. Mineral. 52, 1364-1385
		-8.00	1573	4	Using data from ref. [a]

Notes:
- Thermochemical properties of formation ($\Delta H°_f$, $\Delta G°_f$) are for formation from the *oxides* (not elements) at the temperature indicated.
- All molar properties are computed on the basis of pyroxene molecules having 6 oxygens.
- For the conversions necessary in casting all measurements in common units (J/mole·K or kJ/mole), the factor: 1 thermochemical calorie = 4.1840 Joule, was used. Decimal fractions given may be artifacts of multiplication, and are not necessarily significant.

References:
RW Robie & Waldbaum, 1968
RHF Robie et al., 1978
1 Charlu et al., 1975
2 Charlu et al., 1978
3 Chen, 1975
4 Kitayama, 1970
5 Kleppa & Newton, 1975
6 Navrotsky, 1971b
7 Navrotsky, 1978
8 Navrotsky & Coons, 1976
9 Newton et al., 1979
10 Newton et al., 1977
11 Saxena, 1976a
12 Shearer & Kleppa, 1973
13 Tardy & Garrels, 1977
14 Thompson et al., 1978
15 Torgeson and Sahama, 1948
16 Zen and Chernosky, 1976

* [ ] indicates citation number in the original reference.
** for heat capacity estimator, see ref. 14.

The Macroscopic Approach

A general review. The thermodynamic equation of state for any single solution of c components has the general form given by equation (A-7) of Appendix A:

$$\bar{G}_T = \bar{G}_{ID} + \bar{G}_{EX}. \tag{A-7}$$

Here

$$\bar{G}_{ID} = \sum_{i=1}^{c} \mu_i^o X_i + RT \sum_{i=1}^{c} X_i \ln X_i$$

and $\bar{G}_{EX}$ may be expressed in terms of the rational activity coefficients (functions T, P, and composition) as

$$\bar{G}_{EX} = RT \sum_{i=1}^{c} X_i \ln \gamma_i (T,P,X). \tag{A-7c}$$

Although the activity coefficients in a c-component system are interrelated by means of the Gibbs-Duhem equation,

$$X_k \frac{\partial \ln \gamma_k}{\partial X_k} + \sum_{i, i \neq k}^{c} X_i \frac{\partial \ln \gamma_i}{\partial X_k} = 0, \tag{11}$$

it is generally inconvenient to deal with c-1 independent terms in $\ln \gamma_i$ because the values for γ_i vary in complex ways with composition. It is easier to express the activity coefficients, terms in $\ln \gamma_i$, or the excess free energy itself as an expansion in powers of composition. In a binary system 1-2, for example, one might take

$$\ln \gamma_1 = \frac{1}{RT}(AX_2^2 + BX_2^3); \tag{12a}$$

$$\bar{G}_{EX} = X_1 X_2 (W_{G1} X_2 + W_{G2} X_1); \tag{12b}$$

or

$$\bar{G}_{EX} = aX_1 + (b-2a)X_1^2 + (a-b)X_1^3. \tag{12c}$$

The various coefficients, A, B, W_{G1}, W_{G2}, a and b are binary fitting constants having the dimensions of molar energy.

The principal constraints governing the form of these expansion equations are (1) that the behavior of the $\ln \gamma_i$ terms be consistent with the Gibbs-Duhem equation; (2) that $\bar{G}_{EX}$ be related by equations (A-7c) or (A-32) to $\ln \gamma_i$ terms that are, likewise, consistent with (11); and (3) that the expressions for $\bar{G}_{EX}$ or $\ln \gamma_i$ approach zero as the compositions in

364

the system approach the ideal limits (compositional end-members). Clearly, in equations (12), as $X_2 \to 0$ and $X_1 \to 1$, both $\ln\gamma_1$ and $\bar{G}_{EX}$ approach 0.

King (1969) and Wisniak and Tamir (1978) characterize and discuss many of the numerous expressions for $\ln\gamma_i$ and $\bar{G}_{EX}$ that have been used successfully to treat multicomponent solutions. Examples of formulations used for ternary systems $(X_1-X_2-X_3)$ include:

(a) the "strictly regular" model of Prigogine and Defay (1954, p. 257 ff.),

$$\bar{G}_{EX} = X_1 X_2 W_{G12} + X_2 X_3 W_{G23} + X_1 X_3 W_{G13};$$

(b) a compositionally weighted "subregular" model,

$$\bar{G}_{EX} = X_1 X_2 X_3 [(W_{G12} + W_{G13})X_1 + (W_{G21} + W_{G23})X_2$$
$$+ (W_{G31} + W_{G32})X_3 + C];$$

(c) Kohler's (1960) empirical formulation,

$$\bar{G}_{EX} = \sum_{i=1} \sum_{j>1} (X_i + X_j)^2 \bar{G}_{EX(i,j)},$$

where $G_{EX(i,j)}$ is the value for the excess energy in the corresponding binary system i-j, computed at $X_1/(X_1 + X_2)$ (see Barron, 1976); and

(d) the Redlich-Kister (1948a,b) equation,

$$\bar{G}_{EX} = X_1 X_2 [A_{12} + B_{12}(X_1 - X_2) + C_{12}(X_1 - X_2)^2 + D_{12}(X_1 - X_2)^3 + \ldots]$$
$$+ X_1 X_3 [A_{13} + B_{13}(X_1 - X_3) + C_{13}(X_1 - X_3)^2 + D_{13}(X_1 - X_3)^3 + \ldots]$$
$$+ X_2 X_3 [A_{23} + B_{23}(X_2 - X_3) + C_{23}(X_2 - X_3)^2 + D_{23}(X_2 - X_3)^3 + \ldots]$$
$$+ X_1 X_2 X_3 [A + B_{2,1}(X_1 - X_2) + B_{3,1}(X_1 - X_3) + Z_1(X_2 - X_3)] .$$

Again, the coefficients W_{Gij}, A_{ij}, and so on are fitting parameters with the dimensions of energy.

As the number of components considered is increased, the expressions used to represent $\bar{G}_{EX}$ become more complex (see, for example, the general form for the Redlich-Kister equation, given by Wisniak and Tamir, 1978, p. XXIV). At the same time, the extraction of fitting constants becomes mathematically more difficult as the number of experimental data appropriate for modelling the multicomponent system of interest dwindles.

Each of the numerous models available to represent $\bar{G}_{EX}$ or $\ln\gamma_i$ finds appropriate (or even powerful) application in some system or other. For certain systems, however, the mathematical behavior of particular mixing models makes them totally ineffective in representing the observed phase relations. In choosing a suitable model for a given application, it is useful to consider:

(1) the need for simple, tractable equations;

(2) the number of data available and appropriate for thermodynamic evaluation and massaging;

(3) the need to extrapolate the results obtained for a simple system to a more complex one, or to predict phase relations on the basis of scanty data;

(4) the experience others have had in evaluating the mixing properties of similar systems, and the need for accurate correlations among several systems;

(5) the success that a given model has in corroborating measured thermochemical constants;

(6) the need (or desire) to ascribe physical or crystal-chemical meaning to the fitting parameters;

(7) the need to account for unusual nonideal behavior (ordering, partial miscibility, second-order structural inversions, for example); and

(8) the nature of the phases present in the system (gaseous, aqueous, metallic, mixed) (Wisniak and Tamir, 1978, p. XVI).

Several authors, notably Hardy (1953), Sundquist (1966), Green (1970), Powell (1974), Barron (1976) and Blencoe (1977), have compared the functional form, ease in use, and predictive capacities of the models used most frequently in treating mineral solutions and metallic alloys.

Properties of the Margules equations. Of the many formulations for excess energy, the Margules equations (Thompson, 1967; Grover, 1977) or closely related expansions (Wohl, 1946, 1953; Currie and Curtis, 1976; Brown and Skinner, 1974) have been used most often in modelling binary or multicomponent pyroxene solutions. For a binary system, the Margules expression for $\bar{G}_{EX}$ with two adjustable coefficients (the "asymmetric" case) is given by equation (12b). The fitting parameters W_{G1} and W_{G2} are functions only of temperature and pressure, not composition, and have the dimensions of energy. Like other thermodynamic measures of energy, W_{G1} and W_{G2} may be decomposed into terms representing internal energy, enthalpy and entropy:

$$W_{Gi} = W_{Ui} + PW_{Vi} - TW_{Si}; \quad i = 1,2. \tag{13}$$

When equation (12b) is combined with (A-32),

$$\ln\gamma_1 = \frac{1}{RT} [\bar{G}_{EX} - X_2(\frac{\partial \bar{G}_{EX}}{\partial X_2})] \ ,$$

it is found that

$$\ln\gamma_1 = \frac{1}{RT} [(2W_{G2} - W_{G1})X_2^2 + 2(W_{G1} - W_{G2})X_2^3] \ . \tag{14a}$$

The comparable expression for $\ln\gamma_2$ is

$$\ln\gamma_2 = \frac{1}{RT} [(2W_{G1} - W_{G2})X_1^2 + 2(W_{G2} - W_{G1})X_1^3] \ . \tag{14b}$$

Thus, it follows that equations (12a) and (12b) are identical, provided

$$A = 2W_{G2} - W_{G1} \quad \text{and} \quad B = 2(W_{G1} - W_{G2}) \ .$$

The significance of the W_G terms is made clear when we set $X_2 = 1$ in equation (14a), and $X_1 = 1$ in (14b). For these limiting compositions,

$$W_{G1} = (RT\ln\gamma_1)_{X_2=1} = -(d\bar{G}_{EX}/dX_2)_{X_2=1}$$

$$\text{and } W_{G2} = (RT\ln\gamma_2)_{X_1=1} = (d\bar{G}_{EX}/dX_2)_{X_2=0}$$

(see Grover, 1977, pp. 74-79; Hardy, 1953, p. 209).

If $(1-X_1)$ is substituted for X_2 in equation (12b), we obtain

$$\bar{G}_{EX} = W_{G1}X_1 + (W_{G2} - 2W_{G1})X_1^2 + (W_{G1} - W_{G2})X_1^3 \ , \tag{15}$$

which is identical to (12c) when $a = W_{G1}$ and $b = W_{G2}$. In short, equations (12a), (12b) and (12c) are equivalent expressions for the same (asymmetric Margules) model, cast in different forms!

The analogous expressions for the "symmetric" or regular case can be obtained from (12) by setting $W_{G1} = W_{G2} \equiv W_G$:

$$\ln\gamma_1 = \frac{W_G}{RT} X_2^2 \quad \text{and} \quad \bar{G}_{EX} = X_1 X_2 W_G.$$

Figure 4 shows how the total Gibbs free energy, $\bar{G}_T$, for a binary system 1-2 varies with different values of the asymmetric Margules constants, W_{G1} and W_{G2}. These curves were generated using equations (A-7) and (12b), combined to give:

$$G_T = \mu_1^o + (\mu_2^o - \mu_1^o)X_2 + RT(X_1 \ln X_1 + X_2 \ln X_2)$$
$$+ X_1 X_2 (W_{G1}X_2 + W_{G2}X_1) \ . \tag{16}$$

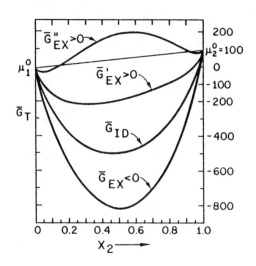

Figure 4. Variation of the total molar Gibbs free energy with composition, X_2, for a binary solution 1-2. The free energy of mechanical mixing is

$$\bar{G}_{ID}^M = X_1\mu_1^o + X_2\mu_2^o = 100X_2,$$

and the molar free energy of ideal mixing at this T and P is represented by the curve $\bar{G}_{ID}$. The other curves show the behavior of the free energy of mixing with either positive ($\bar{G}_{EX} > 0$) or negative ($\bar{G}_{EX} < 0$) departures from ideality. These curves were generated using the asymmetric Margules formulation for $\bar{G}_T$, equation 16, with values for W_{G1} and W_{G2} as given in the text. (From Grover, 1977).

The data used in the calculations are summarized here:

$\mu_1^o = 0$ cal. mol^{-1}; $\mu_2^o = 100$ cal. mol^{-1}

$\bar{G}_{EX} < 0$: $W_{G1} = -1500$ cal. mol^{-1}; $W_{G2} = -1000$ cal. mol^{-1}

$\bar{G}_{EX}' > 0$: $W_{G1} = 1500$ cal. mol^{-1}; $W_{G2} = 1000$ cal. mol^{-1}

$\bar{G}_{EX}'' > 0$: $W_{G1} = 3000$ cal. mol^{-1}; $W_{G2} = 2500$ cal. mol^{-1}

$\bar{G}_{EX} = 0$ ($\bar{G}_{ID}$): $W_{G1} = W_{G2} = 0$

$R = 1.98717$ cal. K^{-1} mol^{-1}; $T = 400°K$

Note that when W_{G1} and W_{G2} are both positive, the solution displays positive deviation from ideality ($\bar{G}_{ID}$); when both Margules constants are negative, the solution shows negative deviation from ideality ($\bar{G}_{EX} < 0$). In the case of extreme positive deviation from ideality ($\bar{G}_{EX}'' > 0$), the free-energy curve shows a pronounced downward concavity in its central part. This represents the condition for phase separation: Any homogeneous (metastable) phase having a composition between approximately $X_2 = 0.06$ and $X_2 = 0.95$ (those points on the curve sharing a common tangent line) should unmix to form two phases.

An expression for ternary mixing, analogous to (12b) is

$$\bar{G}_{EX} = X_1X_2(A_{21}X_1 + A_{12}X_2) + X_1X_3(A_{31}X_1 + A_{13}X_3)$$
$$+ X_2X_3(A_{32}X_2 + A_{23}X_3)$$
$$+ X_1X_2X_3(A_{21} + A_{13} + A_{32} - C) ,$$

where the constants A_{ij} and A_{ji} are obtained from data for the respective binary systems, and the ternary constant C is obtained by fitting composition data in the ternary system 1-2-3.

A complex model presented by Wohl (1946, 1953) has been used increasingly to represent the mixing properties of multicomponent solutions among pyroxene and other mineral components. Several alternative expressions for the excess quantities $\bar{G}_{EX}$ and $ln\gamma_i$, consistent with this model, have been developed and discussed by Currie and Curtis (1976) and King (1969, pp. 327-336, 555-556). The expression for the activity coefficient of component k in a c-component system is:

$$RT ln\gamma_k = \sum_{\substack{i=1 \\ (i\neq k)}}^{c} X_i^2 W_{Gi,k} + \sum_{\substack{i=1 \\ (i,j\neq k; \, i<j)}}^{c-1} \sum_{j=2}^{c} X_iX_j (W_{Gi,k} + W_{Gj,k} - W_{Gi,j})$$

(see Powell and Powell, 1974, who show how this expression can be applied to the case of coexisting olivine and orthopyroxene). For the case of a ternary system, an alternative form generated by T. Gasparik (personal communication) makes the extraction of ternary coefficients $W_{G,ij}$, $W_{G,ji}$, $W_{G,j}$ and $W_{G,ijk}$ especially simple and convenient. The general expression for μ_i^{EX} (= $RT ln\gamma_i$; see equations A-28) is:

$$\mu_i^{EX} = \sum_{\substack{j \ k \ m \\ (i\neq j\neq k\neq m)}} [W_{G,ij}(2X_j^3 - X_j^2 - 0.5X_jX_k + 3X_j^2X_k + X_jX_k^2 - X_jX_kX_m) +$$
$$+ W_{G,ji}(2X_j^2 - 2X_j^3 + 1.5X_jX_k - 3X_j^2X_k - X_jX_k^2 + X_jX_kX_m) +$$
$$+ W_{G,jk}(X_j^2X_k - X_jX_k^2 - 0.5X_jX_k)] + W_{G,ijk}(2X_j^2X_k + 2X_jX_k^2 - y$$

For component 1, for example, this becomes

$$\mu_1^{EX} = W_{G,12}(2X_2^3 - X_2^2 - 0.5X_2X_3 + 3X_2^2X_3 + X_2X_3^2) +$$
$$+ W_{G,21}(2X_2^2 - 2X_2^3 + 1.5X_2X_3 - 3X_2^2X_3 - X_2X_3^2) +$$
$$+ W_{G,13}(2X_3^3 - X_3^2 - 0.5X_2X_3 + 3X_2X_3^2 + X_2^2X_3) +$$
$$+ W_{G,31}(2X_3^2 - 2X_3^3 + 1.5X_2X_3 - 3X_2X_3^2 - X_2^2X_3) +$$

$$+ W_{G,23}(X_2^2X_3 - X_2X_3^2 - 0.5X_2X_3) + W_{G,32}(X_2X_3^2 - X_2^2X_3 - 0.5X_2X_3)$$
$$+ W_{G,123}(2X_2^2X_3 + 2X_3^2X_2 - X_2X_3) \quad .$$

The corresponding expression for $\bar{G}_{EX}$ is

$$\bar{G}_{EX} = \sum_i \sum_j \sum_k [W_{G,ij}X_iX_j(X_j + \frac{1}{2}X_k) + W_{G,ijk}X_iX_jX_k].$$
$$(i{\neq}j{\neq}k)$$

Gasparik and Lindsley (Chapter 7, this volume) show how these equations can be applied in the thermodynamic treatment of systems involving the components $CaMgSi_2O_6$, $CaAl_2SiO_6$ and $Ca_{1/2}AlSi_2O_6$ (or, alternatively, Di, CaTs plus SiO_2 or $CaAl_2Si_2O_8$). For further illustration of the use of expressions of this kind in modelling pyroxene solutions, the reader should consult Currie and Curtis (1976). In this important but difficult paper the authors study reactions involving jadeitic pyroxene, nepheline (both taken as "subregular" multicomponent solutions) and albite (treated as an ideal solution), deriving values for the several adjustable parameters from published experimental data. They find that the limits on the composition of omphacitic pyroxene predicted by the model agree with observations of natural assemblages.

Methods for extracting fitting parameters from phase-equilibrium data. In polynomial or other series-form expansions in *composition* for $\bar{G}_{EX}$ or $ln\gamma_i$, the coefficients of the composition terms are taken to be functions only of temperature and pressure. They are usually evaluated empirically from data on the compositions of coexisting phases at fixed temperature and pressure. To do this, it is necessary to express the condition for chemical equilibrium in a system containing several phases in terms of the series coefficients. This presents no difficulty in principle, because the series coefficients may always be related to the chemical potentials and activity coefficients of individual components by means of equations such as A-18 and A-32.

The several methods for extracting model fitting parameters, discussed in this section, are similar in that each involves equating chemical potentials for specific components in phases that coexist in equilibrium. This approach is general, and does not depend on the mathematical form of the excess free energy, $\bar{G}_{EX}$, or the excess chemical potentials, equation (A-28). I have framed the discussion in terms of the Margules fitting parameters, W_{G1} and W_{G2}, for an asymmetric binary system because (1) the Margules model has become popular in treating geologic systems, and is already well represented in the literature (see Grover, 1977, for an annotated bibliography); (2) much of the discussion given above is based on this model; and (3) the examples of thermodynamics applied to pyroxene solutions, presented below, also favor this approach.

(A) Models for isostructural systems

(i) Thompson (1967, pp. 353-356, especially equation (87)), Luth and Fenn (1973), Saxena (1973) and Blencoe (1977) present algorithms for calculating the Margules coefficients W_{G1} and W_{G2} for a binary system from the compositions of two coexisting phases at fixed temperature and pressure. These formulations assume, necessarily, that the phases share a common equation of state; with this assumption, the standard-state chemical potentials for the end-members can be cancelled when the chemical potentials, or activities, of the coexisting phases are equated. This simplification is by no means appropriate for many pyroxene systems in which chemical unmixing occurs, however.

(ii) Another technique for determining Margules coefficients involves the equation of exchange for two species common to two co-existing phases. When two or more phases coexist in equilibrium at fixed temperature and pressure, the chemical potentials for each of the several components are equal in all the phases. If a and b denote two phases co-existing in equilibrium in the binary system 1-2, the equivalence of chemical potentials, $\mu_1^a = \mu_i^b$, leads to equilibrium expressions having a form

$$\mu_1^a + \mu_2^b = \mu_2^a + \mu_1^b \ . \tag{17}$$

Chemical potentials for individual components can be calculated for our two-coefficient model by combining equation (16) (the expression for total free energy) with (A-18). This gives

$$\mu_1 = \mu_1^o + RT\ell nX_1 + X_2^2[W_{G1} + 2(W_{G2}-W_{G1})X_1] \tag{18a}$$

and

$$\mu_2 = \mu_2^o + RT\ell nX_2 + X_1^2[W_{G2} + 2(W_{G1}-W_{G2})X_2] \ . \tag{18b}$$

Inserting appropriate chemical potentials, (18), for the phases a and b in equation (17), we can derive a general expression for the distribution coefficient K_D:

$$-RT\ln K_D = (\mu_2^{ao} - \mu_1^{ao}) - (\mu_2^{bo} - \mu_1^{bo}) +$$
$$[W_{G2}^a(x_1^a)^2 - W_{G1}^a(x_2^a)^2 + 2(W_{G1}^a - W_{G2}^a)x_1^a x_2^a] -$$
$$[W_{G2}^b(x_1^b)^2 - W_{G1}^b(x_2^b)^2 + 2(W_{G1}^b - W_{G2}^b)x_1^b x_2^b] , \qquad (19)$$

where $K_D = (x_2^a x_1^b)/(x_1^a x_2^b)$ and the two-coefficient Margules model is assumed to apply for both phases, a and b. This equation is valid when a and b do not obey the same equation of state, but it cannot in general be used to extract independent values for the standard-state chemical potentials from two-phase data. Neglecting that problem for a moment, we will look briefly at how equation (19) might be solved for the values of the W_G terms. Equation (19) can be written in an alternate, *linear* form such that each of the four independent Margules parameters appears as the single coefficient of a term in composition. We can take the terms $(\mu_2^{ao} - \mu_1^{ao}) - (\mu_2^{bo} - \mu_1^{bo}) = C(T,P)$, as constant at fixed temperature and pressure. If "phases" a and b were, in fact, two coexisting binary solutions, a(1-2) and b(1-2) in a reciprocal system, it would be possible to carry out a series of isothermal equilibration experiments using starting materials mixed to give different bulk compositions. These experiments would establish sets of tie-lines among coexisting phases of different composition in the two solutions. With a sufficient number of compositional analysis of coexisting phases in solutions a and b, it would be possible to solve equation (19) (in the form $-RT\ln K_D = C + W_{G1}^a f_1(X) + W_{G2}^a f_2(X) + W_{G1}^b f_3(X) = W_{G2}^b f_4(X))$ for the W_G terms, either approximately by the method of linear least squares, or explicitly using determinants (Cramer's Rule).

(iii) If the mixing behavior of one of the constituent mixtures, a or b, is well-known, equation (19) can be used to determine the solution properties of the other, coexisting phase. Given W_{G1}^a and W_{G2}^a, or otherwise presuming that solution *a* is well known, it is convenient to write equation (19) in the form

$$[-RT\ln K_D - f(x_1^a, x_2^a)] = A_o + A_1 x_2^b + A_2 \overline{x}_2^{b2} , \qquad (20a)$$

where

$$A_0 = [(\mu_2^{ao} - \mu_1^{ao}) - (\mu_2^{bo} - \mu_1^{bo}) + W_{G2}^b], \text{ a constant for T,P; (20b)}$$

$$A_1 = 2(W_{G1}^b - 2W_{G2}^b); \text{ and} \tag{20c}$$

$$A_2 = 3(W_{G2}^b - W_{G1}^b). \tag{20d}$$

The expressions for A_1 and A_2 may be solved simultaneously to give:

$$W_{G1}^b = -\frac{1}{2}A_1 - \frac{2}{3}A_2 ; \tag{21a}$$

$$W_{G2}^b = -\frac{1}{2}A_1 - \frac{1}{3}A_2 \tag{21b}$$

Like W_{G1}^b and W_{G2}^b, A_0, A_1 and A_2 are independent of composition. The experimentally-determined compositions for coexisting phases (tie lines) x_1^a, x_2^a, x_1^b and x_2^b, can be used to plot $[-RT \ln K_D - f(x_1^a, x_2^a)]$ *versus* X_2 (*cf.* equation 20). If the resulting curve is linear or parabolic, the coefficients A_0, A_1 and A_2 can be determined from equation (20) by means of a least-squares polynomial fitting technique. W_{G1}^b and W_{G2}^b can then be calculated using equations (21). If the function is linear in X_2, $A_2 = 0$ and $W_{G1}^b = W_{G2}^b$. Again, the standard state chemical potentials μ_2^{ao}, μ_1^{ao}, μ_2^{bo} and μ_1^{bo} cannot be determined individually. If the function is neither parabolic nor linear, a two- (or one-) coefficient Margules expression for the excess free energy does not apply and another solution model should be sought.

(B) Model for non-isostructural systems

For the models discussed previously in this section it was necessary to assume that two phases *coexisting in a binary system* had a common equation of state. Only then could *standard-state* chemical potentials be cancelled when the chemical potentials of the two components in those phases were equated. If, however, one or both of the phases undergoes a polymorphic transformation such that the forms coexisting stably at a particular temperature and pressure have different crystal structures, the free-energy composition relations are more complicated (Fig. 3; see also the discussion of polymorphism). In this case the standard-state chemical potentials for a component in the coexisting phases cannot be cancelled. By treating the polymorphic inversions (reactions A and B, and equation 7) independently, and with a sufficient number of polythermal

373

and polybaric phase-composition data, it is possible to solve for the several Margules fitting parameters for *each* solution, as well as to determine values for the reaction constants, $\Delta \bar{U}^o_\phi$, $\Delta \bar{V}^o_\phi$ and $\Delta \bar{S}^o_\phi$ for both polymorphic inversions. To illustrate this approach, we will return to a discussion of the system Di-En.

For coexisting Opx and Cpx in the system Di-En the reactions relating the chemical potentials of the components in each of the phases are

$$(Mg_2Si_2O_6)^{Opx} \rightleftarrows (Mg_2Si_2O_6)^{Cpx} \qquad (A)$$

and

$$(CaMgSi_2O_6)^{Opx} \rightleftarrows (CaMgSi_2O_6)^{Cpx}. \qquad (B)$$

The corresponding free energies of reaction pertaining to phases of any composition (*cf.* equation 7a) are

$$\Delta \bar{G}_A = \mu_{En}^{Cpx} - \mu_{En}^{Opx} \quad \text{and} \quad \Delta \bar{G}_B = \mu_{Di}^{Cpx} - \mu_{Di}^{Opx}.$$

At equilibrium, $\Delta \bar{G}_A = 0$ and $\Delta \bar{G}_B = 0$. Although the chemical potentials for the Di and En components in either phase are related at equilibrium by means of the Gibbs-Duhem equation,

$$X_{Di}^{Opx} d\mu_{Di}^{Opx} + X_{En}^{Opx} d\mu_{En}^{Opx} = 0$$

(and similarly for Cpx), it is correct explicitly to write:

$$(\mu_{En}^{Cpx} - \mu_{En}^{Opx}) \equiv (\mu_{En}^{o,Cpx} - \mu_{En}^{o,Opx}) + RTln(\frac{X_{En}^{Cpx}}{X_{En}^{Opx}}) + RTln(\frac{\gamma_{En}^{Cpx}}{\gamma_{En}^{Opx}}) = 0 \quad (22a)$$

and

$$(\mu_{Di}^{Cpx} - \mu_{Di}^{Opx}) \equiv (\mu_{Di}^{o,Cpx} - \mu_{Di}^{o,Opx}) + RTln(\frac{X_{Di}^{Cpx}}{X_{Di}^{Opx}}) + RTln(\frac{\gamma_{Di}^{Cpx}}{\gamma_{Di}^{Opx}}) = 0. \quad (22b)$$

The standard-state terms can be expressed in several ways. For reaction A:

$$\Delta \bar{G}^o_A \equiv \Delta \bar{U}^o_A + P\Delta \bar{V}^o_A - T\Delta \bar{S}^o_A = (\mu_{En}^{o,Cpx} - \mu_{En}^{o,Opx}) ; \qquad (23A)$$

and for B,

$$\Delta \bar{G}^o_B \equiv \Delta \bar{U}^o_B + P\Delta \bar{V}^o_B - T\Delta \bar{S}^o_B = (\mu_{Di}^{o,Cpx} - \mu_{Di}^{o,Opx}) . \qquad (23B)$$

The activity coefficient (or "excess") terms are

$$\Delta\mu_{En}^{Ex} \equiv RT\ln(\gamma_{En}^{Cpx}/\gamma_{En}^{Opx}) \tag{24A}$$

and

$$\Delta\mu_{Di}^{Ex} \equiv RT\ln(\gamma_{Di}^{Cpx}/\gamma_{Di}^{Opx}) \ . \tag{24B}$$

For the asymmetric Margules model, the activity coefficients are given by equations (14a) and (14b).

The compositions of natural, calcic orthopyroxenes are restricted to low values of X_{Di}. Consequently, the form of the free-energy curve for Opx is critical in this analysis only in the region where X_{Di} is less than about 0.3. (See the several T-X sections for the Di-En system figured in Lindsley's chapter. For a very wide range of temperatures and pressures, 900° to 1500°C and 0 to 40 kbar, the experimentally determined limits on the Opx solvus are well below $X_{Di} = 0.3$.) Moreover, the absolute value for $\mu_{Di}^{o,Opx}$, the fictive standard-state potential for an Opx with diopside composition, is unknown and, in fact, unimportant. We could take the Opx solution to be ideal, setting $\Delta\mu_{En}^{Ex} = RT\ln\gamma_{En}^{Cpx}$ in equation (24A). The form of the resulting free-energy curve for Opx would surely be geometrically correct (concave upward) in the region $0 < X_{Di} \lesssim 0.3$, and is inconsequential for higher values of X_{Di}. For crystal-chemical reasons, however, it is quite unlikely that the Opx solution is ideal thermodynamically. We will assume, instead, that it is symmetric ($W_{G1}^{Opx} = W_{G2}^{Opx}$), so that equations (14a) and (14b) take the form

$$\ln\gamma_{En}^{Opx} = \frac{W_G^{Opx}}{RT} (X_{Di}^{Opx})^2 \tag{25a}$$

and

$$\ln\gamma_{Di}^{Opx} = \frac{W_G^{Opx}}{RT} (X_{En}^{Opx})^2 \tag{25b}$$

Equation (24a) becomes

$$\Delta\mu_{En}^{EX} = (2W_{G2}^{Cpx} - W_{G1}^{Cpx})(X_{Di}^{Cpx})^2 + 2(W_{G1}^{Cpx} - W_{G2}^{Cpx})(X_{Di}^{Cpx})^3$$
$$- W_G^{Opx} (X_{Di}^{Opx})^2 \tag{26}$$

(and similarly for $\Delta\mu_{Di}^{EX}$, equation 24b). There is no information to support a two-coefficient (asymmetric) or higher-order model for the Opx solution.

The coexistence of $(C2/c)$ iron-free pigeonite (Pig_{ss}) and Di_{ss} (Schweitzer, 1977; Longhi and Boudreau, 1980), indicates that the Cpx solution can undergo phase separation; the asymmetry of the two-phase region Di_{ss} + Pig_{ss} (see Lindsley, this volume, Fig. 8b) requires that a two-coefficient mixing model be used to represent the free energy-composition relations of Cpx solutions. Combining equations (22), (23), (24), (25) and (26) [(14)], we obtain the relations used to express En_{ss}-Di_{ss} equilibria (Lindsley et $al.$, 1981).

$$\mu_{En}^{Cpx} = \mu_{En}^{Opx}: \quad \mu_{En}^{o,Cpx} - \mu_{En}^{o,Opx} = -\Delta\bar{U}_A^o + T\Delta\bar{S}_A^o - P\Delta\bar{V}_A^o$$

$$= RT\ln(X_{En}^{Cpx}/X_{En}^{Opx}) - W_G^{Opx}(X_{Di}^{Opx})^2 + 2W_{G2}X_{En}^{Cpx}(X_{Di}^{Cpx})^2 + W_{G1}(1 - 2X_{En}^{Cpx})(X_{Di}^{Cpx})^2;$$

$$(27A)$$

$$\mu_{Di}^{Cpx} = \mu_{Di}^{Opx}: \quad \mu_{Di}^{o,Opx} - \mu_{Di}^{o,Cpx} = -\Delta\bar{U}_B^o + T\Delta\bar{S}_B^o - P\Delta\bar{V}_B^o$$

$$= RT\ln(X_{Di}^{Cpx}/X_{Di}^{Opx}) - W_G^{Opx}(X_{En}^{Opx})^2 + W_{G2}(1 - 2X_{Di}^{Cpx})(X_{En}^{Cpx})^2 + 2W_{G1}X_{Di}^{Cpx}(X_{En}^{Cpx})^2$$

$$(27B)$$

(subscripts A and B refer to reactions (A) and (B); subscripts En and Di refer to components $Mg_2Si_2O_6$ and $CaMgSi_2O_6$; and W_{G1} and W_{G2} are asymmetric Margules parameters for Cpx).

There are twelve adjustable parameters in each of these equations: $\Delta\bar{U}^o$, $\Delta\bar{V}^o$, $\Delta\bar{S}^o$, W_G^{Opx}, W_{G1} and W_{G2}, where each of the Margules parameters may be presumed to vary with temperature and pressure (see equation (13)). The W_G coefficients are shared between equations, however, and the number of unknowns is reduced from 24 to 15. Equations (27a) and (27b) can be solved simultaneously by a linear least-squares using the compositions of coexisting Opx_{ss} and Cpx_{ss} phases obtained from phase equilibrium experiments at a variety of temperatures and pressures. With some variation in the assumptions made, and using the additional, independent equilibria:

$$\mu_{En}^{Pig} = \mu_{En}^{Di} \quad \text{and} \quad \mu_{Di}^{Pig} = \mu_{Di}^{Di}$$

to constrain the system further at temperatures and pressures where Di_{ss} and Pig_{ss} coexist, Lindsley et $al.$ (1981) obtained a good and useful model by this approach, able to generate phase diagrams compatible with most existing data over a range of 900°C and 40 kbar. [See additional discussion below.]

The Microscopic Approach

It has become increasingly evident that cation ordering among both octahedral and tetrahedral sites needs to be considered in our attempts to devise realistic, predictive solution models for chemically complex pyroxene systems. Some of the thermodynamic consequences of ordering have already been discussed in terms of a model for ideal, nonconvergent ordering. In this section I will present expressions for the free-energy functions of binary solutions in which certain kinds of ordering occur.

The thermodynamics--and kinetics--of both convergent and nonconvergent ordering have been discussed in detail by many authors, particularly in the chemical and metallurgical literature (see, for example, Swalin, 1972, esp. pp. 153-160; Guggenheim, 1952; and Dienes, 1955). Instructive commentaries on the theory of ordering, as applied to pyroxenes and certain other mineral groups have been given by Mueller (1962a,b); Matsui and Banno (1965); Banno and Matsui (1966, 1967); Thompson (1969, 1970); Saxena (1973, esp. pp. 79-112, 138-141, 144-152); Grover (1974); Kerrick and Darken (1975); and Ganguly and Ghose (1979). Mueller (1967, 1969) has treated the kinetics of the ordering process in terms of the thermodynamic properties of intracrystalline exchange equilibria, and Virgo and Hafner (1969; see also Saxena, 1973, esp. pp. 92-106, 148-153) have used this theory to interpret the results from their important experimental study on the kinetics of ordering for natural orthopyroxenes at temperatures between 500° and 100°C. Several other workers, notably Saxena and Ghose (1971) and Saxena et $al.$ (1974), have studied the effects of controlled heating on the distribution of species (Fe^{2+}, Mg^{2+} and Ca) between the M1 and M2 sites in terrestrial, lunar and synthetic pyroxenes, and have interpreted their results using a site-partitioning model. Grover and Orville (1969) showed theoretically that (ideal) cation partitioning over two sites in a single phase can have a profound effect on the exchange reactions--and element distributions--between that phase and a coexisting one. Figure 5 shows five distribution diagrams for isothermal partitioning of two species, A and B, between an ideal single-site phase, I, and a double-site phase (II) that shows ideal ordering on the sites. $\Delta \bar{G}_T^0$ is the standard free energy for the exchange reaction between phases I and II,

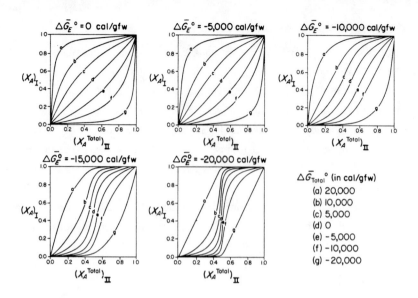

Figure 5. Isothermal distribution diagrams for partitioning of components A and B between an ideal single-site phase (I), and a double-site phase (II) with ideal solution on both sites. X_A^{Total} is the mole fraction of species A in phase II, and is related to the site occupancies

$$X_A^\alpha \text{ and } X_A^\beta \text{ by } X_A^T = (X_A^\alpha + X_A^\beta)/2.$$

$\Delta\bar{G}_E^0$ for each of the five distribution diagrams is fixed, while $\Delta\bar{G}_T^0$ for each curve is given by (a)-(g). The progressive distortion of the partitioning curves with decreasing (negative) values of $\Delta\bar{G}_E^0$ is evident; the diagram with $\Delta\bar{G}_E^0 = 0$ represents the familiar case for partitioning between coexisting single-site phases. These curves were generated for a temperature of 1373°K using R = 1.98717. (From Grover and Orville, 1969).

$$2A^I + (B^\alpha + B^\beta)^{II} \rightleftarrows 2B^I + (A^\alpha + A^\beta)^{II}.$$

The several curves, a–g, represent different (arbitrary) values for $\Delta\bar{G}_T^0$, as given, and α and β denote cation sites in phase II (e.g., M1 and M2).

$\Delta\bar{G}_E^0(T,P)$ is the standard free energy of exchange for the site-interchange reaction,

$$A^\alpha + B^\beta \rightleftarrows B^\alpha + A^\beta. \tag{5a}$$

Recall from equations (1a) and (5c) that for ideal ordering (as represented here),

$$\Delta\bar{G}_E^0 = -RT\ln\left(\frac{X_A^\beta X_B^\alpha}{X_A^\alpha X_B^\beta}\right).$$

$\Delta \bar{G}_E^O$ for each of the five distribution diagrams is fixed. When $\Delta \bar{G}_E^O = 0$, sites α and β are energetically equivalent and there is no ordering of species A and B between them ($s = 0$, equation 3). In this case the familiar distribution relations, given by Kretz (1961, p. 368) and others, for two coexisting single-site phases are obtained. When $\Delta \bar{G}_T^O$ $= \Delta \bar{G}_E^O = 0$, there is no preferential partitioning of A between I and II. That is, since α, β and I are indistinguishable energetically, the mole fraction of A is equal in both phases for all compositions. Maximum (negative) ordering occurs when ΔG_E^O is fixed (here arbitrarily) at $-20,000$ cal (for which $X_B^\alpha \gg X_A^\alpha$, $X_A^\beta \gg X_B^\beta$, and $1 > S \gg 0$). The distortion of individual isotherms, produced because of site partitioning, is particularly evident in this case. See Grover and Orville (1969) for additional discussion.

For a review of published measurements on inter- and intracrystalline cation distributions for pyroxenes and some discussion of the theoretical interpretations given them, see DHZ, 1978. Specifically, for information on [inter- or] intra-phase partitioning- in the following phases see the pages listed opposite:

	pages
augite	[326–330;372]
diopside–hedenbergite	221–223
omphacite	425–426[437–438;445–447;449,454]
orthopyroxene	24–25[63–73;90]
pigeonite	165–166

Nonconvergent ordering. In nonconvergent ordering the geometric and energetic distinctions among the sublattice sites in a phase are maintained throughout the entire range of temperature and pressure where the phase exists. So long as the material retains crystallographic integrity, avoiding structural transformations such as melting, the value for the "exchange potential," $\Delta \bar{G}_E^O$, is never 0. Thus, while $\Delta \bar{G}_E^O$ can be expected to vary as a function of temperature and pressure, the "site potentials" by which it is formally defined (equation 5b) never disappear. In general, $\Delta \bar{G}_E^O$ will decrease with increasing temperature as thermal vibrations obscure progressively (but never completely) the energetic differences among the sites and among the several cations that occupy them.

In nonconvergent ordering the configurational enthalpy is either independent of the degree of disorder--as in the models proposed by Navrotsky and Kleppa (1967) and Navrotsky (1971)--or depends directly on the value of $\Delta \bar{G}_E^O$ (and sometimes additional constants) and the ordering parameter s (equation 3). In general, $|\bar{H}_c|$ increases as s approaches 1 or -1; s = 0 only if disorder is complete. (See Thompson, 1969, p. 363 ff., 1970; Grover, 1974.) By comparison, the configurational enthalpy for a *convergent* ordering model approaches zero as s approaches 1.

In a general model for nonconvergent ordering, the bulk or macroscopic activity for *species* A in a binary phase with cation sites α and β is given by

$$a_A = (X_A^\alpha X_A^\beta)^{1/2} (\gamma_A^\alpha \gamma_A^\beta)^{1/2} \tag{28}$$

(see Grover, 1974, p. 1542). The quantities γ_j^k are "site activity coefficients." These cannot be measured individually, although their product $(\gamma_A^\alpha \gamma_A^\beta)$ has experimental significance. The factor $(\gamma_A^\alpha \gamma_A^\beta)^{1/2}$ is *not* equivalent to the macroscopic activity coefficient because $(X_A^\alpha X_A^\beta)^{1/2}$ is not, in general, equal to X_A (equation 2). For *ideal ordering* $a_A = (X_A^\alpha X_A^\beta)^{1/2} \neq X_A$. Phases that display ordering can never be ideal in the *classic* sense because $H_{mix} \geq H_c \neq 0$. The bulk activity coefficient for the ordered phase is given by

$$\gamma_A = \frac{a_A}{X_A} = \frac{2(X_A^\alpha X_A^\beta)^{1/2} (\gamma_A^\alpha \gamma_A^\beta)^{1/2}}{(X_A^\alpha + X_A^\beta)} , \tag{29a}$$

or

$$ln\gamma_A = ln[\frac{2(X_A^\alpha X_A^\beta)^{1/2}}{(X_A^\alpha + X_A^\beta)}] + Y_A \tag{29b}$$

where $Y_A \equiv ln(\gamma_A^\alpha \gamma_A^\beta)^{1/2}$. This formulation is useful when the total free energy for the ordered binary phase is written as

$$\bar{G}_T = X_A \mu_A + X_B \mu_B . \tag{A-2b}$$

This is equivalent to

$$\bar{G}_T = \mu_A^o X_A + \mu_B^o X_B + \epsilon RT[X_A ln a_A + X_B ln a_B] ,$$

380

where $\mu_A = \mu_A^o + \varepsilon RT \ln a_A$ (Thompson, 1967). With (28) this becomes

$$\bar{G}_T = \mu_A^o X_A + \mu_B^o X_B + \frac{\varepsilon RT}{2} [X_A \ln(X_A^\alpha X_A^\beta) + X_B \ln(X_B^\alpha X_B^\beta)]$$
$$+ \frac{\varepsilon RT}{2} [X_A \ln(\gamma_A^\alpha \gamma_A^\beta) + X_B \ln(\gamma_B^\alpha \gamma_B^\beta)] . \tag{30}$$

In this case, $\varepsilon = 2$ because equations (28) and (29) imply mixing of twice as many moles as are indicated by the binary form of (A-2b), above. Thus, with equation (29), $\bar{G}_T$ becomes

$$\bar{G}_T = \mu_A^o X_A + \mu_B^o X_B + RT[X_A \ln(X_A^\alpha X_A^\beta) + X_B \ln(X_B^\alpha X_B^\beta)]$$
$$+ 2RT [X_A Y_A + X_B Y_B] . \tag{31}$$

We will discuss below several forms suggested for Y.

Another, equivalent form for the free energy of a solution in the system A-B with nonconvergent ordering on two sites is

$$\bar{G}_T = \mu_A^o X_A + \mu_B^o X_B + \frac{s\Delta\bar{G}_E^o}{2} + \bar{H}_C^N + RT \sum_j \sum_k X_j^k \ln X_j^k . \tag{32}$$

Here $\bar{G}_T$ can also be written as

$$\bar{G}_T = \bar{G}_{ID}^M + \bar{H}_C - T\bar{S}_C , \tag{33a}$$

where the free energy of mechanical mixing, $\bar{G}_{ID}^M$, is

$$\bar{G}_{ID}^M \equiv \mu_A^o X_A + \mu_B^o X_B = \mu_B^o + (\mu_A^o - \mu_B^o)X_A \tag{33b}$$

and the configurational enthalpy, $\bar{H}_C$, is

$$\bar{H}_C = \bar{H}_C^{ID} + \bar{H}_C^N$$
$$= \frac{s\Delta\bar{G}_E^o}{2} + \bar{H}_C^N . \tag{33c}$$

$\bar{H}_C^N$ is the contribution to configurational enthalpy due to nonideal site-partitioning (see below). The configurational entropy, $\bar{S}_C$, is

$$\bar{S}_C = -\varepsilon R \sum_j \sum_k X_j^k \ln X_j^k , \tag{6}$$

where, as before, $j = A, B$; $k = \alpha, \beta$; and $\varepsilon = 1$ (cf. equation 6).

This expression for $\bar{G}_T$ can be compared with Thompson's (1969) equations (20) and (46) or his (1970) equation (1a). Specifically, Thompson proposes

$$\bar{G}_T' = \bar{G}^* + 2RT \sum_j \sum_k X_j^k \ln X_j^k \tag{34a}$$

(1969, equation (20) with 1970, equation (1a)), where G* can be written as an expansion in s and X_A:

$$\frac{\bar{G}^*}{RT} = g_o + g_x(2X_A-1) + g_s s + g_{xx}(2X_A-1)^2 + g_{xs}(2X_A-1)s + g_{ss}s^2 + \ldots \tag{34b}$$

It follows that

$$\frac{\bar{G}^*}{2} = \mu_B^o + (\mu_A^o - \mu_B^o)X_A + \frac{s\Delta\bar{G}_E^o}{2} + \bar{H}_C^N , \tag{34c}$$

and therefore that $\bar{H}_C^N$ can be related to specific terms in the expansion for $\bar{G}^*$ (the factor of 1/2 is inserted in equation 34c to correct for a difference in the size of molar units; $\bar{G}_T' = 2\bar{G}_T$). The form of that relation may be complicated algebraically, depending upon the number of terms in the expansion for $\bar{G}^*$ (see Davidson *et al.*, in preparation). For purposes of illustration it is sufficient here to observe that if $\bar{H}_C^N = 0$, and if the expansion in s and $(2X_A-1)$ [Thompson's r] is truncated at the term in s^2, shown, then

$$\mu_B^o = \frac{RT}{2}(g_o - g_x + g_{xx});$$

$$(\mu_A^o - \mu_B^o) = \frac{RT}{2}(2g_x - 4g_{xx});$$

and

$$\Delta\bar{G}_E^o = RT(g_s - g_{xs}) ,$$

which may be solved using equation (32) and its partial derivatives in X_A and s to give

$$g_o = (\frac{\mu_A^o + \mu_B^o}{RT}) ;$$

$$g_x = (\frac{\mu_A^o - \mu_B^o}{RT}) ;$$

$$g_s = \frac{\Delta\bar{G}_E^o}{RT} ;$$

and

$$g_{ss} = g_{xx} = g_{xs} = 0 \ .$$

Although it is not possible in this chapter to discuss the principles and procedures involved in expanding, combining and using equations (32) and (34) to model a complex pyroxene solution, it should be evident that this method is both pertinent and powerful. The form of the expansion, $\bar{G}^*$ (equation 34b) is conducive to a simple and straightforward treatment of phase equilibrium data (provided cation occupancies of sites can be measured or accurately estimated), and suggests, moreover, that an extension of the method to ternary or higher-order systems should not be difficult conceptually. (In a ternary system, for instance, there will be two independent compositional variables, say $\bar{X}$ and $\bar{Y}$, and two independent order paramters, analogous to s (say s and p). The free-energy expression comparable to (29) becomes simply

$$\bar{G}_T = g_o + g_x\bar{X} + g_y\bar{Y} + g_s s + g_p p + g_{xx}\bar{X}^2$$
$$+ g_{xy}\bar{X}\bar{Y} + g_{xs}\bar{X}s + g_{xp}\bar{X}p + g_{yy}\bar{Y}^2$$
$$+ g_{ys}\bar{Y}s + \dots + \varepsilon RT \sum_j \sum_k x_j^k ln x_j^k \ ,$$

omitting four terms in the second-order expansion or 20 terms in the third-order expansion.) This approach is likely to be used extensively in the future.

We will now look specifically at the equations for $\bar{G}_T$, $ln\gamma_A$ and $\bar{G}_{EX}$ in a binary system with nonconvergent ordering:

(A) The ideal case

Many of the thermodynamic properties of the ideal case for nonconvergent ordering have been discussed previously (equations 1-6). In summary, when ordering is ideal $\bar{H}_C^N = 0$ and $Y_A = Y_B = 0$ (these conditons are not independent). We find from equations (28)-(34):

$$\bar{G}_T = \mu_A^o X_A + \mu_B^o X_B + RT \ [X_A ln(X_A^\alpha X_A^\beta) + X_B ln(X_B^\alpha X_B^\beta)] \ ; \tag{35a}$$

$$\bar{G}_T = \mu_A^o X_A + \mu_B^o X_B + \frac{s\Delta\bar{G}_E^o}{2} + RT \sum_j \sum_k x_j^k ln x_j^k \ ; \tag{35b}$$

$$\bar{G}^* - 2\mu_B^o + 2(\mu_A^o - \mu_B^o)X_A + S\Delta\bar{G}_E^o \ ; \tag{35c}$$

$$\bar{G}_{EX} = \frac{s\Delta\bar{G}^O_E}{2} + RT \sum_j \sum_k X^k_j \ln X^k_j - (\mu^O_A X_A + \mu^O_B X_B)$$

$$- 2RT \, (X_A \ln X_A + X_B \ln X_B) \tag{35d}$$

(see Grover, 1974, equation 14; $s = -2X_A\Omega$); and

$$\ln\gamma_A = \ln\left[\frac{2(X^\alpha_A X^\beta_A)^{1/2}}{(X^\alpha_A + X^\beta_A)}\right] \tag{35e}$$

(B) The nonideal case (with two symmetric W terms)

Models presented by Thompson (1969, 1970) and by Saxena and Ghose (1970) predict nonideal "ordering" contributions to the model bulk activity coefficients of the form

$$\ln(\gamma^\alpha_A \gamma^\beta_A) = \frac{1}{2RT} \, [W_\alpha (X^\alpha_B)^2 + \Delta\bar{G}^O_* (X^\alpha_B X^\beta_B) + W_\beta (X^\beta_B)^2] \tag{36a}$$

and

$$\ln(\gamma^\alpha_A \gamma^\beta_A) = \frac{ZL}{RT} \, [W'_\alpha (X^\alpha_B)^2 + W'_\beta (X^\beta_B)^2] \,, \tag{36b}$$

respectively. (Here Z is analogous to a coordination number and L is Avagadro's number.) These expressions may be associated with that portion of the total enthalpy of mixing that is the result of nonideal intracrystalline partitioning, $\bar{H}^N_{mix}$. The similarity between these expressions is evident and we will focus on Thompson's formulation as the more general one. In equation (36a), W_α, W_β and $\Delta\bar{G}^O_*$ are constants at some given temperature and pressure for all bulk compositions of the double-site phase. The parameters W_α and W_β may be identified with nonideality on the α and β sites, respectively. Because one W term is associated with each site, this model may be considered to be a *symmetric approximation* for mixing on individual sites.

$\Delta\bar{G}^O_*$ is a molar free-energy term that describes the difference in energy between the ordered crystal at $X_A = 0.5$ and the crystal at the pure end-member compositions. Specifically,

$$\Delta\bar{G}^O_* = \bar{G}^O_+ + \bar{G}^O_- - (\bar{G}^O_A + \bar{G}^O_B) \,,$$

where $\bar{G}^O_+$ denotes positive order at $X_A = 0.5$;
$\bar{G}^O_-$ denotes negative order at $X_A = 0.5$,

$\bar{G}_A^O$ denotes the pure A end-member; and

$\bar{G}_B^O$ denotes the pure B end-member.

The importance of the cross-term coefficient, $\Delta\bar{G}_*^O$, has been emphasized by Sack (1980). In reviewing published information on the activity-composition relations among coexisting olivines and Fe-Mg orthopyroxenes, he found that ordering models that omitted the cross term were unable to account for (1) the simultaneous temperature and composition dependence observed for the apparent equilibrium constant, $K_D^{ol-opx} = X_{Mg}^{opx}X_{Fe}^{ol}/X_{Fe}^{opx}X_{Mg}^{ol}$; and (2) the magnitude of the positive (macroscopic) deviation from ideality observed for opx at temperatures of from 1100° to 1200°C (Nafziger and Muan, 1967; Kitayama and Katsura, 1968; Williams, 1971).

Expressions analogous to (35) for the nonideal, nonconvergent ordering prescribed by Thompson's formulation (36a) may be assembled quickly from equations (29b) and (31) by inserting values for Y_A and Y_B:

$$Y_A = \frac{1}{4RT} \ [W_\alpha(X_B^\alpha)^2 + \Delta\bar{G}_*^O X_B^\alpha X_B^\beta + W_\beta(X_B^\beta)^2] \ ; \qquad (37a)$$

and

$$Y_B = \frac{1}{4RT} \ [W_\alpha(X_A^\alpha)^2 + \Delta\bar{G}_*^O X_A^\alpha X_A^\beta + W_\beta(X_A^\beta)^2] \ . \qquad (37b)$$

The excess free energy is simply

$$\bar{G}_{EX} = \bar{G}_T - (\mu_A^O X_A + \mu_B^O X_B) - 2RT(X_A ln X_A + X_B ln X_B) \ ; \qquad (38)$$

and $\bar{H}_c^N$ (see equation 32) is

$$\bar{H}_c^N = 2RT(X_A Y_A + X_B Y_B) + \frac{1}{2} \ (W_\alpha - W_\beta)X_A s$$

$$- \frac{1}{4} \ (W_\alpha + W_\beta - \Delta\bar{G}_*^O) \ s^2 \ . \qquad (39a)$$

The *total* configurational enthalpy, $\bar{H}_c$, can be represented using equations (39a) and (33c). Inserting equations (37) in (39a), we obtain

$$\bar{H}_c = \frac{s\Delta G_E^O}{2} + \frac{(W_\alpha + W_\beta + \Delta\bar{G}_*^O)}{2} \ X_A X_B$$

$$+ \frac{(W_\alpha - W_\beta)}{2} \ X_A s - \frac{(W_\alpha + W_\beta - \Delta\bar{G}_*^O)}{8} \ s^2 \ . \qquad (39b)$$

(C) The nonideal case (with one asymmetric W term)

By analogy with equation (14), which represents the asymmetric Margules formulation for the macroscopic activity coefficient, we can write expressions for the nonideal contributions to ordering, Y_A and Y_B, that suppose more complicated (asymmetric) behavior in the W terms associated with individual sites. Let us assume, for instance, that site β (= 2) displays asymmetric behavior, while α retains the simple symmetric form. The equations analogous to (37) become

$$Y_A = \frac{1}{4RT} [(2W_{21} - W_{22})(X_B^\beta)^2 + 2(W_{22} - W_{21})(X_B^\beta)^3 + \Delta\bar{G}_*^o X_B^\alpha X_B^\beta + W_\alpha (X_B^\alpha)^2] \quad (40a)$$

and

$$Y_B = \frac{1}{4RT} [(2W_{22} - W_{21})(X_A^\beta)^2 + 2(W_{21} - W_{22})(X_A^\beta)^3 + \Delta\bar{G}_*^o X_A^\alpha X_A^\beta + W_\alpha (X_A^\alpha)^2] , \quad (40b)$$

where W_{21} and W_{22} are the nonideality parameters for asymmetric behavior in the substitution of species A and B on site β. Expressions for $\bar{G}_T$ and $\ln\gamma_A$ can be formulated easily by combining equations (40a) and (40b) with (31) and (29b), respectively. The expression for $\bar{H}_c^N$ comparable to (39a) is straightforward, but too long to be included here. Davidson, Grover and Lindsley (in preparation) have employed this model successfully in treating phase-relations in the system Di-En. In that case, mixing on the M2 site of Cpx was taken to be asymmetric.

Convergent ordering. In convergent ordering the geometric and energetic distinctions among the considered sublattice sites in a phase disappear with increasing temperature. At a single critical temperature (or over a small range in temperature) the sites become energetically equivalent and experimentally indistinguishable. Prigogine and Defay (1954, pp. 299-305) give a clear and succinct derivation of the free-energy equation for convergent ordering in the compound A_1B_1 of binary system A-B. Assuming no interaction between species A and B, they show (equation 19.54) that the molar configurational enthalpy predicted by the model is

$$\bar{H}_c = \frac{W_c}{4} (1 - s^2) \quad (41a)$$

where s is the order parameter defined previously and W_c is an energy term comparable to W_α or W_β (*cf.* equation 4). Setting the s-partial derivative of the expression for molar free energy equal to zero (the criterion for internal equilibrium with respect to ordering), they derive the condition for equilibrium,

$$ln\left(\frac{1+s}{1-s}\right) = \frac{sW_c}{RT} ,$$

and show that when temperature reaches a critical value,

$$T_c = W_c/2R ,$$

s necessarily becomes 0.

In a similar, but somewhat variant approach Swalin (1972, pp. 153–160) shows that the admission of an interchange energy, W_c^I, between species A and B for composition A_1B_1 gives rise to an additional term in the expression for configurational enthalpy,

$$\bar{H}_c = \frac{W_c}{4}(1-s^2) + \frac{W_c^I}{2}(1+s^2) , \tag{41b}$$

or

$$\bar{H}_c = \frac{1}{4}(W_c + 2W_c^I) + \frac{1}{4}(2W_c^I - W_c)s^2 .$$

This leads to a value for the critical temperature of

$$T_c = \frac{W_c}{2R} - \frac{W_c^I}{R} .$$

(The lower the energy of the A–B pair, relative to the energy required for a simple interchange of species on sites, the higher T_c will be.) It is instructive to compare this formulation for $\bar{H}_c$ with that deriving from Thompson's model, equation (39b). When (39b) is limited to a comparable composition ($X_A = X_B = \frac{1}{2}$), the two expressions for configurational enthalpy are identical in form provided $\Delta\bar{G}_E^O = 0$, $\Delta\bar{G}_*^O = 4W_c^I$, and $W_\alpha = W_\beta = W_c$. Thompson's representation is the more general one, and reduces to (41b) only when the free-energy function is symmetric with respect to positive and negative order.

Navrotsky and Loucks (1977) and Holland et al. (1979) use a solution model for convergent disordering to represent free-energy composition relations in the system Di–En. Their formulation extends the simple model (41a), described previously, to include compositional variations. It predicts that the free energy of mixing due to disordering is

$$\Delta \bar{G}_{(\text{disordering})} = \frac{W_c X_A^2}{4}(1-s^2) + \frac{1}{2} RT \sum_j \sum_k X_j^k \ln X_j^k$$

$$- RT[X_A \ln X_A + (1-X_A) \ln(1-X_A)]$$

(modified from Navrotsky and Loucks, 1977, equation 7). This expression can be compared to the equations given in nonconvergent ordering models for the excess free energy of solution, neglecting mechanical mixing terms (equations 35d and 38). Here, as for the other ordering models discussed, the site fractions X_j^k can be expressed as simple, explicit functions of bulk composition and s. The total free energy of mixing given by this model is

$$\bar{G}_T = D + Cs - BX_A + \bar{H}_c + \frac{1}{2} RT \sum_j \sum_k X_j^k \ln X_j^k \ , \qquad (42)$$

where the value of $\varepsilon = \frac{1}{2}$ implies a pyroxene formula unit $MSiO_3$. The configurational enthalpy is given by

$$\bar{H}_c = \frac{W_c}{4} X_A^2 (1 - s^2) \ ,$$

and the constants B, C and D determine the character of the standard-state terms and non-configurational contributions in the mixing function.

Figure 6 shows how these constants are interrelated for the system En-Wo (*cf.* also Fig. 3). The coexisting phases stable at the prescribed temperature and pressure are (*Pbca*) En and (*C2/c*) Di; their chemical potentials are given by μ_{En}^{Opx} and μ_{Wo}^{Cpx}, respectively. The free-energy curve for Cpx is anchored at $\bar{G}_{f;T,P}^o$(Di) for $X_{Wo} = \frac{1}{2}$ but intersects the free-energy axis $X_{Wo} = 0$ at $\mu_{En}^{o,Cpx}$, the chemical potential for a *C2/c* pyroxene with $MgSiO_3$ composition. According to the theory first presented by Navrotsky and Loucks (1977, p. 116 ff.), the value of s for the end-member magnesium-*Cpx* may be fixed somewhere between 1 (appropriate to a "frozen diopside structure" with radically different Ml and M2 cation sites) and 0 (representing the disordered state of a "relaxed" clinopyroxene whose Ml and M2 sites are energetically equivalent).

While a "true" value for s(T,P) is unknown, Navrotsky and Loucks (1977) contend that its potential for variability can be incorporated in the model by allowing $\mu_{En}^{o,Cpx}$ to vary linearly between the extreme

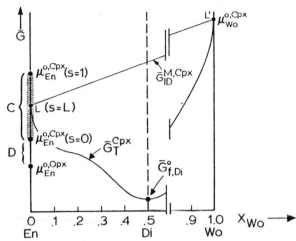

Figure 6. Free energy-composition diagram for the system En - (Di) - Wo showing schematically how the constants C and D of Navrotsky and Loucks, 1977 (equation 42) are related to the standard state chemical potentials for components En and Wo in Cpx. In particular,

$$C = \mu_{En}^{o,Cpx}(s=1) - \mu_{En}^{o,Cpx}(s=0) \quad \text{and} \quad D = \mu_{En}^{o,Cpx}(s=0) - \mu_{En}^{o,Opx},$$

where s is the order parameter defined by equation 3. Line L-L' represents the free energy of a hypothetical mechanical mixture consisting of metastable end-member clinopyroxenes of composition $Mg_2Si_2O_6$ and $Ca_2Si_2O_6$ in proportions defined by X_{Wo}. $\bar{G}_T^{Cpx}$ is the free energy curve for the Cpx solution. It is anchored at $\bar{G}_{f,Di}^{o}$ for $X_{Wo} = \frac{1}{2}$ and at a chemical potential, $\mu_{En}^{o,Cpx}(s=L)$, representing an intermediate state of disorder (s=L) appropriate to the clinopyroxene of composition $X_{Wo} = 0$ "stable" at the temperature and pressure represented.

limits defined at s = 1 and s = 0. Thus, if the relaxed Cpx defined by $\mu_{En}^{o,Cpx}(s=0)$ is less stable than orthorhombic $MgSiO_3$ by the amount

$$D \equiv [\mu_{En}^{o,Cpx}(s = 0) - \mu_{En}^{o,Opx}] , \tag{43a}$$

the constant C is simply

$$C = [\mu_{En}^{o,Cpx}(s = 1) - \mu_{En}^{o,Cpx}(s = 0)] . \tag{43b}$$

The "true" standard-state chemical potential for the Mg end-member of the Cpx solution, $\mu_{En}^{o,Cpx}(s)$, is then

$$\mu_{En}^{o,Cpx}(s) = \mu_{En}^{o,Opx} + D + [\mu_{En}^{o,Cpx}(s = 1) - \mu_{En}^{o,Cpx}(s = 0)]s . \tag{43c}$$

For purposes of illustration in Figure 6 the "true" value for s was arbitrarily taken at s = L; the schematic free-energy curve for Cpx, $\bar{G}_T^{Cpx}$, is thus anchored at that point when $X_{Wo} = 0$.

It is clear that variability in s is in fact a fictive property for *pure* $MgSiO_3$; both sites M1 and M2 contain Mg, and $s = X_{Mg}^{M2} - X_{Mg}^{M1} = 0$ by equation (3). It is possible, however, to define a value for the "macro-scopic property" s using a measured value for the configurational enthalpy and an estimate for W_c in equation (41a). As Navrotsky and Loucks (1977) also point out, it might be feasible to determine values for s in pyrox-enes having compositions *close* to $X_{Wo} = 0$ and to extrapolate those values to the limit where $X_{Wo} \to 0$.

The (-)B term (eqn 42) is the slope of the mechanical mixing line, $\mu_{En}^{o,Cpx}(s) -- \mu_{Wo}^{o,Cpx}$. As such, its value depends strictly on s (equation 43c), although if $|\mu_{Wo}^{o,Cpx}| >> |\mu_{En}^{o,Opx}|$ it may be acceptable to take B as constant, as Navrotsky and Loucks have done. Equations (42) and (43) may now be combined to give the expression for $\bar{G}_T^{Cpx}$ in the system Wo-En:

$$\bar{G}_T^{Cpx} = (\mu_{En}^{o,Opx} + D + [\mu_{En}^{o,Cpx}(s=1) - \mu_{En}^{o,Cpx}(s=0)]s)X_{En}^{Cpx}$$

$$+ \mu_{Wo}^{o,Cpx}X_{Wo}^{Cpx} + \frac{W_c}{4}(X_{Wo}^{Cpx})^2(1-s^2) + \frac{1}{2} RT \sum_j \sum_k X_j^k ln X_j^k , \qquad (44)$$

where j = Ca, Mg; k = M1, M2; and, because $\varepsilon = \frac{1}{2}$, the components Wo and En represent the molecular compositions $CaSiO_3$ and $MgSiO_3$, resepctively. Using this model with a selection of arbitrary but reasonable values for the thermochemical constants B, D ($\mu_{En}^{o,Opx} = 0$ for convenience), C and W_c, and assuming for computational simplicity that the Opx solution is ideal, Navrotsky and Loucks (1977) generated temperature-composition diagrams for the system En-Di (as part of En-Wo) that are topologically consis-tent with published phase diagrams. Although these calculated phase diagrams do not reproduce the positions of experimental phase boundaries, they do show a field for coexisting Pig_{ss} and Di_{ss}. Many models based on macroscopic considerations have been unable to do this.

This approach in treating the system $MgSiO_3-CaSiO_3$ avoids certain theoretical problems involving the stability of pure $CaMgSi_2O_6$, relative to Mg-enriched Di_{ss} plus Wo (see Fig. 2a,b and related discussion). It may create others, however, in assuming that the M1 and M2 sites of Cpx tend to converge at magnesian compositions so that Ca and Mg can parti-tion between them. This assumption requires that for a given temperature, Ca and Mg should be more disordered in Fe-free $Pig_{(ss)}$ than in Di_{ss}. As a consequence, the average size of the M1 site in Cpx should increase as the bulk Ca content decreases (and Ca enters M1). In fact, there is no

390

crystal-chemical evidence to suggest that the average size of M1 increases with increasing Mg content. Concerted studies involving both high-temperature crystallography and solution calorimetry on Fe-free Pig_{ss} phases of various compositions, X_{Wo}, should be made to determine the extent of ordering in Ca-Mg clinopyroxenes.

Hybrid models. We have seen that the thermodynamic analysis of experimental data for crystalline solutions can sometimes be simplified when the mixing model chosen uses components that reflect the known crystal-chemical properties of the system. (See Ganguly and Ghose, 1979, for example.) In a system with cation ordering on distinct structural sites, the macroscopic activity for a given species can sometimes be expressed simply in terms of the site mole fractions for that species. Thus, with ideal ordering, equation (28) predicts that the activity of species A in a phase having two sites, α and β, will be

$$a_A = (X_A^\alpha X_A^\beta)^{1/2} ; \qquad (45a)$$

for the activity of Mg in a pyroxene, this simple formulation gives

$$a_{Mg} = (X_{Mg}^{M1} X_{Mg}^{M2})^{1/2} . \qquad (45b)$$

The comparable expression for the activity of the *component* $Mg_2Si_2O_6$ (En), is

$$a_{En} = (a_{Mg})^2 = X_{Mg}^{M1} X_{Mg}^{M2} . \qquad (45c)$$

For nonideal partitioning, equation (28) is sometimes also written

$$a_A = (a_A^\alpha a_A^\beta)^{1/2} ,$$

implying that the product $\gamma_j^k X_j^k$ can be regarded as a "site activity," a_j^k.

In cases where simple formulations like (45) are not adequate to represent the observed solution behavior, more complicated models for site activities are needed. Beyond the expanded (nonideal) order-disorder models, already discussed, there has been some use of *hybrid* models. These treat the mixing properties of individual cation sites independently, using one or more of the standard formulations for nonideal *macroscopic* mixing behavior to express site activities. Thus, for example, a model

suitable for treating asymmetric mixing in a ternary system might be
applied to the mixing of Fe, Ca and Mg on the M2 site in pyroxene, while
a model representing ideal mixing might be used to treat M1. The models
described above for nonideal, non-convergent ordering differ in that
they incorporate *cross terms*--$\Delta G_*^o X_B^\alpha X_B^\beta$ and $\Delta \bar{G}_*^o X_A^\alpha X_A^\beta$--which allow for the
coupling of individual species between sites. For examples of the use
of such hybrid models, the reader can refer to

where

Saxena (1973):   Wohl's (1946, 1953) expression for activity
                          in an asymmetric ternary solution is used to
                          treat Ca, Fe^{2+}, and Mg mixing on M2 of Cpx
                          and a two-coefficient Margules expression is
                          used to treat mixing on M1;

Powell and Powell (1974): Cpx - olivine equilibria are treated by taking
                          (a) olivine to be ideal at high temperatures;
                          (b) M1 as a ternary regular solution in Mg,
                          Fe^{2+} and "Al" (= $Al+Ti+Cr+Fe^{3+}$); (c) M2 to
                          contain all Ca and Mn, with an (Fe/Mg) ratio
                          governed by the constraint (d) $(Fe/Mg)_{M2}$ =
                          $(Fe/Mg)_{M1}$ = $(Fe/Mg)_{bulk\ Cpx}$. (However, see
                          Wood, 1976);

Saxena (1976b):   Equilibria between coexisting Opx and Cpx are
                          treated by taking M2 in each pyroxene phase
                          as a simple ternary solution of Fe^{2+}, Mg and
                          Ca and M1 in each pyroxene as an asymmetric
                          binary solution of Mg and Fe^{2+}. Site occu-
                          pancies for Cpx are approximated by consid-
                          ering relations among mixing equations for
                          coexisting Ca-pyroxene, Pig and Opx solutions.
                          Saxena also considers the possibility for
                          treating mixing of species Fe(M1), Fe(M2),
                          Mg(M1) and Mg(M2) in each of two coexisting
                          pyroxenes in terms of two regular quaternary
                          models (with 10 fitting constants for each
                          phase: i.e., six binary, three ternary and
                          one quaternary W-terms);

Grove and Lindsley (1981): Equilibria among Cpx phases and coexisting
                          silicate liquids are treated using a symmet-
                          ric ternary model for mixing of Ca, Mg and
                          Fe^{2+} on M2 of Cpx, and an ideal model for
                          mixing of Mg and Fe^{2+} on M1. The activity
                          of silica in coexisting liquids is found to
                          be nearly constant. Published x-ray refine-
                          ments for single crystals are used to solve
                          for site occupancies.

392

To illustrate a simple hybrid approach, applicable to QUAD pyroxenes and involving the manipulation of material-balance equations in ways common to many site-mixing models, we will consider the problem of "Ca-blocking" in a Ca-Fe^{2+}-Mg clinopyroxene. If the partitioning of Fe and Mg on sites M1 and M2 in this phase is ideal, and constrained only by the requirement that Ca cannot occupy M1, we can express a_{Mg} in terms of the bulk composition and the partition coefficient K_E (equation 1a). Specifically, for ideal partitioning the macroscopic activity a_{Mg} is given by (45b), and we will use the relations among site fractions and macroscopic mole fractions to cast X_{Mg}^{M1} and X_{Mg}^{M2} in terms of bulk composition and K_E:

For Ca: In general, $X_{Ca} = \dfrac{(X_{Ca}^{M1} + X_{Ca}^{M2})}{2}$.

Because $X_{Ca}^{M1} = 0$,

$$X_{Ca}^{M2} = 2XCa, \qquad (46a)$$

where $2X_{Ca} \leq 1$.

For Fe: Similarly, $X_{Fe}^{M1} + X_{Fe}^{M2} = 2XFe$.

Moreover, $X_{Fe}^{M1} + X_{Mg}^{M1} = 1$,

so that $X_{Fe}^{M1} = 1 - X_{Mg}^{M1}$; $\qquad (46b)$

and $X_{Fe}^{M2} = 2XFe - X_{Fe}^{M1}$

$$= 2X_{Fe} - 1 + X_{Mg}^{M1} \qquad (46c)$$

$$= 1 - 2X_{Ca} - X_{Mg}^{M2} .$$

For Mg: $X_{Mg}^{M1} + X_{Mg}^{M2} = 2XMg$; $X_{Mg}^{M2} = 2X_{Mg} - X_{Mg}^{M1}$. $\qquad (46d)$

From equation (1a), we have for K_E:

$$K_E = \frac{X_{Fe}^{M2} X_{Mg}^{M1}}{X_{Fe}^{M1} X_{Mg}^{M2}} , \qquad (46e)$$

or, with (46b,c) and (d)

$$K_E = \frac{(2X_{Fe} - 1 + X_{Mg}^{M1}) X_{Mg}^{M1}}{(1 - X_{Mg}^{M1})(2X_{Mg} - X_{Mg}^{M1})} . \qquad (46f)$$

(Although X_{Ca} does not occur in this equation, it would appear if $(2X_{Fe}-1)$ were written in the form $(1-2X_{Mg}-2X_{Ca})$. The influence of X_{Ca} is significant in (46f) inasmuch as $X_{Fe} + X_{Mg} \neq 1$. If $X_{Fe} + X_{Mg}$ is taken to be 1, the expression reduces to

$$K_E = \frac{(1 - X_{Mg}^{M2})X_{Mg}^{M1}}{(1 - X_{Mg}^{M1})X_{Mg}^{M2}} = \frac{X_{Fe}^{M2}X_{Mg}^{M1}}{X_{Fe}^{M1}X_{Mg}^{M2}} \quad (cf. \text{ equation 1).})$$

Equation (46f) can be expanded to give a quadratic expression in X_{Mg}^{M1}:

$$(K_E-1)(X_{Mg}^{M1})^2 - (2X_{Fe} - 1 + K_E + 2X_{Mg})X_{Mg}^{M1} + 2K_E X_{Mg} = 0$$

The value obtained for X_{Mg}^{M1} from the solution of this equation can be used to determine a value for X_{Mg}^{M2} (equation 46d). Taken together, X_{Mg}^{M1} and X_{Mg}^{M2} fix a_{Mg} or a_{En} (equation 45). The problem of "Ca-blocking" is treated theoretically by Blander (1972); and Fleet (1974) applies this theory in analyzing Mg and Fe^{+2} site occupancies for M1 and M2 in coexisting natural pyroxenes.

THE SYSTEM DIOPSIDE-ENSTATITE: A REVIEW

In this section we will review several thermodynamic solution models that have been presented to treat published phase equilibrium data in the system Di-En. If we omit the cooperative disordering model of Navrotsky and Loucks (1977), which was discussed previously, we find that these models fall into five categories. These are listed in Table 4 (Lindsley *et al.*, 1981), along with the values given by the author(s) for the standard reaction constants of (A) and (B) (see Table 4, 'conventions') and fitting parameters for the assumed equation(s) of state. All parameters are referred to a common nomenclature and converted to energies in kJoules gfw^{-1} as necessary, permitting direct comparison of the analogous values predicted by each model. Following closely the treatment given by Lindsley, Grover and Davidson (1981), I will discuss the assumptions and limitations of the models in each category in light of the conceptual problems and thermodynamic approaches already considered in this chapter.

The Asymmetric Margules Formulation that Treats Opx and Cpx as a Single Solution

In this model (WL in Table 4) the *Pbca* and *C2/c* structures are treated as having a single G curve, so that the reaction energies $\Delta \bar{G}_A^o$ and $\Delta \bar{G}_B^o$ (equation 23) are equal to zero. Warner and Luth (1974) recognized the limitations in this assumption and contended in their use of the asymmetric Margules model, equation (16), that the fitting parameters, $W_{G(En)}$ and $W_{G(Di)}$ should be regarded merely as empirical constants, without thermodynamic significance. Their approach has not been followed by later workers.

Models Assuming Ideal Mixing of Ca and Mg on the M2 Sites

Because they also treat the effects of Fe on Di-En solutions, most of the models in category II begin by considering cation mixing on both the M1 and M2 sites. For the system Di-En these models can be simplified as follows. A general expression for the activity of Mg in a solution phase, ϕ, which has mixing of both Ca and Mg on crystallographic sites M1 and M2, is given by

$$ln\,a_{Mg}^\phi = ln(a_{En}^\phi)^{1/2} = ln\,\gamma_{Mg}^\phi X_{Mg}^\phi \quad .$$

From equation (29),

$$ln\,a_{Mg}^\phi = ln(X_{Mg}^{M1}X_{Mg}^{M2})^{1/2} + Y^\phi \tag{47}$$

where $X_{Mg}^\phi = (X_{Mg}^{M1} + X_{Mg}^{M2})^\phi/2 = 1 - X_{Ca}^\phi$, and Y^ϕ represents the contribution to the activity coefficient, γ_{Mg}^ϕ, that results from nonideal partitioning of Ca and Mg. (Recall that one useful formulation for Y is given by equation (37). For the system Di-En this becomes

$$Y_T^\phi = \frac{1}{4RT} [W_{M1}^\phi (X_{Ca}^{M1})^2 + \Delta \bar{G}_*^{o,\phi} X_{Ca}^{M1} X_{Ca}^{M2} + W_{M2}^\phi (X_{Ca}^{M2})^2] \quad .) \tag{48}$$

The models in this category proceed from equations (23) and (47), taking $Y^\phi = 0$, so that

$$ln\,a_{En}^\phi \equiv 2(ln\,a_{Mg}^\phi) = ln(X_{Mg}^{M1}X_{Mg}^{M2})^\phi$$

and therefore,

$$\Delta \bar{G}_A^o = (\Delta \bar{U}_A^o + P\Delta \bar{V}_A^o - T\Delta \bar{S}_A^o) = -RT\,ln\,\frac{(X_{Mg}^{M1}X_{Mg}^{M2})^{Cpx}}{(X_{Mg}^{M1}X_{Mg}^{M2})^{Opx}} \quad . \tag{49}$$

Table 4. Comparisons among fitting parameters for the equations of state used to describe phase relations in the system $Mg_2Si_2O_6$-$CaMgSi_2O_6$

Category	Reference	Note	For reaction $\underline{A}$: $\Delta\bar{U}^o_{\underline{A}}(H)$	$\Delta\bar{V}^o_{\underline{A}}$	$\Delta\bar{S}^o_{\underline{A}}$	For reaction $\underline{B}$: $\Delta\bar{U}^o_{\underline{B}}(H)$	$\Delta\bar{V}^o_{\underline{B}}$	$\Delta\bar{S}^o_{\underline{B}}$	For the monoclinic crystalline solution: $W_1(U,G,H)$	$W_2(U,G,H)$	W_{V1}	W_{V2}	W_{S1}	W_{S2}	Orthorhombic Solution W^{opx}_G
I: Asymmetric Margules formulation, treating Opx and Cpx as a single solution.														
	WL 1974	1	---	---	---	---	---	---	100.550	50.999	---	0.1531	0.0511	0.0101	--see W_1--
II: Formulation for the activities of En in Opx and Cpx, assuming ideal mixing of Ca and Mg on the M2 site.														
	WB 1973	2	84.822	---	0.0445	---	---	---	---	---	---	---	---	---	---
	NW 1974	3	41.621	---	0.0143	---	---	---	---	---	---	---	---	---	---
	M 1977	4	61.035	(0)	0.0279	---	---	---	---	---	---	---	---	---	---
III: Symmetric model for Opx and Cpx solutions, based upon Opx-Cpx exchange equilibria.														
	F 1977	5	--cannot be determined from an exchange reaction--						24.7		---	---	---	---	106.7
IV: Symmetric ('regular') Margules formulation for independent Opx and Cpx solutions.														
	SN 1975	6	---	---	---	(-4.184)	---	---	(2.092)	27.326	---	---	---	---	34.68
	P 1978	7	---	---	---	---	---	0.00682	11.90	33.5	---	---	---	---	(0)
	HNN 1979	8	---	---	---	---	---	0.00275	6.80	24.47	---	---	---	---	(34.00)
V: Asymmetric Margules formulation for the Cpx solution; the independent Opx solution is symmetric.														
	GLS 1976	9	(0.398)	(-0.007)	(-0.00154)	-13.535	---	---	26.740	30.359	0.105		0.0		18.20
	LGD 1981		3.561	0.0355	0.00191	-21.178	-0.0908	-0.00816	25.484	31.216	0.0812	-0.0061	0.0	0.0	(25)

Conventions:

--The dimensions of terms or coefficients are, for molar energy: kJoule gfw⁻¹; molar volume: kJoule gfw⁻¹ kbar⁻¹; and for molar entropy: kJoule gfw⁻¹ Kelvin⁻¹. All equations of state refer to the molecular components $MgMgSi_2O_6$ (En) and $CaMgSi_2O_6$ (Di).

--Reaction ($\underline{A}$) is $(Mg_2Si_2O_6)_{opx} \rightleftharpoons (Mg_2Si_2O_6)_{cpx}$. Reaction ($\underline{B}$) is $(CaMgSi_2O_6)_{opx} \rightleftharpoons (CaMgSi_2O_6)_{cpx}$; the absence of an entry for $\Delta\bar{U}^o_{\underline{B}}$ indicates that the author(s) did not apply the constraint of equilibrium implied by reaction ($\underline{B}$). There is a relation,

$$(\mu^{o,cpx}_{\underline{N}} - \mu^{o,opx}_{\underline{N}}) = \Delta\bar{U}^o_{\underline{N}} = \Delta\bar{U}^o_{\underline{N}} + P\Delta\bar{V}^o_{\underline{N}} - T\Delta\bar{S}^o_{\underline{N}},$$

where $\underline{N}$ refers to reaction $\underline{A}$ or $\underline{B}$.

Thus, if $\Delta\bar{V}^o_{\underline{N}}$ is not given, the first column represents $\Delta\bar{U}^o_{\underline{N}}$. If neither $\Delta\bar{V}^o_{\underline{N}}$ nor $\Delta\bar{S}^o_{\underline{N}}$ is given, the first column represents $\Delta\bar{G}^o_{\underline{N}}$.

--For the monoclinic solution,

$$\bar{G}_{EX} = X_{En}X_{Di}(X_{En}W_{G2} + X_{Di}W_{G1}),$$

so that $W_{G2} = W_G(Di)$ and $W_{G1} = W_G(En)$, according to the convention of Thompson (1967). Because

$$W_G = W_U + PW_V - TW_S,$$

the values entered for W_1 and W_2 represent either internal energies (W_{U1} or W_{U2}, if values are given) or enthalpies (W_{H1} or W_{H2}, if values are given only for W_{S1} or W_{S2}, respectively); or Gibbs energies (W_{G1} or W_{G2}, if no volume or entropy terms are given).

--Parentheses, ( ), denote an assumed value; the other data given in the body of the table represent values of best-fit, computed by the author(s) according to the model specified, and converted by us to kJoules as necessary.

[1.] Warner and Luth (1974). The authors state explicitly that W_{G1} and W_{G2} are fitting parameters, empirically determined and without thermodynamic significance.

[2.] Wood and Banno (1973). This formulation incorporates all deviations from the ideal condition in one constant, and one temperature-dependent term:

$$(\mu_{En}^{o,Cpx} - \mu_{En}^{o,Opx}) \equiv -RT\ln(a_{En}^{Cpx}/a_{En}^{Opx}) = \Delta\bar{H}_A^o - T\Delta\bar{S}_A^o.$$

The authors' initial model for ideal partitioning, $a_{En}^{px} = (a_{Mg}^{px})^2 = (X_{Mg}^{M1}X_{Mg}^{M2})$, reduces to $a_{En}^{px} = X_{En}^{px}$ for the system En-Di because of the assumption that $X_{Ca}^{M1} = 0$. If $X_{Ca}^{M1} \neq 0$, this model cannot be limited to the system En-Di, but requires end-member component $Mg_2Si_2O_6$-$Ca_2Si_2O_6$.

[3.] Nehru and Wyllie (1974). Uses data from unreversed dissolution experiments at 30 kbar.

[4.] Wells (1977). Very similar in approach to Wood and Banno (1973), but using more recent experimental data.

[5.] Finnerty (1977). The equation representing exchange equilibrium between En and Di components in Opx and Cpx is a linear combination of the equations in chemical potential for reactions (A) and (B). Without studying (A) or (B) independently, it is not possible to extract singular values for the reaction coefficients (e.g., $\Delta\bar{U}_A^o$ or $\Delta\bar{U}_B^o$).

[6.] Saxena and Nehru (1975). An important early demonstration that equations treating the solution properties of coexisting Opx and Cpx can be derived and manipulated independently. The values given here for W_G^{Opx} and W_G^{Cpx} represent averages, based on data for three temperatures: 1100°, 1200°, and 1300°C.

[7.] Powell (1978). When applied to the system En-Di, the author's "site population" model reduces to a simple regular (symmetric) equation of state. (Cf. the analogous reduction for Wood and Banno's (1973) model, discussed in note 2.) This simplification does not, of course, apply when Fe or other components are present in the system.

[8.] Holland, Navrotsky and Newton (1979). These authors constrain the value for ΔU^o by considering available calorimetric data for the Ca clinopyroxenes. Their value for W_G^{Opx} was taken from Saxena and Nehru (1975).

[9.] Grover, Lindsley and Schweitzer (1976). Although the published abstract suggests that an equation representing the exchange reaction between Opx and Cpx was used to compute fitting parameters, the equations of state for Opx and Cpx solutions were determined independently. For the isobarically invariant condition, where the three phases Di_{ss}, Pig_{ss} and ortho En_{ss} coexist in equilibrium (taken here to be at 1713 K and 15 kbar), it is possible to derive an explicit solution for W_1(Cpx), W_2(Cpx) and W_G(Opx) (but not for asymmetric behavior in Opx), provided the compositions of the coexisting phases are well-known. ΔU_A^o, ΔV_A^o and ΔS_A^o were computed (not fitted) using the reversed low-CEn ↔ orthoEn inversion curve of Grover (1972a,b).

For each pyroxene solution, Cpx and Opx, the authors then assume that $X_{Ca}^{M1} = 0$. Because $X_{Ca} = (X_{Ca}^{M1} + X_{Ca}^{M2})^{\phi}/2 = (1 - X_{Mg}^{\phi})$, this gives $2X_{Ca}^{\phi} \equiv X_{Di}^{\phi} = (X_{Ca}^{M2})^{\phi}$, and establishes Di and En as the appropriate components for the system. (The mole fractions X_{Mg} and X_{Ca} imply components En and Wo.) The site-fractions for Mg are related to X_{Di}^{ϕ}:

$$(X_{Mg}^{M2})^{\phi} = (1 - X_{Ca}^{M2})^{\phi} = 1 - X_{Di}^{\phi} \; ;$$

and

$$(X_{Mg}^{M1})^{\phi} = 2X_{Mg}^{\phi} - (X_{Mg}^{M2})^{\phi} = 2 - 2X_{Ca}^{\phi} - (1 - X_{Di}^{\phi})$$

$$= 2 - X_{Di}^{\phi} - 1 + X_{Di}^{\phi}) = 1 \; .$$

Substituting these relations in (49) we obtain,

$$\Delta \bar{G}_{A}^{o} = -RT \mathit{l}n (X_{En})^{Cpx} / (X_{En})^{Opx}$$

Thus for the system Di-En, if Ca is prohibited from entering M1 and Y is taken to be 0, the model for "ideal mixing of Mg and Ca on M2" is equivalent to macroscopic ideality of the En component in both the Cpx and Opx solutions.

The several proponents of this model (WB, NW, W; Table 4) use equation (49) to fit experimental data on the compositions of coexisting Opx and Cpx phases as a function of $1/T$, presuming no pressure effect. Thus, because $\Delta \bar{G}_{A}^{o} = \Delta \bar{H}_{A}^{o} - T\Delta \bar{S}_{A}^{o}$, their equation of state takes the form:

$$\mathit{l}n \frac{(a_{En})^{Cpx}}{(a_{En})^{Opx}} = \mathit{l}n \frac{(X_{En})^{Cpx}}{(X_{En})^{Opx}} = - \frac{\Delta \bar{H}_{A}^{o}}{RT} + \frac{\Delta \bar{S}_{A}^{o}}{R} \; . \tag{50}$$

There are several serious problems inherent in this formulation: (1) By neglecting reaction (B), the authors using this model lose an important thermodynamic constraint on the solution properties of the phases in question (namely, that $\mu_{Di}^{Opx} = \mu_{Di}^{Cpx}$). (2) By neglecting Y, the authors force all nonideal effects inherent in the Cpx and Opx solutions to be amalgamated within a single constant ($-\Delta \bar{H}_{A}^{o}$) and a single temperature-dependent term ($\Delta \bar{S}_{A}^{o}$). This, of course, explains why $\Delta \bar{H}_{A}^{o}$ for these models is so large in comparison to that for models that treat the solutions as nonideal (Table 4, categories III, IV and V). It is clear, in considering Figure 3, that the procedure for fitting phase-equilibrium data by means of equation (50) involves nothing more than repositioning two *ideal* free-energy curves in free energy-composition-temperature^{-1} space by varying the distance between them until the prescribed compositions for the coexisting phases En$_{ss}$ and Di$_{ss}$ have been intersected.

Because it was designed for use in geothermometry, this ideal model for the mixing properties of the Opx and Cpx solutions is incapable of predicting the phase relations in the system Di-En.

Symmetric Models for Opx and Cpx Solutions, Based upon Opx-Cpx Exchange Equilibria

This model (F) is based upon the exchange reaction,

$$(En)^{Cpx} + (Di)^{Opx} \rightleftarrows (Di)^{Cpx} + (En)^{Opx}, \qquad (C)$$

which is equivalent to the *difference* between reactions (B) and (A),

$$(C) = (B) - (A) .$$

At equilibrium,

$$\Delta \bar{G}_C = (\Delta \bar{G}_B - \bar{G}_A) = (\mu_{Di}^{Cpx} + \mu_{En}^{Opx} - \mu_{Di}^{Opx} - \mu_{En}^{Cpx}) = 0 \qquad (51)$$

and the equation for $\Delta \bar{G}_C^o$ comparable to equations (22) and (23) becomes

$$\Delta \bar{G}_C^o = \mu_{Di}^{o,Cpx} + \mu_{En}^{o,Opx} - \mu_{Di}^{o,Opx} - \mu_{En}^{o,Cpx},$$

or

$$(\Delta \bar{U}_C^o + P\Delta \bar{V}_C^o - T\Delta \bar{S}_C^o) = -RT ln \frac{x_{Di}^{Cpx} x_{En}^{Opx}}{x_{Di}^{Opx} x_{En}^{Cpx}} - RT ln \frac{\gamma_{Di}^{Cpx}}{\gamma_{En}^{Cpx}} - RT ln \frac{\gamma_{Di}^{Opx}}{\gamma_{En}^{Opx}} , \qquad (52)$$

where

$$\Delta \bar{G}_C^o = (\Delta \bar{G}_B^o - \Delta \bar{G}_A^o) = \Delta \bar{U}_C^o + P\Delta V_C^o - T\Delta \bar{S}_C^o .$$

Finnerty (1977) took both the Opx and Cpx solutions to be symmetric in composition $(W_{G(En)}^{\phi} = W_{G(Di)}^{\phi})$, so that equation (15) reduces to

$$\bar{G}_{Ex}^{\phi} = W_G^{\phi} x_{En}^{\phi} x_{Di}^{\phi} . \qquad (53)$$

He then used equations (52) and (53), transformed appropriately by means of the relation

399

$$RT\ell n \; \frac{\gamma_{Di}^{\phi}}{\gamma_{En}^{\phi}} = \frac{\partial \bar{G}_{Ex}^{\phi}}{\partial X_{Di}} = W_G^{\phi} \frac{\partial}{\partial X_{Di}} (X_{Di}^{\phi} [1 - X_{Di}^{\phi}]) \; ,$$

to fit the reversed data of Lindsley and Dixon (1976). Note that a model of this type requires only that $(\mu_{Di}^{Cpx} - \mu_{En}^{Cpx}) = (\mu_{Di}^{Opx} - \mu_{En}^{Opx})$ at equilibrium. It is sufficient by this criterion that the tangents to the free-energy curves at the points representing the compositions of coexisting phases be merely parallel and *not coincident*. This model thereby loses the power of the independent constraints

$$\mu_{En}^{Cpx} = \mu_{En}^{Opx} \quad \text{and} \quad \mu_{Di}^{Cpx} = \mu_{Di}^{Opx} \; .$$

Moreover, the reaction constants $\Delta\bar{G}_A^o$ and $\Delta\bar{G}_B^o$ (see equations 23A and 23B) cannot be extracted independently. Thus, this model is not appropriate for treating two-phase data.

Symmetric Margules Formulation for Independent Cpx and Opx Solutions

In the symmetric Margules approximation, $W_{G(En)}^{\phi} = W_{G(Di)}^{\phi}$ and equation (15) reduces to $G_{Ex}^{\phi} = W_G^{\phi} X_{En} X_{Di}$. Although the three papers, (SN,P, HNN in Table 4) listed under this category presented different approaches in the treatment of two-phase equilibria by means of the symmetric approximation, they have in common (1) recognition of the importance of equation (A), with the computation or assignment of a value of $\Delta\bar{u}_A^o$, and (2) computation of W_G^{Cpx} in conjunction with a calculated or assigned value for W_G^{Opx}. Only Saxena and Nehru (1975), however, recognized the important constraints implied by the use of both equations (A) and (B) in treating two-phase equilibria, employing the criteria for equilibrium $\mu_{En}^{Cpx} = \mu_{En}^{Opx}$ and $\mu_{Di}^{Cpx} = \mu_{Di}^{Opx}$.* Because there are two independent free-energy functions for the system, $\bar{G}_{Cpx}$ and $\bar{G}_{Opx}$, it is entirely possible that *one* of these equilibrium conditions could be met while the other was not. Thus, for example, the appropriate tangents at or near the correct compositions for each G-curve might meet a single point on the $X_{Di} = 0$ axis ($\mu_{En}^{Opx} = \mu_{En}^{Cpx}$), yet diverge markedly at $X_{Di} = 1$ ($\mu_{Di}^{Opx} \neq \mu_{Di}^{Cpx}$). To guarantee that the

---

*If both the thermodynamic model and the experimental data were perfect, these two constraints would no longer be independent, inasmuch as they are related through the Gibbs-Duhem equation. But since a major purpose in fitting a model to the data is to *test* the model, it is misleading to ignore the second constraint.

fitting process will produce values for the W_G terms appropriate to the condition of complete internal equilibrium, it is necessary to employ *both* equations (A) and (B). Among the several models listed in Table 4, only those of SN (1975) and GLS (1976) and LGD (in Category V) have been appropriately constrained through the combined use of equilibrium reactions (A) and (B).

The model presented by Powell (1978) is interesting as an extension of the "activity" approximation discussed in category II. Powell combined equation (22a) with expressions, like (47), for the activity of Mg in Cpx and Opx solution phases. He made the additional approximation that $X_{Ca}^{M1} = 0$, so that the expression for Y_T^{ϕ}, equation (48), reduces directly to:

$$Y_T^{\phi} = \frac{1}{4RT} [W_{M2}^{\phi} (X_{Ca}^{M2})^2] .$$

Together, these give:

$$-\Delta \bar{G}_A^o = RT \ln \frac{(X_{Mg}^{M1} X_{Mg}^{M2})^{Cpx}}{(X_{Mg}^{M1} X_{Mg}^{M2})^{Opx}} + \frac{1}{2} (W_{M2}^{Cpx}[X_{Ca}^{M2,Cpx}]^2 - W_{M2}^{Opx}[X_{Ca}^{M2,Opx}]^2). \quad (54)$$

Powell considered W_{M2}^{Opx} to be negligible and wrote $\Delta \bar{G}_A^o$ as $\Delta \bar{H}_A^o - T\Delta \bar{S}_A^o$. However, if one assumes $X_{Ca}^{M1} = 0$, then $X_{Mg}^{M1} = 1$, and equation (54) reduces to

$$-\Delta \bar{H}_A^o + T\Delta \bar{S}_A^o = RT \ln \frac{X_{En}^{Cpx}}{X_{En}^{Opx}} + \frac{1}{2} W_{M2}^{Cpx}(X_{Di}^{Cpx})^2$$

which is, of course, equivalent to the "regular solution" approximation with $W_G^{Opx} = 0$, and $W_{M2}^{Cpx} = 2W_G^{Cpx}$.

The HNN regular solution model (Holland *et al.*, 1979) is remarkably successful in reproducing the temperatures of the experimental data. It suffers, however, from its inability both to account for equilibria involving Pig and (because the constants for reaction (B) were not refined) to permit the calculation of phase diagrams.

Asymmetric Margules Formulation for the Cpx Solution, with a Symmetric Opx Solution

Models that account for equilibria involving Pig necessarily treat Cpx as an asymmetric solution. GLS (Table 4) attempted to fit En_{ss} +

Pig + Di$_{ss}$ data at a single temperature and pressure where the three
phases coexist (Grover *et al.*, 1976). Although the compositions of the
coexisting phases were not closely constrained experimentally, these
authors were successful in deriving the asymmetric fitting parameters
W_{G1}^{Cpx} and W_{G2}^{Cpx} for the Cpx solution as well as an independent (rather
than assumed) value for the symmetric Margules coefficient for the Opx
solution, W_G^{Opx}. The reaction constants $\Delta \bar{U}_A^o$, $\Delta \bar{V}_A^o$, $\Delta \bar{S}_A^o$ and $\Delta \bar{G}_B^o$ had, however, to be computed independently from phase equilibrium measurements
(see the notes for Table 4). Furthermore, the propagated error for
thermochemical quantities derived by means of the fitting procedure
were large, owing primarily to the uncertainties in composition and
temperature characteristic of a single set of experiments on phases
whose free energies are not greatly different (see Fig. 7b).

The effect of pronounced uncertainties in composition for experiments conducted at a single temperature and pressure can be minimized
by fitting polythermal and polybaric data. The results obtained by
Lindsley *et al.* (1981; LGD) represent a least-squares fit for bracketed
(or half-bracketed) polythermal and polybaric data in the system Di-En.
These authors used a method of simultaneous equations to fit the 10
thermodynamic parameters given in Table 4, assuming the value for W_G^{Opx}.
(For reasons discussed previously the shape of the Opx G-curve is critical only at compositions near $X_{Di} = 0$; it is not markedly sensitive in
that region to the value taken for W_G^{Opx}.) The procedures used by Lindsley *et al.* have been reviewed earlier and are explained in detail in
their paper (1981). The expressions from LGD for the excess Gibbs free
energy of mixing for the Opx and Cpx solutions, are respectively,

$$\bar{G}_{EX}^{Opx} = W_G^{Opx} X_{En}^{Opx} X_{Di}^{Opx} = 25 X_{En}^{Opx} X_{Di}^{Opx} \ kJ \qquad (55a)$$

and

$$\bar{G}_{EX}^{Cpx} = W_{G1} X_{En}^{Cpx} (X_{Di}^{Cpx})^2 + W_{G2} X_{Di}^{Cpx} (X_{En}^{Cpx})^2$$

$$= (25.484 + .0812P) X_{En}^{Cpx} (X_{Di}^{Cpx})^2 + (31.216 - .0061P) X_{Di}^{Cpx} (X_{En}^{Cpx})^2 \quad (55b)$$

(*cf.* equations A-7, 9 and 10). In this treatment Pig and Di are taken
to be part of the same solid solution series at high T (Cpx; space group
$C2/c$). Thus, because the compositions of coexisting Pig and Di$_{ss}$ are not

symmetric with respect to $X_{Di} = \frac{1}{2}$, the expression for the free energy of Cpx must include an asymmetric excess energy.

This model takes the theoretical problem of polymorphism into consideration by treating reactions (A) and (B) independently. This approach allows the model to be used both to calculate phase diagrams (by means of equations 16 and 55) and to extract equilibration temperatures, given data on the compositions of coexisting Opx-Cpx pairs. The expressions given by Lindsley et $al.$ for estimating the equilibration temperature of Opx-Cpx pairs were obtained by rearranging the equilibrium expressions (55):

$$T,°C(A) = \frac{3.561 + .0355P + 2W_{G2}X_{En}^{Cpx}(X_{Di}^{Cpx})^2 + W_{G1}(X_{Di}^{Cpx})^2(1-2X_{En}^{Cpx}) - W_{G}^{Opx}(X_{Di}^{Opx})^2}{.00191 - R\ln(X_{En}^{Cpx}/X_{En}^{Opx})}$$

$$- 273.15$$

$$T,°C(B) = \frac{-21.178 - .0908P + W_{G2}(X_{En}^{Cpx})^2(1-2X_{Di}^{Cpx}) + 2W_{G1}X_{Di}^{Cpx}(X_{En}^{Cpx})^2 - W_{G}^{Opx}(X_{En}^{Opx})^2}{-.00816 - R\ln(X_{Di}^{Cpx}/X_{Di}^{Opx})}$$

$$- 273.15$$

where R = 0.0083143, P is in kbar, W_{G1} = 25.484 + 0.0812P and W_{G2} = 31.216 - 0.0061P. The agreement between the model predictions for temperature and the experimental data is excellent in most cases. The model value for $\Delta \bar{H}_A^0$ (Table 4) is about 3kJ mole^{-1} lower than the value inferred by Holland et $al.$ (1979) from the calorimetric data of Newton et $al.$ (1979), however.

Lindsley (Chapter 6, this volume) presents the phase diagrams computed using this model (equations 16 and 55), in relation to the experimental brackets used to constrain the fit for the Margules parameters. Some typical free-energy curves calculated from the model (Figs. 7a-d; compare Fig. 3) illustrate the various equilibria encountered in the calculated phase diagrams and the behavior of the free-energy function with changing temperature and pressure.

CONCLUSIONS

The successful thermodynamic treatment of natural pyroxenes will depend both on the thoughtful formulation of realistic (and sometimes innovative) solution models capable of treating some difficult conceptual problems, and the continued accumulation of precise compositional data

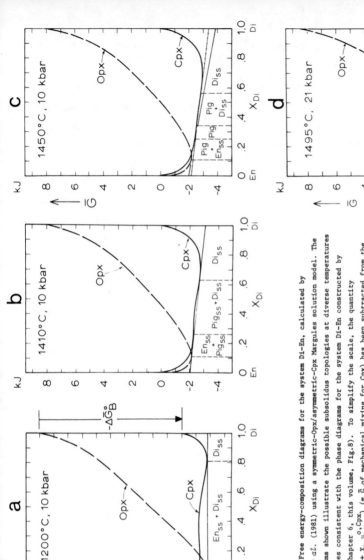

Figure 7. Free energy-composition diagrams for the system Di-En, calculated by Lindsley *et al.* (1981) using a symmetric-Opx/asymmetric-Cpx Margules solution model. The four diagrams shown illustrate the possible subsolidus topologies at diverse temperatures and pressures consistent with the phase diagrams for the system Di-En constructed by Lindsley (Chapter 6, this volume, Fig.8). To simplify the scale, the quantity $(X_{En}\mu^{o,Cpx}_{En} + X_{Di}\mu^{o,Cpx}_{Di})$ (= $\bar{G}$ of mechanical mixing for Cpx) has been subtracted from the total free energy of solution, so that the $\bar{G}$-curve for Cpx is $\bar{G}^{Cpx}_{mixing}$ whereas that for Opx is $\bar{G}^{Opx}_{soln} - \bar{G}^{Cpx}_{mech. mix.}$

(a) $En_{ss} + Di_{ss}$ stable; Pig + Di_{ss} metastable. Applies to low-temperature portions of Figures 8a-e, Chapter 6 (this volume).

(b) En_{ss}, Pig (low-Ca Cpx), and Di_{ss} have a mutual tangent and thus co-exist. Applies to three-pyroxene assemblages in Figures 8a-c, Chapter 6.

(c) The phases and assemblages En_{ss}, En_{ss} + Pig, Pig, Pig + Di_{ss}, and Di_{ss} are stable, depending on bulk Ca/Mg ratio. Applies to Figures 8a-c (Chapter 6) at temperatures above that of the three-pyroxene line but below that of the Pig-Di_{ss} consolute point.

(d) En_{ss} coexists with a singular Cpx: the consolute point on the Pig-Di_{ss} miscibility gap. The assemblage Pig + Di_{ss} will not be stable at any pressure higher than 21 kbar.

from reversed experiments on systems of increasing complexity. While we are still at an early stage in that process, the fits and starts, failures and successes of the past 10 years have given us some useable data and considerable understanding. Thermodynamic analysis in the future will involve vast amounts of algebra on conceptually simple expressions of egregious length, and an increased use of computer and matrix techniques for the rapid solution of simultaneous equations of state.

ACKNOWLEDGEMENTS

It is a special pleasure for me to acknowledge Donald Lindsley for the generous assistance and kind advice and criticism he offered me during the preparation of these notes. He and others at the State University of New York, Stony Brook, made it possible for me to spend a most pleasant and productive sabbatical year. I also thank the University of Cincinnati for financial support during that time. Paula Davidson helped me with her insightful comments on diverse theoretical questions, and Clara Podpora assisted in collecting and putting in order the references and data for Table 3. This work was supported in part by the National Science Foundation, Grant EAR7622129 (to D. H. Lindsley).

THE THERMODYNAMICS OF MULTI-COMPONENT SOLUTIONS, BRIEFLY REVIEWED

Because we are so often concerned with temperature and pressure as natural variables, or because temperature and pressure are relatively easy to manage experimentally, the Gibbs free energy, G, is the state variable most commonly used as the basis for constructing solution models. Infinitesimal changes in the Gibbs free energy of a c-component crystalline solution are given by the relation

$$dG = -SdT + VdP + \sum_i^c \mu_i dn_i \qquad \text{(A-1)}$$

where S is entropy; V is volume; μ_i is the chemical potential of component i and n_i is the number of moles of i. At constant temperature and pressure, the total *molar* free energy for the solution, $\bar{G}_T$, is

$$\bar{G}_T \equiv \frac{G}{\sum_i n_i} = \frac{1}{\sum_i n_i} \int \sum_i^c \mu_i dn_i = \sum_i^c \mu_i \left(\frac{n_i}{\sum_i n_i}\right) \qquad \text{(A-2a)*}$$

or

$$\bar{G}_T = \sum_i \mu_i X_i \qquad \text{(A-2b)}$$

where the mole fraction $X_i = (n_i / \sum_i n_i)$. Equation (A-2b), which may also be written

$$\bar{G}_T = \sum_i \left(\frac{\partial G}{\partial n_i}\right)_{T, P n_{j \neq i}} X_i ,$$

has analogs in terms of each of the other thermodynamic variables of state:

$$\bar{V}_T = \sum_i \bar{v}_i X_i ; \quad \bar{S}_T = \sum_i \bar{s}_i X_i ,$$

and so on. The lower-case variables

----

*An infinitesimally small change in composition is accomplished when dn_i moles are added to the solution. $\bar{G}_T$ is thus evaluated at a fixed composition, so that the constant of integration in equation (A-2a) is equal to zero. Moreover,

$$\mu_i = (\partial G/\partial n_i)_{T, P, n_j}$$

is an *intensive* variable that does not change with dn_i and is therefore constant in integration. (See Denbigh, 1966, pp. 92-93.) Hereafter, unspecified limits in summation will denote sums taken over all i, i = 1, ..., c.

$$\bar{v}_i \equiv (\partial V / \partial n_i)_{T,P,n_j}$$

and

$$\bar{s}_i \equiv (\partial S / \partial n_i)_{T,P,n_j} = -(\partial \mu_i / \partial T)_{P,n_i,n_j}$$

are the partial molar volume and partial molar entropy, respectively; these are interrelated through equations analogous to those pertaining to the corresponding molar variables of state. For example,

$$\mu_i = \bar{e}_i + P\bar{v}_i - T\bar{s}_i$$

(see Denbigh, 1966, pp. 100–104).

The Molar Gibbs Free Energy of Mixing as an Equation of State

The activity, a_i, of component i in a solution is related to the chemical potential of i, $\mu_i(T,P,X)$, and a standard-state (or ideal *reference*) chemical potential, $\mu_i^o(T,P)$, by

$$RT\ln a_i = \mu_i(T,P,X_i,X_j) - \mu_i^o(T,P) \qquad (A-3)$$

(Prigogine and Defay, 1954, p. 88; parentheses here denote functional dependence). If we define the rational activity coefficient, γ_i, as

$$\gamma_i \equiv (a_i/X_i) \ , \qquad (A-4)$$

then (A-3) may be rewritten in the form:

$$\mu_i = \mu_i^o(T,P) + RT\ln X_i \gamma_i \ . \qquad (A-5)$$

In general, γ_i is a function of temperature, pressure and composition. By combining equations (A-2b) and (A-5), and by associating like terms, we have, for the total molar free energy of mixing,

$$\bar{G}_T = \sum_i \mu_i^o X_i + RT \sum_i X_i \ln X_i + RT \sum_i X_i \ln \gamma_i (T,P,X) \ . \qquad (A-6)$$

The Gibbs free energy of mixing for a solution may also be written as the sum of an ideal and a nonideal (or *excess*) contribution:

$$\bar{G}_T = \bar{G}_{ID} + \bar{G}_{EX} \ . \qquad (A-7a)$$

$\bar{G}_{ID}$ can be decomposed into terms representing the free energy of *mechanical mixing* and the free energy of *ideal chemical mixing*:

$$\bar{G}_{ID} = \sum_i \mu_i^o(T,P)X_i + RT \sum_i X_i \ln X_i \ , \qquad (A-7b)$$

respectively. The *excess* Gibbs free energy may then be associated with
that part of equation (A-6) that involves activity coefficients:

$$\overline{G}_{EX} = RT \sum_i X_i \ln\gamma_i (T,P,X) \ . \qquad \text{(A-7c)}$$

Derivation of Equations relating μ_i and $\overline{G}_T$, γ_i and $\overline{G}_{EX}$ for Multicomponent Solutions

It follows from equation (A-2b) that an infinitesimal change in $\overline{G}_T$
is given by

$$d\overline{G}_T = \sum_i \mu_i dX_i + \sum_i X_i d\mu_i \ . \qquad \text{(A-8)}$$

The Gibbs-Duhem equation requires that

$$\sum_i X_i d\mu_i = 0$$

at fixed temperature and pressure, however, and (A-8) becomes

$$d\overline{G}_T = \sum_i \mu_i dX_i = \mu_1 dX_1 + \sum_{i=2}^{c} \mu_i dX_i \ . \qquad \text{(A-9)}$$

The explicit separation of one term from the summation in (A-9) is necessary in order to isolate a single component to serve as the dependent
compositional variable. The specific $\mu_i dX_i$ term selected is arbitrary
(here i = 1), and affects only the notation involved in the derivation,
not the form of the final equations or the development that leads to them.
For complete generality, one could label the specific term selected with
an integer variable k and take the remaining sum over i = 1, 2, ..., (k-1),
(k+1), ..., c.

Because $\sum_{i=1}^{c} X_i = 1$, we have

$$X_1 = 1 - \sum_{i=2}^{c} X_i \qquad \text{(A-10a)}$$

and

$$dX_1 = - \sum_{i=2}^{c} dX_i \ , \qquad \text{(A-10b)}$$

which, when combined with (A-9), gives

$$d\overline{G}_T = -\mu_1 \sum_{i=2}^{c} dX_i + \sum_{i=2}^{c} \mu_i dX_i \ ,$$

or

$$d\bar{G}_T = \sum_{i=2}^{c} (\mu_i - \mu_1)\,dX_i \; . \tag{A-11}$$

It follows directly from equation (A-11) that

$$\left(\frac{\partial \bar{G}_T}{\partial X_i}\right)_{T,P,X_j} = \mu_i - \mu_1 \qquad (j \neq 1, i) \quad , \tag{A-12}$$

and the combination of equation (A-12) with (A-11) gives

$$d\bar{G}_T = \sum_{i=2}^{c} \left(\frac{\partial \bar{G}_T}{\partial X_i}\right)_{T,P,X_j} dX_i \; . \tag{A-13}$$

Equation (A-12) is an important and powerful tool in the derivation of additional relations pertinent to a system of c components. If, for example, the term $\mu_1 X_1$ is isolated in equation (A-2b), we have

$$\bar{G}_T = \mu_1 X_1 + \sum_{i=2}^{c} \mu_i X_i$$

substitution of the expression for X_1 from equation (A-10A) in this equation gives

$$\bar{G}_T = \mu_1\left(1 - \sum_{i=2}^{c} X_i\right) + \sum_{i=2}^{c} \mu_i X_i$$

or

$$\bar{G}_T = \mu_1 + \sum_{i=2}^{c} (\mu_i - \mu_1) X_i \tag{A-14}$$

Combining equations (A-12) and (A-14), we find, then, that

$$\bar{G}_T = \mu_1 + \sum_{i=2}^{c} \left(\frac{\partial \bar{G}_T}{\partial X_i}\right)_{T,P,X_j} X_i \tag{A-15}$$

For a two-component system, this reduces to the familiar form

$$\bar{G}_T = \mu_1 + \left(\frac{\partial \bar{G}_T}{\partial X_2}\right)_{T,P} X_2 \tag{A-16a}$$

or, with (A-12),

$$\bar{G}_T = \mu_1 + (\mu_2 - \mu_1) X_2 \tag{A-16b}$$

which is the equation for the straight line tangent to the free-energy curve $\bar{G}_T(X_2)$ at bulk composition X_2. μ_1 and μ_2 are the $\bar{G}_T$-intercepts for this tangent at $X_2 = 0$ and $X_2 = 1$, respectively, and represent the partial molar free energies of the solution for the composition of interest. In a three-component system, the analogous equation,

$$\bar{G}_T = \mu_1 + (\mu_2 - \mu_1)X_2 + (\mu_3 - \mu_1)X_3 \qquad \text{(A-17)}$$

represents a *plane* tangent to the free-energy surface in $\bar{G}_T - X_2 - X_3$ space. The free-energy intercepts for this plane are μ_3, μ_2 and μ_1 at compositions $X_3 = 1$ ($X_2 = 0$), $X_2 = 1$ ($X_3 = 0$), and $X_2 = X_3 = 0$, respectively. These intercepts represent the partial molar free energies for the composition (X_2, X_3) at the point of tangency.

Rearranging terms in equation (A-15), we have

$$\mu_1 = \bar{G}_T - \sum_{i=2}^{c} X_i \left(\frac{\partial \bar{G}_T}{\partial X_i} \right)_{T,P,X_{j\neq1,i}} \qquad \text{(A-18)}$$

In the case of a binary system, equation (A-18) also takes a familiar form:

$$\mu_1 = \bar{G}_T - X_2 \left(\frac{\partial \bar{G}_T}{\partial X_2} \right)_{T,P}$$

(see Guggenheim, 1959, p. 213). Equation (A-18) takes a somewhat less familiar form for the case of a ternary mixture:

$$\mu_1 = \bar{G}_T - X_2 \left(\frac{\partial \bar{G}_T}{\partial X_2} \right)_{T,P,X_3} - X_3 \left(\frac{\partial \bar{G}_T}{\partial X_3} \right)_{T,P,X_2} \qquad \text{(A-19a)}$$

Analogous expressions for $\mu_2 (\bar{G}_T, X_{i\neq2})$ and $\mu_3 (\bar{G}_T, X_{i\neq3})$ in the ternary system can be obtained by permuting the subscripts. Additional equations giving μ_2 and μ_3 as functions only of $\bar{G}_T$, X_2 and X_3 can be obtained by combining equations (A-19a) and (A-12), with $i = 2$ or 3:

$$\mu_2 = \bar{G}_T + (1-X_2) \left(\frac{\partial \bar{G}_T}{\partial X_2} \right)_{T,P,X_3} - X_3 \left(\frac{\partial \bar{G}_T}{\partial X_3} \right)_{T,P,X_2} \qquad \text{(A-19b)}$$

$$\mu_3 = \bar{G}_T - X_2 \left(\frac{\partial \bar{G}_T}{\partial X_2} \right)_{T,P,X_3} + (1-X_3) \left(\frac{\partial \bar{G}_T}{\partial X_3} \right)_{T,P,X_2} \qquad \text{(A-19c)}$$

Equations (A-19) confirm that there are only two independent composition variables in a ternary system. Our choice of X_2 and X_3 as independent variables is arbitrary.

Equations having the general form of (A-18) are useful in establishing conditions for the *coexistence of stable phases* in c-component systems. Although explicit solutions for such equations, including values for the compositions of the coexisting phases, are usually not possible to obtain, the equations themselves can be set up with little difficulty. They may be amenable to approximate solution using iterative or other techniques (*e.g.*, Brown and Skinner, 1974). To write the coexistence equations we need only recognize that the chemical potential of each component, μ_i, is the same in each of the several phases containing that component, so long as the system is in equilibrium. Thus, for example,

$$\mu_i^A = \mu_i^B = \ldots = \mu_i^\phi \qquad (A-20)$$

where A, B and ϕ denote phases. Moreover, because of equation (A-12),

$$\left(\frac{\partial \bar{G}_T}{\partial X_i}\right)_{T,P,X_j}^A = \mu_i^A - \mu_1^A \qquad (A-21a)$$

$$\vdots$$

$$\left(\frac{\partial \bar{G}_T}{\partial X_i}\right)_{T,P,X_j}^\phi = \mu_i^\phi - \mu_1^\phi \ ; \qquad (A-21\phi)$$

and it follows directly from (A-20) that

$$\left(\frac{\partial \bar{G}_T}{\partial X_i}\right)_{T,P,X_j}^A = \left(\frac{\partial \bar{G}_T}{\partial X_i}\right)_{T,P,X_j}^B = \ldots = \left(\frac{\partial \bar{G}_T}{\partial X_i}\right)_{T,P,X_j}^\phi . \qquad (A-22)$$

Thus, for any number of phases (within the limits imposed by the Phase Rule) coexisting in equilibrium in a c-component system, we can combine equations (A-18), (A-20) and (A-22) to obtain

$$\mu_k = \bar{G}_T^A - \sum_{i,i \neq k}^c X_i^A \left(\frac{\partial \bar{G}_T}{\partial X_i}\right)_{T,P,X_{j \neq i,k}}^A = \ldots = \bar{G}_T^\phi - \sum_{i,i \neq k}^c X_i^\phi \left(\frac{\partial \bar{G}_T}{\partial X_i}\right)_{T,P,X_{j \neq i,k}}^\phi \qquad (A-23)$$

in which the specific compositional subscript 1 has been replaced by a general subscript k. (See Prigogine and Defay, 1954, p. 254 for the example of a ternary system.)

The separability of the ideal and excess portions of $\bar{G}_T$ (equation A-7a) is an extremely important property in that it permits expressions in $\bar{G}_T$ to be treated *by parts*. Thus, for example, equation (A-18) becomes

$$\mu_1 = (\bar{G}_{ID} + \bar{G}_{EX}) - \sum_{i=2}^{c} X_i [(\frac{\partial \bar{G}_{ID}}{\partial X_i}) + (\frac{\partial \bar{G}_{EX}}{\partial X_i})]_{T,P,X_{j\neq 1,i}} \quad , \qquad (A-24)$$

and in accord with Guggenheim (1959, p. 242) and others, we can define

$$\mu_1^{ID} \equiv \bar{G}_{ID} - \sum_{i=2}^{c} X_i (\frac{\partial \bar{G}_{ID}}{\partial X_i})_{T,P,X_{j\neq 1,i}} \qquad (A-25a)$$

and

$$\mu_1^{EX} \equiv \bar{G}_{EX} - \sum_{i=2}^{c} X_i (\frac{\partial \bar{G}_{EX}}{\partial X_i})_{T,P,X_{j\neq 1,i}} \quad , \qquad (A-25b)$$

so that

$$\mu_1 = \mu_1^{ID} + \mu_1^{EX} \; . \qquad (A-26)$$

It follows from equations (A-25) and (A-26) that for component i

$$\mu_i^{ID} = \mu_i^o(T,P) + RT ln X_i \qquad (A-27a)$$

and

$$\bar{G}_{ID} = \sum_i \mu_i^{ID} X_i \; ; \qquad (A-27b)$$

and that

$$\mu_i^{EX} = RT ln \gamma_i \qquad (A-28a)$$

and

$$\bar{G}_{EX} = \sum_i \mu_i^{EX} X_i \qquad (A-28b)$$

(*cf.* equations A-2b, 5, 7a and 7b). By analogy, equation (A-15) gives

$$\bar{G}_{EX} = \mu_1^{EX} + \sum_{i=2}^{c} X_i (\frac{\partial \bar{G}_{EX}}{\partial X_i})_{T,P,X_{j\neq 1,i}} \quad , \qquad (A-29)$$

and equation (A-12) gives

$$(\frac{\partial \bar{G}_{EX}}{\partial X_i})_{X_{T,P,X_j}} = \mu_i^{EX} - \mu_1^{EX} \quad (j\neq 1) \; . \qquad (A-30)$$

412

Combining equations (A-28a) and (A-30), we obtain the important result:

$$\frac{1}{RT} \left(\frac{\partial \bar{G}_{EX}}{\partial X_i}\right)_{T,P,X_j} = \ln\left(\frac{\gamma_i}{\gamma_1}\right) . \qquad (A-31)$$

Substitution of equation (A-28a) in a generalized form of (A-25b), where the specific compositional subscript has been replaced by the general subscript k, yields an equally important result:

$$RT\ln\gamma_k = \bar{G}_{EX} - \sum_{i,i\neq k}^{c} X_i \left(\frac{\partial \bar{G}_{EX}}{\partial X_i}\right)_{T,P,X_{j\neq k,i}} \qquad (A-32)$$

Aldridge, L. P., Bancroft, G. M., Fleet, M. E. and Herzberg, C. T. (1978) Omphacite studies, II. Mössbauer spectra of C2/c and P2/n omphacites. Amer. Mineral. 63, 1107-1115.

Banno, S. and Matsui, Y. (1966) Intracrystalline exchange equilibrium in orthopyroxene. Proc. Japan Acad. 42, 629-633.

Banno, S. and Matsui, Y. (1967) Thermodynamic properties of intracrystalline exchange solid solution. Proc. Japan Acad. 43, 762-767.

Barrer, R. M. and Klinowski, J. (1972) Ion exchange involving several groups of homogeneous sites. J. Chem. Soc., Faraday Trans. I 68, 73-87.

Barrer, R. M. and Klinowski, J. (1977) Theory of isomorphous replacement in alumino-silicates. Phil. Trans. Roy Soc. London, A. Math. Phys. Sci. 285, 637-680.

Barrer, R. M. and Klinowski, J. (1979) Order-disorder model of cation exchange in silicates. Geochim. Cosmochim. Acta 43, 755-766.

Barron, L. M. (1976) A comparison of two models of ternary excess free energy. Contr. Mineral. Petrol. 57, 71-81.

Blander, M. (1972) Thermodynamic properties of orthopyroxenes and clinopyroxenes based on the ideal two-site model. Geochim. Cosmochim. Acta 36, 787-799.

Blencoe, J. G. (1977) Computation of thermodynamic mixing parameters for isostructural, binary crystalline solutions using solvus experimental data. Computers Geosci. 3, 1-18.

Broecker, W. S. and Oversby, V. M. (1971) *Chemical Equilibria in the Earth.* McGraw-Hill, New York, 318 p.

Brown, T. H. and Skinner, B. J. (1974) Theoretical prediction of equilibrium phase assemblages in multicomponent systems. Amer. J. Sci. 274, 961-986.

Carmichael, I. S. E., Turner, F. J. and Verhoogen, J. (1974) Igneous Petrology. McGraw-Hill, New York, 739 p.

Charlu, T. V., Newton, R. C. and Kleppa, O. J. (1975) Enthalpies of formation at 970 K of compounds in the system $MgO-Al_2O_3-SiO_2$ from high temperature solution calorimetry. Geochim. Cosmochim. Acta 39, 1487-1497.

Charlu, T. V., Newton, R. C. and Kleppa, O. J. (1978) Enthalpy of formation of some lime silicates by high-temperature solution calorimetry, with discussion of high pressure phase equilibria. Geochim. Cosmochim. Acta 42, 367-375.

Chen, C.-H. (1975) A method of estimation of standard free energies of formation of silicate minerals at 298.15°K. Amer. J. Sci. 275, 801-817.

Chernoski, J. V., Jr. (1976) Gibbs free energy of enstatite, clinochlore, and hydrous Mg-cordierite evaluated from phase equilibrium data. Trans. Amer. Geophys. Union 57, 1201.

Currie, K. L. and Curtis, L. W. (1976) An application of multicomponent solution theory to jadeitic pyroxenes. J. Geol. 84, 179-194.

Darken, L. S. and Gurry, R. W. (1953) *Physical Chemistry of Metals.* McGraw-Hill, New York, 535 p.

Davidson, P. M., Engi, M. and Lindsley, D. H. (1979) The nature of clinopyroxene (cpx) solutions near $Di(CaMgSi_2O_6)$ in the system $En-Wo(Mg_2Si_2O_6 - Ca_2Si_2O_6)$. II. Geol. Soc. Amer. Abstr. with Programs 11, 409.

Deer, W. A., Howie, R. A. and Zussman, J. (1978) *Rock-forming Minerals*, Vol. 2A, Single-chain Silicates (second ed.), Wiley, New York, 668 p.

Denbigh, K. (1964) *The Principles of Chemical Equilibrium.* Cambridge Univ. Press, New York, 491 p.

Dienes, G. J. (1955) Kinetics of order-disorder transformations. Acta Metall. 3, 549-557.

Ernst, W. G. (1976) *Petrologic Phase Equilibria.* Freeman, San Francisco, 333 p.

Finnerty, T. A. (1977) The regular solution model and exchange equilibria: methodology and application to enstatite--diopside equilibrium. Carnegie Inst. Wash. Yearbook, 76, 550-553.

Fleet, M. E. (1974) Mg, Fe^{2+} site occupancies in coexisting pyroxenes. Contr. Mineral. Petrol. 47, 207-214.

Fraser, D. G., ed. (1977) *Thermodynamics in Geology.* (Proc. N.A.T.O. Adv. Study Inst., Oxford, 1976). D. Reidel, Boston, 410 p.

Ganguly, J. and Ghose, S. (1979) Aluminous orthopyroxene: order--disorder, thermodynamic properties, and petrologic implications. Contr. Mineral. Petrol. 69, 375-385.

Gasparik, T. and Lindsley, D. H. (1980) Experimental study of pyroxenes in the system $CaMgSi_2O_6-CaAl_2SiO_6-Ca_{0.5}AlSi_2O_6$. Trans. Amer. Geophys. Union 61, 402-403 (abstract).

Ghose, S. (1965) $Mg^{2+}-Fe^{2+}$ order in an orthopyroxene $Mg_{0.93}Fe_{1.07}Si_2O_6$. Z. Kristallogr. 122, 81-99.

Goldman, D. S. and Rossman, G. R. (1979) Determination of quantitative cation distribution in orthopyroxenes from electronic absorption spectra. Phys. Chem. Minerals 4, 43-53.

Green, E. J. (1970) Predictive thermodynamic models for mineral systems. I. Quasichemical analysis of the halite-sylvite subsolidus. Amer. Mineral. 55, 1692-1713.

Greenwood, H. J., ed. (1977) *Short Course in Application of Thermodynamics to Petrology and Ore Deposits*. Mineral. Assoc. Canada Short Course Vol. 2, Vancouver, 231 p.

Grove, T. L. and Lindsley, D. H. (1981) An experimental study of clinopyroxene/melt equilibria. Geochim. Cosmochim. Acta. (in press).

Grover, J. E. (1972a) *Two problems in pyroxene mineralogy: a theory of partitioning cations between coexisting single- and multi-site phases; and a determination of the stability of low-clinoenstatite under hydrostatic pressure*. Ph. D. Thesis, Yale Univ. (Univ. Microfilms no. 72-22387).

Grover, J. E. (1972b) The stability of low-clinoenstatite in the system $Mg_2Si_2O_6-CaMgSi_2O_6$. Trans. Amer. Geophys. Union 53, 539 (abstract).

Grover, J. (1974) On calculating activity coefficients and other excess functions from the intracrystalline exchange properties of a double-site phase. Geochim. Cosmochim. Acta 38, 1527-1548.

Grover, J. (1977) Chemical mixing in multicomponent solutions: an introduction to the use of Margules and other thermodynamic excess functions to represent nonideal behavior; pp. 67-97 in *Thermodynamics in Geology* (D. G. Fraser, ed.), D. Reidel, Boston, 410 p.

Grover, J. E., Lindsley, D. H. and Schweitzer, E. L. (1976) Calculation of thermodynamic excess parameters for the invariant three-phase assemblage diopside$_{(ss)}$-pigeonite$_{(ss)}$- orthoenstatite$_{(ss)}$. Geol. Soc. Amer. Abst. with Programs 8, 895.

Grover, J. E. and Orville, P. M. (1969) The partitioning of cations between coexisting single- and multi-site phases with application to the assemblages: orthopyroxene--clinopyroxene and orthopyroxene--olivine. Geochim. Cosmochim. Acta 33, 205-226.

Guggenheim, E. A. (1952) *Mixtures*. Clarendon Press, Oxford, 270 p.

Guggenheim, E. A. (1959) *Thermodynamics* (fourth ed.), North-Holland Pub. Co., Amsterdam, 476 p.

Haas, J. L., Jr. and Fisher, J. R. (1976) Simultaneous evaluation and correlation of thermodynamic data. Amer. J. Sci. 276, 525-545.

Hardy, H. K. (1953) A "sub-regular" solution model and its application to some binary alloy systems. Acta Metall. 1, 202-209.

Helgeson, H. C., Delany, J. M., Nesbitt, H. W., and Bird, D. K. (1978) Summary and critique of the thermodynamic properties of rock-forming minerals. Amer. J. Sci. 278-A, 1-229.

Holland, T. J. B., Navrotsky, A. and Newton, R. C. (1979) Thermodynamic parameters of $CaMgSi_2O_6-MgSi_2O_6$ pyroxenes based on regular solution and cooperative disordering models. Contr. Mineral. Petrol. 69, 337-344.

Kerrick, D. M. and Darken, L. S. (1975) Statistical thermodynamic models for ideal oxide and silicate solid solutions, with application to plagioclase. Geochim. Cosmochim. Acta 39, 1431-1442.

King, M. B. (1969) *Phase Equilibrium in Mixtures*. Pergamon Press, Oxford, 584 p.

Kitayama, K. (1970) Activity measurements in orthosilicate and metasilicate solid solutions. II. $MgSiO_3-FeSiO_3$ at 1154, 1204, and 1250°C. Bull. Chem. Soc. Japan 43, 1390-1393.

Kitayama, K. and Katsura, T. (1968) Activity measurements in orthosilicate and metasilicate solid solutions. I. $Mg_2SiO_4-Fe_2SiO_4$ and $MgSiO_3-FeSiO_3$ at 1204°C. Bull. Chem. Soc. Japan 41, 1146-1151.

415

Kleppa, O. J. and Newton, R. C. (1975) The role of solution calorimetry in the study of mineral equilibria. Fortschr. Mineral. 52, 3-20.

Ko, H. C., Ferrante, M. J., and Stuve, J. M. (1978) Thermophysical properties of acmite. *Proc. 7th Symposium on Thermophysical Properties,* Washington, D. C., 10-12 May, 1977. Amer. Soc. Mech. Engineers, New York, pp. 392-395.

Kohler, F. (1960) Zur Berechnung der Thermodynamischen Daten eines ternären Systems aus dem zugehörigen binären System. Monatsh. Chem. 91, 738-740.

Kretz, R. (1961) Some applications of thermodynamics to coexisting minerals of variable composition. Examples: orthopyroxene--clinopyroxene and orthopyroxene--garnet. J. Geol. 69, 361-387.

Kushiro, I. (1964) The system diopside-forsterite-enstatite at 20 kilobars. Carnegie Inst. Wash. Year Book 63, 101-108.

Lindsley, D. H. and Dixon, S. A. (1976) Diopside--enstatite equilibria at 850° to 1400°C, 5 to 35 kb. Amer. J. Sci. 276, 1285-1301.

Lindsley, D. H. and Grover, J. E. (1978) The nature of clinopyroxene (Cpx) solutions near $CaMgSi_2O_6$ in the system $Mg_2Si_2O_6$-$Ca_2Si_2O_6$. Geol. Soc. Amer. Abst. with Programs 10, 445.

Lindsley, D. H., Grover, J. E. and Davidson, P. M. (1981) The thermodynamics of the $Mg_2Si_2O_6$-$CaMgSi_2O_6$ join: a review and a new model. *Advances in Physical Geochemistry, Vol. 1* (R. C. Newton, A. Navrotsky and B. J. Wood, eds.), Springer-Verlag, New York, pp. 149-175.

Longhi, J. and Boudreau, A. E. (1980) The orthoenstatite liquidus field in the system forsterite - diopside - silica at one atmosphere. Amer. Mineral. 65, 563-573.

Luth, W. C. and Fenn, P. M. (1973) Calculation of binary solvi with special reference to the sanidine--high albite solvus. Amer. Mineral. 58, 1009-1015.

Matsui, Y. and Banno, S. (1965) Intracrystalline exchange equilibrium in silicate solid solutions. Proc. Japan Acad. 41, 461-466.

Muan, A. (1967) Determination of thermodynamic properties of silicates from locations of conjugation lines in ternary systems. Amer. Mineral. 52, 797-804.

Mueller, R. F. (1962a) Energetics of certain silicate solid solutions. Geochim. Cosmochim. Acta 26, 581-598.

Mueller, R. F. (1962b) Theory of homologous sublattices and intracrystalline equilibria. Science 137, 540-541.

Mueller, R. F. (1967) Model for order-disorder kinetics in certain quasi-binary crystals of continuously variable composition. J. Phys. Chem. Solids 28, 2239-2243.

Mueller, R. F. (1969) Kinetics and thermodynamics of intracyrstalline distributions. Mineral. Soc. Amer. Spec. Paper 2, 83-93.

Mueller, R. F. and Saxena, S. K. (1977) *Chemical Petrology.* Springer-Verlag, New York, 394 p.

Nafziger, R. H. and Muan, A. (1967) Equilibrium phase compositions and thermodynamic properties of olivines and proxenes in the system MgO-'FeO'-SiO_2. Amer. Mineral. 52, 1364-1385.

Navrotsky, A. (1971a) The intracrystalline cation distribution and the thermodynamics of solid solution formation in the system $FeSiO_3$-$MgSiO_3$. Amer. Mineral. 56, 201-211.

Navrotsky, A. (1971b) Thermodynamics of formation of the silicates and germanates of some divalent transition metals and of magnesium. J. Inorg. Nucl. Chem. 33, 4035-4050.

Navrotsky, A. (1977) Geological applications of high temperature reaction claorimetry; pp. 1-10 in Thermodynamics in Geology (D. G. Fraser, ed.), D. Reidel, Boston, 410 p.

Navrotsky, A. (1978) Thermodynamics of element partitioning: (1) systematics of transition metals in crystalline and molten silicates and (2) defect chemistry and "the Henry's Law problem." Geochim. Cosmochim. Acta 42, 887-902.

Navrotsky, A. and Coons, W. E. (1976) Thermochemistry of some pyroxenes and related compounds. Geochim. Cosmochim. Acta 40, 1281-1288.

Navrotsky, A. and Kleppa, O. J. (1967) The thermodynamics of cation distributions in simple spinels. J. Inorg. Nucl. Chem. 29, 2701-2714.

Navrotsky, A. and Loucks, D. (1977) Calculation of subsolidus phase relations in carbonates and pyroxenes. Phys. Chem. Minerals 1, 109-127.

Nehru, C. E. and Wyllie, P. J. (1974) Electron microprobe measurement of pyroxenes coexisting with H_2O-understaurated liquid in the join $CaMgSi_2O_6-MgSi_2O_6-H_2O$ at 30 kilobars, with applications to geothermometry. Contr. Mineral. Petrol. 48, 221-228.

Newton, R. C., Charlu, T. V., Anderson, P. A. M., and Kleppa, O. J. (1979) Thermochemistry of synthetic clinopyroxenes on the join $CaMgSi_2O_6-Mg_2Si_2O_6$. Geochim. Cosmochim. Acta 43, 55-60.

Newton, R. C., Charlu, T. V. and Kleppa, O. J. (1977) Thermochemistry of high pressure garnets and clinopyroxenes in the system $CaO-MgO-Al_2O_3-SiO_2$. Geochim. Cosmochim. Acta 41, 369-377.

Powell, M. and Powell, R. (1974) An olivine--clinopyroxene geothermometer. Contr. Mineral. Petrol. 48, 249-263.

Powell, R. (1974) A comparison of some mixing models for crystalline silicate solid solutions. Contr. Mineral. Petrol. 46, 265-274.

Powell, R. (1978) The thermodynamics of pyroxene geotherms. Phil. Trans. R. Soc. Lond. A 288, 457-469.

Prigogine, I. and Defay, R. (1954) *Chemical Thermodynamics*. (D. H. Everett, transla.) Longmans, London, 543 p.

Redlich, O. and Kister, A. T. (1948a) Thermodynamics of nonelectrolyte solutions. x-y-t relations in a binary system. Indus. Eng. Chem. 40, 341-345.

Redlich, O. and Kister, A. T. (1948b) Thermodynamics of nonelectrolyte solutions. Algebraic representation of thermodynamic properties and the classification of solutions. Indus. Eng. Chem. 40, 345-348.

Robie, R. A., Hemingway, B. S. and Fisher, J. R. (1978) Thermodynamic properties of minerals and related substances at 298.15 K and 1 bar (10^5 Pascals) pressure and at higher temperatures. U. S. Geol. Survey Bull. 1452, 456 p.

Robie, R. A. and Waldbaum, D. R. (1968) Thermodynamic properties of minerals and related substances at 298.15°K (25.0°C) and one atmosphere (1.013 bars) pressure and at higher temperatures. U. S. Geol. Surv. Bull. 1259, 256 pp.

Sack, R. O. (1980) Some constraints on the thermodynamic mixing properties of Fe-Mg orthopyroxenes and olivines. Contr. Mineral. Petrol. 71, 257-269.

Sato, M. (1971) Electrochemical measurements and control of oxygen fugacity and other gaseous fugacities with solid electrolyte sensors. pp 43-99 in *Research Techniques for High Pressure and High Temperature* (G. C. Ulmer, ed.), Springer-Verlag, New York, 367 p.

Saxena, S. K. (1973) *Thermodynamics of Rock-Forming Crystalline Solutions*. Springer-Verlag, New York, 188 p.

Saxena, S. K. (1976a) Entropy estimates for some silicates at 298°K from molar volumes. Science 193, 1241-1242.

Saxena, S. K. (1976b) Two-pyroxene geothermometer: a model with an approximate solution. Amer. Mineral. 61, 643-652.

Saxena, S. K. and Ghose, S. (1970) Order--disorder and the activity--composition relation in a binary crystalline solution. I. Methamorphic orthopyroxene. Amer. Mineral. 55, 1219-1225.

Saxena, S. K. and Ghose, S. (1971) $Mg^{2+}-Fe^{2+}$ order-disorder and the thermodynamics of the orthopyroxene-crystalline solution. Amer. Mineral. 56, 532-559.

Saxena, S. K., Ghose, S. and Turnock, A. C. (1974) Cation distribution in low-calcium pyroxenes: dependence on temperature and calcium content and the thermal history of lunar and terrestrial pigeonites. Earth Planet. Sci. Lett. 21, 194-200.

Saxena, S. K. and Nehru, C. E. (1975) Enstatite--diopside solvus and geothermometry. Contr. Mineral. Petrol. 49, 259-267.

Schweitzer, E. (1977) *The reaction: pigeonite = diopside + enstatite at 15 kbar*. M. S. Thesis, State Univ. of New York at Stony Brook, 25 p.

Shearer, J. A. and Kleppa, O. J. (1973) The enthalpies of formation of $MgAl_2O_4$, $MgSiO_3$, Mg_2SiO_4 and Al_2SiO_5 by oxide melt solution calorimetry. J. Inorg. Nucl. Chem. 35, 1073-1078.

Sundquist, B. E. (1966) The calculation of thermodynamic properties of miscibility--gap systems. Trans. Met. Soc. AIME 236, 1111-1122.

Swalin, R. A. (1972) *Thermodynamics of Solids* (second ed.), Wiley-Interscience, New York, 387 p.

Tardy, Y. and Garrels, R. M. (1977) Prediction of Gibbs energies of formation of compounds from the elements-II. Monovalent and divalent metal silicates. Geochim. Cosmochim. Acta 41, 87-92.

Thompson, A. B., Perkins, D., III, Sonderegger, U., and Newton, R. C. (1978) Heat capacities of synthetic $CaAl_2SiO_6$-$CaMgSi_2O_6$-$Mg_2Si_2O_6$ pyroxenes. Trans. Amer. Geophys. Union 59, 395.

Thompson, J. B., Jr. (1967) Thermodynamic properties of simple solutions. *Researches in Geochemistry*, Vol. 2 (P. H. Abelson, ed.), pp. 340-361, Wiley, New York.

Thompson, J. B., Jr. (1969) Chemical reactions in crystals. Amer. Mineral. 54, 341-375.

Thompson, J. B., Jr. (1970) Chemical reactions in crystals: corrections and clarification. Amer. Mineral. 55, 528-532.

Torgeson, D. R. and Sahama, Th. G. (1948) A hydrofluoric acid solution calorimeter and the determination of the heats of formation of Mg_2SiO_4, $MgSiO_3$, and $CaSiO_3$. J. Amer. Chem. Soc. 70, 2156-2160.

Turner, F. J. and Verhoogen, J. (1960) *Igneous and Metamorphic Petrology*, 2nd edition, McGraw-Hill, New York, 694 p.

Virgo, D. and Hafner, S. S. (1969) Fe^{2+}, Mg order-disorder in heated orthopyroxenes. Mineral. Soc. Amer. Spec. Paper 2, 67-81.

Warner, R. D. and Luth, W. C. (1974) The diopside-orthoenstatite two-phase region in the system $CaMgSi_2O_6$-$Mg_2Si_2O_6$. Amer. Mineral. 59, 98-109.

Wells, P. R. A. (1977) Pyroxene thermometry in simple and complex systems. Contr. Mineral. Petrol. 62, 129-139.

Williams, R. J. (1971) Reaction constants in the system $Fe-MgO-SiO_2-O_2$ at 1 atm between 900° and 1300°C: experimental results. Amer. J. Sci. 270, 334-360.

Wisniak, J. and Tamir, A. (1978) *Mixing and Excess Thermodynamic Properties*. Elsevier, New York, 935 p.

Wohl, K. (1946) Thermodynamic evaluation of binary and ternary liquid systems. Trans. Amer. Inst. Chem. Eng. (Chem. Eng. Progress) 42, 215-249.

Wohl, K. (1953) Thermodynamic evaluation of binary and ternary liquid systems. Chem. Eng. Progress 49, 218-219.

Wood, B. J. (1976) An olivine-clinopyroxene geothermometer: a discussion. Contr. Mineral. Petrol. 56, 297-303.

Wood, B. J. and Banno, S. (1973) Garnet--orthopyroxene and orthopyroxene--clinopyroxene relationships in simple and complex systems. Contr. Mineral. Petrol. 42, 109-124.

Wood, B. J. and Fraser, D. G. (1976) *Elementary Thermodynamics for Geologists*. Oxford Univ. Press, 303 p.

Wood, B. J. and Henderson, C. M. B. (1978) Compositions and unit-cell parameters of synthetic non-stoichiometric tschermakitic clinopyroxenes. Amer. Mineral. 63, 66-72.

Zen, E-An and Chernosky, J. V., Jr. (1976) Correlated free energy values of anthophyllite, brucite, clinochrysotile, enstatite, forsterite, quartz, and talc. Amer. Mineral. 61, 1156-1166.

Chapter 9

THE COMPOSITION SPACE of TERRESTRIAL PYROXENES –
INTERNAL and EXTERNAL LIMITS Peter Robinson

INTRODUCTION

Understanding of the composition space of pyroxenes is central to studies of a very wide spectrum of terrestrial rocks. This is so because pyroxene compositions, like feldspar compositions, occupy a position that is chemically central to the composition realm of rocks. Thus, there are relatively few rock bulk compositions in which pyroxene cannot be formed under any set of P, T, and volatile conditions that could be found within the crust or mantle of the earth.

Quite clearly understanding of pyroxene composition space and its limits as a function of environmental conditions must be based on the crystal chemistry, thermodynamic properties, and experimental data on the pyroxenes themselves as covered elsewhere in this volume. It must also be based equally on understanding of these factors for all of the other phases with which pyroxenes interact in the terrestrial realm. Because of the chemically central location of pyroxene composition space, the number of such other interacting phases is large, leading to endless complexity. What follows is an attempt to outline the various kinds of limits and controls on pyroxene composition space that appear to exist in the terrestrial environment and to give specific examples of a few of them. A much more encyclopedic but differently organized coverage is presented in Deer *et al.* (1978). These limits and controls can be broadly classified as follows:

I   Internal limits of pyroxene composition space governed by coexistence of more than one pyroxene.

II   External limits governed by coexistence with other solid phases with pyroxene-like stoichiometry including other single chain silicates, garnet, etc.

III External limits governed by coexistence with two or more other phases connected by tie lines that pass through composition space of pyroxene-like stoichiometry.

IV Controls on pyroxene compositions exerted by combinations of phases all of which lie on the same side of pyroxene composition space.

Phases involved in III and IV may include melt or hydrous, carbonate and other phases whose presence is controlled by the activity of a volatile constituent.

In its very simplest form, ignoring minor or trace elements, pyroxene composition space must exist in eight to ten dimensions, thus ideally lending itself to algebraic or vectorial manipulation. There are undoubtedly great advances to be made by workers with talent in this area, though they will also require special talent in communicating their conclusions. Fortunately or not, the present author has little facility in this field[1] and will rely more heavily on graphical representations that are more closely related to the tool kit of the average geologist. These two- and quasi-three-dimensional images of the ten-dimensional pyroxene composition elephant help us to understand the truth in much the same way that a two-dimensional photograph helps with the real elephant, even though certain features are lost.

GENERAL REVIEW OF PYROXENE COMPOSITION SPACE

The Pyroxene Formula

The presentation of any pyroxene composition must be based on a sound structural formula. The structural formula is only as good as the analysis that backs it up. In my opinion it is the responsibility of any mineralogist who is publishing pyroxene data to measure his analysis against a structural formula and discuss the result.

Elsewhere in this volume Cameron and Papike have outlined the criteria they used to evaluate 405 chemical analyses[2], mostly based on wet chemistry, published in Deer *et al.* (1978). They accepted an analysis if:

(1) Si + Al in tetrahedral positions is greater than 1.979 and less than 2.021.

(2) The sum of permissible octahedral (M1) cations, $^{VI}Al + Fe^{3+} + Cr^{3+} + Ti^{4+} + Mg + Fe^{2+} + Mn$, is greater than 0.98.

----

[1]Possibly a result of youthful rebellion because both his parents were mathematicians!

[2]Deer *et al.* (1978) list 406 analyses excluding spodumene; however, one analysis is listed twice as Titanaugite, Table 36, Analysis 8; and as Fassaite, Table 39, Analysis 16. In succeeding pages all analyses from Deer *et al.* (1978) will be referred to by page and number, thus 320-8 and 406-16.

(3) The sum of cations assigned to M2 is greater than 0.979 and less than 1.021.

(4) The charge imbalance ("residual") of $X + Z \simeq Y$ is less than 0.030, where $X = $ Na in M2, $Z = $ Al in tetrahedral positions, and $Y = {}^{VI}Al + Cr + Fe^{3+} + 2Ti$ in M1, and where site excesses and vacancies are ignored.

On this basis Cameron and Papike accepted 175 analyses and rejected 230. In my own evaluation, based on computer print-out kindly supplied by these authors, 25 additional analyses were accepted by permitting substitution of Fe^{3+} in tetrahedral positions as suggested by Huckenholz *et al.* (1969), and an additional 92 analyses were accepted by permitting unlimited vacancies[3] in M2 (i.e., occupancy less than 0.979 provided above) as suggested by experimental work by Kushiro (1969), Wood (1976) and Gasparik and Lindsley (1980). Surprisingly, 44 out of 72 "accepted" Na pyroxenes are among those accepted under this cation-deficient provision.

The bulk of recent pyroxene analyses have been done by electron probe, in which there is no direct measurement of the oxidation state of Fe. The importance of both Fe^{2+} and Fe^{3+} is shown by the wet chemical analyses, and an assumption that all Fe is Fe^{2+} is clearly unfounded except in the most special circumstances. Barring the possibility of obtaining a small separate for wet determination of FeO, the best procedure seems to be to estimate the composition on the assumption of a perfectly stoichiometric formula with four cations and six oxygens. This ignores the possibility of cation deficiency and also the possibility of H_4 substitution for Si as proposed by Sclar *et al.* (1968). While this calculation is a very simple matter of arithmetic, it seems so prone to mistake and misunderstanding that it is worth reviewing here.

The first step is to normalize the analysis to four cations. Then oxygen is assigned to each cation according to valence, two for R^{4+} cations, 1.5 for R^{3+} cations, one for R^{2+} cations, and 0.5 for R^{1+} cations. In this Fe is treated as R^{2+}. The oxygen is now summed. If the oxygen sum is less than six, then an appropriate amount of oxygen is *added* to bring the sum to six. For each amount added, *twice* that amount of Fe is

---

[3] In terms of charge balance a vacancy has the same effect as the substitution of 2Na for Ca in M2. The charge "imbalance" criterion as expressed in (4) above must also be relaxed.

converted from Fe^{2+} to Fe^{3+}. If not enough Fe is available for this, it most probably indicates analytical error unless there is reason to suspect the presence of Mn^{3+}. If the oxygen sum is greater than six for terrestrial pyroxene, this also probably indicates analytical error. In the case of a lunar or meteoritic pyroxene, where Ti^{3+} is probable (Dowty and Clark, 1973), oxygen is subtracted to bring the sum to six. For each such amount subtracted *twice* that amount of Ti is converted from Ti^{4+} to Ti^{3+} (Tracy and Robinson, 1977)[4].

For a perfectly stoichiometric pyroxene and a perfectly accurate analysis, the above method of calculating Fe^{3+} should be perfectly accurate. Unfortunately, the method is *extremely* sensitive to errors in analysis of major elements, in particular SiO_2, so that it has given frustrating results in many cases. Cawthorn and Collerson (1974) gave it up for a more empirical method for Na pyroxenes because many wet analyses of Na pyroxenes show apparent cation deficiency. Mitchell (1980) assumed that Fe^{3+} substitution is present only to the extent of Na substitution in a series of pyroxenes from carbonatites, thus arbitrarily ruling out the recognition of a coupled Fe^{3+} ^{VI}Al substitution that is apparent in many wet analyses (see below; also Cameron and Papike, this volume).

Most workers agree on a reasonable order for filling sites; tetrahedral, then M1, then M2; and an order for placing cations in sites; Si, Al, Fe^{3+}, Ti^{4+}, Cr, Mg, Fe^{2+}, Mn, Ca, then Na. Obviously, distribution of ions between sites, particularly Fe^{2+} and Mg, can be partly a function of temperature, but crystal structure studies show the above sequence has a strong qualitative basis.

It has become common practice to recast pyroxene structural formulae in terms of percentages of ideal end members. As will be shown below, the pyroxene composition space is a complex reciprocal space in which the choice and order of choice of ideal end members is quite arbitrary. For this reason it is recommended that emphasis be placed on the

----

[4]If a pyroxene is truly cation deficient, then normalization to four cations would result in a calculation of more oxygen than is actually present. Adjustment to six would then result in an overestimate of the amount of Ti^{3+} present or an underestimate of the amount of Fe^{3+}. If the titanaugites from Allende meteorite are in fact cation deficient, their Ti^{3+} contents may be overestimated. However, in this case there is independent spectroscopic evidence for the presence of Ti^{3+} (Dowty and Clark, 1973).

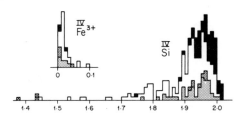

Figure 1. Histogram of inferred tetrahedral Si and Fe^{3+} in "accepted" pyroxene analyses from Deer, Howie and Zussman (1978). Tetrahedral Al is equal to 2-Si except where there is tetrahedral Fe^{3+}. Stippled pattern represents orthopyroxenes; lined pattern, pigeonites; unpatterned, Ca pyroxenes; solid, Na pyroxenes. Crossed boxes indicate selected electron probe analyses listed in Table 1.

formula and on actual cation content rather than on a series of arbitrarily chosen end members.

The Composition Range of Terrestrial Pyroxenes

A simple way to explore pyroxene compositions, without becoming involved in multidimensional space is through a set of single-element histograms based on a suite of accepted analyses. Those processed from Deer *et al.* (1978) serve as the best available example. The histograms were prepared for each of the three cation sites using the site assignment scheme outlined above. In the histograms the four broad classes--orthopyroxenes, pigeonites, Ca pyroxenes, and Na pyroxenes--are indicated separately.

For the tetrahedral sites (Fig. 1) only Si and Fe^{3+} are graphed because ^{IV}Al is the reciprocal of Si except in those few cases where Fe^{3+} is assigned to the tetrahedral site. The maximum Si is, of course, when the tetrahedral site is completely filled as is common in many pyroxenes. The minimum Si in the representative sample is 1.443 (320-8, 406-16) in a titanaugite, but recent probe analyses yielded 1.434 in a fassaite from Soufriere (Devine and Sigurdsson, 1980), and 1.430 and 1.376 in titanaugite from Tahiti (Tracy and Robinson, 1977 and unpublished data). None of these approaches the extreme low Si of 1.196 in a Ti^{3+}-rich augite from Allende meteorite (Mason, 1974; Tracy and Robinson, 1977). In spite of these extremes, there are very few pyroxenes in the representative sample with Si less than 1.7, and 1.762 is the extreme low for orthopyroxene (42-9). The maximum estimated Fe^{3+} content in the tetrahedral site is 0.092 for a titanaugite (320-11). There is at present no spectroscopic proof of this substitution in this specimen, but the experimental work of Huckenholz *et al.* (1969) strongly indicates the possibility of this substitution in Al-poor environments.

In the M1 octahedral site (Fig. 2), Al naturally reaches a maximum of 0.977 in a jadeite (464-1). In Ca pyroxene the maximum in the representative sample is the much lower value of 0.345 (403-4). The maximum for orthopyroxene is 0.150 (49-6).

423

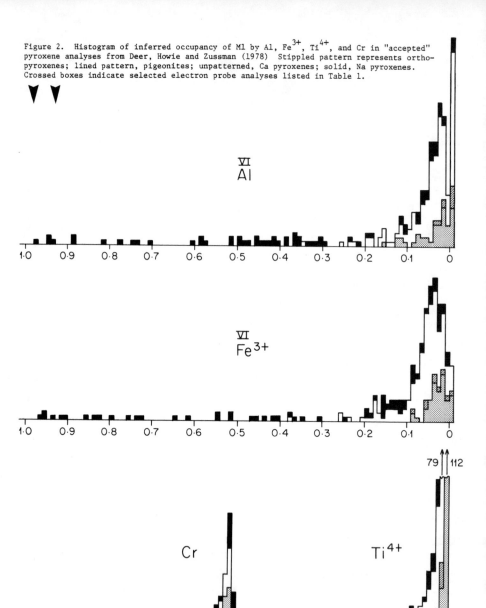

Figure 2. Histogram of inferred occupancy of M1 by Al, Fe^{3+}, Ti^{4+}, and Cr in "accepted" pyroxene analyses from Deer, Howie and Zussman (1978) Stippled pattern represents orthopyroxenes; lined pattern, pigeonites; unpatterned, Ca pyroxenes; solid, Na pyroxenes. Crossed boxes indicate selected electron probe analyses listed in Table 1.

Figure 3. Histogram of inferred occupancy of M1 and M2 by Mg, Fe^{2+}, Mn, Ca, and Na in "accepted" pyroxene analyses from Deer, Howie and Zussman (1978). Histograms, except for Na, are arranged back to back in such a way as to reflect the site assignment scheme: Mg is assigned to M2 only when Fe^{2+} in M1 is less than zero; Fe is assigned to M2 only when Mn in M1 is less than zero; Mn, if any, is assigned to M2 in the usual case where Ca in M1 is less than zero; and Na is assigned to M2. Electron probe analyses indicated by crossed squares include eulites and orthoferrosilites from Jaffe et al. (1975), in the Fe and Mg portions, and Mn pyroxenes from Table 1.

424

FIGURE 3

Legend at bottom of opposite page.

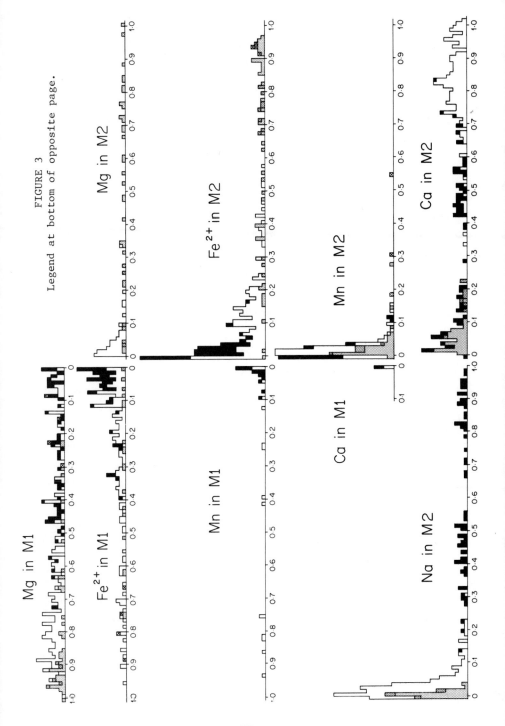

The situation for Fe^{3+} in Ml is similar to Al. The highest value of 0.966 is in acmite (487-1). In Ca pyroxene the maximum is the much lower value of 0.370 (321-5). In orthopyroxene the maximum is 0.085 (42-9). There is an interesting contrast between ^{VI}Al and $^{VI}Fe^{3+}$ in Ca pyroxenes. For Al the most frequent value is 0.000, but for Fe^{3+} it is 0.030, supporting the preponderance of $^{VI}Fe^{3+}$ ^{IV}Al over ^{VI}Al ^{IV}Al coupled substitution in many terrestrial pyroxenes (see also Cameron and Papike, this volume).

Ti^{4+} also reaches a maximum in Na pyroxene, in the representative sample in this case 0.266 (488-9), which is about half of that to be expected in the idealized $NaR^{2+}_{0.5}Ti^{4+}_{0.5}Si_2O_6$ end member. In Ca pyroxenes of the representative sample Ti reaches 0.165 (320-8, 406-16) in the same sample that has minimum tetrahedral Si. However, recent probe analyses give 0.252 and 0.282 (Tracy and Robinson, 1977 and unpublished data) in the same analyses that give Si of 1.430 and 1.376, or about half of an idealized $CaR^{2+}_{0.5}Ti^{4+}_{0.5}AlSiO_6$ end member. Despite these extremes the most common Ca pyroxene has Ti^{4+} of about 0.005. Orthopyroxene reaches 0.043 (40-30).

In spite of the name "chrome diopside" and a table of 11 analyses, the maximum Cr content in the representative terrestrial sample is 0.058 (207-11). A "chrome omphacite" has 0.033 Cr, but an $NaCrSi_2O_6$ end member has little importance in terrestrial pyroxenes. The maximum Cr in orthopyroxene is 0.012.

Absolute occupancy of Mg, Fe^{2+}, and Mn in Ml is of somewhat secondary interest because all of these ions may also appear in M2, because the distribution of ions between these sites is a function of temperature and cooling rate, and because the site assignment scheme is somewhat arbitrary. Figure 3 is a set of histograms that shows these assigned occupancies, arranged in a way that emphasizes the site assignment scheme.

The maximum amount of Mg in the Ml site requires that there be a minimum amount of ^{VI}Al, Fe^{3+}, Cr, and Ti and also so little Fe, Mn, and Ca that it can all be accommodated in M2. In a meteoritic enstatite (50-1) the ideal value of 1.000 is achieved, but 0.983 is the largest value for a terrestrial orthopyroxene (41-1). A terrestrial diopside, however, achieves 0.999 (202-1).

The minimum Mg assigned to Ml in orthopyroxenes is governed mainly by replacement of Mg by Fe^{2+}, reaching 0.223 in the representative sample (47-34) and as low as 0.088 in some probe analyses. The failure to reach zero in orthopyroxene is related to the instability of orthoferrosilite. In calcic and sodic pyroxenes there are Mg-free end members where Mg in M1 does reach zero.

In order for Mg to be assigned to M2, the amount of Fe^{2+} assigned to M1 must be less than zero, and to achieve large amounts of Mg in M2 there can be little or no Ca and Mn, as in orthopyroxene. The maximum is 1.005 in a terrestrial enstatite (41-1, see above). The maximum in Ca pyroxene is 0.337 (317-1) in Mg-rich sub-calcic augite, but recent probe analyses would undoubtedly extend this limit because it is merely a measure of the amount of Mg-rich pigeonite in solid solution.

Fe^{2+} in M1, naturally requires Al, Fe^{3+}, Cr, Ti, and Mg to be minimal. The highest value in orthopyroxene in the representative sample is 0.757 (47-33), but recent probe analyses of orthoferrosilite yield 0.880 (Jaffe $et\ al.$, 1978). In a hedenbergite Fe^{2+} in M1 reaches 0.955 (220-13). Fe^{2+} in M2 is maximal in Fe-rich orthopyroxenes and pigeonites reaching 0.961 (47-33). The maximum Fe^{2+} in Ca pyroxene is 0.575 (318-7) but this is arbitrary because under high T conditions there is probably complete solid solution between Ca pyroxene and iron-rich pigeonite.

Mn is assigned to M1 only when there is too little Fe to assign any to M2. As can be seen from the graph, this is impossible in orthopyroxene and is only of minor importance in Ca and Na pyroxenes except johannsenites where it reaches 0.931 (417-2). Maximum Mn in M2 is apparently in kanoite, $MnMgSi_2O_6$, an Mn analog of pigeonite (Kobayashi, 1977). Maximum in an "accepted" analysis is 0.54 from Gordon $et\ al.$ (1980) in a solid solution between kanoite and diopside. Maximum Mn in M2 in orthopyroxene reaches 0.273 (45-21) in the representative sample, and up to 0.301 in recent probe analyses (Huntington, 1975). Most frequent Mn content of pyroxenes is about 0.005.

Ca is assigned to M1 in only six accepted analyses, in no case in amounts greater than 0.010, and thus is not really demonstrable within analytical error, in spite of the suggestions of Huckenholz $et\ al.$ (1969). Ca content of M2 reflects the type of pyroxene, being minimal in orthopyroxenes and Na pyroxenes, slightly higher in pigeonites, intermediate in Na-Ca pyroxenes, and maximal in Ca pyroxenes attaining 1.016 (204-7) near the maximum allowed for an accepted analysis. The maximum in an orthopyroxene is 0.163 in a bronzite (41-5).

Table 1. Some extreme compositions of pyroxenes. Numbers such as 320-8, etc. indicate page and analysis number from Deer, Howie, and Zussman, 1978, except orthopyroxene Po-17 from Jaffe et al., 1975. Other references are in text. * indicates estimated Ti^{3+} from pyroxene in Allende meteorite.

		Na	Ca	Mn	Fe^{2+}	Mg	Ca	Mn	Fe^{2+}	Mg	Ti^{4+}	Cr	Fe^{3+}	Al	Fe^{3+}	Al	Si
TETRAHEDRAL Si	Titanaugite 406-16,320-8	.007	.992	.005	.002			.227	.385	.165	.003	.128	.091		.557	1.443
	Fassaite D+S'80	.002	.979	.005	.014			.046	.408	.022		.218	.306		.566	1.434
	Titanaugite T+R'77	.074	.886	.007	.033			.171	.433	.252	.001	.098	.045		.570	1.430
	" T+RT11	.067	.889	.004	.040			.169	.422	.282	.001	.105	.021		.624	1.376
	" T+RAllende	1.000					.021		.289	.111			.394*.186		.804	1.196
	Orthopyroxene 42-9	.007	.014	.006	.625	.349			.772	.014			.085	.128		.238	1.762
TETRAHEDRAL Fe^{3+}	Augite 320-11	.016	.547		.384	.058			.877	.106			.017		.092	.133	1.775
VI Al	Jadeite 464-1	.980	.013	.001	.003			.002	.009	.002			.010	.977		.004	1.996
	Fassaite 403-4	.132	.748	.002	.072	.029			.603	.014			.037	.345		.245	1.755
	Orthopyroxene 49-6	.012	.072	.004	.212	.695			.817	.006	.010	.017	.150			.169	1.831
VI Fe^{3+}	Acmite 487-1	.971	.013	.006	.006			.016	.005	.012			.966		.004	.006	1.989
	Ferrian Augite 321-5		.777		.023	.195			.575	.014			.370	.041		.428	1.572
	Orthopyroxene 42-9	.007	.014	.006	.625	.349			.772	.014			.085	.128		.238	1.762
Ti^{4+}	Aegerine 488-9	.933	.067				.003	.107	.150	.266			.459		.025	.032	1.943
	Titanaugite 320-8	.007	.992	.005	.002			.227	.385	.165	.003	.128	.091		.557	1.443
	" T+R'77	.074	.886	.007	.033			.171	.433	.252	.001	.098	.045		.570	1.430
	" T+R T11	.067	.889	.004	.040			.169	.422	.282	.001	.105	.021		.624	1.376
	Orthopyroxene 40-30	.031	.062	.010	.895			.441	.406	.043			.039	.071		.161	1.839
Cr	Chrome Diopside 207-11	.097	.821	.003	.045	.037			.881	.008	.058	.020	.033		.035	1.965
	Chrome Omphacite 428-1	.190	.731	.002	.042			.011	.702	.011	.033	.048	.195		.038	1.962
Mg in M1	Max. in 50-1		.011		.011	.970			1.000								2.004
	Orthopyroxene 41-1	.002	.004			1.005			.983				.017		.002	.036	1.962
	Min. in 47-34			.066	.182	.749			.746	.223	.003		.010	.017		.028	1.972
	Orthopyroxene Po-17	.003	.039	.041	.918				.880	.088	.004	.002	.014		.001	.019	1.980
	Max. in Ca Pyx 202-1	1.001	.001	.002	.012				.999						.018	.003	1.976
Fe^{2+} in M1	Orthopyroxene 47-33	.002	.035	.005	.961				.757	.235	.001		.006		.012	1.988
	" Po-17	.003	.039	.041	.918				.880	.088	.004	.002	.014		.001	.019	1.980
	Ca Pyx 220-13	.021	.836	.009	.140				.955	.009	.030		.006		.012	.046	1.943
Mn in M1	Johannsenite 417-2		.981	.030				.931	.024	.033			.001	.011		.034	1.966
Ca in M1	Diopside 203-6	.013	.982				.009		.047	.884	.001		.016	.043		.039	1.961
Mg in M2	Orthopyroxene 41-1	.002	.004			1.005			.983				.017		.002	.036	1.962
	Ca Pyx 317-1	.057	.438	.005	.158	.337			.814	.008	.018	.017	.143		.126	1.874
Fe^{2+} in M2	Orthopyroxene 47-33	.002	.035	.005	.961				.757	.235	.001		.006		.012	1.988
	Ca Pyx 318-7	.018	.361	.027	.575				.264	.641	.017		.049	.029		.054	1.946
Mn in M2	Orthopyroxene 45-21			.020	.273	.704			.290	.668	.009		.033		.013	.033	1.954
	Huntington,1975	.006	.028	.301	.667				.758	.236	.004	.002			.006	1.994
	Ca Pyx,Tracy et al.'80	.020	.806	.170				.112	.495	.385				.008			2.000
	Kanoite,Gordon et al.'80	'	.44	.54				.16	.02	.82							2.00
Ca in M2	Diopside 204-7		1.016	.001					.032	.830	.004		.034	.100		.177	1.823
	Orthopyroxene 41-5	.013	.163	.009	.395	.414			.921	.023	.002	.025	.030		.075	1.925
Na in M2	Jadeite 464-1	.980	.013	.001	.003			.002	.009	.002			.010	.977		.004	1.996
	Orthopyroxene 38-16	.034	.106	.015	.530	.335			.975	.006			.019		.018	.020	1.963

428

Na content of M2 reflects substitution of Na pyroxene end members and reaches 0.980 (464-1) in jadeite. The most frequent Na content of orthopyroxene and Ca pyroxene is about 0.005, and the maximum Na content of orthopyroxene is 0.034 (38-16).

In the frequency discussion a series of extreme compositions has been described that crudely outlines the extreme limits of terrestrial pyroxene space. The structural formulae of these compositions are given in Table 1. There are undoubtedly other extremes in the literature that the author has missed, but this listing will serve as a starting point for the research of pyroxene extremists.

Terrestrial Pyroxene Composition Space and Its Population

A two-dimensional view of pyroxene composition space may be obtained in a triangular plot of quadrivalent *vs* trivalent *vs* divalent oxides (Fig. 4). The quadrivalent apex is SiO_2 (quartz). The trivalent apex includes Al_2O_3, Fe_2O_3 and Cr_2O_3 (corundum, hematite, eskolaite). TiO_2 might be successfully included with SiO_2, but the crystal chemistry of pyroxenes shows that it is inextricably associated with Al_2O_3 substitution. For purposes of this study it was convenient to create a "mean trivalent ion" and oxide by combining equal parts of R^{2+} and Ti^{4+} to give the oxide $R^{2+}Ti^{4+}O_3$. Tracy and Robinson (1977) have shown a relationship between Fe^{2+} and Ti^{4+} in some titanaugites, suggesting this oxide may conveniently be thought of as $Fe^{2+}Ti^{4+}O_3$ (ilmenite). As a practical matter in plotting an analysis, the amount of this oxide is equal to two times the amount of Ti^{4+}, hence "2Ti" in the plotting parameter. At the same time an amount of R^{2+} equivalent to the amount of Ti must be subtracted from the R^{2+} apex, hence "-Ti" in the plotting parameter.

The divalent apex includes CaO, MnO, FeO, and MgO. The important monovalent ion Na^+ is seemingly left out in this process, but in fact its substitution is *always* linked with substitution of trivalent ions and hence can be combined with R^{3+} to give a "mean divalent ion" $Na_{0.5}R^{3+}_{0.5}O$. For the practical purpose of plotting, the amount of this ion is equal to two times the amount of Na, hence "2Na" in the plotting parameter. At the same time an amount of R^{3+} equivalent to the amount of Na must be subtracted from the R^{3+} apex, hence "-Na" in the plotting parameter. Li^+ substitution can also be accommodated in the same manner if desired.

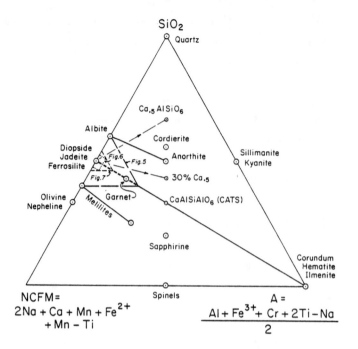

Figure 4. Composite triangular projection of the major oxide composition of pyroxenes and other minerals (see text). Heavy dashed line within inset triangle labelled "Fig. 5" indicates composition range of natural terrestrial pyroxenes.

In summary, the plotting parameters used to convert a structural formula into a plotted point are:

For SiO_2 (abbreviated S): Si

For R_2O_3 (abbreviated NCFM): $\dfrac{Al+Fe^{3+}+Cr+2Ti-Na}{2}$[5]

For RO (abbreviated NCFM): 2Na+Ca+Mn+Fe+Mg-Ti

For RO without Na component (abbreviated CFM): Ca+Mn+Fe+Mg-Ti

For Na component (abbreviated N): 2Na

In Figure 4 any pyroxene conforming to a stoichiometric formula must lie on a line that extends from halfway between S and NCFM toward A. This line exactly cuts the chemical composition space in half, explaining simply

----

[5]Division by two is needed to convert from the cation basis to the traditional oxide basis. For each Al there are half as many Al_2O_3, etc. For some purposes it may be an advantage to remain on the cation basis, in which case this division by two is eliminated. All other parameters remain unchanged.

why pyroxenes are bound to occur in such a wide variety of rocks. Since half the pyroxene cation sites are tetrahedral, the only way for an analysis to drop below this line is for there to be tetrahedral vacancies, which seem to be ruled out on crystal chemical grounds. For an analysis to rise above the line, there must be vacancies in the M1 or M2 sites. The latter has been proposed (Kushiro, 1969; Bell and Mao, 1971; Wood, 1976; Gasparik and Lindsley, 1980) to explain high SiO_2 pyroxenes synthesized at high temperature, and Gasparik and Lindsley have synthesized several compositions between diopside and about 30% of a $Ca_{0.5}AlSi_2O_6$ "end member" (see Fig. 4).

As will be considered in more detail shortly, all natural terrestrial pyroxenes fall within the larger inset triangle (labelled "Fig. 5") within Figure 4. This is so even though the synthesis and stability relations of the pyroxene $CaAlSiAlO_6$ have been fairly extensively studied (Hays, 1966). Other minerals that fall on the pyroxene line include pyroxenoids, garnet (which can be thought of as a "polymorph" of some extreme pyroxene compositions), corundum, hematite, and ilmenite. In this light it is also convenient to consider suggested and synthesized high P polymorphs of simple (NCFM + S) pyroxene compositions with garnet and ilmenite structures, and also to add perovskite to the A apex. Experimental and analytical studies of phase relations within this "line" are crucial to an extensive segment of geothermometry and geobarometry, and a cornerstone of our understanding of relations in the upper mantle as exemplified by the imaginative pioneering studies of F. R. Boyd (1970, 1973).

Important volatile-free phases below the pyroxene line include olivine, nepheline, all spinel group minerals, melilites, and sapphirine. Important phases above the pyroxene line include quartz, all plagioclase feldspars, cordierite, and Al-silicate polymorphs. Obviously pyroxene compositions and stabilities must be restricted by the coexistence of mineral pairs on opposite sides of the pyroxene line. For example: coexistence of quartz and fayalite restricts the stability of ferrosilite. Coexistence of albite and nepheline restricts the stability of jadeite. Coexistence of plagioclase and olivine restricts the amount of Al substitution in pyroxene as does the coexistence of sapphirine and cordierite (Anastasiou and Seifert, 1972) or plagioclase and spinel. Sets of minerals all on one side of the pyroxene line exert control on the composition of the pyroxene without placing a limit on its stability. Examples would be

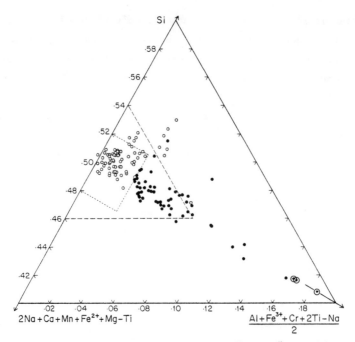

Figure 5. Blow-up of inset in Figure 4 showing "accepted" compositions of Na pyroxenes (open symbols) and highly A-substituted Ca pyroxenes (closed symbols). Dotted circles indicate extreme probe analyses listed in Table 1.

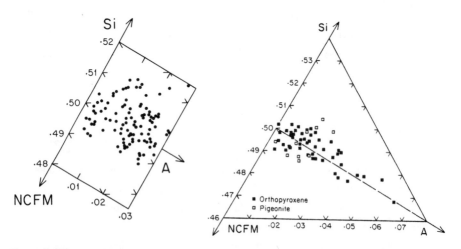

Figure 6. Blow-up of inset in Figure 4 showing "accepted" analyses of weakly A-substituted Ca pyroxenes from Deer, Howie and Zussman (1978).

Figure 7. Blow-up of inset in Figure 4 showing "accepted" analyses of orthopyroxenes (closed symbols) and pigeonites (open symbols) from Deer, Howie and Zussman (1978).

the Al content of pyroxenes in equilibrium with olivine and spinel, the jadeite content of pyroxenes in equilibrium with albite and quartz, or the Al content of orthopyroxenes in equilibrium with cordierite and quartz. Finally, the composition of silicate melt with respect to the pyroxene line and to other phases can exert control on the limits of stability and composition variations of pyroxenes. Examples of some of these relations will be considered below as time and space permits.

Figure 5 is the inset in Figure 4 that shows all Ca pyroxenes with extreme A substitutions and all Na pyroxenes (with N/NCFM > 0.2) in the representative sample. The remaining A-poor Ca pyroxenes are shown in a rectangular inset in Figure 6, and all orthopyroxenes are shown in a triangular inset in Figure 7. All "accepted" terrestrial pyroxene analyses fall well within the space of Figure 5. The distance that analyses plot below the pyroxene line is restricted by the criterion that ions assigned to M2 cannot exceed 1.020. There is no such limit to analyses above the line because cation deficiencies in M2 were permitted. As pointed out previously, most of the analyses exhibiting this deficiency are Na rich. This apparent dependence is explored in Figure 8. The relations shown are fairly likely to be a function of analytical difficulty, but the whole question of such postulated vacancies needs to be explored by crystallographers. The amount of vacancy indicated by the most "vacant" Na pyroxene is comparable to that in the most "vacant" compositions produced by Gasparik and Lindsley (1980).

The four pyroxenes with the largest amount of A in Figure 5 happen to be the first four in Table 1 and are in this position because of extreme tetrahedral Al substitution. In three of these (320-8, 406-16, A = 15.95; T + R '77, A = 16.69; and T + R unpublished, A = 18.48) the tetrahedral Al is largely compensated by Ti^{4+}, and these three are also listed as the most Ti-rich augites. The fourth (D + S '80, A = 16.48) is compensated by nearly equal amounts of Al and Fe^{3+} but is maximal in neither. It does have the highest ^{VI}Al of any of the eight analyses closest to the A apex. The augite with largest amount of ^{VI}Al (Table 1, 403-4) has A = 7.03. What this data points to is that extreme values of A substitution are favored by extreme Ti^{4+} and Fe^{3+}, and that such extreme substitutions with pure ^{VI}Al are ruled out in most environments by stable coexistence of olivine plus Ca-plagioclase of equivalent compositions.

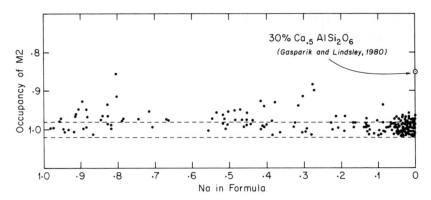

Figure 8. Plot of occupancy of M2 *vs* Na in formula for "accepted" analyses from Deer, Howie and Zussman (1978). Dashed lines are limits of 1.020 and 0.980 set for "accepted" analyses by Cameron and Papike (this volume). The limit at 0.980 was relaxed in this study to permit larger vacancies as proposed in several studies.

"A" substitutions in orthopyroxenes are much less extensive than in Ca pyroxenes and reach a maximum of 6.27 in analysis 42-9 (Table 1). In this analysis ^{VI}Al does predominate over Fe^{3+}, but Fe^{3+} is also significant. Pigeonites, as expected, do not have significant A substitutions.

An important expansion of the pyroxene line is into a triangular plane with the important N component treated separately. This, of course, expands the triangle of Figure 4 into a tetrahedron as shown in Figure 9, allowing separate portrayal of nepheline and olivine and better portrayal of the plagioclase series. The pyroxene plane itself is also separated out in Figure 9. It is important to note that for correct portrayal of this plane in the same terms that it appears in the tetrahedron, appropriate amounts of Si must be added to the N and CFM apices. Within the plane, a heavy line from $CaAlSiAlO_6$ to the N apex marks a stoichiometric limit beyond which pyroxene compositions would be possible only by substituting Al and other small R^{3+} cations into M2. A more lightly dashed line outlines the limit of most natural terrestrial samples. A few natural compositions are indicated including the extreme titanaugite compositions from Tahiti that exceed the generalized limit.

Within this simple triangle in Figure 9 are illustrated the major complex compositional variations of stoichiometric pyroxenes. In the lower right are illustrated the purely divalent or "quadrilateral" pyroxenes ("QUAD" of Cameron and Papike, this volume). Toward the top are the substitutions permitted by placing Al^{3+} and in rare cases Fe^{3+} for Si^{4+} in tetrahedral positions. Toward the left are the substitutions permitted

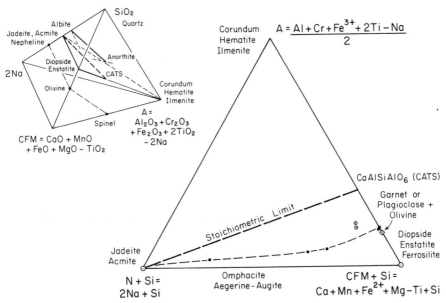

Figure 9. Expansion of composite triangular projection into a tetrahedron (upper left) from which the pyroxene triangular plane (lower right) has been extracted. For key to plotting parameters see text. Tetrahedron shows relation of pyroxene composition plane to various minerals or mineral groups with higher and lower SiO_2 content. Heavy dashed line in triangular plane shows stoichiometric limit for pyroxenes beyond which R^{3+} substitution in M2 would be required. Composition of garnet, which is also the piercing point for the anorthite plus olivine tie line is shown by small square. Light-dashed line is approximate upper limit of A substitution in terrestrial samples. Circled points are probe analyses of extreme titanaugites from Tahiti (Table 1).

by placing Na^{1+} for R^{2+} in M2 positions. Together these two substitutions produce what Cameron and Papike call "others" end members. Both substitutions are charge-reducing substitutions and must be compensated by charge-increasing substitutions. Fortunately, there is only a single type of charge-increasing substitution, namely, that of R^{3+} (or mean R^{3+}) ions for R^{2+} in M1. In this, the pyroxenes are fortunately much simpler than the amphiboles.

Figure 10 is a detailed plot, reoriented to orthogonal coordinates (see inset), of the representative sample within the pyroxene plane. Except for extreme A-rich compositions, no attempt was made to plot analyses with N/NCFM less than 0.01. As expected, there is a very high concentration of analyses near the CFM corner, and a nearly complete spread of analyses toward the jadeite-acmite apex with a possible gap between 0.6 and 0.7. Note that elevation of points above the base line indicates the amount of their R^{3+} substitution compensated by tetrahedral Al, because Na compensation has been subtracted from the plotting parameter. Numerous Na-rich pyroxenes have substantial amounts of the A

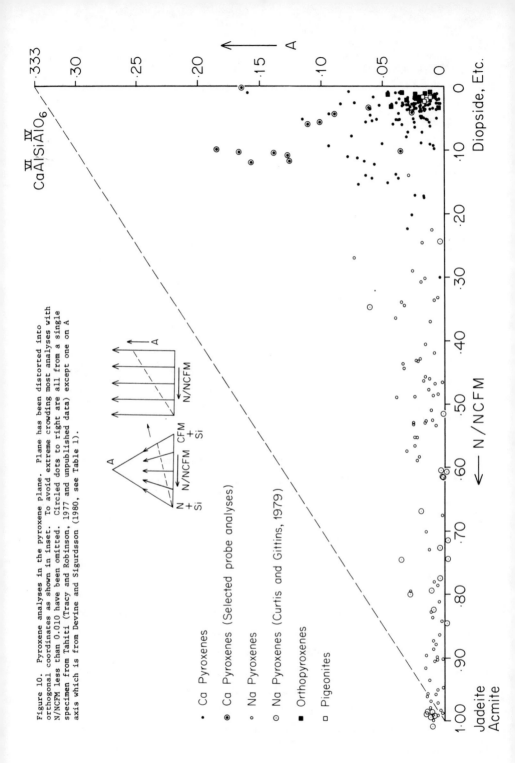

Figure 10. Pyroxene analyses in the pyroxene plane. Plane has been distorted into orthogonal coordinates as shown in inset. To avoid extreme crowding most analyses with N/NCFM less than 0.010 have been omitted. Circled dots to right are all from a single specimen from Tahiti (Tracy and Robinson, 1977 and unpublished data) except one on A axis which is from Devine and Sigurdsson (1980, see Table 1).

- Ca Pyroxenes
◉ Ca Pyroxenes (Selected probe analyses)
∘ Na Pyroxenes
⊙ Na Pyroxenes (Curtis and Gittins, 1979)
■ Orthopyroxenes
□ Pigeonites

substitution. However, the absolute substitution in the A direction is not impressive although a few analyses exceed half of the permissible limit. The long string of probe analyses leading to the highest A value are all from a single probe section of Tahitian basanite (Tracy and Robinson, 1977 and unpublished data) and cover about 80% of the known terrestrial range.

In treating pyroxenes dominated by quadrilateral components, Cameron and Papike (this volume) have narrowed down considerations to 59 accepted analyses with "other" substitutions less than 10%. In the traditional method of plotting quadrilateral pyroxenes in terms of Ca, Mg, and Fe, any non-quadrilateral component causes substitution of R^{3+} ions for Mg and Fe in M1. This in turn causes the ratio Ca/Ca+Mg+Fe to increase, and in some cases to show "$CaSiO_3$" content greater than 50%. Since the fundamental relationship that distinguishes Ca-rich and Na-rich pyroxenes from Ca-poor clinopyroxenes and orthopyroxenes has to do with occupancy of the large M2 site, it seems more rational to consider the occupancy by large cations of the M2 site directly. In this case the choice was made to consider Ca+Na in M2 as a plotting parameter, although an argument could alternatively be made to include Mn with these.

In the first plot (Fig. 11) this parameter is plotted against R^{3+} ions in the M1 site. Within modest limits between 0.08 and 0.3, substitution of R^{3+} in M1 seems to have no systematic effect on the Ca content of Ca pyroxenes coexisting with low Ca pyroxenes. As expected, many orthopyroxenes and pigeonites contain 10-20% Ca+Na in M2 and less than 0.1 R^{3+} ions. What is perhaps most revealing is that there are only *three* Ca pyroxene analyses (318-6 is the most extreme) in the representative sample between Ca+Na = 0.3 and 0.90 that have R^{3+} less than 0.07. In effect there are practically no Ca pyroxenes except in the diopside-hedenbergite series that are *really* in the pyroxene quadrilateral! To illustrate this R^{3+} effect, some iron-rich metamorphic pyroxene pairs are plotted (Jaffe *et al.*, 1975). In these pairs, particularly the most Fe-rich, the augites are notably low in Al_2O_3 even though saturated, as shown by the presence of garnet. However, none of these has R^{3+} less than 0.07. The author was singularly unsuccessful in locating "acceptable" analyzed pairs of high Al omphacites coexisting with orthopyroxene from hypersthene eclogites. A single pair from an inclusion at Salt Lake Crater, Oahu, is plotted (Yoder and Tilley, 1962).

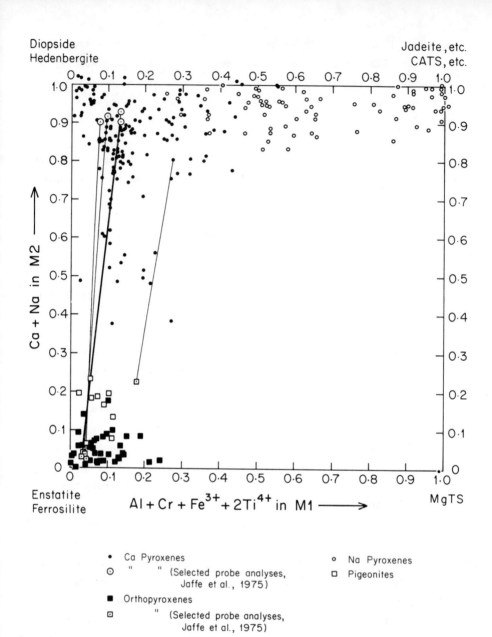

Diopside
Hedenberg ite

Jadeite, etc.
CATS, etc.

Ca + Na in M2 →

$Al + Cr + Fe^{3+} + 2Ti^{4+}$ in M1 →

Enstatite
Ferrosilite

MgTS

- • Ca Pyroxenes
- ⊙ " " (Selected probe analyses, Jaffe et al., 1975)
- ■ Orthopyroxenes
- ◻ " (Selected probe analyses, Jaffe et al., 1975)
- ○ Na Pyroxenes
- ▫ Pigeonites

Figure 11. Plot of Ca + Na in M2 versus Al + Cr + Fe^{3+} + 2Ti^{4+} in M1. Illustrates the principal parameter controlling mutual solubility in the "quadrilateral" *vs* the extent of non-quadrilateral substitution. Analyses connected by tie lines are coexisting Fe-rich metamorphic pyroxenes from the Adirondacks (Jaffe *et al.*, 1975) and one pyroxene pair from a Hawaiian hypersthene eclogite nodule (Yoder and Tilley, 1962). They illustrate the very strong preference for non-quadrilateral substitution in Ca pyroxenes.

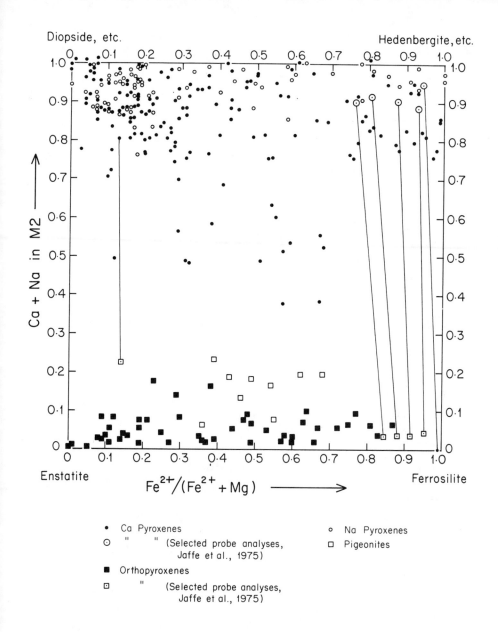

Figure 12. Plot of Ca + Na in M2 *vs* $Fe^{2+}/(Fe^{2+}+Mg)$. This illustrates the principal parameter controlling mutual solubility in the "quadrilateral" *vs* the Fe^{2+} ratio. Analyses connected by tie lines are co-existing Fe-rich metamorphic pyroxenes and one Ca pyroxene-fayalite pair from the Adirondacks (Jaffe *et al.*, 1975), and one pyroxene pair from a Hawaiian hypersthene eclogite nodule (Yoder and Tilley, 1962).

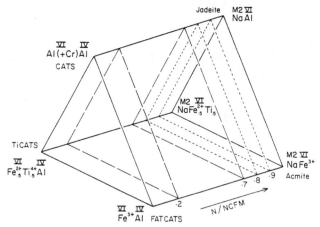

Figure 13. Triangular reciprocal prism showing non-quadrilateral substitutions in pyroxenes. Position in triangular coordinates indicates proportions of R^{3+} charge-increasing substitutions (Cr^{3+} is lumped with ^{VI}Al). At the front end of the prism these are balanced by ^{IV}Al. At the back end they are balanced by Na in M2.

In the second plot (Fig. 12), Ca+Na in M2 is plotted against Fe^{2+}/ (Fe^{2+}+Mg). This is directly analogous to the pyroxene quadrilateral except it is more realistic. It is nearly identical to a figure of Cameron and Papike (this volume) except all accepted analyses are plotted, not just the 59 with quadrilateral components greater than 90%. The same pairs as in Figure 11 are shown here, also.

All of the non-quadrilateral substitutions in pyroxenes are portrayed in a triangular prism in Figure 13. The proportions of the three principal R^{3+} substitutions in M1 are portrayed by the triangular cross section. These are Al^{3+}, Fe^{3+} and $2Ti^{4+}$. Cr^{3+}, which is generally very minor, is lumped with Al. These R^{3+} charge-increasing substitutions must be compensated by some proportion of the two charge-decreasing substitutions, $Al^{3+}(+Fe^{3+})$ for Si^{4+} in tetrahedral sites and Na^+ for R^{2+} in M2. The front end of the prism represents pure tetrahedral compensation. The back end of the prism represents pure Na compensation. The prism should be thought of as the target *toward which* pyroxenes trend as they deviate from pure quadrilateral compositions. It is an expansion of the "others" triangle of Cameron and Papike (this volume). It should also be thought of as a quarternary reciprocal slice through a quinary system. This prism helps to illustrate the illusory nature of end-member calculation schemes. Any such scheme involves dividing the prism into three four-pointed (tetrahedral) volumes. There are *six* alternate ways to do this, each equally valid.

For this study the prism has been sectioned perpendicular to the length. Figure 14 is a projection of analyses within such a section.

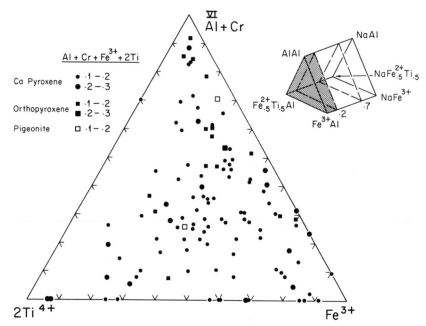

Figure 14. Distribution of 'R^{3+}' substitutions in moderately substituted Ca pyroxenes and Fe-Mg pyroxenes. Includes all "accepted" analyses from Deer, Howie and Zussman (1978) with N/NCFM less than 0.2 (see inset) and with total 'R^{3+}' between 0.1 and 0.3.

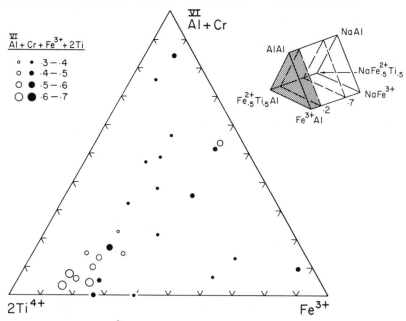

Figure 15. Distribution of 'R^{3+}' substitutions in highly substituted Ca pyroxenes. Includes all "accepted" analyses from Deer, Howie and Zussman (1978) plus selected probe analyses from Tracy and Robinson (1977 and unpublished data) and Devine and Sigurdsson (1979) (open symbols) with N/NCFM less than 0.2 (see inset) and with total 'R^{3+}' greater than 0.3.

441

It includes analyses with N/NCFM less than 0.2, and in which the R^{3+} substitutions are greater than 0.1 and less than 0.3. In essence it covers Ca pyroxenes, orthopyroxenes, and pigeonites with modest amounts of R^{3+} substitutions. Symbols are larger for analyses with R^{3+} between 0.2 to 0.3. The broad variety of substitution in this range is notable with analyses near all three apices, and with no striking differences between the three types of pyroxenes. There is, however, a great scarcity of analyses near the Al-2Ti edge.

Figure 15 covers the same range of N/NCFM but shows only analyses with larger amounts of R^{3+} substitution. Size of symbols shows analyses between 0.3 and 0.4, 0.4 and 0.5, 0.5 and 0.6, and 0.6 and 0.7. Open symbols are selected probe analyses, those to lower left from Tracy and Robinson (1977); the one to upper right from Devine and Sigurdsson (1980). With the exception of the latter, all analyses with R^{3+} greater than 0.5 are rich in Ti, and most of the extremes are listed in Table 1. Overall distribution is similar to that in Figure 14.

Figure 16 covers the middle portion of the prism with N/NCFM between 0.2 and 0.7. Size of symbols is again related to the absolute amount of R^{3+} in M1. Al-rich omphacites and Fe^{3+}-rich aegerine-augites are clearly grouped, and intermediate and Ti-rich compositions are only shown by the analyses of Curtis and Gittins (1979).

Figure 17 covers the Na-rich portion of the prism with N/NCFM greater than 0.7. Here symbols are grouped according to how close they come to the pure Na end of the prism, 0.7-0.8, 0.8-0.9, and 0.9-1.0. In the representative sample, analyses are separately grouped into jadeites, aegerines, and one analysis (488-9), a Ti-aegerine with greater than 50% of an unnamed $Fe^{2+}_{0.5}Ti^{4+}_{0.5}$ end member. The analyses of Curtis and Gittins (1979) fill in the gap with aegerine-jadeites and other Ti-bearing intermediate members.

Na and R^{3+} variations are shown simultaneously in two "side-views" of the prism, Figures 18 and 19. In Figure 18, the Fe^{3+} and $2Ti^{4+}$ edges of the prism are collapsed on each other, essentially lumping these two components together. The result is a reciprocal square with the components $NaAl$, $NaFe^{3+}$(+Ti), $AlAl$, and Fe^{3+}(+Ti)Al. Only analyses with R^{3+} greater than 0.1 are shown. On the Na-poor end there is a thick array of augites with the greatest abundance toward the $Fe^{3+}Ti^{4+}$ end. There are nearly continuous arrays of analyses from augites to acmite and from

442

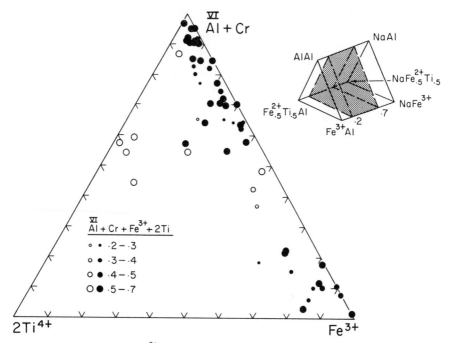

Figure 16. Distribution of 'R³⁺' substitutions in Na-Ca pyroxenes with N/NCFM between 0.2 and 0.7 (see inset). Includes all "accepted" analyses from Deer, Howie and Zussman (1978) plus analyses from Curtis and Gittins (1979) (open symbols).

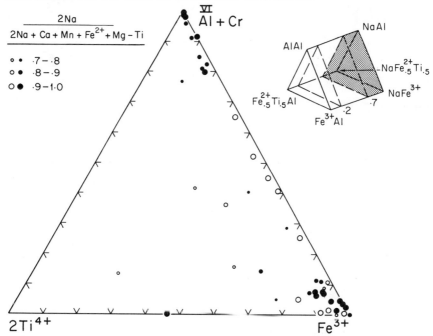

Figure 17. Distribution of 'R³⁺' substitutions in Na pyroxenes with N/NCFM greater than 0.7 (see inset). Includes all "accepted" analyses in Deer, Howie and Zussman (1978) plus analyses from Curtis and Gittins (1979) (open symbols).

443

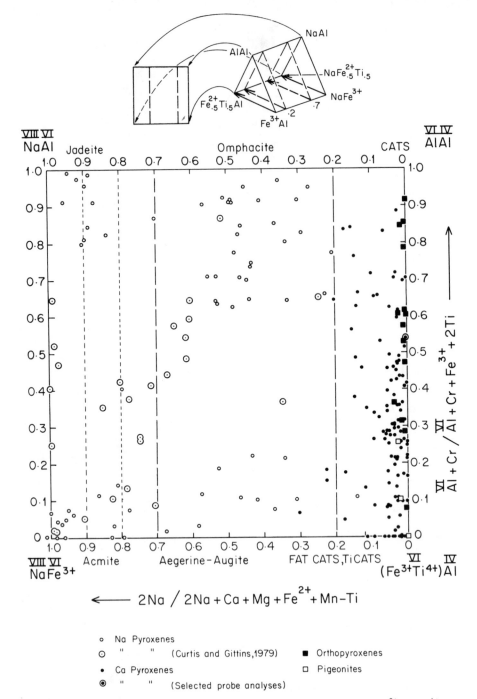

Figure 18. Side view of non-quadrilateral reciprocal prism (see inset). Fe^{3+} and Ti^{4+} substitutions are combined. Shows proportions of these substitutions vs ^{VI}Al $(+Cr)$ for ^{IV}Al-compensated pyroxenes to right and ^{M2}Na-compensated pyroxenes to left. Includes all "accepted" analyses from Deer, Howie and Zussman (1978) with 'R^{3+}' greater than 0.1 plus analyses from Curtis and Gittins (1979, circled points) plus the analyses from Devine and Sigurdsson (1980, circled dot).

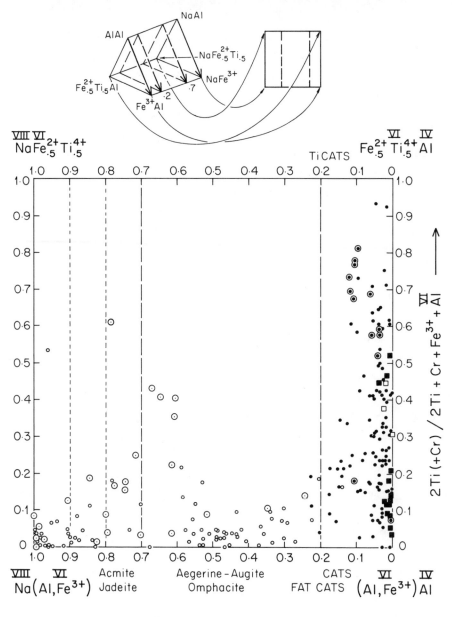

Figure 19. Side view of non-quadrilateral reciprocal prism (see inset). ^{VI}Al and Fe^{3+} substitutions are combined. Shows proportion of these *vs* $2Ti^{4+}$ (+Cr) for ^{IV}Al-compensated pyroxenes to right and for ^{M2}Na-compensated pyroxenes to left. Includes all "accepted" analyses from Deer, Howie and Zussman (1978) with 'R^{3+}' greater than 0.1 plus analyses from Curtis and Gittins, titanaugites from Tracy and Robinson (1977 and unpublished data) and low Ti augite from Devine and Sigurdsson (1980).

augites to jadeite with a possible hiatus around N/NCFM = 0.6. A major
gap in the left central part of the figure is partly filled by probe
analyses from Curtis and Gittins (1979). In addition, these seem to
provide a linear array from omphacite to acmite. Obviously, the controls
on these compositions are a fruitful field for study.

In Figure 19 the ^{VI}Al and Fe^{3+} edges of the prism have been collapsed
on each other in order to see the effect of Na on Ti^{4+} substitution. (In
this case Cr is lumped with Ti for numerical expediency, but is trivial.)
The Ti end member is most important in some augites, though it reaches substantial proportions also in several Na pyroxenes. The limits of such Na
substitutions need exploration.

INTERNAL LIMITS - PYROXENE-PYROXENE INTERACTIONS

Introduction

For the pyroxene purist, nothing can be so exciting as coexisting pyroxenes. Because the equilibria involved are a function of element distribution between phases of the same or very similar structure, they have
the potential for being extremely sensitive to environmental conditions
and thermal histories. Two-pyroxene geothermometry has been the most
widely and fairly successfully applied geothermometer in terrestrial rocks
of high temperature origin, particularly those of deep-seated origin.
However, the very similarity of the two structures means there are small
free energy differences, rampant possibilities for development of metastability and for recrystallization during cooling, as well as for the
development of microstructures covered at length by Buseck and others
(this volume). In order to use coexisting pyroxenes as geothermometers,
it is commonly necessary to see through later transformations, both
chemical and physical, in order to understand the pyroxenes under the
P-T conditions for which a temperature assessment is to be made. For
these reasons, both in assessments of natural occurrences and of laboratory experiments, it is commonly necessary to go beyond routine optical
petrography and electron probe analyses and use expertise in the fields
of single crystal x-ray crystallography and electron microscopy. However, enough experience has already been gained in these fields to make
certain generalizations possible, and thus permit relatively routine
optical and electron probe interpretations in many cases. A further

point in favor of sound optical petrography and careful electron probe analysis is that many of the effects of interest occur on a fairly coarse scale at high temperature where diffusion rates are high, whereas many features observed in x-ray single crystal and electron microscope studies are related to the low temperature part of the cooling range.

The Pyroxene Quadrilateral at P and T

Assemblages of coexisting pyroxenes that fall fairly close to the quadrilateral in composition are by far the most common and most commonly studied. This is so because such pairs can occur in a wide variety of igneous rocks ranging from tholeiitic basalts to granites as well as ultramafic rocks, and the metamorphosed equivalents of all of these. Numerous views of the quadrilateral have been published for a variety of rocks, and a variety of P-T conditions. For igneous differentiation series, pyroxene compositions are commonly presented in relation to a cooling sequence and may cover one or two hundred degrees centigrade. Pyroxene pairs from areas of regional metamorphism have been presented in such a way as to give essentially an isothermal section of the quadrilateral. In all of these cases the Ca-pyroxene is a $C2/c$ clinopyroxene of variable composition but apparently monotonous structure. The Ca-poor pyroxene, however, may have been a $C2/c$ high-T pigeonite that has suffered one or more displacive or reconstructive transformations during cooling, or may have been an orthopyroxene to start with. The crucial difference between the Ca-rich and the Ca-poor quadrilateral pyroxenes is clearly the nature of the dominant cations in the M2 positions: Ca in the former, Fe^{2+} and Mg in the latter. From the point of view of thermodynamic equations, it would be nicer if low Ca pyroxenes retained their high-T $C2/c$ structure throughout. However, it is the small Fe and Mg ions in M2 that permit low Ca pyroxenes to "drop" into the low-T primitive and orthorhombic states, and this is quite frankly what makes the petrography and interpretation of coexisting pyroxenes "interesting[6]."

In rocks such as lunar basalts, formed at very high temperatures, it appears that all coexisting pyroxenes originally had the high-T $C2/c$ structure, although the pigeonite underwent displacive transformation to

----

[6]"Interesting" is used here in the mountaineering sense, to describe a climbing route filled with challenge and surprises.

$P2_1/c$ on cooling. Such an example is shown in Figure 20. The two-pyroxene solvus is extremely narrow and probably reaches a critical point not far to the right of the middle of the quadrilateral, reflecting the very similar behavior of Ca and Fe^{2+} in M2 at high T. Unfortunately, peritectic reactions with liquid do not permit us to examine the critical pyroxene in this case.

In the case of the terrestrial Picture Gorge basalt flow, Smith and Lindsley (1971) have shown that pyroxene compositions are related to the rate of cooling and possibly also to pressure of volatiles (Fig. 21). Within the slowly-cooled flow interior, augite compositions seem to trend parallel to a probable solvus although no primary pigeonites are present, whereas in the quenched exterior, some compositions fall well within the solvus limits. Several of the low Ca augites illustrated in Figure 12 are probably similar to such quench pyroxenes.

The Bushveld Complex of South Africa is probably the largest and most slowly-cooled basaltic intrusion on earth, and one of the most thoroughly studied. Bulk compositions of coexisting pyroxenes from this intrusion are shown in Figure 22. In this case the temperature was low enough for orthopyroxenes to precipitate directly from the most Mg-rich magmas, but because the pigeonite-orthopyroxene transformation declines more steeply with Fe than does the crystallization temperature, the low Ca pyroxene becomes pigeonite near Fe_{30}, by the reaction Opx + Liq = Aug + Pig. Pigeonite is later lost by another peritectic reaction related to narrowing of the solvus--Pig + Liq = Aug. Within such layered intrusions it is always important to distinguish, if possible, between primary precipitate pyroxenes, and pyroxenes crystallized from later interstitial liquids that may give lower crystallization temperatures.

In pyroxene syenites and granites, as exemplified by the White Mountain Batholith, New Hampshire, crystallized under relatively low P conditions, the liquid temperatures were much lower, permitting orthopyroxene crystallization far across the quadrilateral, until the pigeonite-in reaction was finally reached at Fe_{75} (Fig. 23). In this case pigeonite only lasted to Fe_{82} where all Ca-poor pyroxene was replaced by olivine plus quartz.

In dioritic to granitic rocks of the Nain Province, Labrador, intruded at 3-6 kbar under essentially anhydrous conditions, the primary

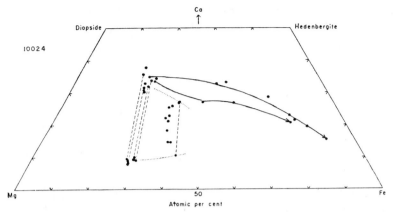

Figure 20. Pyroxene quadrilateral showing compositions of clinopyroxenes in Apollo 11 basalts 10024-23 from Kushiro and Nakamura (1970). Solid lines indicate ranges of continuous zoning within single crystals, dashed lines are tie lines between directly contacting augite and pigeonite, and dotted lines indicate a possible solvus between augite and pigeonite. Compositions between solvus limbs may be quench crystals.

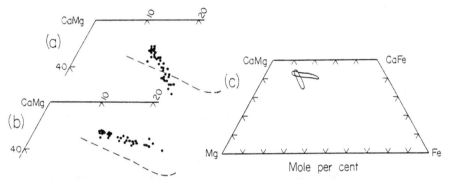

Figure 21. Compositions of augites from the Picture Gorge Basalt flow, from Smith and Lindsley (1971), plotted on the pyroxene quadrilateral. (a) From the chilled basal margin of the flow. (b) From the central portion of the flow 45 meters above the base. (c) Full quadrilateral showing contrasting quench trend from chilled margin and equilibrium trend from flow interior. Dashed lines in (a) and (b) represent the early part of the Skaergaard augite trend.

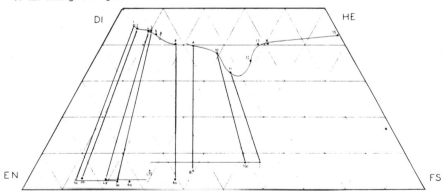

Figure 22. Pyroxene quadrilateral showing compositions from the eastern limb of the Bushveld complex, South Africa, from Buchanan (1979). Most analyses are from Atkins (1969) with analysis numbers corresponding to those in the original publication. L77 (solid triangle) is an inverted pigeonite from Gruenewalt (1970). Analyses A and B (open triangles) are microprobe determinations by Buchanan.

449

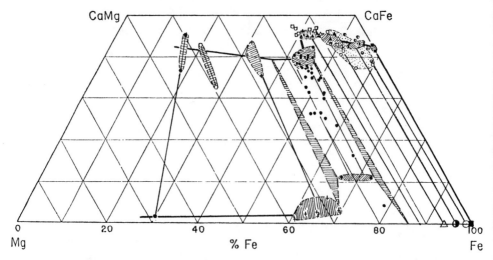

Figure 23. Pyroxene quadrilateral showing pyroxene and fayalite compositions from gabbros, diorites, syenites and granites from the White Mountain Batholith, New Hampshire, adapted from Creasy (1974). Analysis points represent averages of numerous analyses of single pyroxene grains which show considerable variation within each rock unit. Augites within the two most magnesian rocks show a continuous range between the two analysis points plotted. Numerous analysis points within the augite-pigeonite field are averages of two to five determinations on augite and pigeonite hosts with abundant exsolution lamellae 10 micrometers or less in thickness that could not be resolved by the electron beam. Analyses in the Fe corner are fayalite. Tie lines, oval outlines around analyses, hypothetical three-phase fields, and pyroxene trends were added by the present author. Creasy (1974) estimates the conditions of crystallization of the pigeonite syenite at 825°C and 1.6 kbar.

intermediate to Fe-rich pyroxenes were pigeonites and augites very close to each other in composition (Fig. 24) and possibly in one or two cases actually hypersolvus. These have produced some of the most spectacular examples of exsolved pigeonites, in some cases without the complete reconstructive transformation to inverted pigeonite, so common elsewhere.

In metamorphic terrains pyroxenes may have been produced by breakdown of hydrous minerals or pre-existing igneous pyroxenes may have been recrystallized by metamorphic events. The latter seems to have been largely the case in the Mt. Marcy area of the Adirondacks (Jaffe et al., 1975) as exemplified in Figure 25. Here much of the quadrilateral is filled with orthopyroxene-augite tie lines including an extreme extension of the orthopyroxene field against olivine plus quartz caused by pressure of about 7-8 kbar (Bohlen and Boettcher, 1980). However, a few rocks in this terrain retain relict igneous textures and evidence of premetamorphic plutonic crystallization of pigeonite (Jaffe et al., 1977; Bohlen and Essene, 1978; Ashwal, 1979). In local high-T contact aureoles pyroxenes may also contain features indicating the former presence of pigeonite (Bonnichsen, 1969; Simmons et al., 1974).

450

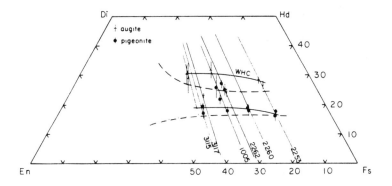

Figure 24. Quadrilateral showing reconstructed compositions of primary augites and pigeonites from the Wyatt Harbor Complex, Labrador, diorites, monzodiorites and granites from Huntington (1980). Compositions were obtained by modal analysis of coarsely exsolved grains and electron probe analyses of coarse lamellae and host grains (see Figure 33). In most cases it was possible to estimate whether the primary host was pigeonite (circles) or augite (crosses) but some grains with intermediate compositions could represent hypersolvus crystallization. The vertical axis represents the variation in estimated Wo content. Solid lines represent a possible solvus for pyroxenes of the Wyatt Harbor Complex. Dashed lines are the intersection of the solidus and solvus for lunar basalt 12021 as determined by Ross *et al.* (1973). Similar pyroxenes from another part of the Nain Province have been described by Ranson (1978, 1979).

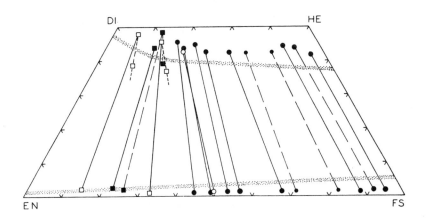

Figure 25. Pyroxene quadrilateral showing compositions of augite and orthopyroxene equilibrated under granulite facies metamorphic conditions from the Mt. Marcy area of the Adirondacks (closed circles), the Hudson Highlands (open circles), the Cortlandt Complex (closed squares), and the Belchertown Intrusive Complex (open squares). From Jaffe *et al.* (1975). All compositions determined by electron probe except small closed circles which are optical estimates. Solid tie lines indicate specimens for which x-ray single crystal data have been obtained, long-dashed tie lines, only optical data on exsolution lamellae. Short dashed lines indicate zoning trends in recrystallized igneous augite. Stippled pattern indicates limits of mutual solid solution of Ca-Mg-Fe augite and orthopyroxene at 810°C as determined by Lindsley *et al.* (1974).

451

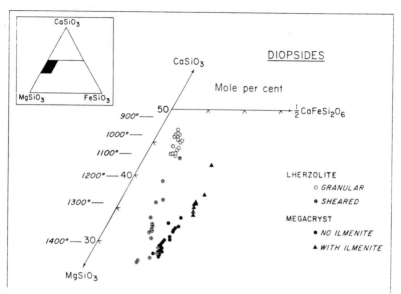

Figure 26. Portion of pyroxene quadrilateral showing composition of diopsides in equilibrium with orthopyroxene from garnet lherzolites and also diopsides from megacrysts, all from Lesotho kimberlites (from Fig. 7 of Boyd and Nixon, 1973).

The greatest and most successful recent explosion in pyroxene research has come from studies of inclusions in kimberlite pipes, particularly of garnet lherzolites (Boyd, 1973; Boyd and Nixon, 1973). For the most part these inclusions appear to have been captured deep in the mantle at very high temperatures and pressures, and transported very rapidly to the earth's surface. For this reason they contain very few of the exsolution features found in pyroxenes from other environments. In addition, these mantle compositions are Mg-rich so that the coexisting pyroxenes are diopside and orthopyroxene fairly close to the experimentally studied system $CaMgSi_2O_6$-$Mg_2Si_2O_6$, and the compositions of the diopsides provide a reasonable temperature interpretation (Fig. 26).

These few examples represent but a tiny fragment of the amount of available information on quadrilateral pyroxenes and their crystallization in igneous and megamorphic environments. I have tried to emphasize the compositions of coexisting pyroxenes at their thermal peak. We will now examine some of the features formed during cooling that aid the optical petrographer in interpreting thermal history.

Optical Petrographer's Guide to Quadrilateral Pyroxene Exsolution

Exsolution processes. The character of exsolution lamellae in pyroxenes is a complex function of bulk composition, maximum temperature and

452

cooling rate; hence, characterization of the lamellae may provide qualitative if not quantitative information on these parameters. Pyroxenes
quenched from high temperature such as those in basalts or in mantle
nodules emplaced in kimberlite diatremes tend not to have optically
observable exsolution lamellae. Pyroxenes from large layered high temperature intrusions tend to show the most complex arrays of lamellae because the pyroxenes crystallized at high temperatures, generally over
1000°C, and then underwent very slow cooling during which exsolution
lamellae could form at several stages. Similar pyroxenes can be found
in the few, very high temperature contact aureoles. Pyroxenes subjected
to regional metamorphism and equilibrated around 750-800°C have far less
mutual solid solution as well as much narrower cooling range. In these
the lamellae are finer and less abundant, and the patterns of lamellae
tend to be simpler.

As discussed in more detail elsewhere in this volume (Buseck *et al.*),
lamellae can be initiated by one of two processes, spinodal decomposition
involving the growth and enlargement of compositional waves, and heterogeneous nucleation and growth. It is probably true that most of the pyroxene lamellae that are optically observable formed by the latter process,
but the principles concerning *orientation* of lamellae are probably equally
valid for either process of initiation.

Growth of lamellae clearly requires solid state diffusion and diffusion rates are very sensitive to temperature. Hence, it is to be expected
that high temperature lamellae will be large and few in number, whereas
low temperature lamellae formed under conditions of slow diffusion will
initiate at many locations and grow to very small size. Diffusion for
lamellar growth implies chemical gradients and inhomogeneity. Portions
of pyroxene host next to early high temperature lamellae may become depleted in lamella-forming constituents so that when a new lower temperature set of lamellae is initiated it precipitates in portions of the host
well away from the high temperature set of lamellae. This is fortunate
because it tends to separate the lamellae into reasonably distinct optically separable sets. Although size of lamellae can often be used to
determine relative age, this is safest with lamellae in approximately
the same orientation. Because pyroxenes are distinctly anisotropic with
respect to diffusion, it is possible in some cases to have finer lamellae
in one direction that are actually older than coarser lamellae in

another direction. At the same time it is quite probable that localized
ion exchange processes occur between hosts and lamellae that further
confuse the chemical picture.

Types of lamellae.   In orthopyroxene host grains, only one orientation of exsolution lamellae is possible, namely clinopyroxene on (100).
Although this is not a plane of exact fit between the two dissimilar
structures, it is by far the best dimensional match and the only plane
in the two structures that is at all similar. In clinopyroxene host
grains it is possible to have orthopyroxene lamellae parallel to (100)
governed by the same criteria. In clinopyroxene host grains it is also
possible to have exsolution lamellae of another clinopyroxene, either
augite or pigeonite. For many years optical petrographers (see Poldervaart and Hess, 1951) described these as being parallel to (001) and
(100). This was agreed to by x-ray crystallographers, who, however, did
not actually observe the phase boundaries, but only the relative orientation of *host and lamellar lattices*, which do, indeed, in most cases,
have either (001) or (100) nearly in common. It is now belatedly recognized (Robinson *et al.*, 1971; Jaffe *et al.*, 1975; Robinson *et al.*,
1977) that, in the general case, clinopyroxene lamellae in a clinopyroxene host are oriented in irrational planes close to (001) and (100)
but deviating from these orientations by as much as 20° and 25°, respectively. These irrational planes related to (001) and (100) are termed
"001" and "100," and are related to the relative lattice parameters (a,
b, and β) of host and lamellae under the conditions of initiation of the
lamellae. Such parameters are a function of chemical composition and of
temperature and pressure; hence, the exact orientation of the lamellae
may be used under favorable circumstances to make estimates of chemical
composition or of temperature.

To this author it is curious that the phenomenon of irrational
lamellae in pyroxenes was not found decades earlier during the period
of intense optical examination of these minerals from the Skaergaard,
Stillwater, and Bushveld intrusions. It is even more curious that it
was first described from the structurally related amphibole system,
where the phenomenon is less pronounced. Irrational lamellae in horn-
blende-cummingtonite pairs were first reported by Callegari (1966) and
Jaffe *et al.* (1968). More striking examples were found by Jaffe

in pigeonite lamellae in augite from the Hudson Highlands, by Klein in a peculiar sample of coexisting Mn-cummingtonite and Mg-arvedsonite, and by Ross in the first example of two sizes of pigeonite "001" lamellae with different orientations in a specimen from the Duluth Gabbro. These and the basic lattice-fit theory were presented by Robinson *et al.* (1971). Subsequent papers dealt with the relation between pigeonite lamellar orientation and composition in metamorphic augites (Jaffe *et al.*, 1975) and on interpretation of lamellae in high T-augites (Robinson *et al.*, 1977). These ideas have been expanded in studies of augite from the Stillwater Complex (McCallum, 1974), studies of augites and pigeonites in Labrador (Ranson, 1978; Huntington, 1980), and studies of pyroxenes from iron-rich igneous rocks in southwest Norway (Rietmeijer, 1979).

Exact phase boundary theory. The theory states that for any two monoclinic pyroxenes with different values of a, c, and β, and with essentially identical values of b, there are two different planes of exact *dimensional* fit, one close to (001) termed "001" and one close to (100) termed "100." Such planes of fit are illustrated in scale drawings for two imaginary pyroxene pairs in Figure 27. In each case the phase boundary lies at a considerable angle to the rational (001) or (100) plane, even though the lattices are within 1° of being parallel and differ only by the very small "phase-boundary lattice rotation." The relative size of the a dimension of PIG (the phase with the larger angle β) vs that of AUG (the phase with the smaller angle β) is the main determinant of the orientation of "001" lamellae. The relative values of c are the main determinant of the orientation of "100" lamellae. The lamellae can be strictly rational only in special cases where $a_{AUG} = a_{PIG}$ or $c_{AUG} = c_{PIG}$. These concepts are illustrated schematically in Figure 28, with its parallels to Orwell's *Animal Farm*. They would be of limited utility for temperature interpretation if it were not for the striking difference between the thermal expansion curves for augite and for pigeonite (Fig. 29). Pigeonite, in fact, goes through a very large change in lattice parameters in the vicinity of the displacive, nonquenchable $C2/c$ to $P2_1/c$ transition. Furthermore, above the transition $a_{AUG} < a_{PIG}$ and $c_{AUG} < c_{PIG}$ whereas below $a_{AUG} > a_{PIG}$ and $c_{AUG} > c_{PIG}$. Thus, in many cases it is possible to make a unique determination as to whether a pigeonite lamellae in augite, or an augite lamella in pigeonite was initiated above or below the $C \rightarrow P$ inversion. Figure 30 illustrates an

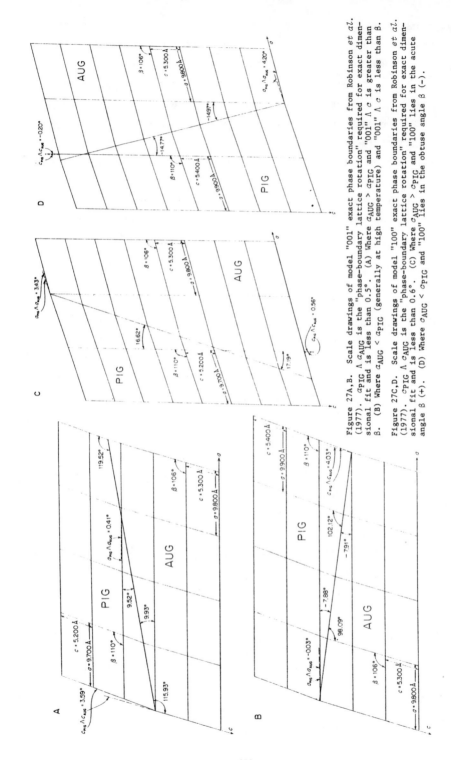

Figure 27A,B. Scale drawings of model "001" exact phase boundaries from Robinson *et al.* (1977). $\alpha_{PIG} \wedge \alpha_{AUG}$ is the "phase-boundary lattice rotation" required for exact dimensional fit and is less than 0.5°. (A) Where $\alpha_{AUG} > \alpha_{PIG}$ and "001" $\wedge$ c is greater than β. (B) Where $\alpha_{AUG} < \alpha_{PIG}$ (generally at high temperature) and "001" $\wedge$ c is less than β.

Figure 27C,D. Scale drawings of model "100" exact phase boundaries from Robinson *et al.* (1977). $c_{PIG} \wedge c_{AUG}$ is the "phase-boundary lattice rotation" required for exact dimensional fit and is less than 0.6°. (C) Where $c_{AUG} > c_{PIG}$ and "100" lies in the acute angle β (+). (D) Where $c_{AUG} < c_{PIG}$ and "100" lies in the obtuse angle β (−).

456

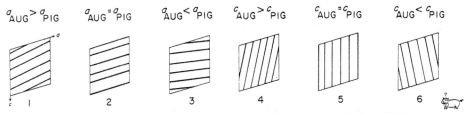

Figure 28. Schematic relationship between relative a and c dimensions and the orientation of exsolution lamellae in clinopyroxenes ($C2/c$ or $P2_1/c$) and clinoamphiboles ($I2/m$ or $P2_1/m$). Rhombs represent schematic crystals bounded by (001) and (100) planes. From Robinson *et al.* (1977).

interpolated semi-quantitative model for patterns of lamellae to be expected from three different cycles of nucleation and growth of lamellae from a Bushveld augite.

High temperature exsolution. Pyroxene precipitated from very magnesian liquids may lie below the $C2/c \rightarrow$ Opx reconstructive transformation. In this case coexisting augite and orthopyroxene usually

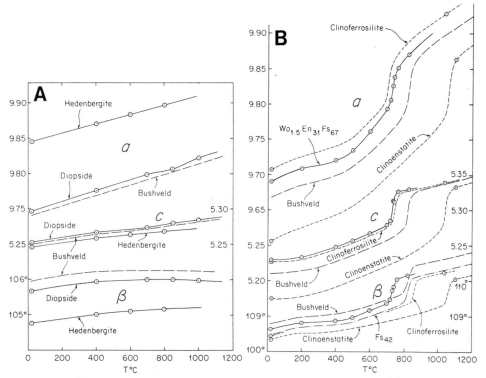

Figure 29. (A) Lattice parameters of Ca-rich pyroxenes measured at temperature (T). Dashed line is interpolated curve for Bushveld composition. Experimental points are from Cameron *et al.* (1973). (B) Lattice parameters of Ca-poor clinopyroxenes (pigeonites) measured at temperature (T). Dash-dot curve is interpolated for pigeonite Fs$_{42}$ and the curve with long dashes is interpolated for Bushveld composition. Experimental points for Wo$_{1.5}$En$_{31}$Fs$_{67}$ are from Smyth (1974). Points for clinoenstatite and clinoferrosilite are from J.V. Smith (1969) and J.J. Papike (pers. comm., 1973). Abrupt changes in slope near 700°C indicate region of symmetry change, $P2_1/c$ below, $C2/c$ above. From Robinson *et al.* (1977).

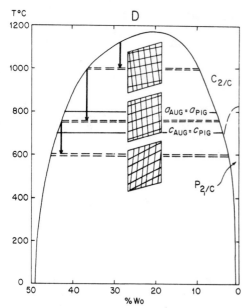

Figure 30. Temperature-composition diagram for model Bushveld augite and pigeonite on the join $Di_{72}He_{28}$-$En_{58}Fs_{42}$ showing temperatures where $a_{AUG} = a_{PIG}$ and $c_{AUG} = c_{PIG}$ and ideal patterns of "001" and "100" lamellae to be expected in each of three cycles of nucleation of lamellae. Estimated position of $C2/c \rightarrow P2_1/c$ transition is also shown. From Robinson *et al.* (1977).

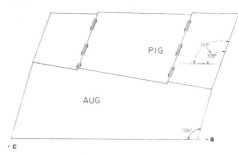

Figure 31. Diagrammatic view of model "001" interface between high temperature pigeonite lamella and augite host after cooling. (100) normal faults in pigeonite lamellae permit β of pigeonite to decrease from 110° to 109° during cooling without changing the overall orientation of the interface. The greater the amount that pigeonite β must decrease during cooling, the greater the number of faults produced. This process results in marked misorientation of (001) of augite and pigeonite lattices in x-ray single crystal photographs. From Robinson *et al.* (1977).

can exsolve lamellae of each other parallel to (100). For other compositions the stable high temperature phases are $C2/c$ augite and $C2/c$ pigeonite, and each tends to exsolve the other, most commonly on "001" planes at an angle to the c axis less than β or about 103° in augite or 106° in pigeonite. What happens next to these pyroxenes is heavily dependent upon cooling rate and possibly to other environmental conditions, in particular availability of H_2O.

Under relatively rapid cooling rates the reconstructive transformation to orthopyroxene can be missed entirely, but at the very least all of the pigeonite must undergo the reversible $C2/c \rightarrow P2_1/c$ displacive transformation. This transformation and other cooling requires the β angle of pigeonite to decrease from about 110° to 108.5° (1.5°) whereas augite decreases by 0.75° or less. Pigeonite "001" lamellae confined in an augite host accomplish this major change in β by development of a series of (100) normal faults (Fig. 31) so that individual fault blocks can decrease their β without changing the overall position and orientation of the lamella. The lattice twisting caused by this phenomenon can give up to 1.5° of (001) misorientation in x-ray single

458

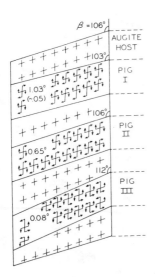

Figure 32. Observed phase boundaries and observed lattice rotations ($a_{PIG} \wedge a_{AUG}$) for three sets of "001" pigeonite lamallae in Bushveld augite crystal 1A. PIG I and PIG II are "fascist pigs" in which (001) has been twisted by 1.03° and 0.65° by the means illustrated in Figure 31. PIG III had an original "phase boundary lattice rotation" of about 0.1 to 0.2° which has been reduced toward 0° by very slight twisting since exsolution in the vicinity of 600°C. From Robinson *et al.* (1977).

crystal photographs and in the opposite sense from the relatively minor "phase-boundary lattice rotation." The sense of this lattice twisting as illustrated in Figure 32 suggests the name "fascist pigs." Although the stacking faults are largely submicroscopic in pigeonite lamellae in augite except for a few lamellae of orthopyroxene developed on them, they seem to have much more prominent development where the host is pigeonite (Fig. 33). In pigeonite host the stacking faults seem to have been important in the process of inversion to orthopyroxene, and such (100) planes can form boundaries between inverted and uninverted pigeonite (Ranson, 1978). Such (100) faults also serve to inhibit any further growth of "001" lamellae in the pigeonite host while further such exsolution continues in adjacent augite.

Relatively slower cooling and/or greater access to volatiles permits pigeonite to undergo reconstructive transformation to stable orthopyroxene. This may take an orientation with the *c*-axis parallel to that of the former pigeonite, or as a completely disoriented replacement of one or more pre-existing pigeonite grains. In this case only the old coarse augite lamellae give a clue to the former presence of pigeonite. Commonly, the replacement orthopyroxene has subsequently exsolved augite on its (100) planes, which bear no intelligent relation to the previous augite lamellae. Once orthopyroxene is stable it appears to form (100) exsolution lamellae in augite (Fig. 34) but this fact in no way inhibits further precipitation of presumably metastable pigeonite on "001" and "100," both in augite host and within coarse augite lamellae in former pigeonite host. The orientations of these lamellae continue to change to accommodate changes in relative lattice parameters. "001" lamellae change from angles less than β

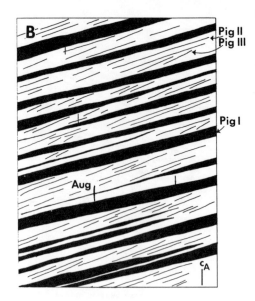

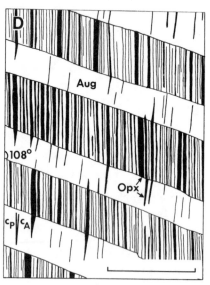

Figure 33. Photomicrographs and drawings of high temperature augites and inverted pigeonites from the Wyatt Harbor Complex, Labrador (Huntington, 1980).
Scale bar is 100 μm. (A) Photomicrograph of augite with three sets of "001"
pigeonite lamellae. (B) Drawing of same grain. (C) Photomicrograph of
pigeonite host with coarse "001" augite lamellae. Apparently as a result of
reduction of β of pigeonite, the entire grain is pervaded by (100) faults
along which both oriented inversion to orthopyroxene and exsolution of (100)
lamellae took place. (D) Detailed drawing of a portion of same crystal. Note
that (100) of augite "001" lamellae in pigeonite host "lean forward" whereas
(100) of pigeonite "001" lamellae in augite would "lean backward" (see Fig.
34B).

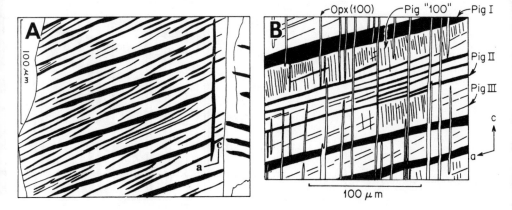

Figure 34. (A) Pattern of pigeonite exsolution lamellae in an (010) section of host augite from chill zone gabbro, Stillwater Complex, Montana. Two sets of "001" lamellae, one at about 106° and the other at 115° to the c-crystallographic axis, each show curvature at the ends. The curvature is believed to be due to longitudinal growth during cooling and/or composition change that gradually changed the lattice parameters. (B) Pattern of exsolution lamellae in (010) section of Bushveld augite (SA-1019) showing three generations of "001" pigeonite lamellae including coarse PIG I at 103° to c axis of augite, medium-sized PIG II at 106° to the c axis, and fine PIG III at 112° to the c axis. Also present are pigeonite (PIG) "100" lamellae at -6° and orthopyroxene (Opx) (100) lamellae (unshaded) parallel to the c axis. Where orthopyroxene lamellae pass through PIG I lamellae they bend 3-4° so that they are parallel to (100) of the pigeonite of the lamella.

through β to angles much larger. There are also pigeonite "100" lamellae. Lamella orientations are changed in two ways: (1) by change of orientation as lamellae propogate longitudinally into unexsolved material, giving rise to curved lamellae; and (2) by initiation of a new set of lamellae, at a different angle, usually in a region far from pre-existing lamellae. Both of these features are illustrated in Figure 34. Numerous examples have now been reported with three different angular sets of "001" pigeonite lamellae in augite. In each case the last and finest set of lamellae unquestionably precipitated well below the $C \rightarrow P$ inversion and has orientations fairly close to lamellae observed in metamorphic rocks.

Low temperature exsolution. In pyroxenes recrystallized during regional metamorphism all Ca-poor pyroxene hosts are orthopyroxene. Orthopyroxenes commonly contain (100) lamellae of augite and augites contain (100) lamellae of orthopyroxene. In addition, augites commonly contain "001" and "100" lamellae of pigeonite (Fig. 35). Lamellar orientations clearly show that exsolution took place where $a_{AUG} > a_{PIG}$ and $c_{AUG} > c_{PIG}$, i.e., well below the $C \rightarrow P$ inversion. In fact, the correspondence between calculated exact phase boundaries based on room temperature lattice parameters and the observed phase boundaries is excellent, suggesting the parameters have not changed greatly since exsolution took place.

461

Figure 35. Tracings from photomicrographs of augites with different iron-magnesium ratios showing patterns and angles of exsolution lamellae. Scale and directions of augite a and c crystallographic axes indicated. FE indicates value of the ratio $100(Fe+Mn)(Fe+Mn+Mg)$. (A) Specimen 447 from Belchertown Complex. Crystal twinned on (100) with single orthopyroxene lamella on twin plane. Thin and thick stubby lamellae are pigeonite "001" lamellae at 122° to the c axis. Thin abundant lamellae forming herringbone pattern across twin plane are pigeonite "100" lamellae at 22° to the c axis. (The b axis is nearly normal to plane of paper, but slightly misoriented, so that angles shown are slightly less than maximum observed angles of 122° and 22°. (B) Specimen J223 from Hudson Highlands. Coarse vertical lamellae are (100) orthopyroxene. Thick stubby lamellae are "001" pigeonite at 116° to the c axis. Very thin lamellae are "100" pigeonite at 6° to the c axis. (C) Specimen Ca-17 from Adirondacks. Vertical lamellae are (100) orthopyroxene. Thick stubby lamellae are "001" pigeonite at 112.5° to the c axis. Thin lamellae are "100" pigeonite at 3° to the c axis. (D) Specimen Go-2 from Adirondacks. Thick vertical lamellae are (100) orthopyroxene. Thick tapered lamellae are "001" pigeonite at 111.5° to the c axis. Very thin lamellae are "100" pigeonite at 0.5° to the c axis. From Jaffe et al. (1975).

462

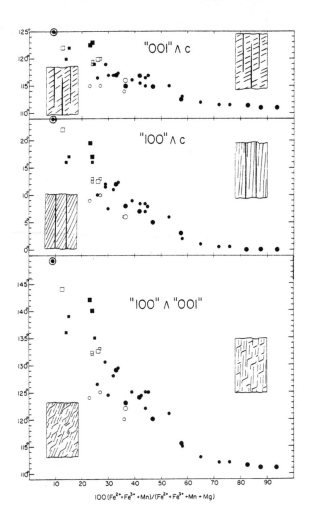

Figure 36. Angles of pigeonite exsolution lamellae in metamorphic augites plotted against $100 \times (Fe^{2+} + Fe^{3+} + Mn)/(Fe^{2+} + Fe^{3+} + Mn + Mg)$ of augite determined by electron probe analyses (large symbols) or by measurement of the gamma index of refraction (small symbols). Insets to left and right show some exsolution patterns of Mg-rich and Mg-poor augites, respectively. Closed circles: Mt. Marcy area, Adirondacks. Open circles: Monroe quadrangle, Hudson Highlands. Closed squares: Cortland Complex, New York. Open squares: Belchertown Intrusive Complex, central Massachusetts. From Jaffe et al. (1975). An addition determination (circled dot) is from fine lamellae in augite from a coarse augite-orthopyroxene intergrowth in a megacryst from a Moses Rock spinel-websterite 1307 kindly supplied by T. R. McGetchin (1968 and pers. comm., 1975). The data here support McGetchin's independent conclusion that groundmass equilibration and fine pyroxene exsolution took place in the upper mantle below 900°C, where Mg-pigeonite would have been in the primitive state (see Fig. 29B), before it was entrained and erupted in the kimberlite.

The exact angular orientations of "001" and "100" pigeonite lamellae in augite are a function of Δa and Δc of the lattice parameters of host and lamellae. These differences are largest in Mg-rich pyroxene pairs and decrease as the small Mg ion is replaced by the larger Fe^{2+} ion. Thus the orientations of lamellae in metamorphic pyroxenes, all of which exsolved at roughly the same conditions around 600°C, can be measured to obtain a rough thin-section estimate of composition. As can be seen in Figure 36, this can be done by measuring one of three different angles depending on the relative abundances of lamellae. The angle "100" Λ "001" has the most sensitivity to composition.

Exsolution lamellae and mineral analyses. An obvious objective of mineral analysis is to obtain compositions that represent distinct stages of the cooling process including stages representing the highest temperatures attained. Mineral separation and wet chemical analysis have the obvious disadvantages of (a) allowing contamination by fragments of the other host phase, or (b) allowing loss of lamellae composition from the host of interest during purification. Electron microprobe analysis has the great advantage of being non-destructive and permitting observation of what is being analyzed. The bulk composition of coarsely exsolved primary grains can be determined either by broad beam analysis (Ashwal, 1978, Fig. 37; Bohlen and Essene, 1978) or by narrow beam analysis of host and lamellae, and calculation of bulk composition from modal data (Ranson, 1979; Huntington, 1980). Small lamellae, however, can cause difficulties in narrow beam analysis of coarse lamellae or host. Buchanan (1979) has explored and approved the validity of broad beam analysis using both electron probe and electron microscopic analytical technique (EMMA-4) on very fine lamellae. He obtained bulk compositions for primary igneous precipitates that are equal to or better than those obtained by wet chemical analyses. Clearly, the *patterns* of lamellae should be used as an adjunct in trying to decide what broad beam analyses represent. For example, it may be pointless to use broad beam analysis to obtain a bulk composition on a pyroxene aggregate that is probably a metamorphic intergrowth and contains only low T metamorphic pigeonite lamellae. Relict high T lamellae would be helpful to give a high T interpretation to such a bulk analysis. One of the most comprehensive studies of the patterns of lamellae has been carried out by Rietmeijer (1979).

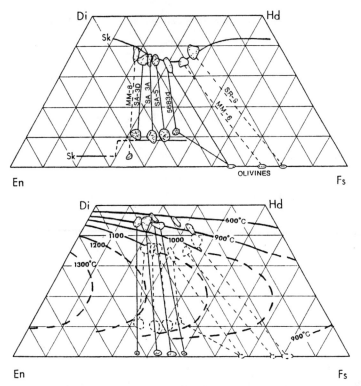

Figure 37. Pyroxene and olivine compositions from the Mt. Marcy anorthosite-mafic gabbro suite, Adirondack Mountains. From Ashwal (1978). TOP: Primary pyroxene compositions determined using broad beam energy-dispersive techniques. Solid tie lines: mafic-enriched cumulate layers; dashed tie lines: mafic enriched dikes. Olivine compositions determined by wavelength dispersive analyses. Skaergaard trend (Sk) shown for comparison. BOTTOM: Compositions of primary igneous pyroxene and pyroxene-olivine pairs (short-dashed lines) and reconstituted metamorphic pairs (solid lines) plotted in terms of the augite-pigeonite solvus (dashed contours) and the augite-orthopyroxene solvus (solid contours) of Ross and Huebner (1975).

The analyses of Buchanan (1979) (Fig. 37) are also a superb example of the change in element fractionation between pyroxenes at (a) the temperature of primary precipitation as represented by bulk analyses, and (b) the temperature of exsolution represented by host and lamella compositions. In Figure 37, the tie line between the bulk compositions is steep, indicating low fractionation between coexisting primary pyroxenes at high temperature. The tie lines between hosts and lamellae are flatter by contrast, but as expected, host-lamella tie lines pass through the bulk compositions.

Other Coexisting Pyroxenes

As pointed out above (see Fig. 12), hypersthene eclogites contain coexisting orthopyroxene and omphacite of relatively low jadeite content.

The extent of solid solutions in such pairs has not been well explored in published analyses. It was recognized early on (see, for example, O'Hara, 1960) that omphacites enriched in Na and Al occur in kyanite and corundum-bearing eclogites, rather than in hypersthene eclogites, and the single pair shown in Figure 12 may be fairly close to the limit for hypersthene-omphacite pairs in eclogites. It is possible that the search for R^{3+}-enriched pairs for Figure 12 would be better carried out in rocks of lower pressure derivation where pyroxene substitutions are not restricted by coexistence with garnet (see below).

The question of miscibility gaps in the system diopside-hedenbergite-jadeite-acmite has been summarized by Onuki and Ernst (1969). A number of workers have postulated a broad gap between diopside and jadeite at low temperature and high pressure, but others have found evidence for complete solid solution. There seems to be little disagreement that the series diopside-acmite and hedenbergite-acmite are complete. I am not aware of any reports of coarse coexisting calcic and sodic pyroxenes. At a submicroscopic scale, however (Carpenter, 1980), there is ample evidence for first order exsolution between diopside-rich and jadeite-rich solid solutions, which is grossly complicated by the appearance of the ordered $P2$ omphacite (see Buseck *et al.*, this volume) in the middle of the gap.

Ginzburg (1969) proposed a miscibility gap between Ca pyroxenes and more aluminous and ferric compositions, but Shedlock and Essene (1979) have found apparently complete solid solution within this range in a tactite from Montana and from a compilation of other natural occurrences (Fig. 39).

In summary, there is potential for finding coarse coexisting pyroxenes of non-quadrilateral compositions, particularly in the structurally favorable diopside-jadeite series. The fact of finding a complete series of compositions formed under one set of conditions does not preclude a miscibility gap under other conditions. The most convincing proof in favor of a miscibility gap would be the occurrence of two pyroxenes in textural relationships suggesting mutual growth under equilibrium conditions.

Pyroxenes *vs* Pyroxenoids

In the pyroxene quadrilateral it is generally recognized that when the amount of Ca exceeds that of stoichiometric $CaMgSi_2O_6$, the single

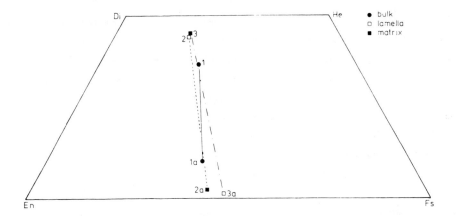

Figure 38. Analytical data on pyroxenes from a single Bushveld sample (KLG/2-3530) from Buchanan (1979): 1 and 1a are bulk compositions of primary coexisting augite and pigeonite; 2a is the orthopyroxene matrix of the inverted pigeonite and 2 represents coarse augite lamellae in inverted pigeonite; and 3 is the matrix of the augite host and 3a represents coarse "001" pigeonite lamellae in augite. The relations shown reflect an increase in fractionation of Fe and Mg from high to low temperature conditions. Similar relations have been shown in detail for the Skaergaard intrusion by Coleman (1978).

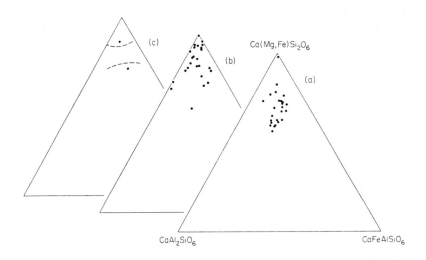

Figure 39. Composition plots of metamorphic Ca-pyroxenes. (a) Pyroxenes from the Knopf tactite, Montana, (b) pyroxenes from other published sources, (c) clinopyroxene analyses given in Ginzburg (1969) and proposed solvus (dashed lines) represented in these pyroxenes. From Figure 2 of Shedlock and Essene (1979).

chain silicate wollastonite appears. The wollastonite structure and other pyroxenoid structures are capable of taking in a higher proportion of large cations such as Ca, Mn, Na (plus OH), and even Fe^{2+} (Liebau, 1959; Belov, 1961; Peacor and Prewitt, 1963; Ohashi and Finger, 1975, 1978) than is the pyroxene structure. The pyroxenoids include wollastonite, bustamite, and pectolite made of tetrahedral chains with a three-unit repeat, and various hybrids that include segments of wollastonite-like chain and segments of pyroxene-like chain with a two-unit repeat. These are: rhodonite (3 Wo, 2Pyx) five tetrahedron repeat; pyroxmangite (also Lunar Pyroxferroite) (3 Wo, 4 Pyx) seven tetrahedron repeat; ferrosilite III (synthetic only) (3 Wo, 6 Pyx) nine tetrahedron repeat.

Clearly, cations of intermediate size such as Fe^{2+} and Mn^{2+} may behave more like small cations at low temperature and high pressure, more like large cations at high temperature and low pressure. Thus, pyroxenoids commonly turn out to be high T, low P polymorphs of pyroxenes of identical composition. An instructive set of experiments on the composition $MnSiO_3$ was performed by Akimoto and Syono (1972) who, at constant temperature and increasing pressure, successively produced rhodonite, pyroxmangite, clinopyroxene, and finally garnet at about 120 kbar. Elsewhere Lamb *et al.* (1972) have studied the stability of the rhodonite $CaMnSi_2O_6$ and its high P polymorph, the pyroxene johannsenite. All of this work illustrates the fact that any "limit" on the composition of pyroxene in equilibrium with pyroxenoid or pyroxenoids is going to be very much a function of pressure and temperature. Further complicating the issue is the fact that Mn-bearing pyroxenoids and pyroxenes are commonly stable in relatively low T environments where the CaFeMg silicates are commonly amphiboles (Huntington, 1975; Tracy *et al.*, 1980; Brown *et al.*, 1980).

Perhaps the most celebrated case of such polymorphism occurs in the Skaergaard Intrusion (Lindsley *et al.*, 1969; Burnham, 1976) where a ferrobustamite of composition close to $CaFeSi_2O_6$ appears to have crystallized from late liquid fractions, and then inverted to hedenbergite pseudomorphs. Experimental work on this, coupled with data on the quartz-tridymite inversion, suggest late Skaergaard liquids crystallized at 600 ± 100 bars and a temperature between 980 and 950°C. Absence of ferrobustamite pseudomorphs in the Bushveld Intrusion is attributed to higher pressure crystallization.

468

Brown *et al.* (1980) have recently compiled analytical data on co-existing pyroxenes and pyroxenoids in or close to the systems $CaSiO_3$-$FeSiO_3$-$MnSiO_3$ (Fig. 40) and $CaSiO_3$-$MgSiO_3$-$MnSiO_3$ (Fig. 41) for middle- to upper-amphibolite facies conditions. In the Fe-bearing system Ca pyroxenes coexisting with bustamites and rhodonites show nearly 50% solid solution toward the johannsenite end member which would itself only be stable at higher pressure. Orthopyroxene, which is not strictly stable under these conditions without some Mg, shows solid solution toward $MnSiO_3$ limited by coexistence with pyroxmangite. Elsewhere Huntington (1975) has suggested that Fe-Mn-Mg orthopyroxene (see Table 1) may form by dehydration of the assemblage Fe-Mg-Mn grunerite + Fe-Mn-Mg olivine + Fe-Mn-Mg pyroxmangite. In the Mg-bearing system, Figure 41, small Mg ions permit an apparently complete solid solution from diop-side to kanoite, $MnMgSi_2O_6$, the Mn analogue of $FeMgSi_2O_6$ high-T pigeonite (Kobayashi, 1977). This lies between the fields of orthopyroxene and pyroxmangite, and is consistent with experimental work of Lindsley *et al.* (1974) showing preference of Mn for pigeonite over coexisting ortho-pyroxene. $C2/c$ pyroxenes that formed under amphibolite facies condi-tions at Balmat, New York, and with intermediate compositions between diopside and kanoite (Table 1) have broken down at lower temperatures to submicroscopic mixtures of $C2/c$ Ca-pyroxene and $P2_1/c$ primitive kanoite. Tracy *et al.* (1980) have reported Mn-ferrosalite (Table 1) in the kyanite zone in central Massachusetts coexisting with both rho-donite and pyroxmangite (Fig. 42). The pyroxene contains optically visible "001" lamellae that were too scarce to identify in x-ray single crystal photographs, but may be primitive Mn pigeonite. These authors also suggest the former presence of Mn-Fe-Mg orthopyroxene in this as-semblage, later retrograded to a grunerite-magnetite-quartz intergrowth.

The foregoing summary serves to illustrate the sensitivity of the limits of pyroxene solid solutions toward pyroxenoids both to bulk com-position (particularly in Mn^{2+}) and to environmental conditions. These limits relate to pyroxene crystal chemistry, but even more to the com-plex crystal chemistry of the pyroxenoids.

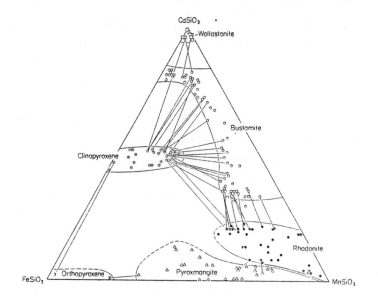

Figure 40. Compilation of analyses for Ca-Fe-Mn single chain silicates estimated to have formed between 4 and 8 kbar and temperatures of 600 ± 100°C. Tie lines join coexisting phases. From Brown, Essene and Peacor (in press).

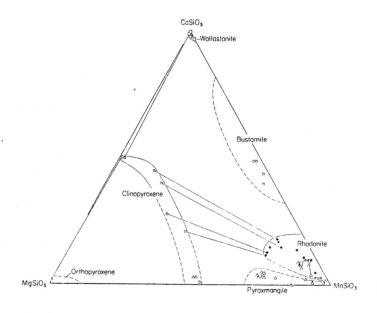

Figure 41. Compilation of analyses for Ca-Mg-Mn single chain silicates estimated to have formed between 3 and 6 kbar and temperature of 600 ± 100°C. From Brown, Essene and Peacor (in press).

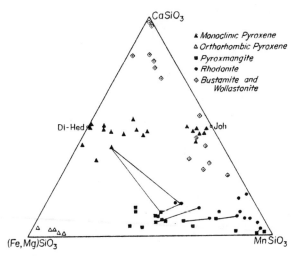

Figure 42. Compilation of analyses for Ca-Mg-Fe-Mn single chain silicates of intermediate Fe/Mg ratio formed at moderate to high pressure metamorphic conditions. Three
phase pyroxene-pyroxmangite-rhodonite assemblage is from the kyanite zone in the Pelham
dome, Massachusetts (Tracy *et al.*, 1980).

EXTERNAL LIMITS CONTROLLED BY MINERALS OF "IN PLANE" COMPOSITION

General Statement

Limits of pyroxenes toward pyroxenoids have been covered under internal limits because of the crude similarity of pyroxene and pyroxenoid
crystal chemistries. Otherwise, they might have been covered here. The
other important minerals that have "in-plane" compositions, namely garnets and rhombohedral oxides, are present because of large amounts of
"A-substitution" components, and these minerals are nearly always richer
in the "A" components than the coexisting pyroxenes. This tends to simplify consideration of the equilibria. A further simplifying factor in
natural pyroxenes is that the A-substitutions are facilitated mainly by
substitution of tetrahedral Al which is usually most significant in SiO_2-
poor bulk compositions. The other A-components can only be substituted
in major amounts when they can substitute in octahedral positions and be
compensated by tetrahedral Al. The main exception to this is in the case
of tetrahedral Fe^{3+} which does reach modest amounts in a few natural pyroxenes. It is natural then to commence this short review with consideration of pure Al substitutions.

Pyroxenes with Aluminous Garnets and Corundum

Saturation of pyroxenes with these minerals mainly involves consideration of combinations of the CFM components and Al_2O_3, yielding
grossular-rich garnet, pyrope-almandine-spessartine-rich garnet solid

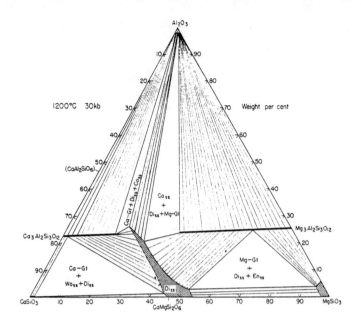

Figure 43. Summary diagram for phase relations in the system Ca-SiO$_3$-MgSiO$_3$-Al$_2$O$_3$ at 1200°C and 30 kbar. From Boyd (1970).

solutions, and corundum. Of key importance in modeling the petrology of garnet lherzolites and of making pressure estimates for mantle samples has been the system CaSiO$_3$-MgSiO$_3$-Al$_2$O$_3$ (Fig. 43) in which fairly extensive high pressure experimental work has now been done (Boyd, 1970; Wood, 1974; Akella, 1976; Lindsley and Dixon, 1976; Lindsley, this volume). Alan Thompson (1979) has outlined the theoretical basis for consideration of equilibria within this plane and presented several alternative topologies for the plane under different P-T conditions. He has also considered equilibria that pierce the plane or involve phases outside the plane, in a manner similar to the general outline expressed at the beginning of this paper and in Figures 4 and 9[7]. To the pure CaSiO$_3$-MgSiO$_3$-Al$_2$O$_3$ system various experimentalists and theoreticians have added components such as FeO, Na$_2$O, TiO$_2$, and Cr$_2$O$_3$ to more closely model natural rock compositions, particularly under mantle conditions, and considerable experimental work has also been done using natural garnet lherzolites themselves.

----

[7]Unfortunately, Thompson's key figure is too small to be reproduced here and the reader is referred to the original.

In Figure 43, at simulated mantle conditions, pyroxenes are limited
by four three-phase assemblages, clinopyroxene-orthopyroxene-pyrope-rich
garnet appropriate to garnet lherzolites; clinopyroxene-pyrope-rich garnet-corundum, clinopyroxene-grossular-rich garnet-corundum, and clinopyroxene-grossular-rich garnet-wollastonite. The "gap" in the garnet
solid solution is broached by clinopyroxenes rich in AlAl substitution.
Under lower-pressure, higher-temperature conditions the Ca-pyroxene solid
solution can extend to $CaAlSiAlO_6$ (Hays, 1966), and could exsolve corundum
on cooling. At lower pressures pyrope-rich garnet becomes metastable
with respect to orthopyroxene-corundum (Schreyer and Seifert, 1969; Morse
and Talley, 1971) or some combination of phases above and below the pyroxene plane (see below). At higher pressures and/or lower temperatures the
grossular-pyrope solution may become complete or there may be coexisting
garnets, effectively cutting off pyroxenes from corundum.

The solubility of Al_2O_3 in clinopyroxene and orthopyroxene in equilibrium with garnet is a clear function of pressure, decreasing markedly
with increasing pressure at constant temperature, because of the smaller
partial molar volume of Al_2O_3 in solution in garnet as compared to pyroxene.
Other things being equal, which they are commonly not, the Al_2O_3 content
of pyroxenes in this assemblage serves as an excellent geobarometer. In
the region where the clinopyroxene and garnet solid solutions cross, relations must be much more complex. In general, however, it is to be expected that high T pyroxenes, particularly those crystallized at intermediate to high pressures, should be capable of exsolving garnet if ex-
posed to the proper cooling conditions. Garnet lamellae, usually on
(100), have indeed been found in both clinopyroxene and orthopyroxene
(Beeson and Jackson, 1970; Dickey, 1970; Jaffe and Jaffe, 1979) in
settings where the garnet composition appears to have been derived exclusively from pre-existing homogeneous pyroxene. By careful probe work
and modal analyses Beeson and Jackson (1970) reconstructed the original
bulk compositions of pyroxene that exsolved garnet and have compared
these (Fig. 44) to the equilibrium compositions in the pure Mg system
at 1200°C and 30 kbar determined by Boyd (1970). Although the quantitative effect of Fe is not certain it appears that the pyroxenites from
Salt Lake Crater originally crystallized at pressures less than 30 kbar,
and were subsequently held in the mantle at somewhat lower temperatures

473

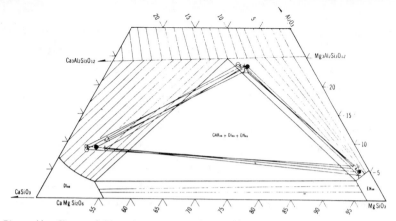

Figure 44. Phase relations in a portion of the system CaSiO$_3$-MgSiO$_3$-Al$_2$O$_3$ at 30 kbar and 1200°C after Boyd (1970). Large symbols with tie lines are reconstructured compositions of primary orthopyroxene, augite, and garnet from garnet pyroxenites at Salt Lake Crater, prior to cooling and exsolution. Numbers indicate weight %; from Beeson and Jackson (1970).

where garnet exsolution took place before they were incorporated in the Honolulu Series basalts. Fe-rich analogues of these assemblages were found in Adirondack granitic gneisses (Jaffe *et al.*, 1978) believed to have been metamorphosed at about 750°C and 7-8 kbar. Here the effect of Fe and Mn in lowering the free energy of garnet is clear, and the amount of Al substitution in the pyroxenes is very low (see Fig. 11) reaching only 1 wt % in the most iron-rich clinopyroxenes as compared to 8-11% in the reconstructed clinopyroxene compositions from Salt Lake Crater.

Although their "A-substitutions" are generally small, it should be noted that omphacites have been found coexisting with corundum in a number of eclogites (Sobolev, 1968). This is understandable in that there is no Na equivalent of garnet to intervene between Na-rich pyroxene and corundum.

Pyroxenes with Ilmenite or Perovskite

Equilibria of pyroxenes with ilmenite are of major interest in a variety of terrestrial settings, but very few definitive experimental and theoretical studies have been completed. As shown above, the most extreme "A-substitutions" in any natural pyroxenes involve Ti in M1 coupled with tetrahedral Al. Further, exsolution lamallae of ilmenite are extremely common (as well as spinels, rutile, and other out-of-plane oxides) in former high T pyroxenes (Morse, 1970; Ashwal, 1974; Grapes, 1975). Spectacular pyroxene-ilmenite intergrowths in nodules from kimberlite were once ascribed to breakdown of pre-existing Ti-rich garnet or ilmenite-structured silicate, but this has been discounted by many (Rawlinson and Dawson, 1979).

Akella and Boyd (1972) examined the compositions of pyroxenes in equilibrium with ilmenite, garnet, liquid and metallic iron under simulated lunar mantle conditions. Under these conditions, however, solubility of Al_2O_3 in pyroxenes is low, which permits only very low coupled substitutions of $FeTiO_3$. Yagi and Onuma (1967) performed experiments on the join $CaMgSi_2O_6-CaTiAl_2O_6$ at 1 atmosphere and found 3.7 wt % TiO_2 in diopside coexisting with perovskite and corundum. Yang (1973) reports experiments in the system $CaMgSi_2O_6-CaTiAl_2O_6-CaAlSiAlO_6$ in which augites were produced with up to 16 wt % TiO_2 and containing exsolutions of perovskite and corundum. The effects of $Fe^{2+}Ti^{4+}O_3$ component as suggested by Tracy and Robinson (1977) have yet to be extensively explored experimentally. Further, it is fairly obvious that redistribution of Al should take place within the pyroxene left over following exsolution of lamellae. The importance of Ti substitutions in augite suggests that experimental theoretical and analytical exploration would yield valuable results.

Cr Substitutions in Augites

The substitution of Cr in augite on the join $CaMgSi_2O_6-CaCrSiAlO_6$ was studied by Dickey et $al.$ (1971) and Dickey and Yoder (1972). In the subsolidus at 1200°C and 1 atmosphere, the bulk composition of 20 wt % $CaCrSiAlO_6$ consists of chrome diopside with 14 wt % of the chrome pyroxene component (13 mole % $CaCrSiAlO_6$) plus spinel plus possibly Cr_2O_3 (eskolaite). Since spinel composition lies outside the pyroxene plane it appears either that the bulk composition was pulled out of the plane by reaction with the crucible or that the pyroxene produced was non-stoichiometric with Ca vacancies and excess SiO_2. Solubility of (Cr,Al_2O_3) appears to be slightly less at 10 and 20 kbar than at 1 atmosphere.

Pyroxenes with Hematite and Andradite Garnet

As part of a systematic study of phase relations among rock-forming oxides Huckenholz et $al.$ (1969) studied the system $CaSiO_3-CaMgSi_2O_6-Fe_2O_3$ at 1 atmosphere between 1050 and 1400°C (Fig. 45). The bulk compositions are probably realistic only for extremely Al-poor skarns, but they do illustrate the potential of substitution of the $CaFe^{3+}Fe^{3+}SiO_6$ component

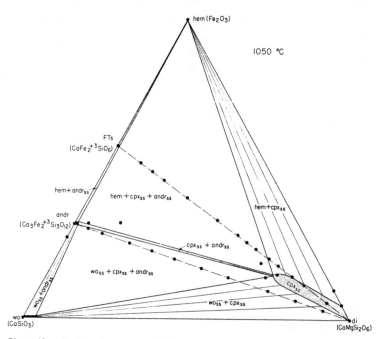

Figure 45A. Isothermal section at 1050°C in the plane CaSiO₃-CaMgSi₂O₆-Fe₂O₃ from Huckenholz *et al.* (1969). Shows the saturation of Al-free pyroxene with hematite and andradite garnet.

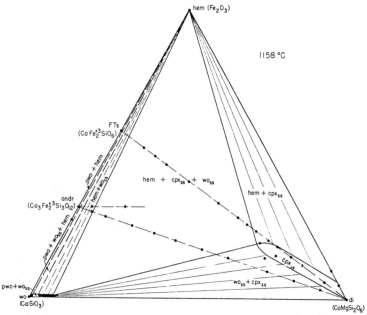

Figure 45B. Isothermal section at 1158°C in the plane CaSiO₃-CaMgSi₂O₆-Fe₂O₃ from Huckenholz *et al.* (1969). Shows the saturation of Al-free pyroxene with hematite and wollastonite.

476

in diopside up to about 30% either in equilibrium with andradite and hematite or, above 1150°C, with hematite and wollastonite. Preliminary data suggest these pyroxenes may also have a small amount of Ca replacing Mg in M1. Replacement of Mg by Fe^{2+} would eventually lead to incompatibility of Ca pyroxene and hematite in favor of andradite-magnetite-wollastonite and quartz-magnetite-wollastonite. The absence of an Fe^{3+}-rich silicate analogous to plagioclase probably means such ferric pyroxenes can be stable with quartz.

The substitution of Fe^{3+} in M1 coupled with tetrahedral Al needs to be explored experimentally. As shown above, such substitutions are widely important in a variety of terrestrial pyroxenes. A range of such pyroxenes in aluminous-ferric skarns from Montana, containing various assemblages with grossular garnet, forsterite, spinel and hydrous minerals, were explored by Shedlock and Essene (1979) (see Fig. 39). They pointed out that some of the pyroxene compositions are equivalent to garnet of intermediate grossular-andradite-pyrope composition, whereas the garnets actually present in the skarns are nearly pure grossular-andradite garnets. They also showed that, although the "A-substitutions" in these pyroxenes are high, none of them coexists with corundum or hematite in a combination with garnet that actually provides a natural limit to the amount of A-substitution under the prevailing conditions.

EXTERNAL LIMITS CONTROLLED BY MINERAL TIE LINES
PASSING THROUGH THE PYROXENE PLANE

General Statement

Under many conditions compositions within the "pyroxene plane" are metastable with respect to a mineral pair of equivalent composition on a tie line that passes through the "plane." The classic example is the "Forbidden Zone" of the pyroxene quadrilateral (Lindsley and Munoz, 1969) where Fe^{2+}-rich pyroxenes are metastable with respect to fayalite plus quartz. This and other less obvious examples are reviewed briefly. Obviously, if a melt composition lying on one side of the pyroxene-plane coexists with a mineral on the other, this also will provide a limit on pyroxene composition. Melt compositions are considered too complex to cover in the time and space available here.

477

Instability of Quadrilateral Pyroxenes

The history and phase relations of the "Forbidden Zone" of the pyroxene quadrilateral are reviewed by Lindsley elsewhere (this volume).
Recent experimental refinements by Bohlen *et al.* (1980a,b) and Bohlen
and Boettcher (1980) allow fairly precise pressure estimates, taking into
account effects of Mg and Mn, where Ca-poor Fe-rich pyroxenes coexist with
olivine plus quartz, and where pyroxene is stabilized by higher pressure
and lower temperature. Pyroxene compositions alone (Jaffe *et al.*, 1978;
Klein, 1978) can be used to estimate minimum pressures of equilibration
at known temperatures. Olivine plus quartz can be used to estimate maximum pressures.

Instability of Jadeite and Omphacite

The reaction by which the NaAl pyroxenes become stable in tholeiitic
gabbro compositions and many other common rock compositions involve the
breakdown of plagioclase on its own composition. These are *not* the reactions that govern the ultimate stability of jadeite and omphacite. As
Curtis and Gittins (1979) point out, in their study of the Red Wine Complex, Labrador, jadeite and omphacite are first stabilized by breakdown
of the association of nepheline plus albitic plagioclase in strongly
silica-undersaturated bulk compositions. These reactions appear to take
place under fairly normal amphibolite facies conditions rather than the
higher pressure conditions required for feldspar breakdown and the production of eclogite.

Limits on A-Substitution

A great deal of recent experimental endeavor has gone into characterization of the conditions for formation of lherzolites, particuarly
as modelled in the system $MgO-CaO-Al_2O_3-SiO_2$. At high pressure, as shown
above, lherzolites consist of olivine, two pyroxenes, and garnet, the last
providing Al saturation for both pyroxenes. At intermediate pressures
garnet and olivine react to form pyroxenes plus spinel, creating spinel
lherzolites in which there is no phase that provides complete Al-satura-
tion for the pyroxenes, although the Al-content of the pyroxenes in
equilibrium with olivine and spinel is a function of P and T (Obata, 1976;
Danckwerth and Newton, 1978, see below). Under P-T conditions of spinel

lherzolites, maximum Al content of pyroxenes is to be found in other rock bulk compositions, in some cases, still limited by garnet or corundum, in others by cordierite plus sapphirine (Anastasiou and Siefert, 1972), quartz plus sapphirine (Caporuscio and Morse, 1978), quartz plus spinel, sillimanite plus spinel, anorthite plus spinel, and so on. It is presumably in this range that many garnet compositions become metastable with respect to other combinations of phases, and in which pyroxenes with maximum amounts of pure Al-Al substitutions are to be found. Interestingly, the hypersthene of Caporuscio and Morse (1978), growing in Al-saturated equilibrium with either sapphirine and quartz or spinel and quartz was also apparently Fe_2O_3-$FeTiO_3$-saturated by virtue of presence of a 50% hematite-ilmenite solid solution. Octahedral R^{3+} values of 0.180 for Al, 0.032 for Fe^{3+}, and 0.002 for Ti^{4+} should be compared with limiting values from Deer *et al.* (1978) in Table 1.

At still lower pressures in lherzolites, pyroxenes and spinel react to produce olivine plus plagioclase, and saturation with this mineral pair places fairly severe limits on the amount of pure Al substitution that is possible in pyroxenes of low pressure origin. However, there is no analog to the olivine-plagioclase pair in the comparable Fe^{3+} and $Fe^{2+}Ti^{4+}$ systems, so a coupling of these components in octahedral positions with tetrahedral Al still permits very large "A-substitutions" in near-surface rocks, as exemplified by the skarn inclusions in Soufriere Volcano (Devine and Sigurdsson, 1980, see Table 1) and in the Boulder Batholith (Shedlock and Essene, 1979), and by the extreme substitutions in some titanaugites (Tracy and Robinson, 1977, see Table 1).

Exsolution lamellae of plagioclase parallel to (100) have been described in orthopyroxenes and less commonly clinopyroxenes from a number of localities (Dickey, 1970; Emslie, 1975; Morse, 1975) in what were probably originally homogeneous aluminous pyroxenes. In some of the examples described by these authors it appears that the amount of plagioclase exsolved is roughly balanced by exsolution of a low silica phase, either olivine or a spinel group mineral. In one variety Morse suggests that oxidation of Fe to produce magnetite lamellae provided the SiO_2 needed for plagioclase exsolution. In other cases it appears the original pyroxenes had compositions above the pyroxene plane, i.e., with cation vacancies, so that they were capable of exsolving plagioclase.

This conclusion is apparently to be drawn from recent detailed work by Brett and Emslie (1979). The fact that the lamellae in orthopyroxene studied by Emslie (1975) and Morse (1975) contain andesine cores and anorthite rims, suggests early exsolution of sodic component (Na-deficient jadeite) followed by successive exsolution of a calcic component (Ca-deficient CATS), and is consistent with progressive unloading. Orthopyroxenes of similar primary bulk composition could exsolve both plagioclase and garnet, as well as ilmenite, if exsolved under somewhat lower T or higher P conditions as reported by Jaffe and Jaffe (1979) in the Adirondacks.

CONTROLS ON PYROXENE COMPOSITIONS EXERTED BY COMBINATIONS OF PHASES ALL ON ONE SIDE OF PYROXENE SPACE

General Statement

Equilibria within pyroxene composition space or on tie lines passing through it determine whether a given pyroxene will or will not be stable on its own composition. Equilibria on one side or the other of the space determine which stable pyroxenes can occur in which bulk compositions. These equilibria can be conveniently grouped into those on the SiO_2-rich side of the plane and those on the SiO_2-poor side.

Equilibria in SiO_2-rich Compositions

In this composition range a chief question is: Which pyroxene compositions can be stable with quartz? The answer is that any pyroxene can be stable with quartz that does not have a more siliceous mineral between it and quartz or a tie line from a more siliceous phase to another pyroxene between it and quartz.

In Al-rich systems the more siliceous phases (see Fig. 4) include plagioclases, cordierite, and Al-silicate minerals. Albite prevents jadeite from occurring with quartz under moderate pressure conditions where jadeite is stable on its own composition. Anorthite, cordierite, and Al-silicate prevent quartz from occurring with Al-rich pyroxenes. These relations have been studied experimentally on the joins diopside-albite, diopside-anorthite, and diopside-$An_{49}Ab_{51}$ by Kushiro (1969) and others. Results in diopside-albite are portrayed in diopside-jadeite-quartz in Figure 46. Omphacites in equilibrium with albite plus

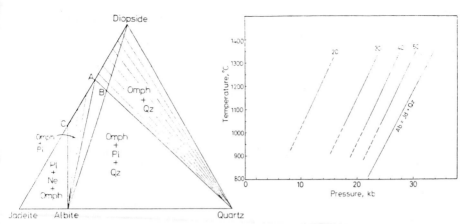

Figure 46. Hypothetical isothermal-isobaric section of the system diopside-jadeite-quartz. It is assumed for convenience that plagioclase is pure albite, and all omphacites lie on the join jadeite-diopside. Point C is the limiting composition of omphacite in equilibrium with albite and nepheline, and should move with increasing pressure to the limiting reaction albite + nepheline → jadeite. Point A is the limiting composition of omphacite in equilibrium with albite and quartz and should move with increasing pressure to a higher pressure limiting reaction albite → jadeite + quartz. From Kushiro (1969).

Figure 47. P-T diagram showing isopleths of omphacite composition (point A in Fig. 46) in equilibrium with albite and quartz. Isopleths are in mole % of jadeite in omphacite. From Kushiro (1969).

nepheline (C) are much more jadeitic than those with albite and quartz (A). P-T dependence of the latter is shown in Figure 47. In diopside-anorthite (Fig. 48, upper right edge of each triangle) pyroxene with quartz and plagioclase becomes more aluminous with increasing pressure until pyroxene plus plagioclase react to give garnet (approx. $Gros_2Pyrope_1$) plus quartz. At higher pressures than this Al content of Ca pyroxenes is reduced in equilibrium with garnet plus quartz. Overall relations in the pseudo-ternary quartz projection are illustrated at four pressures in Figure 48. At low pressures (15 kb) the pyroxene field is limited to low Na content by quartz-albite but reaches higher ^{VI}Al content in equilibrium with quartz and anorthite. By 20 kbar AlAl substitution is being reduced by the garnet-quartz equilibrium whereas NaAl substitution is increasing at the expense of albite-quartz. This process continues through 35 kb where the diopside-jadeite series is complete and no plagioclase is stable. Meanwhile, anorthite has broken down at about 26 kbar on its own composition to grossular + kyanite + quartz.

In the Fe^{3+}-rich analog of the system just discussed there are no Fe^{3+} analogs of plagioclase, cordierite, or Al-silicate. The result is

481

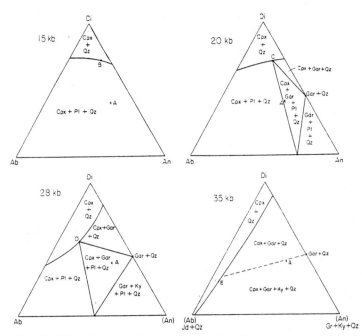

Figure 48. Simplified isothermal-isobaric projections from quartz in the system
$CaMgSi_2O_6$–$NaAlSi_3O_8$–$CaAl_2Si_2O_8$ at 1150°C and 15, 20, 28, and 35 kbar. Abbreviations:
Ky, kyanite; Gr, grossular; Qz, quartz; Cpx, clinopyroxene; Pl, plagioclase; Gar,
garnet. Composition of garnet is assumed to be Pyrope, Grossular 2. A is a mixture
of 40·weight % diopside and 60% plagioclase ($An_{75}Ab_{25}$ mole %). From Kushiro (1969).

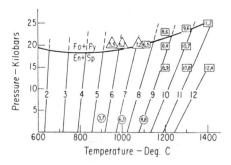

Figure 49. Calculated Al_2O_3 isopleths of enstatite, in weight percent, in the presence of
forsterite plus spinel, assuming ideal pyroxene
solution of the $Mg_2Si_2O_6$ and $MgAl_2SiO_6$ components and the parameters $\Delta H° = 9.0$ kcal, $\Delta S° =$
3.60 cal/K. Triangles are experimental data of
Danckwerth and Newton (1978), rectangles are
data of Fujii (1976), and circles are data of
Anastasiou and Seifert (1972). Plotted points
are at centers of symbols. From Danckwerth and
Newton (1978).

that quartz is stable with $NaFe^{3+}$ pyroxene at 1 atmosphere as well as with
hematite and andradite that would
limit the amount of Fe^{3+} substitution
in Ca pyroxenes. A similar situation
probably holds for the $Fe^{2+}Ti^{4+}$ system
where quartz is certainly stable with
ilmenite, but more realistic combinations of Al, Fe^{3+} and Ti^{4+} should relate to much more complex equilibria.

Equilibria in SiO_2-poor Compositions

The reaction of garnet plus olivine to yield pyroxenes plus spinel,
which creates the distinction between
garnet and spinel lherzolites has been
mentioned above. The fine structure

482

of this reaction is exceedingly complex because it depends on the Fe/Mg as well as Cr, Ti, and Fe^{3+} of coexisting phases. Danckwerth and Newton (1978) have completed the most recent calibration of the Al content of orthopyroxene in equilibrium with olivine and spinel in a pure $MgO-Al_2O_3-SiO_2$ "lherzolite" composition (Fig. 49). Clearly, there are numerous other equilibria that control pyroxene compositions on the SiO_2-poor side of the "plane."

EXTERNAL LIMITS GOVERNED BY EQUILIBRIA WITH MINERALS
BEARING VOLATILE CONSTITUENTS

A full discussion of this topic would involve fully one-third or more of the entire subject of metamorphic petrology, and would involve the introduction of a vast number of new phases. Many of these new phases, with the addition of H_2O, CO_2, F or Cl, would plot somewhere in Figure 4, and would be involved in the same types of equilibria vis-a-vis the pyroxenes as discussed previously, except that a fluid phase would usually be involved in the reactions. Another group of equilibria involving orthopyroxenes with biotite and K feldspar require the addition of K_2O as well as H_2O, even though the pyroxenes themselves contain no K_2O.

The carbonate minerals of principal importance plot at the CFM apex of Figures 4 and 9 and CFM pyroxenes can be formed by direct decarbonation of assemblages of these carbonates plus quartz in contact or regional metamorphism. More commonly, where metamorphic fluids contain even small amounts of H_2O, the decarbonation process involves production of hydrous minerals that are subsequently devolatilized to produce pyroxenes by mixed volatile CO_2-H_2O equilibria (Greenwood, 1962; Skippen, 1974; Kerrick, 1974). Reactions involving melting or devolatilization of carbonate-bearing peridotite with or without a hydrous phase also appear to be of major importance in the mantle at pressures greater than 23 kbar (Eggler, 1976; Wyllie, 1979).

The production of pyroxenes by dehydration of amphiboles is of particular interest because it is common, and also because of the crystal-chemical fact that amphibole compositions can be characterized as a combination of pyroxene plus talc or Na biotite. Thus, all amphibole compositions in which the sheet silicate component is talc and which vary only in the pyroxene component form a line in Figure 4 or a plane in

483

Figure 9 that is exactly parallel to the "pyroxene plane" but with a higher SiO_2 content. On the other hand, amphiboles in which Na substitutes in the empty "A site" with coupled substitution of tetrahedral Al, in effect Na biotite instead of talc, drop to lower Si content and may lie in or well below the pyroxene plane. Thus, the production of pyroxenes by devolatilization of amphiboles has as much or more to do with the complex crystal chemistry of amphiboles as it does with the pyroxenes.

Although the production of pyroxenes from amphiboles usually takes place with increasing temperature, it may take place at constant temperature with increasing pressure under mantle conditions where amphibole has a *higher* molar volume that its anhydrous products (usually including garnet) plus dense high pressure aqueous fluid (Lambert and Wyllie, 1968; Mysen and Boettcher, 1975). There also are some curious anomalies where pyroxene is actually destroyed by prograde dehydration. For example, Trommsdorff and Evans (1974) have shown that a common low grade assemblage in the Malenco Serpentinite is antigorite-olivine-diopside. Diopside in this assemblage breaks down to tremolite by the prograde dehydration diopside + antigorite → tremolite + olivine + H_2O, commonly followed by antigorite → olivine + talc + H_2O to yield talc-olivine-tremolite rock.

The dehydration of biotite plus quartz to yield orthopyroxene plus K feldspar is the chief reaction by which pyroxene appears in prograde metamorphism of rocks of granitic composition, and is commonly cited as the key reaction leading to the true granulite facies. This reaction is sensitive to Fe/Mg ratio, as well as activity of H_2O, occurring first in more Fe-rich compositions, but has not yet been well calibrated experimentally. By this stage pyroxenes have appeared at the expense of amphiboles in most mafic and calcareous compositions, although some hornblende compositions may still persist. The fact that Fe-Mg amphiboles break down to orthopyroxenes before the dehydration of biotite to orthopyroxene plus K feldspar explains why assemblages of Fe-Mg amphiboles with K feldspar are unknown.

I have mentioned previously the possibility of creating conditions for spinel exsolution from pyroxene by oxygen metasomatism. This is akin to the better known oxidation-exsolution processes in the opaque oxides (see Reviews of Mineralogy 3, *Oxide Minerals*, 1976). The same notion may be extrapolated to the possibility of hydration-exsolution in which

amphibole is produced from pyroxene. Veblen *et al.* (1977) and Veblen
and Buseck (1977) have reviewed some of the structural complexities involved with this process. In certain favorable cases all of the necessary constituents except H_2O may be housed in the pre-existing pyroxene.
More commonly, slight additions or removals of SiO_2, Al_2O_3, Na_2O, K_2O
or other locally available constituents would be needed for amphibole
formation. Some workers (Papike *et al.*, 1969; Yamaguchi *et al.*, 1978),
however, have suggested a natural amphibole-pyroxene compound (in effect
an ordered submicroscopic amphibole-pyroxene intergrowth) that could be
recrystallized to separate pyroxene and amphibole lamellae. The plane
of best structural fit between amphibole and pyroxene is normally (010)
(Veblen *et al.*, 1977), and examples of oriented microscopic and submicroscopic intergrowths on this plane are well known (Papike *et al.*, 1969;
Yamaguchi *et al.*, 1978). For example, in the Belchertown Intrusive Complex, Massachusetts (Ashwal *et al.*, 1979), where amphiboles are replacing
pyroxenes on a broad front, (010) lamellae of hornblende are common in
partially replaced augite (personal observation, 1970). Yamaguchi *et al.*
(1978) also report (100) amphibole lamellae in diopside, for which the
structural fit is less satisfactory. In addition, there are countless
examples of exsolved pyroxenes in which the lamellae appear to have been
subsequently altered to amphibole to give extremely complex and confusing
patterns. It is, of course, the determination of the true sequence of
events that is the key to genetic interpretations.

Finally, I should mention the world's best known occurrence of pyroxene formed under low temperature diagenetic conditions, acmite in the
lake deposits of the Eocene Green River Formation in Wyoming and Utah
(Milton and Eugster, 1959; Clark *et al.*, 1959), where it occurs together
with magnesioriebeckite, calcite, dolomite, albite, and a variety of
matrix materials. Although the saline waters may have had an effect in
lowering activity of H_2O to promote pyroxene crystallization, acmite,
unlike many pyroxenes, does not have a closely similar amphibole analog
free of FeO and MgO.

SOME PYROXENE-LIQUID FRACTIONATION TRENDS: A SHORT NOTE ADDED IN PROOF

An objective of igneous petrology is to trace the evolution of igneous
liquids either during fractional crystallization, fractional fusion, or any

485

other dynamic process. In such studies pyroxenes play a central role because they are such an extensive solid solution, and igneous liquids, especially under relatively H_2O-poor conditions, stay saturated with one or more pyroxenes through much of their crystallization history. Thus, the study of the evolution of pyroxenes is central to the petrology of igneous complexes and the tracing of the liquid line of descent (or ascent). Detailed coverage of this topic is beyond the scope of this report, but a summary of a few of the more obvious terrestrial trends seems warranted.

Igneous fractionation trends in pyroxenes relate to fractionation of various ions between pyroxenes and liquids, and also to the various other phases the liquid is precipitating and their relations with liquids. A nearly universal relationship in ferromagnesian silicates is the enrichment of Mg relative to Fe^{2+} in crystals, and thus the enrichment of liquids in Fe^{2+} in fractional crystallization. This has been alluded to above under quadrilateral pyroxenes including those of the Skaergaard, Bushveld and other tholeiitic layered intrusions. In basaltic liquids of lower silica content not saturated with orthopyroxenes or pigeonite, but with Ca-pyroxene and olivine (Smith and Lindsley, 1971; Morse, 1972) the Ca-pyroxene tends to follow a similar Fe-enrichment trend nearly parallel to but at a higher Ca content, consistent with tie lines from the Si-under-saturated surface of the pyroxene field to coexisting olivines (see Fig. 21 part b). Interestingly, Ca contents of Ca pyroxenes also rise in equilibrium with extremely Fe-rich compositions of tholeiitic liquids also saturated with fayalite and quartz. In this case the low free energy of fayalite plus quartz is effective in removing "Fe-pigeonite component" from the coexisting Ca-pyroxene (see Figs. 22 and 23).

It has long been known (see Lindsley, and Huebner, this volume) that extreme fractionation toward Fe^{2+}-rich liquids requires relatively low oxygen fugacity. Crystallization under more oxidizing conditions induces the early crystallization of Fe^{3+}-rich magnetite, and massive early removal of Fe^{2+} from the liquid. Thus, there are numerous examples of complexes that are chemically evolved in other respects that precipitated very Mg-rich pyroxenes due to high oxygen fugacity. The Belchertown Quartz Monzodiorite, Massachusetts, for example (Ashwal *et al.*, 1979), contains coexisting orthopyroxene and augite ($Wo_{3.5}$, $En_{65.5}$, Fs_{31}, and $Wo_{43.9}$, $En_{41.2}$, $Fs_{14.8}$, see Fig. 25) with Fe/Mg ratios fairly typical of Bushveld main zone gabbros (Fig. 22).

Large amounts of "A-substitutions" in igneous pyroxenes are clearly favored by physical or chemical conditions under which a saturating solid phase or combination of phases does not precipitate. Tracy and Robinson (1977) show a continuous increase in Ti^{4+} and Al with Fe/Mg ratio in Tahiti augites to the highest Ti contents, beyond which Ti and Al drop with further increase in Fe/Mg ratio. They suggest that the extreme Ti enrichment and final drop is related to the very late precipitation of ilmenite, part of which occurs as skeletal precipitates in interstitial brown glass. Such late precipitation of ilmenite could be an equilibrium phenomenon related to an unusual bulk composition, or could be a result of metastability and failure to precipitate until a very late stage. Morse (1975) has suggested metastable failure of plagioclase precipitation to account for extreme Al-enrichment in orthopyroxenes that produced exsolution lamellae of plagioclase on cooling. It should also be emphasized that in rapidly cooled pyroxenes the complex chemistry of different pyroxene surfaces, commonly leading to sector zoning, can greatly confuse the question of what an equilibrium pyroxene composition is (Hollister and Hargraves, 1970; Bence *et al.*, 1971; Hollister and Gancarz, 1971; Dowty, 1976)*.

The fractionation of Ca into plagioclase and Na into liquid is known to every petrology student. Less well advertised is the fractionation of Ca into pyroxene and Na into liquid for liquids of appropriate chemistry (available Fe^{3+} and molecular $(Na_2O + K_2O) > Al_2O_3$, i.e. peralkaline) that are in equilibrium with members of the augite - aegerine-augite - acmite series (Yagi, 1953; Larsen, 1976)*. In liquids where there is coexisting plagioclase this trend is not pronounced until plagioclase fractionation has come fairly close to albite composition, and albitic plagioclase coexists with Ca augite (commonly hedenbergite). The end process of this peralkaline crystallization trend would be a liquid that would crystallize acmite, albite, fayalite (or riebeckite), quartz or nepheline, and potassic feldspar. That feldspar crystallization is not an essential part of this trend is shown by the pyroxene trends of the Fen Complex carbonatites and feldspar-free nepheline-rich rocks (Mitchell, 1980).

---

* See REFERENCES FOR NOTES ADDED IN PROOF below.

Studies of zoned pyroxenes with the electron probe have great potential for understanding complex liquid lines of descent. Much of this potential has been realized for the Moon (see Papike, this volume) where bulk compositions are relatively limited, but such work is only beginning for terrestrial rocks, particularly those of more unusual bulk compositions. There is the potential, in pyroxenes as well as other minerals, not only to follow lines of descent, but to recognize momentary reversals of the process caused by injections of fresh magma (Buchanan, 1979), magma mixing, or changes in pressure or volatile conditions. The augites of St. Dorotheé Sill, Quebec (Robert O'Laskey, pers. comm., 1980), for example, have green cores rich in Fe^{3+} Al substitution which grade outward toward more normal pale green hedenbergite. These are sharply overgrown by pale brown more Mg-rich titanaugite which then increases in Ti^{4+} Al substitution to a very high value. The extreme outermost edge of the augites shows a steep drop in Ti Al substitution coupled with a large increase in Na Fe^{3+} substitution toward acmite. These augites, thus, have the record of a complex history of liquid-crystal fractionation and reaction, that will be very interesting to translate.

REFERENCES FOR NOTE ADDED IN PROOF

Bence, A.E., Papike, J.J., and Lindsley, D.H. (1971) *Proceedings of the Second Lunar Science Conference,1,* 559-574.

Dowty, E. (1976) Crystal structure and crystal growth II. Sector zoning in minerals. Am. Mineral., 61, 460-469.

Hollister, L.S., and Hargraves, R.B. (1970) Compositional zoning and its significance in pyroxenes from two coarse-grained Apollo 11 samples. Apollo 11 Lunar Sci. Conf., 1, 541-550.

————, and Gancarz, A.J. (1971) Compositional sector-zoning in clinopyroxene from the Narce area, Italy. Am. Mineral., 56, 959-979.

Larsen, L.M. (1976) Clinopyroxenes and coexisting mafic minerals from the alkaline Ilimaussaq intrusion. J. Petrol., 17, 258-290.

Yagi, K. (1953) Petrochemical studies of the alkalic rocks of the Morotu district, Sakhalin. Bull. Geol. Soc. Am., 64, 769-810.

ACKNOWLEDGMENTS

Due to teaching and other commitments, this project was undertaken seriously only in the final months before the August 1 deadline. The scope of the assigned topic of this chapter could have involved many months of work, but due to other obligations I have been unable to give it full treatment. Thus my objective has been to provide a brief survey heavily based on published data, a summary of pyroxene-pyroxene exsolution phenomena on a microscopic scale, and a rough chemical framework for future compilations.

The presentation of data from Deer, Howie and Zussman (1978) would have been impossible without the timely contribution of computer print-out and a manuscript by Maryellen Cameron and J.J. Papike. Programs for further computations were written by Joel Sparks and run at the University of Massachusetts X-ray Fluorescence Analytical Facility. John C. Schumacher assisted with computer input. Schumacher and Catherine Hodgkins helped assemble references. Fred Luckey plotted several of the figures. These students along with the author might better have spent the time in the field! Figures 1 - 19 were drafted from rough copy by Marie Litterer. J.B. Thompson, Jr., was helpful in obtaining the material for Figure 23 from Harvard University. The originals for Figures 24 and 33 were provided directly by Hope D. Huntington. Figure 37 was taken from L.D. Ashwal's Ph.D. dissertation at Princeton University. Figures 40 and 41 were kindly provided from an unpublished manuscript by P.E. Brown, E.J. Essene, and D.R. Peacor. R.J. Tracy prepared Figure 42 for this chapter. Various drafts of the text were typed by Camille St. Onge. Joe Kwiecinski and Colleen O'Laskey. My wife and two sons had to endure several hectic summer weekends of manuscript drafting, when it would have been more fun to go swimming! To each of these persons I express my grateful acknowledgment.

Akella, J. (1976) Garnet pyroxene equilibria in the system CaSiO₃-MgSiO₃-Al₂O₃ and in a natural mineral mixture. Am. Mineral., 61, 589-598.

_____ and Boyd, F. R. (1972) Partitioning of Ti and Al between pyroxenes, garnets, and oxides. Carnegie Inst. Washington Year Book, 71, 378-384.

Akimoto, S. and Syono, Y. (1972) High pressure transformation in MnSiO₃. Am. Mineral., 57, 76-84.

Anastasiou, P. and Seifert, F. (1972) Solid solubility of Al₂O₃ in enstatite at high temperatures and 1-5 kb water pressure. Contrib. Mineral. Petrol., 34, 272-287.

Ashwal, L. D. (1974) *Metamorphic hydration of augite-orthopyroxene monzodiorite to hornblende granodiorite gneiss, Belchertown Batholith, west-central Massachusetts.* Contribution No. 18 (M.S. thesis), Univ. Massachusetts, Amherst, MA, 117 p.

_____ (1978) *Petrogenesis of massif-type anorthosites: Crystallization history and liquid line of descent of the Adirondack and Morin complexes.* Ph.D. thesis, Princeton Univ., Princeton, New Jersey.

_____, Leo, G. W., Robinson, P., Zartman, R. E., and Hall, D. J. (1979) The Belchertown Quartz Monzodiorite pluton, west-central Massachusetts: A syntectonic Acadian intrusion. Am. J. Sci., 279, 936-969.

Atkins, F. B. (1969) Pyroxenes of the Bushveld intrusion, South Africa. J. Petrol., 10, 222-249.

Beeson, M. H. and Jackson, E. D. (1970) Origin of the garnet pyroxenite xenoliths of Salt Lake Crater, Oahu. Mineral. Soc. Am. Spec. Paper, 3, 95-112.

Bell, P. M. and Mao, H. K. (1971) Composition of clinopyroxene in the system NaAlSi₂O₆-CaAl₂Si₂O₈. Carnegie Inst. Washington Year Book, 70, 31.

Belov, N. V. (1961) *Crystal Chemistry of Large-cation Silicates.* Consultants Bureau, New York, 162 p.

Bohlen, S. R. and Boettcher, A. L. (1980) The effect of magnesium on orthopyroxene-olivine-quartz stability: orthopyroxene geobarometry (abstr.). EOS, 61, 393.

_____, _____, Dollase, W., and Essene, E. J. (1980a) The effect of manganese on olivine-quartz-orthopyroxene stability. Earth Planet. Sci. Letters, 47, 11-20.

_____, Essene, E. J., and Boettcher, A. L. (1980b) Reinvestigation and application of olivine-quartz-orthopyroxene barometry. Earth Planet. Sci. Letters, 47, 1-10.

_____ and Essene, E. J. (1978) Igneous pyroxenes from metamorphosed anorthosite massifs. Contrib. Mineral. Petrol., 65, 433-442.

Bonnichsen, B. (1969) Metamorphic pyroxenes and amphiboles in the Biwabik Iron Formation, Dunka River area, Minnesota. Mineral. Soc. Am. Spec. Paper, 2, 217-239.

Boyd, F. R. (1970) Garnet peridotites and the system CaSiO₃-MgSiO₃-Al₂O₃. Mineral. Soc. Am. Spec. Paper, 3, 63-75.

_____ (1973) A pyroxene geotherm. Geochim. Cosmochim. Acta, 37, 2533-2546.

_____ and Nixon, P. H. (1973) Structure of the upper mantle beneath Lesotho. Carnegie Inst. Washington Year Book, 72, 431-445.

Brett, C. P. and Emslie, R. F. (1979) Orthopyroxene megacrysts in anorthosites of the Mealy Mountains. Prog. with Abstr., Ann. Meeting Geol. Assoc. Canada, 41.

Brown, P. E., Essene, E. J., and Peacor, D. R. (1980) Phase relations inferred from field data for Mn pyroxenes and pyroxenoids. Contrib. Mineral. Petrol. (in press).

Buchanan, D. L. (1979) A combined transmission electron microscope and electron microprobe study of Bushveld pyroxenes from the Bethal area. J. Petrol., 20, 327-354.

Burnham, C. W. (1976) Ferrobustamite: the crystal structure of two Ca,Fe bustamite-type pyroxenoids. Z. Kristallogr., 142, 450-462.

Callegari, E. (1966) Osservazioni su alcune cummingtonite del massiccio dell'Adamello. Mem. Accad. Patavina Sci., Lett., Arti, Padua, 78, 273-310.

Cameron, M., Sueno, S., Prewitt, C. T., and Papike, J. J. (1973) High temperature crystal chemistry of acmite, diopside, hedenbergite, jadeite, spodumene, and ureyite. Am. Mineral., 58, 594-618.

Caporuscio, F. A. and Morse, S. A. (1978) Occurrence of sapphirine plus quartz at Peekskill, New York. Am. J. Sci., 278, 1334-1342.

Carpenter, M. A. (1980) Sodic pyroxenes from Mybo, Norway: solid solution and cation ordering at high temperatures (abstr.). EOS, 61, 403.

Cawthorn, R. G. and Collerson, K. D. (1974) The recalculation of pyroxene end-member parameters and the estimation of ferrous and ferric iron content from electron microprobe analyses. Am. Mineral., 59, 1203-1208.

Clark, J. R., Appleman, D. E., and Papike, J. J. (1969) Crystal-chemical characterization of clinopyroxenes based on eight new structure refinements. Mineral. Soc. Am. Spec. Paper, 2, 31-50.

Coleman, L. C. (1978) Solidus and subsolidus compositional relationships of some coexisting Skaergaard pyroxenes. Contrib. Mineral. Petrol., 66, 221-227.

Creasy, J. W. (1974) *Mineralogy and petrology of the White Mountain batholith, Franconia and Crawford Notch quadrangles, New Hampshire*. Ph.D. thesis, Harvard Univ., Cambridge, MA.

Curtis, L. W. and Gittins, J. (1979) Aluminous and titaniferous clinopyroxenes from regionally metamorphosed agpaitic rocks in central Labrador. J. Petrol., 20, 165-186.

Danckwerth, P. A. and Newton, R. C. (1978) Experimental determination of the spinel peridotite-garnet peridotite reaction in the system $MgO-Al_2O_3-SiO_2$ in the range $900°-1100°C$ and Al_2O_3 isopleths of enstatite in the spinel field. Contrib. Mineral. Petrol., 66, 189-201.

Deer, W. A., Howie, R. A., and Zussman, J. (1978) *Rock Forming Minerals*, Volume 2A, *Single-Chain Silicates* (2nd Ed.), Longmans, London.

Devine, J. D. and Sigurdsson, H. (1980) Garnet-fassaite calc-silicate nodule from La Soufrière, St. Vincent. Am. Mineral., 65, 302-305.

Dickey, Jr., J. S. (1970) Partial fusion products in alpine-type peridotites: Serrania de la Ronda and other examples. Mineral. Soc. Am. Spec. Paper, 3, 33-49.

_____ and Yoder, Jr., H. S. (1972) Partitioning of chromium and aluminum between clinopyroxene and spinel. Carnegie Inst. Washington Year Book, 71, 384-392.

_____, and Schairer, J. F. (1971) Chromium in silicate-oxide systems. Carnegie Inst. Washington Year Book, 70, 118-122.

Dowty, E. and Clark, J. R. (1973) Crystal structure refinement and optical properties of a Ti^{3+} fassaite from the Allende meteorite. Am. Mineral., 58, 230-242.

Eggler, D. H. (1976) Does CO_2 cause partial melting in the low-velocity layer of the mantle? Geology, 4, 69-72.

Emslie, R. F. (1975) Pyroxene megacrysts from anorthositic rocks: new clues to the sources and evolution of the parent magmas. Canadian Mineral., 13, 138-145.

Gasparik, T. and Lindsley, D. H. (1980) Experimental study of pyroxenes in the system $CaMgSi_2O_6-CaAl_2SiO_6-Ca_{0.5}AlSi_2O_6$ (abstr.). EOS, 61, 402.

Ginzburg, I. V. (1969) Immiscibility of the natural pyroxenes diopside and fassaite and the criterion for it. Doklady Akad. Nauk, 186, 423-426.

Gordon, W. A., Peacor, D. R., Brown, P. E., Essene, E. J., and Allard, L. F. (1980) Exsolution relationships in a clinopyroxene of average composition $Ca_{.43}Mn_{.69}Mg_{.82}Si_2O_6$ from Balmat, New York: X-ray diffraction and scanning transmission electron microscopy. Am. Mineral. (in press).

Grapes, R. H. (1975) Petrology of the Blue Mountain Complex, Marlborough, New Zealand. J. Petrol., 16, 371-428.

Greenwood, H. J. (1962) Metamorphic reactions involving two volatile components. Carnegie Inst. Washington Year Book, 61, 82-85.

Gruenewaldt, G. von (1970) On the phase-change orthopyroxene-pigeonite and resulting textures in the Main and Upper Zones of the Bushveld Complex in Eastern Transvaal. Geol. Soc. South Africa Special Publ., 1, 67-73.

Hays, J. F. (1966) Lime-alumina-silica. Carnegie Inst. Washington Year Book, 65, 234-239.

Huckenholz, H. F., Schairer, and Yoder, Jr., H. S. (1969) Synthesis and stability of ferri-diopside. Mineral. Soc. Am. Spec. Paper, 2, 163-177.

Huntington, H. D. (1980) *Anorthositic and related rocks from Nukasorsuktokh Island, Labrador*. Ph.D. thesis, Univ. Massachusetts, Amherst, MA.

Huntington, J. C. (1975) *Mineralogy and petrology of metamorphosed iron-rich beds in the Lower Devonian Littleton Formation, Orange area, Massachusetts.* Contrib. No. 19 (M.S. thesis), Geol. Dept., Univ. Massachusetts, Amherst, MA.

Jaffe, H. W. and Jaffe, E. B. (1979) The old order changeth, yielding place to new, in *Feldspars and Pyroxenes* (abstr.). First Annual Five College Geology Symposium, Mt. Holyoke College, November 30, 1979.

———, ———, and Ashwal, L. D. (1977) Structural and petrologic relations in the High Peaks region, northeastern Adirondacks. Geol. Soc. Am. Abstr. with Progr., 9, 279–280.

———, Robinson, P., and Klein, Jr., C. (1968) Exsolution lamellae and optic orientation of clinoamphiboles. Science, 160, 776–778.

———, ———, and Tracy, R. J. (1978) Orthoferrosilite and other iron-rich pyroxenes in microperthite gneiss of the Mount Marcy area, Adirondack Mountains. Am. Mineral., 63, 1116–1136.

———, ———, ———, and Ross, M. (1975) Orientation of pigeonite exsolution lamellae in metamorphic augite: correlation with composition and calculated optimal phase boundaries. Am. Mineral., 60, 9–28.

Kerrick, D. M. (1974) Review of metamorphic mixed volatile (H_2O-CO_2) equilibria. Am. Mineral., 59, 729–762.

Klein, Jr., C. (1978) Regional metamorphism of Proterozoic iron-formation, Labrador Trough, Canada. Am. Mineral., 63, 898–912.

Kobayashi, H. (1977) Kanoite ($Mn^{2+},Mg)_2Si_2O_6$, a new clinopyroxene in the metamorphic rocks from Tatehira, Oshima Peninsula, Hokkaido, Japan. J. Geol. Soc. Japan, 83, 537–542.

Kushiro, I. (1969) Clinopyroxene solid solutions formed by reactions between diopside and plagioclase at high pressures. Mineral. Soc. Am. Spec. Paper, 2, 179–191.

——— and Nakamura, Y. (1970) Petrology of some lunar crystalline rocks. Proc. Apollo 11 Lunar Science Conf., 1, 607–626.

Lamb, C. L., Lindsley, D. H., and Grover, J. E. (1972) Johannsenite-bustamite: inversion and stability range. Geol. Soc. Am. Abstr. with Progr., 4, 571–572.

Lambert, I. B. and Wyllie, P. J. (1968) Stability of hornblende and a model for the low velocity zone. Nature, 215, 1240–1241.

Liebau, F. (1959) Über die Kristallstruktur des Pyroxmangits (Mn,Fe,Ca,Mg)SiO₃. Acta Crystallogr., 12, 177–181.

Lindsley, D. H., Brown, G. M., and Muir, I. D. (1969) Conditions of the ferrowollastonite-ferrohedenbergite inversion in the Skaergaard intrusion, east Greenland. Mineral. Soc. Am. Spec. Paper, 2, 193–201.

——— and Dixon, S. A. (1976) Diopside-enstatite equilibria at 850° to 1400°C, 5 to 35 Kb. Am. J. Sci., 276, 1285–1301.

———, King, Jr., H. E., and Turnock, A. C. (1974) Compositions of synthetic augite and hypersthene coexisting at 810°C: Application to pyroxenes from lunar highlands rocks. Geophys. Res. Letters, 1, 134–136.

——— and Munoz, J. L. (1969) Subsolidus relations along the join hedenbergite-ferrosilite. Am. J. Sci., 267A, 295–324.

———, Tso, J., and Heyse, J. V. (1974) Effect of Mn on the stability of pigeonite. Geol. Soc. Am. Abstr. with Progr., 6, 846–847.

Maresch, W. V. and Mottana, A. (1976) The pyroxmangite-rhodonite transformation for the MnSiO₃ composition. Contrib. Mineral. Petrol., 55, 69–76.

Mason, B. (1974) Aluminum-titanium-rich pyroxenes, with special reference to the Allende meteorite. Am. Mineral., 59, 1198–1202.

McCallum, I. S. (1974) Exsolution and inversion in Stillwater pyroxenes (abstr.). EOS, 55, 468.

McGetchin, T. R. (1968) *The Moses Rock Dike: Geology, petrology and mode of emplacement of a kimberlite-bearing breccia pipe, San Juan County, Utah.* Ph.D. thesis, California Institution of Technology, Pasedena, CA.

Milton, C. and Eugster, H. P. (1959) Mineral assemblages in the Green River Formation. In Abelson, P. H., Ed., *Researches in Geochemistry*, Wiley, New York, p. 118–150.

492

Mitchell, R. H. (1980) Pyroxenes of the Fen alkaline complex, Norway. Am. Mineral., 65, 45-54.

Morse, S. A. (1970) Preliminary chemical data on the augite series of the Kiglapait layered intrusion (abstr.). Am. Mineral., 55, 303-304.

_____ (1975) Plagioclase lamellae in hypersthene, Tokkoatokhakh Bay, Labrador. Earth Planet. Sci. Letters, 26, 331-336.

_____ and Talley, J. H. (1971) Sapphirine reactions in deep-seated granulites near Wilson Lake, Central Labrador, Canada. Earth Planet. Sci. Letters, 10, 325-328.

Mysen, B. O. and Boettcher, A. L. (1975) Melting of a hydrous mantle: I. Phase relations of natural peridotite at high pressures and temperatures with controlled activities of water, carbon dioxide, and hydrogen. J. Petrol., 16, 520-548.

Obata, M. (1976) The solubility of Al_2O_3 in orthopyroxenes in spinel and plagioclase peridotites and spinel pyroxenite. Am. Mineral., 61, 804-816.

O'Hara, M. J. (1960) A garnet-hornblende-pyroxene rock from Glenelg, Inverness-shire. Geol. Mag., 97, 145-156.

Ohashi, Y. and Finger, L. W. (1975) Pyroxenoids: a comparison of refined structures of rhodonite and pyroxmangite. Carnegie Inst. Washington Year Book, 74, 564-569.

_____ and _____ (1978) The role of octahedral cations in pyroxenoid crystal chemistry. I. Bustamite, wollastonite and the pectolite-schizolite-serandite series. Am. Mineral., 63, 274-288.

Onuki, H. and Ernst, W. G. (1969) Coexisting sodic amphiboles and sodic pyroxenes from blueschist facies metamorphic rocks. Mineral. Soc. Am. Spec. Paper, 2, 241-250.

Papike, J. J., Ross, M., and Clark, J. R. (1969) Crystal chemical characterization of clinoamphiboles based on five new structure refinements. Mineral. Soc. Am. Spec. Paper, 2, 117-136.

Peacor, D. R. and Prewitt, C. T. (1963) Comparison of crystal structures of bustamite and wollastonite. Am. Mineral., 48, 588-596.

Poldervaart, A. and Hess, H. H. (1951) Pyroxenes in the crystallization of basaltic magma. J. Geol., 59, 472-489.

Ranson, W. A. (1978) Exsolution in Ca-poor pyroxenes from anorthositic rocks of the Nain complex, Labrador. Geol. Soc. Am. Abstr. with Progr., 10, 476.

_____ (1979) *Anorthosites of diverse magma types in the Puttuaaluk Lake area, Nain Complex, Labrador.* Ph.D. thesis, Univ. Massachusetts, Amherst, MA.

Rawlinson, P. J. and Dawson, J. B. (1979) A quench pyroxene-ilmenite xenolith from kimberlite: implications for pyroxene-ilmenite intergrowths. In Boyd, F. R. and Meyer, H. O. A., Eds., *The Mantle Sample: Inclusions in Kimberlites and Other Volcanics.* Am. Geophys. Union, Washington, DC, p. 292-299.

Rietmeijer, F. J. M. (1979) *Pyroxenes from iron-rich igneous rocks in Rogaland, S. W. Norway.* Ph.D. thesis, Univ. Utrecht, Netherlands.

Robinson, P., Jaffe, H. W., Ross, M., and Klein, Jr., C. (1971) Orientation of exsolution lamellae in clinopyroxenes and clinoamphiboles: consideration of optimal phase boundaries. Am. Mineral., 56, 909-939.

_____, Ross, M., Nord, Jr., G. L., Smyth, J. R., and Jaffe, H. W. (1977) Exsolution lamellae in augite and pigeonite: fossil indicators of lattice parameters at high temperature and pressure. Am. Mineral., 62, 857-873.

Ross, M. and Huebner, J. S. (1975) A pyroxene geothermometer based on composition-temperature relationships of naturally occurring orthopyroxene, pigeonite and augite. Intern. Conf. Geothermometry Geobarometry (abstr.), Pennsylvania State Univ., University Park, PA.

Ross, M., Huebner, J. S., and Dowty, E. (1973) Delineation of the one atmosphere augite-pigeonite miscibility gap for pyroxenes from lunar basalt 12021. Am. Mineral., 58, 619-635.

Schreyer, W. and Seifert, F. (1969) Compatibility relations of the aluminum silicates in the systems $MgO-Al_2O_3-SiO_2-H_2O$ and $K_2O-MgO-Al_2O_3-SiO_2-H_2O$ at high pressures. Am. J. Sci., 267, 371-388.

Schlar, C. B., Carrison, L. C., and Stewart, O. M. (1968) High-pressure synthesis and stability of hydroxylated orthoenstatite in the system $MgO-SiO_2-H_2O$ (abstr.). EOS, 49, 356.

Shedlock, R. J. and Essene, E. J. (1979) Mineralogy and petrology of a tactite near Helena, Montana. J. Petrol., 20, 71-97.

Simmons, E. C., Lindsley, D. H., and Papike, J. J. (1974) Phase relations and crystallization sequence in a contact-metamorphosed rock from the Gunflint Iron Formation, Minnesota. J. Petrol., 15, 539-565.

Skippen, G. (1974) An experimental model for low pressure metamorphism of siliceous dolomitic marble. Am. J. Sci., 274, 487-509.

Smith, D. and Lindsley, D. H. (1971) Stable and metastable augite crystallization trends in a single basalt flow. Am. Mineral., 56, 225-233.

Smith, J. V. (1969) Crystal structure and stability of the $MgSiO_3$ polymorphs; physical properties and phase relations of Mg,Fe pyroxenes. Mineral. Soc. Am. Spec. Paper, 2, 3-29.

Smyth, J. R. (1974) The high temperature crystal chemistry of clinohypersthene. Am. Mineral., 59, 1069-1082.

Sobolev, N. V. (1968) Eclogite clinopyroxenes from the kimberlite pipes of Yaketia. Lithos, 1, 54-57.

Thompson, A. B. (1970) Metamorphism in a model mantle I. Predictions of P-T-X relations in $CaO-Al_2O_3-MgO-SiO_2$. In Boyd, F. R. and Meyer, H. O. A., eds., *The Mantle Sample: Inclusions in Kimberlites and Other Volcanics*. Am. Geophys. Union, Washington, DC, p. 15-28.

Tracy, R. J. and Robinson, P. (1977) Zoned titanian augite in alkali olivine basalt from Tahiti and the nature of titanium substitutions in augite. Am. Mineral., 62, 634-645.

_____, Lincoln, T. N., Robinson, P., and Asheuden, D. D. (1980) Rhodonite-pyroxmangite-pyroxene-amphibole assemblages (abstr.). EOS, 61, 390-391.

Trommsdorff, V. and Evans, B. W. (1974) Alpine metamorphism of peridotitic rocks. Schweiz Mineral. Petrogr. Mitt., 54, 333-355.

Veblen, D. R. and Buseck, P. R. (1977) Petrologic implications of hydrous biopyriboles intergrown with igneous pyroxenes (abstr.). EOS, 58, 1242.

_____, _____, and Burnham, C. W. (1977) Asbestiform chain silicates: new minerals and structure groups. Science, 198, 359-365.

Wood, B. J. (1974) The solubility of alumina in orthopyroxene coexisting with garnet. Contrib. Mineral. Petrol., 46, 1-15.

_____ (1976) On the stoichiometry of clinopyroxenes in the system $CaO-MgO-Al_2O_3-SiO_2$. Carnegie Inst. Washington Year Book, 75, 741-742.

Wyllie, P. J. (1979) Magmas and volatile components. Am. Mineral., 64, 469-500.

Yagi, K. and Onuma, K. (1967) The join $CaMgSi_2O_6-CaTiAl_2O_6$ and its bearing on the titanaugites. J. Fac. Science, Hokkaido Univ., Series IV, 13, 463-483.

Yamaguchi, Y., Akai, J., and Tomita, K. (1978) Clinoamphibole lamellae in diopside of garnet lherzolite from Alpe Arami, Bellinzona, Switzerland. Contrib. Mineral. Petrol., 66, 263-270.

Yang, H.-Y. (1973) Synthesis of an Al- and Ti-rich clinopyroxene in the system $CaMgSi_2O_6-CaSl_2SiO_6-CaTiAl_2O_6$. Trans. Am. Geophys. Union, 54, 478.

Yoder, H. S. and Tilley, C. E. (1962) Origin of basalt magmas: an experimental study of natural and synthetic rock systems. J. Petrol., 3, 342-532.

Chapter 10

PYROXENE MINERALOGY of the MOON and METEORITES
James J. Papike
INTRODUCTION

The main objective of this presentation is to show how pyroxenes
reflect differences in planetary environments in which they crystallized.
To carry out this exercise we must select host rocks from different planetary bodies that are similar in chemistry and that had similar petrogenetic
histories. An obvious host rock choice is basalt because it occurs on the
three (or more) parent bodies that have been sampled, the Earth, its moon,
and the parent body (bodies) of basaltic meteorites.

Basalt magmas, derived by partial melting of planetary interiors,
have compositions that reflect (1) the pre-accretionary history of the
material from which the planet formed, (2) the planet's subsequent evolutionary history, (3) the chemistry and mineralogy of the source regions,
and (4) the intensive thermodynamic parameters operating at the source and
emplacement sites. Studies of basalt suites from the Earth, its moon, and
the eucrite parent body (EPB) reveal compositional differences intrinsic
to their source regions which are, in turn, a characteristic of the planet
and its formational and evolutionary history. Because pyroxenes are the
most abundant ferromagnesian phase in most basalts and since their crystal
structures accommodate all of the major elements that occur in basalts
(except potassium) they, like their host basalts, are potentially powerful
planetary probes.

Because basalts are key indicators of planetary evolution, an experimental project entitled, "Basaltic Volcanism on the Terrestrial Planets"
was organized by the Lunar and Planetary Institute, Houston, Texas. The
project, funded in part by NASA, is a pilot project in comparative plane-
tology involving approximately eighty scientists representing a variety
of disciplines. The project was initiated in 1977 and will culminate in
1981 with a book entitled "Basaltic Volcanism on the Terrestrial Planets"
which will summarize the results of the study.

I was given the major responsibility for synthesizing the silicate
mineralogy of the planetary basalt suite and much of what I report here
is a result of my interaction with the project. Here I will consider
pyroxene chemistry based on approximately 1,200 high-quality pyroxene
microprobe analyses that were selected from the following basalt suites:
Archean, Columbia Plateau, deep sea, basaltic meteorite, Hawaiian, island

Table 1. Planetary basalt comparisons (from Bence et al., 1980).

	Mare Basalts	Eucrites	Subalkaline Terrestrial Basalts
Age	3.8→3.2→2.6?	∿4.5	∿3.8→0
% Crust	<1	?	>50(<180 MY)
Tectonic Style	static; diapiric?	?	dynamic; diapiric
fO_2 1150° (atm.)	$10^{-12}-10^{-14}$	$10^{-13}-10^{-14}$	$10^{-8}-10^{-11}$
Gas Phase	CO_2, CH_4, H_2S	?	H_2O, CO_2
Mg# (non-cumulates)	60–30	65–30	88–40
TiO_2 (%)	0.5–1; 1.5–5; 8–14	0.5–1	<3
Al_2O_3 (%)	8–14	10–13	10–18
Na_2O (%)	0.2–0.6	0.4–0.6	1–4
Cr (ppm)	$2-6\cdot10^3$	$2-3\cdot10^3$	$0.1-3\cdot10^3$
Ni (ppm)	<5–90	<30	30–1000
Sr, Ba	depleted	not depleted	not depleted
REE	5–100 x cc	8–15 x cc	8–50 x cc
Eu/Eu*	-ve (intrinsic)	?	none
K/U	∿1-5·10^3	∿3-6·10^3	1-2·10^4
Multiple Saturation (km)	150–400	0	25–100?
Residual Phases	olivine, pyroxene	olivine, pyroxene, plagioclase	olivine, orthopyroxene + clinopyroxene

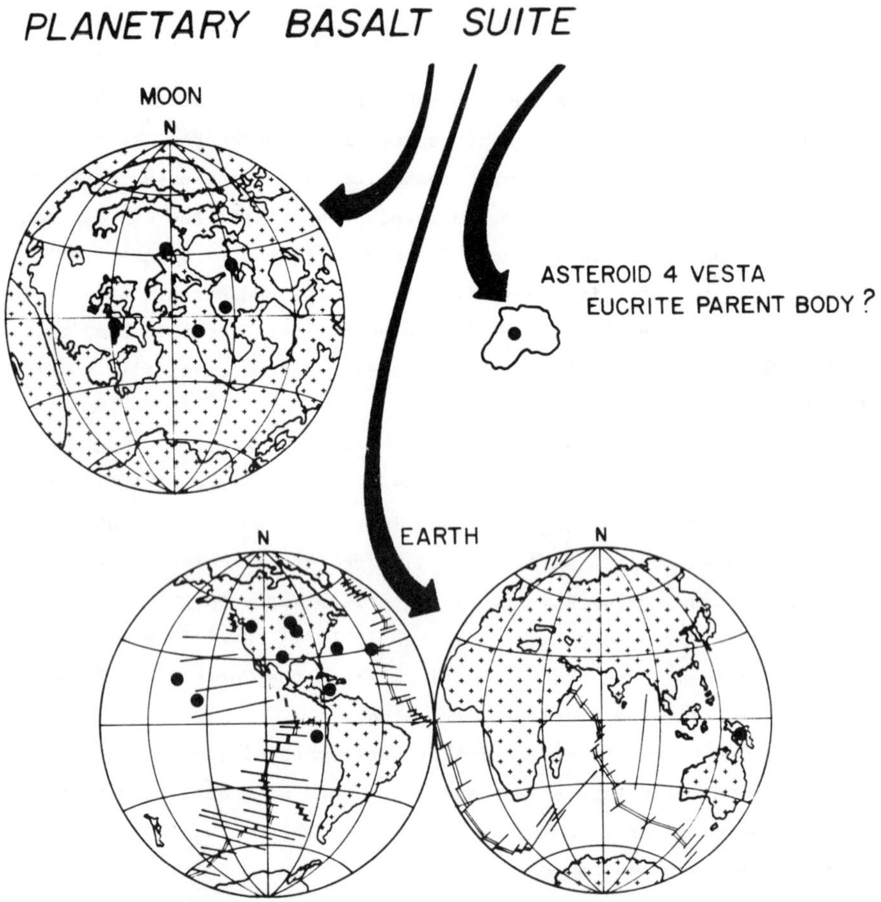

PLANETARY BASALT SUITE

Figure 1. Schematic diagram showing the locations of the planetary basalt suites. The major lunar locations include Apollo 11, 12, 15 and 17 for the mare basalts and Apollo 16 for the lunar highland melt rocks. The identification of the terrestrial suites is discussed in the text.

arc, Keweenawan, lunar highland melt rocks, lunar mare, and Rio Grande.
I emphasize that in this report the mineral chemistry and basalt suites
are considered in overview fashion. Detailed discussions concerning
the suites will be included in the book "Basaltic Volcanism on the
Terrestrial Planets" (1981, in preparation). This summary emphasizes
characteristics of pyroxene chemistry that reflect planetary constraints
on the host basalts from which the pyroxene crystallized.

Some references on planetary basalts that are not included in the
aforementioned book include Papike *et al.* (1980). Figure 1 shows in car-
toon fashion the distribution of the planetary basalt suites and Table 1
taken from Bence *et al.* (1980) lists some planetary comparisons for lunar
mare basalts, eucrites and subalkaline terrestrial basalts.

A SIMPLIFIED EXAMPLE OF PLANETARY COMPARISONS: LUNAR MARE VERSUS TERRESTRIAL MIDOCEAN RIDGE BASALTS

Introduction

A dominant surface lithology on Earth is mid-ocean ridge basalt (MORB)
which covers approximately 60% of the Earth's surface area; in contrast,
the lunar basalts comprise approximately 17% of the Moon's surface area
(Head *et al.*, 1977). Characteristics in common are: (1) both are ex-
truded by fissure type eruptions and (2) both are endogenous melts. There
are also some striking differences: (1) Mare basalt volcanism spans the
range $\sim$3.1-3.9 b.y. compared to 0-200 m.y. for MORB's; (2) MORB's (relative
to mare basalts) are enriched in SiO_2, Al_2O_3, Na_2O, K_2O and depleted in
"Cr_2O_3", "TiO_2", "FeO"; (3) the least fractionated MORB's have higher Mg
values (where Mg value = $Mg/(Mg + Fe^{2+})$ atomic) than the least fractionated
mare basalts; and (4) the mare basalts are anhydrous.

There are many reasons for variability in basalt rocks, e.g., different degrees of partial melting and partial melting mechanisms, diverse
source regions, fractionation at depth in planetary interiors, fractionation en route to the surface, fractionation at the surfaces, etc. However,
the major differences between mare and MORB's emphasized here are due primarily to major differences in planetary characteristics inherent in the
Earth versus the Moon.

Major Element Chemistry

The three major mare basalt groups (high Ti, low Ti and very low Ti
or VLT) are effectively separated on the basis of their TiO_2 contents

497

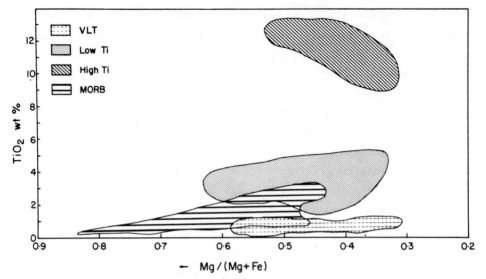

Figure 2. TiO$_2$ (wt %) versus Mg/(Mg + Fe^{2+}) (atomic) variation diagram for MORB and mare basalts. Numbers of analyses for each group in parentheses. Mg value for MORB's calculated assuming Fe^{3+}/(Fe^{2+} + Fe^{3+}) = 0.1 (atomic).

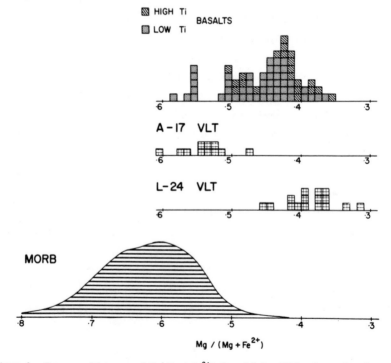

Figure 3. Frequency histogram of Mg/(Mg + Fe^{2+}) (atomic) for MORB and mare basalts.

498

(Papike and Vaniman, 1978). MORB's have TiO_2 contents that overlap two of the mare basalt groups but which are considerably lower than those of the high-Ti basalts (Fig. 2). An additional striking difference involves the much higher Mg values displayed by many of the MORB's.

Important differences in Mg-value between mare basalts and MORB's are illustrated in Figure 3. It is thought that Mg values of 0.5 represent the most primitive of the low and high Ti mare basalts sampled; basalts with higher Mg values are believed to be cumulates (see, for example, Papike *et al.*, 1976). However, basalts from the Apollo 17 VLT suite and the Apollo 15 Emerald Green Glass (see, for example, Taylor, 1975) have Mg values as high as 0.6 and some of these Mg-rich compositions may represent relatively unfractionated basalt liquids. MORB's extend to much higher Mg values, although values greater than $\sim$0.70-0.72 represent cumulates. High-pressure melting experiments on the low Ti mare basalts (Kesson and Lindsley, 1976) and the MORB's (see, for example Bender *et al.*, 1978) indicate comparable residuum mineralogies (but different mineral chemistries).

The Al_2O_3 contents of mare basalts are significantly lower than of the MORB's (Fig. 4), a feature which probably reflects a lower alumina source region for mare basalts. This is consistent with an evolutionary scenario that attracts many lunar scientists. This scenario, developed by Taylor and Jakes (1974), suggests that shortly after accretion, the Moon melted to depths of several hundred kilometers. Cooling of the Moon-wide magma ocean resulted in the formation of Al_2O_3-depleted cumulates and a plagioclase-rich crust. Mare basalts were subsequently derived by partial melting of these Al_2O_3-depleted cumulates during the period 3.1-3.9 b.y. A similar depletion of CaO in the mare basalts is not observed; a possible explanation of this is a higher clinopyroxene content in the mare basalt source region.

The first analyses of lunar rocks in 1969 showed that the Moon is significantly depleted in the alkali elements, e.g., K and Na, relative to the Earth; this is illustrated in Figures 5 and 6 with respect to basaltic rocks. Figure 5 shows a major difference in the K_2O content of MORB's versus mare basalts. Although some of the very high K_2O contents of the MORB's represent sea water alteration, 967 fresh MORB glasses contain up to 1 wt % K_2O and average $\sim$0.25 wt %. The average K_2O content of 1100 crystalline MORB's is $\sim$0.29 wt % (Bence *et al.*, 1978).

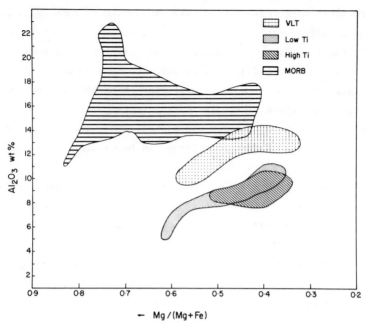

Figure 4. Al$_2$O$_3$ (wt %) versus Mg/(Mg + Fe^{2+}) (atomic) variation diagram for MORB and mare basalts.

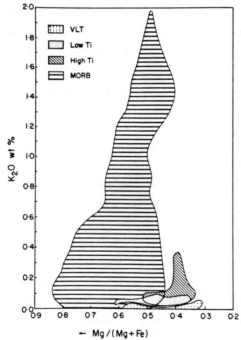

Figure 5. K$_2$O (wt %) versus Mg/(Mg + Fe^{2+}) (atomic) variation diagram for MORB and mare basalts.

The differences in alkali content are even more clearly demonstrated by comparisons of Na_2O. Fresh MORB glasses have Na_2O ranging from 1 to 4 wt % (Fig. 6) and average 2.65 wt %, whereas the 1100 crystalline MORB's average 2.31 wt % Na_2O (Bence *et al.*, 1978).

Inherent differences in oxygen fugacity exist between the two basalt suites. Haggerty (1978) in a summary of the redox state of planetary basalts noted that the oxygen fugacities for terrestrial basalts follow the fayalite-magnetite-quartz (FMQ) buffer curve very closely. In contrast, mare basalts are far more reduced with oxygen fugacities several orders of magnitude lower and which fall below the wüstite-iron (WI) buffer curve (Fig. 7). The low lunar oxygen fugacities result in reduced valence states of several cations. Ferric ion is virtually absent on the Moon and most iron is in the ferrous state with a subordinate amount reduced to Fe^0. In addition, some Ti is reduced from Ti^{4+} to Ti^{3+} and some Cr from Cr^{3+} to Cr^{2+}. Differences in f_{O_2} may be responsible for differences observed in the concentration of these elements. Chromium is markedly depleted in MORB's relative to mare basalts (Fig. 8). Three possibilities for this depletion are: (1) all MORB's experienced early fractionation of Cr spinel and/or Cr spinel is an important phase left behind in the source region, (2) the source regions for MORB's are depleted in Cr relative to mare source regions, (3) the low oxygen fugacity on the Moon results in a significant reduction of Cr^{3+} to Cr^{2+} and this in turn, affects the partition coefficients for mineral-melt equilibria, such that Cr^{2+} favors the melt relative to Cr^{3+} (Huebner *et al.*, 1976; Schreiber and Haskin, 1976).

Mineral Chemistry

The discussion above documents differences in the major element chemistries of these two basalt suites that reflect their individual planetary characteristics. Although the mineral assemblages in these basalts are similar (clinopyroxene + plagioclase + FeTi oxides ± olivine) their mineral chemistries are significantly different.

Pyroxenes from the MORB's are more magnesian than those from the mare basalts (Fig. 9). The open circles represent specific pyroxene analyses for MORB's while the shaded area illustrates the range of compositions for mare pyroxenes. Two features are evident on this diagram; first, the mare pyroxenes are displaced compositionally toward the Fe^{2+}-rich side of the quadrilateral and, second, mare basalt pyroxenes span

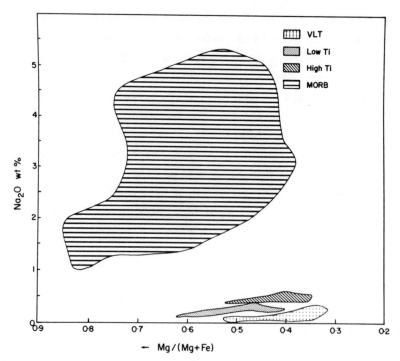

Figure 6. Na$_2$O (wt %) versus Mg/(Mg + Fe^{2+}) (atomic) variation diagram for MORB and mare basalts.

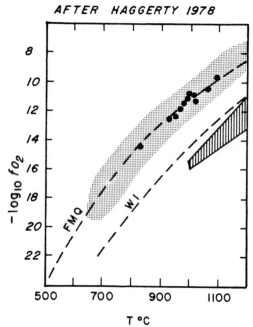

Figure 7. Temperature versus oxygen fugacity diagram from Haggerty (1978).

a greater compositional range. The more magnesian nature of MORB pyroxenes reflects the high Mg values of MORB liquids and presumably a source
region significantly enriched in magnesium relative to the source region
for mare basalts. Although not illustrated in Figure 9, another important
compositional difference between MORB and mare basalt pyroxenes concerns

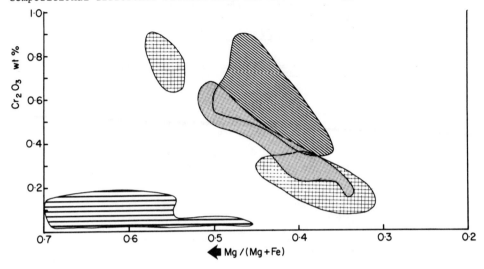

Figure 8. "Cr$_2$O$_3$" (wt %) versus Mg/(Mg + Fe^{2+}) (atomic) variation diagram for MORB and mare basalts.

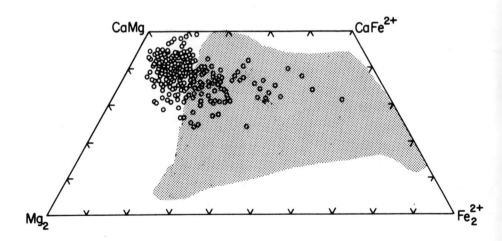

Figure 9. Pyroxene quadrilateral showing data for MORB (open circles) and mare basalts (shaded area).

503

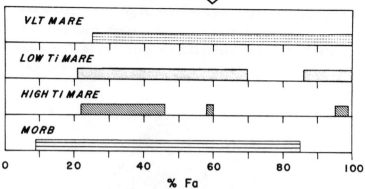

Figure 10. Range of olivine compositions for MORB and mare basalts. Fa = fayalite component = $Fe_2^{2+}SiO_4$. Zero % Fa = 100% Fo = forsterite = Mg_2SiO_4.

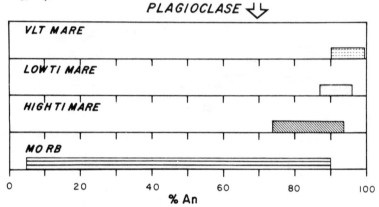

Figure 11. Range of feldspar compositions for MORB and mare basalts. An = anorthite component = $CaAl_2Si_2O_8$. Zero % An = 100% Ab = albite = $NaAlSi_3O_8$.

the pyroxene components other than Mg, Fe and Ca. These include such elements as Na, Cr, Fe^{3+}, Al and Ti. Schweitzer *et al.* (1978) found that the two most important coupled "other" substitutions in MORB pyroxenes are Fe^{3+} (octahedral)-Al (tetrahedral) or $^{VI}Fe^{3+}-{}^{IV}Al$ and Ti^{4+} (octahedral)- Al (tetrahedral) or $^{VI}Ti^{4+}-2^{IV}Al$. These two couples are of roughly equal importance in MORB pyroxenes. Since the low oxygen fugacities on the Moon preclude the possibility of $^{VI}Fe^{3+}-{}^{IV}Al$ as being a substitutional couple in mare basalt pyroxenes, their crystal chemistry and thus the pyroxene- melt interactions are significantly different from those in MORB's. The most important "other" substitutional couples in mare basalt pyroxenes are $^{VI}Ti^{4+}-2^{IV}Al$ and $^{VI}Al-{}^{IV}Al$.

Olivine compositions are illustrated in Figure 10. Two differences
between MORB and mare olivines are apparent. First, although significant
chemical zoning is present in both olivine groups, the mare olivines are
zoned to more iron-rich compositions, and second, the most magnesian MORB
olivines are more magnesian than the most magnesian mare olivines. Again,
this almost certainly is a reflection of MORB magmas with higher Mg values.

The most significant feature of the feldspar chemistry (Fig. 11) is
the striking sodium enrichment of MORB versus mare feldspars. This is
a reflection of the alkali depletion of the Moon relative to the Earth.
The albite content ($NaAlSi_3O_8$) of the feldspar correlates positively with
the Ti content of the mare basalts.

Conclusions

The data and arguments presented above demonstrate that major differences which exist between MORB's and mare basalts are largely a consequence of differing planetary characteristics of the Moon and the Earth:

(1) the higher $Mg/(Mg + Fe^{2+})$ ratio and Al_2O_3 contents of the mantle
source for MORB's relative to the source regions for mare basalts.

(2) the alkali depletion of the Moon relative to the Earth.

(3) the much lower oxygen fugacities that obtained during lunar mare
versus MORB petrogenesis.

THE PLANETARY SUITE: PYROXENE CHEMICAL TRENDS

Chemical Variations

Of the major silicates in basaltic rocks, pyroxenes are the most
chemically complex and are powerful recorders of the evolutionary history
of these rocks. The pyroxenes contain most of the major elements of their
host rocks. Ca, Na, Mn, Fe^{2+} and Mg are accommodated in the large distorted 8 to 6 coordinated M2 sites, Mn, Fe^{2+}, Mg, Fe^{3+}, Cr^{3+}, Cr^{2+}, Ti^{4+},
Ti^{3+}, and Al may substitute in the octahedral M1 site and Al and Si and
sometimes Fe^{3+} can occupy the tetrahedral site. The only major element
that cannot be accommodated in the pyroxene structure is K and in the
basalts considered here this element resides either in feldspar or in
the mesostasis.

Figure 12 presents a series of chemical variation diagrams that are
all normalized to a common analysis listed as the grand average in Tables
2 and 3. The plotted value for each oxide and for each group is obtained

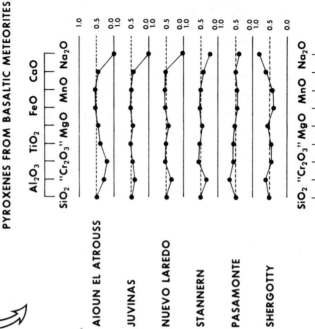

PYROXENES

Figure 12. Variation diagrams for ten oxides for the pyroxene mineral suites. The actual numerical values plotted, which vary between zero and one, are derived from the ratio oxide/(oxide + average). "Oxide" represents the mean for each pyroxene suite (note Table 2) and "average" is the grand average for all of the pyroxene suites (note Table 2).

Figure 13. Variation diagrams for nine oxides for the pyroxenes from the six individual basaltic meteorites. The actual numerical values plotted, which vary between zero and one, are derived from the ratio oxide/(oxide + average). "Oxide" represents the mean for each meteorite suite (note Table 3) and "average" is the grand average for all six meteorites (note Table 3).

PYROXENES FROM BASALTIC METEORITES

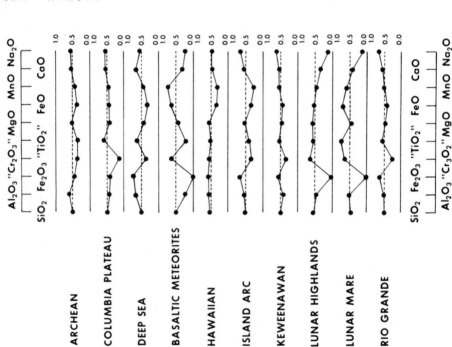

506

Table 2. Pyroxene—mean values.

	Archean	Col. Plat.	Deep Sea	Basaltic Meteorites	Hawaiian	Island Arc
SiO_2	51.68	50.30	50.42	50.17	52.25	51.31
Al_2O_3	3.92	2.00	3.01	0.75	2.82	2.83
FeO	9.60	12.51	10.60	27.17	8.45	9.54
Fe_2O_3	0.70	0.62	1.61	0.00	1.02	1.79
MgO	16.75	13.05	14.89	13.36	21.05	16.02
MnO	0.24	0.27	0.28	0.79	0.14	0.19
TiO_2	0.46	1.27	0.90	0.24	0.85	0.47
Cr_2O_3	0.20	0.05	0.20	0.52	0.38	0.18
CaO	16.82	19.23	17.86	6.81	12.84	17.30
Na_2O	0.17	0.16	0.19	0.04	0.13	0.29
	100.54	99.46	99.96	99.85	99.93	99.92

	Keween.	Lunar High.	Lunar Mare	Rio Grande	(Grand Average)
SiO_2	51.31	52.12	49.00	50.93	50.95
Al_2O_3	1.79	1.98	2.55	2.68	2.43
FeO	13.21	15.62	22.15	10.30	13.91
Fe_2O_3	1.02	0.00	--	1.43	1.17
MgO	15.37	21.18	13.06	15.73	16.05
MnO	0.34	0.24	0.36	0.29	0.31
TiO_2	0.88	0.92	1.44	0.95	0.65
Cr_2O_3	0.15	0.56	0.54	0.12	0.29
CaO	16.02	7.36	10.51	17.02	14.18
Na_2O	0.24	0.02	0.02	0.31	0.16
	100.33	100.00	99.63	99.76	100.10

Table 3. Pyroxenes. Basaltic meteorites—means for individual meteorites.

	Aioun el Atrouss	Juvinas	Nuevo Laredo	Stannern	Pasamonte	Shergotty	(Average)
SiO_2	49.99	49.36	49.00	49.22	50.11	51.87	49.93
Al_2O_3	0.26	0.42	0.33	0.30	1.32	0.98	0.60
FeO	31.57	30.63	32.63	32.85	25.90	19.02	28.75
Fe_2O_3	---	---	---	---	---	---	---
MgO	11.25	12.25	9.58	11.43	14.48	16.48	12.58
MnO	0.94	0.89	0.96	0.83	0.75	0.55	0.84
TiO_2	0.16	0.24	0.24	0.24	0.29	0.22	0.23
Cr_2O_3	0.12	0.46	0.51	0.61	0.65	0.44	0.47
CaO	5.70	5.26	6.80	4.44	6.32	10.11	6.44
Na_2O	0.00	0.00	0.00	0.01	0.02	0.14	0.03
	99.99	99.51	100.05	99.93	99.84	99.82	99.87

by the calculation x/(x + grand average) where x equals the value for
each oxide listed in Tables 2 and 3. This type of diagram, which is similar in some ways to chondrite-normalized REE plots, has the advantage
of portraying similarities and differences of concentrations for a large
number of oxides on one plot.

A considerable range of Fe_2O_3 values is observed for these pyroxenes
(Fe_2O_3 contents were estimated by the method of Papike *et al.*, 1974).
Pyroxenes from basaltic meteorites, lunar highland melt rocks and mare
basalts show essentially no Fe_2O_3 reflecting the low oxygen fugacities
that obtained during the extrusion of the parent basalts. Some Fe_2O_3 in
pyroxenes from the Shergotty meteorite is suggested by the calculation
method of Papike *et al.*, (1974)*. This is consistent with the estimate
of higher oxygen fugacities for Shergottites than eucrites. Relatively
high values of Fe_2O_3 are estimated for pyroxenes from deep sea (Schweitzer
et al., 1978, 1979), Hawaiian, island arc and Rio Grande basalts. It is
interesting to note the overall similarity of the "patterns" for the
lunar highland and mare basalt pyroxenes even though these rock suites
have very different bulk chemistries. These appears to be a distinct
lunar signature on these pyroxenes.

Figure 13 shows the intragroup comparisons for the basaltic meteorites. On this plot it is apparent that Juvinas, Nuevo Laredo and Stannern
have very similar pyroxenes. Aioun el Atrouss is also similar but appears
to be relatively depleted in Al_2O_3, Cr_2O_3 and TiO_2 compared to the other
equilibrated eucrites in the suite. Pyroxenes from the two unequili-
brated meteorites in the basaltic meteorite suite, Pasamonte and Shergotty,
have similarities in their major oxide patterns even though the host rock
chemistries are quite different. This implies a cooling rate control
especially for Al, where pyroxenes from Pasamonte and Shergotty show
significant Al enrichments relative to the average. An obvious difference in the major oxide patterns for Shergotty and Pasamonte pyroxenes
is that Shergotty pyroxenes show Na enrichment. This is a reflection of
the higher Na concentration in the Shergotty host basalt.

- - - - - - - - -

* However, in the comparative mineralogy discussed below all Fe is considered as Fe^{2+} for Shergotty pyroxenes.

PYROXENES
PLANETARY BASALTS

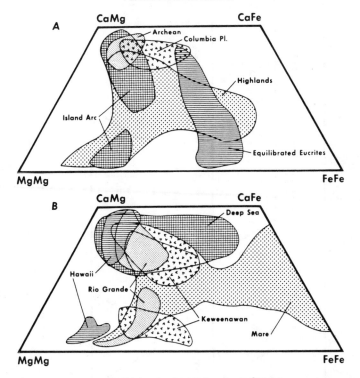

Figure 14. A. Pyroxene Wo (CaSiO₃) - En (MgSiO₃) - Fs (Fe²⁺SiO₃) composite variation diagrams for five of the ten mineral suites. Note that the meteorite pyroxenes are only for the "equilibrated" eucrites. See text for discussion. B. Pyroxene Wo (CaSiO₃) - En (MgSiO₃) - Fs (Fe²⁺SiO₃). Composite variation diagrams for five of the ten mineral suites.

Some of the major features of the "Quad" chemical systematics are illustrated in Figures 14a,b and more detailed chemical trends are shown in Figures 15 and 16. The two lunar suites (highlands and mare) show the greatest range of chemical variability of these diagrams. The suites that have extreme magnesian compositions and most closely approach the CaMg (diopside) - MgMg (enstatite) sideline include the lunar highland, island arc, Archean, deep sea and Hawaiian suites. Pyroxenes from the Columbia Plateau, Rio Grande, and Keweenawan suites are more evolved (i.e., less magnesian); this is consistent with the olivine and feldspar data from these rocks. It should be noted that the basaltic meteorite pyroxene assemblages plotted on the composite diagram 14a are only for the "equilibrated" groups. The Pasamonte and Shergotty pyroxene chemical

509

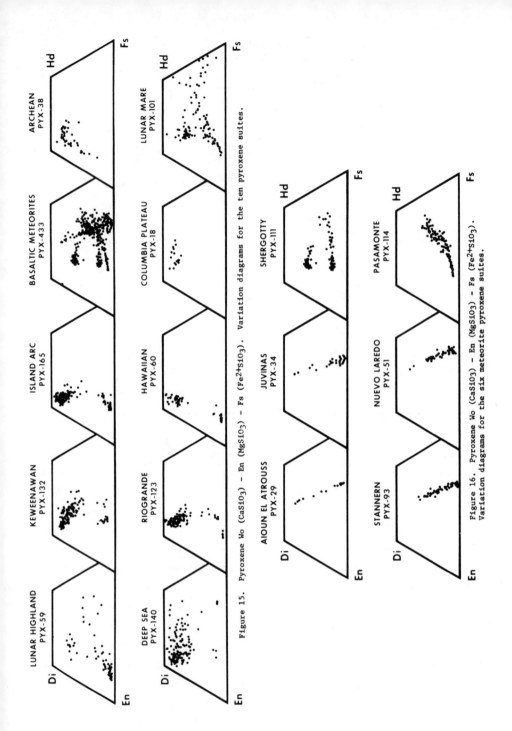

Figure 15. Pyroxene Wo ($CaSiO_3$) – En ($MgSiO_3$) – Fs ($Fe^{2+}SiO_3$). Variation diagrams for the ten pyroxene suites.

Figure 16. Pyroxene Wo ($CaSiO_3$) – En ($MgSiO_3$) – Fs ($Fe^{2+}SiO_3$). Variation diagrams for the six meteorite pyroxene suites.

LUNAR HIGHLAND
PYX-59

KEWEENAWAN
PYX-132

ISLAND ARC
PYX-165

BASALTIC METEORITES
PYX-433

ARCHEAN
PYX-38

DEEP SEA
PYX-140

RIOGRANDE
PYX-123

HAWAIIAN
PYX-60

COLUMBIA PLATEAU
PYX-18

LUNAR MARE
PYX-101

AIOUN EL ATROUSS
PYX-29

JUVINAS
PYX-34

SHERGOTTY
PYX-111

STANNERN
PYX-93

NUEVO LAREDO
PYX-51

PASAMONTE
PYX-114

510

trends show significant zoning consistent with rapid cooling histories (Fig. 16). The mare basalt pyroxenes are displaced towards the $CaFe^{2+}$ (hedenbergite) - $Fe^{2+}Fe^{2+}$ (ferrosilite) sideline. This Fe-enrichment of mare basalt pyroxenes has been interpreted as reflecting an iron-enriched lunar mantle source relative to the earth (see for example, Papike and Bence, 1978).

The plots of $Na + {}^{IV}Al$ against ${}^{VI}Al + 2Ti + Cr$ (Fig. 17) elucidate the oxidation state of these elements. If all the Fe in the pyroxenes were Fe^{2+}, and Ti and Cr were in the 3+ and 2+ states, respectively, the data would plot on the 45° line on such diagrams. Chemical data for pyroxenes from the Archean, Columbia Plateau, deep sea, Hawaiian, island arc, Keweenawan and Rio Grande plot towards the $Na + {}^{IV}Al$ axis indicating significant Fe^{3+} in these pyroxenes. The Fe^{3+} content reported in the tables for terrestrial assemblages was estimated by the method of Papike et $al.$ (1974). In contrast, pyroxenes from lunar highland and lunar maria plot on the other side of the 45° line towards ${}^{VI}Al + 2Ti + Cr$, indicating the presence of Ti^{3+} and/or Cr^{2+} and reflecting the low oxygen fugacity that obtained on the moon relative to the earth. These diagrams also give an indication of the amount of pyroxene components other than $Wo(CaSiO_3)$ - $En(MgSiO_3)$ - $Fs(Fe^{2+}SiO_3)$. These so called "Others" include Na, Al, Cr, Ti, Fe^{3+}. The highest upper limits for "Others" are found in lunar mare and Archean basalts, probably reflecting very rapid cooling histories. Intermediate concentrations are found in pyroxenes from Columbia Plateau, deep sea, Hawaiian, island arc, Keweenawan, lunar highlands, and Rio Grande suites. The lowest concentrations of "Others" are found in pyroxenes from the "equilibrated" eucrites, Juvinas, Stannern, Nuevo Laredo, and Aioun el Atrouss (Fig. 18). The "Others" content of Pasamonte pyroxenes probably reflects a more rapid cooling history, while the "Others" content of Shergotty pyroxenes reflects both a more rapid cooling history and a different bulk host rock composition (e.g., higher Na).

The most important substitutional couples involved in pyroxene "Others" components are: ${}^{VI}R^{3+} - {}^{IV}Al$, ${}^{VI}R^{3+} - Na$, ${}^{VI}Ti - {}^{IV}Al$, ${}^{VI}Ti - Na$ where $R^{3+} = Fe^{3+} + Al + Cr^{3+}$. Plots of Ti against ${}^{IV}Al$ (Figs. 19 and 20) give some insight into the relative importance of these couples;

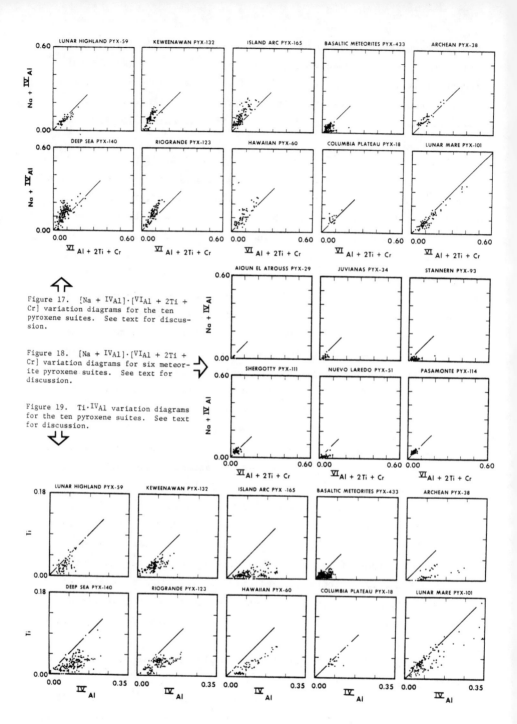

Figure 17. [Na + ^{IV}Al]·[^{VI}Al + 2Ti + Cr] variation diagrams for the ten pyroxene suites. See text for discussion.

Figure 18. [Na + ^{IV}Al]·[^{VI}Al + 2Ti + Cr] variation diagrams for six meteorite pyroxene suites. See text for discussion.

Figure 19. Ti·^{IV}Al variation diagrams for the ten pyroxene suites. See text for discussion.

512

however, a more rigorous treatment of this subject is given below in connection with the factor analysis. If the data plots on the Ti/Al = 1/2 line, a major "Others" substitutional couple is $^{VI}Ti^{4+} - 2^{IV}Al^{3+}$; this couple is important, for example, in pyroxenes from mare basalts.

A useful equation, based on charge balance, describes the permissible cation substitutional couples "Other" than those related to the pyroxene quadrilateral "Quad" ($CaMgSi_2O_6 - CaFe^{2+}Si_2O_6 - Fe_2Si_2O_6 - Mg_2Si_2O_6$) (Papike and Cameron, 1976). It is:

Charge Excess Relative to "Quad" = Charge Deficiency Relative to "Quad" $^{VI}Al + ^{VI}Fe^{3+} + ^{VI}Cr^{3+} + 2^{VI}Ti^{4+} = ^{IV}Al + ^{M2}Na$. When ^{VI}Al, $^{VI}Fe^{3+}$, or $^{VI}Cr^{3+}$ substitutes in the pyroxene M1 site, a charge excess of 1+ results *relative* to the (Mg, Fe^{2+}, Mn^{2+}) occupancy of this site in "Quad" pyroxene components. By similar reasoning, a Ti^{4+} substitution in M1 causes a charge excess of 2+ relative to "Quad". Thus, for charge balance to be maintained in the pyroxene structure, these site charge excesses must be compensated by site charge deficiencies. The substitution of Al for Si in the tetrahedral site causes a deficiency of -1. Similarly, the substitution of Na in the M2 site for (Ca, Fe^{2+}, Mg, Mn^{2+}) also results in a site charge deficiency of -1 relative to "Quad". Since the Na content of the pyroxenes in the reference suites is quite low (note Table 2), most of the charge deficiencies in these pyroxenes are accounted for by ^{IV}Al. Thus, the most important pyroxene "Others" components in these pyroxenes are: $^{VI}Al - ^{IV}Al$, $^{VI}Fe^{3+} - ^{IV}Al$, $^{VI}Cr^{3+} - ^{IV}Al^{3+}$, and $^{VI}Ti^{4+} - 2^{IV}Al$. We will discuss the relative importance of these substitutional couples in detail below. The amount of ^{IV}Al gives us a first approximation of the % "Others" in the pyroxenes or, in other words, the amount of compositional deviation from the pure pyroxene quadrilateral. Since the maximum amount of ^{IV}Al permissable in the pyroxene structure is 1 atom per formula unit, $^{IV}Al \times 100$ is a good estimate of % Others. Thus, Figures 21 and 22, which are variation diagrams of ^{IV}Al vs X_{Fe} where $X_{Fe} = Fe^{2+}/(Fe^{2+} + Mg)$ atomic, give a good estimate of the realtive amount of "Others" in the pyroxene suites. The maximum amount of "Others" are found in mare pyroxenes followed by Archean, island arc, deep sea, and Hawaiian pyroxenes. The lowest "Others" contents are found in the pyroxenes from basaltic meteorites. In fact, it we omit the "relatively" high ^{IV}Al values of Shergotty and

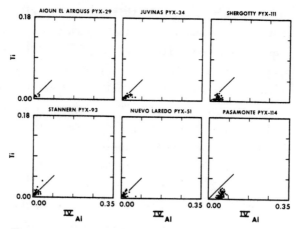

Figure 20. Ti·^{IV}Al variation diagrams for six meteorite pyroxene suites. See text for discussion.

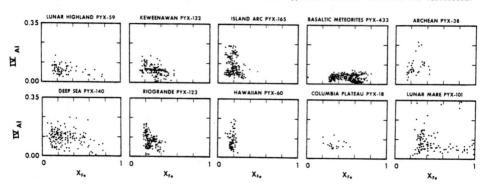

Figure 21. ^{IV}Al·X_{Fe} variation diagrams for the ten pyroxene suites.

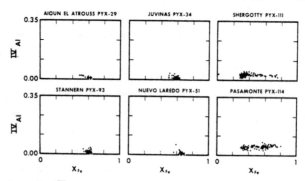

Figure 22. ^{IV}Al·X_{Fe} variation diagrams for six meteorite pyroxene suites.

PYROXENES

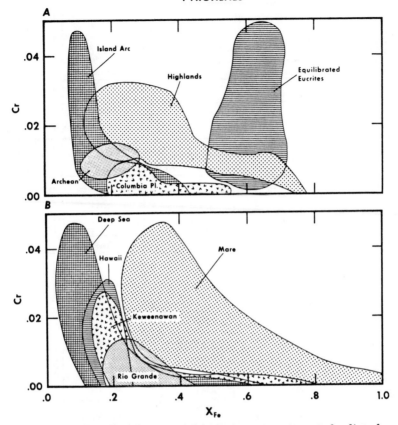

Figure 23. A. Pyroxene Cr·X_{Fe} composite variation diagrams for five of the ten mineral suites. Note that the meteorite pyroxenes are only for the "equilibrated" eucrites. See text for discussion. B. Pyroxene Cr·X_{Fe} composite variation diagrams for five of the ten mineral suites.

Pasamonte pyroxenes, the pyroxenes from "equilibrated" eucrites are exceedingly low (Fig. 22). The significance of this observation is that the many studies of pyroxene phase equilibria for the pure "Quad" are very applicable to pyroxenes from "equilibrated" eucrites.

Figures 23a,b and 24 show the Cr vs X_{Fe} variations for pyroxenes from the basalt suites Pyroxenes from deep sea, island arc, Keeweenawan and Hawaiian basalts show similar behavior in that they all show high Cr concentrations at low X_{Fe}. Pyroxenes from the "equilibrated" eucrites (Aioun el Atrouss, Juvinas, Nuevo Laredo, and Stannern) show anomalous systematics in that they have both high and low Cr concentrations at high X_{Fe}. It is quite possible that a significant portion of the Cr in the eucrites is in the reduced valance state Cr^{2+} because of the low

515

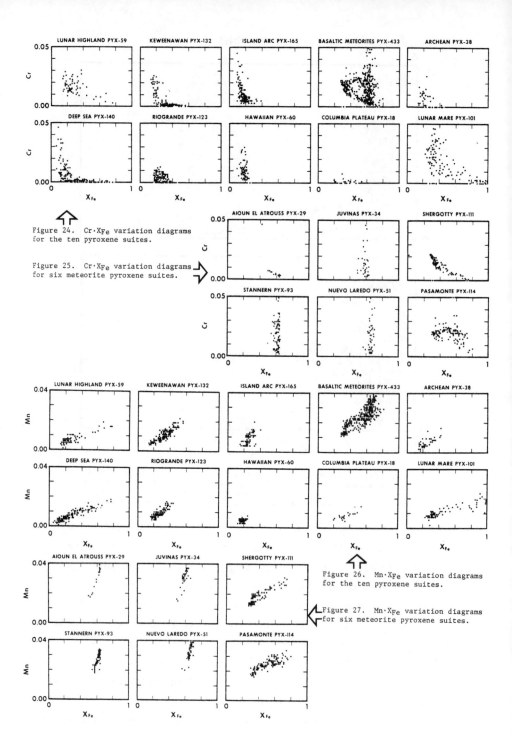

Figure 24. Cr·X_{Fe} variation diagrams for the ten pyroxene suites.

Figure 25. Cr·X_{Fe} variation diagrams for six meteorite pyroxene suites.

Figure 26. Mn·X_{Fe} variation diagrams for the ten pyroxene suites.

Figure 27. Mn·X_{Fe} variation diagrams for six meteorite pyroxene suites.

516

oxygen fugacities they experienced.

The Cr contents of basaltic meteorite pyroxenes require more discussion. Reference to Figure 25 shows a very unusual set of chemical
systematics. Aioun el Atrouss and Shergotty show relatively well-behaved
trends of Cr decreasing with increasing X_{Fe}. Pasamonte also shows a
crude trend of Cr versus X_{Fe}. However, the Cr systematic for Juvinas,
Nuevo Laredo, and Stannern make no sense at all. The Cr does not appear
to be correlated with Ca as one might expect. It is possible that during
the pyroxene equilibration, phase separation took place, involving not
only two pyroxenes, but also a Cr-containing phase, perhaps Cr-spinel.
This possibility requires a close examination of the pyroxenes with TEM
and electron diffraction techniques. The microprobe data indicate that
the Cr is not evenly distributed in these pyroxenes.

The Mn versus X_{Fe} systematics for the pyroxenes from the suites are
illustrated in Figures 26 and 27. In most of the diagrams for the individual suites, Mn and X_{Fe} show a high positive correlation (see discussion below); however, the slopes are different. The lowest Mn/X_{Fe}
values are found in lunar pyroxenes, intermediate values in terrestrial
basalt pyroxenes, and the highest values in meteoritic pyroxenes. However, the systematics have interesting complexities. For example, differences in Mn/X_{Fe} are found among the different suites for one planetary
body; e.g., Keweenawan basalt pyroxenes show higher Mn/X_{Fe} than deep sea
basalt pyroxenes.

Even more interesting features are observed for pyroxenes from
basaltic meteorites. Shergotty pyroxene systematics (Fig. 27) show two
trends of increasing Mn with X_{Fe}, one for high-Ca pyroxene and one for
low-Ca pyroxenes. The trends are parallel to each other and show the
expected positive correlation of Mn and X_{Fe}. However, the trend for
low-Ca pyroxenes is displaced towards higher Mn values. The crystal
chemical interpretation of this feature is that in the low-Ca pyroxenes,
the Mn^{2+} cations are strongly ordered into the pyroxene M2 site. Even
more striking is the significantly different slope of the $Mn \cdot X_{Fe}$ trends
for the equilibrated Aioun el Atrouss, Juvinas, Nuevo Laredo, and
Stannern pyroxenes compared with pyroxenes from Shergotty and Pasamonte.
The Mn/X_{Fe} values for the pyroxenes from equilibrated meteorites are
considerably higher than those for Shergotty and Pasamonte. The reason

for these differing systematics appears to reflect the subsolidus exsolution of the pyroxenes in the equilibrated eucrites. As we mentioned above in reference to the Shergotty meteorite, Mn enters the low-calcium preferentially. It is also observed that low-calcium pyroxenes are enriched in Fe^{2+} relative to the high-Ca pyroxenes. Thus, the diagrams of Mn versus X_{Fe} are simply showing that low-Ca pyroxenes are enriched in both Mn and Fe^{2+} relative to the high-Ca pyroxenes. The crystal chemical explanation for this is straight-forward. During the subsolidus re-equilibration which causes exsolution in these pyroxenes, Ca is fractionated into the M2 sites of the augites while Mn and Fe^{2+} are fractionated into the M2 site of pigeonite. The site preference for the pyroxene M2 site is Ca > Mn^{2+} > Fe^{2+} > Mg. It is clear from the combined chemical systematics for pyroxenes from the equilibrated eucrite pyroxenes that the chemical trends on the pyroxene quadrilaterals (Fig. 16) are simply "mixing" lines caused by lack of microprobe resolution of very fine exsolution lamellae of augite and pigeonite. This is consistent with the results of several x-ray diffraction studies by Takeda (for example, see Takeda *et al*., 1976).

Figures 28 and 29 illustrate Na·X_{Fe} systematics. The Na contents of the pyroxenes show considerable variation. Low values are found in pyroxenes from eucrites, lunar highland, and lunar mare: Intermediate values are found in pyroxenes from the Archean, Columbia Plateau, deep sea, Hawaiian, and the Keweenawan basalts. Higher values are found in pyroxenes from island arc and Rio Grande basalts (Table 2). Although the Na contents of basaltic meteorite pyroxenes are low, variations are found among individual meteorites (Fig. 29). The Na contents of pyroxenes from Aioun el Atrouss, Juvinas and Nuevo Laredo are negligible. Those from Stannern and Pasamonte are higher, but the highest Na contents are found in pyroxenes from Shergotty. This is consistent with high Na in the bulk meteorite and is also reflected in the high albite contents of feldspars from Shergotty. The data for Shergotty also show that Na preferentially enters the M2 site of high-Ca pyroxenes relative to low-Ca pyroxenes.

The Ti·X_{Fe} systematics for the pyroxenes from the various suites are illustrated in Figures 30 and 31. The highest concentrations are found in mare basalts, intermediate concentrations in Columbia Plateau, deep sea, Keweenawan, Hawaiian, lunar highlands and Rio Grande basalts and lower values

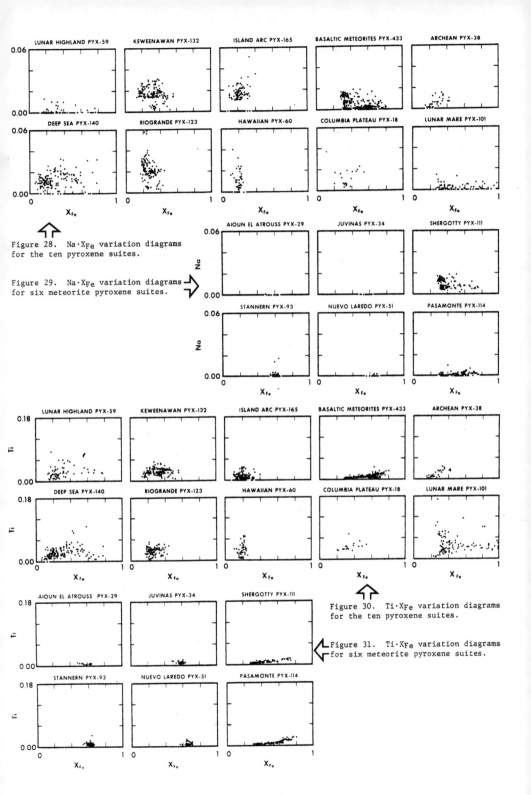

Figure 28. Na·X_{Fe} variation diagrams for the ten pyroxene suites.

Figure 29. Na·X_{Fe} variation diagrams for six meteorite pyroxene suites.

Figure 30. Ti·X_{Fe} variation diagrams for the ten pyroxene suites.

Figure 31. Ti·X_{Fe} variation diagrams for six meteorite pyroxene suites.

in island arc, Archean and basaltic meteorites (Table 2). Within the basaltic meteorite suite, there are also variations (Fig. 31). The equilibrated eucrites, Aioun el Atrouss, Juvinas, Nuevo Laredo, and Stannern have pyroxenes with extremely low-Ti contents. The Ti concentrations in pyroxenes from Shergotty and Pasamonte are higher. In these samples a positive correlation between Ti and X_{Fe} is observed.

Substitutional Couples

To gain further insight into the relative importance of the pyroxene substitutional couples, the cation contents for twelve subgroups of the pyroxenes were treated statistically by a factor analysis program BMDP4M (Dixon, 1975). The twelve groups include Archean, Columbia Plateau, deep sea, equilibrated eucrites, Pasamonte, Shergotty, Hawaiian, island arc, Keweenawan, lunar highlands, lunar mare, and Rio Grande.

Results of the calculations are presented in Tables 4-7. Table 4 illustrates the relative importance of the "Others" substitutional couples. The Archean pyroxenes have $^{VI}Al-^{VI}Al$ and $^{VI}Ti^{4+}-^{IV}Al$ as the two most important "Others" couples. These couples are consistent with the high Al concentrations reported for these pyroxenes. In pyroxenes from the Columbia Plateau suite $^{VI}Fe^{3+}-^{IV}Al$ and $^{VI}Ti^{4+}-^{IV}Al$ are the two most important couples. Fe^{3+} is clearly important in these pyroxenes. In pyroxenes from the deep sea basalts the same two couples are important, indicating similar petrogenetic scenarios for these two rock suites. In the equilibrated eucrites the significance of the correlation coefficients is equivocal because their values are low and the concentration of the "Others" components are also low. Nevertheless, $^{VI}Ti^{4+}-^{IV}Al$ and $^{VI}Al-^{IV}Al$ appear to be present. An equally high correlation coefficient is indicated for $^{VI}Cr^{3+}-^{IV}Al$, but there are problems with the reported Cr contents for these pyroxenes, as discussed above. In Pasamonte, which is virtually identical in bulk composition to the equilibrated eucrites, the most important substitutional couple is $^{VI}Ti^{4+}-^{IV}Al$. In Shergotty, which has a different bulk chemistry (e.g. higher Na), the two most important "Others" couples are $^{VI}Ti^{4+}-^{IV}Al$ and $Na-^{VI}Al$ (jadeite), the latter reflecting higher Na concentrations in the host basalt. Pyroxenes from the Hawaiian suite show an extremely high correlation coefficient (0.932) for the $^{VI}Ti-^{IV}Al$ couple,

Table 4. Pyroxene "Other" – "Other" cation correlations[a].

Group	^{VI}Al–^{IV}Al	$^{VI}Fe^{3+}$–^{IV}Al	$^{VI}Cr^{3+}$–^{IV}Al	$^{VI}Ti^{4+}$–^{IV}Al	Na–^{VI}Al	Na–$^{VI}Fe^{3+}$	Na–$^{VI}Cr^{3+}$	Na–$^{VI}Ti^{4+}$
Archean	0.527	0.203	0.360	0.527	0.427	-0.319	-0.353	-0.282
Columbia Plateau	-0.436	0.568	0.072	0.566	-0.027	0.202	-0.121	0.395
Deep Sea	0.232	0.584	0.068	0.590	-0.051	0.453	-0.137	0.349
Equilibrated Eucrites	0.270	-	0.342	0.328	0.021	-	-0.032	-0.078
Pasamonte	0.033	-	0.234	0.488	-0.059	-	-0.054	0.088
Shergotty	-0.100	-	0.129	0.260	0.234	0.608	0.173	0.160
Hawaiian	0.495	0.436	0.375	0.932	0.178	-0.223	0.069	0.537
Island Arc	0.588	0.486	0.213	0.355	0.378		-0.156	0.382
Keweenawan	0.091	0.794	0.360	0.635	-0.042	0.562	0.159	0.568
Lunar Highlands	0.285	-	0.562	0.642	-0.136	-	0.050	0.611
Lunar Mare	0.675	-	0.300	0.833	0.158	-	0.068	0.598
Rio Grande	0.304	0.818	0.207	0.642	0.571	0.463	-0.210	0.001

[a] Correlation coefficients > 0.6 are underlined.

Table 5. Pyroxene "Other" – "Quad" cation correlations.

Group	Ti–Ca	Ti–Mg	Ti–Fe^{2+}	^{VI}Al–Ca	^{VI}Al–Mg	^{VI}Al–Fe^{2+}	Fe^{3+}–Ca	Fe^{3+}–Mg	Fe^{3+}–Fe^{2+}	Cr–Ca	Cr–Mg	Cr–Fe^{2+}
Archean	0.519	-0.785	0.645	-0.180	-0.020	-0.198	0.266	-0.086	-0.106	0.035	0.150	-0.363
Columbia Plateau	0.405	-0.541	0.357	-0.441	0.146	0.008	0.468	0.234	-0.546	-0.269	0.210	-0.116
Deep Sea	-0.014	-0.486	0.405	0.152	0.142	-0.306	0.342	-0.166	-0.196	0.284	0.382	-0.586
Equilibrated Eucrites	0.159	-0.110	-0.147	0.083	-0.136	-0.107	-	-	-	0.043	-0.086	-0.009
Pasamonte	0.783	-0.841	0.808	-0.234	0.288	-0.370	-	-	-	-0.224	0.218	-0.225
Shergotty	0.323	-0.768	0.372	0.351	0.064	0.301	0.278	-0.249	-0.522	0.639	0.292	-0.792
Hawaiian	0.737	-0.783	-0.550	0.494	-0.533	-0.290	0.278	-0.249	-0.522	0.194	-0.184	-0.358
Island Arc	0.101	-0.505	0.243	0.322	-0.687	0.008	0.009	0.064	-0.233	0.469	-0.153	-0.508
Keweenawan	0.450	-0.529	-0.177	0.132	0.198	0.378	0.558	-0.382	-0.451	0.374	0.133	-0.621
Lunar Highlands	0.760	-0.493	-0.090	-0.357	0.392	-0.285	-	-	-	0.201	0.233	-0.603
Lunar Mare	0.713	-0.213	-0.273	0.135	-0.187	-0.025	-	-	-	-0.033	0.645	-0.680
Rio Grande	0.372	-0.522	-0.105	0.123	-0.013	-0.411	0.637	-0.574	-0.669	0.167	-0.052	-0.225

illustrating the dominance of this couple in these pyroxenes. The Na-containing couples $Na-^{VI}Fe^{3+}$ and $Na-^{VI}Ti^{4+}$ are also important. In the island arc suite, the dominant "Others" components are $^{VI}Al-^{IV}Al$ and $^{VI}Fe^{3+}-^{IV}Al$. In the pyroxenes from the Keweenawan suite, the three most important couples are $^{VI}Fe^{3+}-^{IV}Al$, $^{VI}Ti^{4+}-Al$ and $Na-^{VI}Ti^{4+}$. In the two lunar suites, highlands and mare, $Ti-^{IV}Al$ is the dominant "Others" couple. Lastly, pyroxenes from the Rio Grande basalts show $Fe^{3+}-^{IV}Al$ as the dominant couple followed by $^{VI}Ti^{4+}-^{IV}Al$. Table 7 summarizes the two most important "Others" couples for each of these groups. With the data so summarized several things are apparent. First, the $^{VI}Ti-^{IV}Al$ couple is by far the most important "Others" couple, ranking in the top two for all except the island arc suite. Second, Fe^{3+} is important in all of the terrestrial suites with the possible exception of the Archean; it is usually present as the $^{VI}Fe^{3+}-^{IV}Al$ couple. This emphasizes the importance of f_{O_2} in pyroxene-liquid elemental partitioning. At the relatively high oxygen fugacities of terrestrial basalt petrogenesis, iron enters the pyroxene both as Fe^{3+} and Fe^{2+}, whereas in lunar and most basaltic meteorite pyroxenes, iron is only present as Fe^{2+}.

Tables 5 and 6 list the "Other"-"Quad" cation correlations. These correlation coefficients enable us to ascertain if there is any preferential linking of the "Others" couples with a specific quadrilateral component. We will simplify the discussion by considering only the two most important "Others" couples $^{VI}Ti^{4+}-^{VI}Al$ and $^{VI}Fe^{3+}-^{IV}Al$. By thus observing the correlation coefficients in Tables 5 and 6 we can decide which of the following pyroxene components are most important:

$$CaTiAl_2O_6 \qquad CaFe^{3+}SiAlO_6$$
$$MgTiAl_2O_6 \qquad MgFe^{3+}SiAlO_6$$
$$Fe^{2+}TiAl_2O_6 \qquad Fe^{2+}Fe^{3+}SiAlO_6$$

We can thus see whether these couples are correlated with augitic pyroxenes (high correlation with Ca), whether they occur early in the crystallization sequence (high correlation with Mg), or late in the crystallization sequence (high correlation with Fe^{2+}). As a result of such an analysis we see that the Ti-Al couple is correlated with the calcium content of the pyroxene and with the degree of fractionation in Archean, Columbia Plateau, deep sea, Pasamonte, and Shergotty basalts, while it is essentially only correlated

Table 6. Pyroxene "Other" - "Quad" cation correlations.

Group	Na-Ca	Na-Mg	Na-Fe^{2+}	^{IV}Al-Ca	^{IV}Al-Mg	^{IV}Al-Fe^{2+}	Mn-Ca	Mn-Mg	Mn-Fe^{2+}	Cr-XFe	Mn-XFe	Ti-XFe
Archean	-0.409	0.258	0.208	0.346	-0.498	0.131	0.272	-0.373	0.431	-0.316	0.434	0.773
Columbia Plateau	0.534	-0.555	0.225	0.726	-0.092	-0.331	0.041	-0.509	0.524	-0.137	0.514	0.438
Deep Sea	0.335	-0.469	0.114	0.359	-0.227	-0.203	-0.345	-0.675	0.938	-0.572	0.919	0.446
Equilibrated Eucrites	0.000	-0.015	0.009	0.458	-0.154	-0.501	-0.664	0.175	0.729	0.052	0.646	-0.099
Pasamonte	0.091	-0.135	0.163	0.445	-0.391	0.287	0.521	-0.646	0.699	-0.239	0.660	0.844
Shergotty	0.461	-0.149	-0.252	0.148	-0.044	-0.066	0.020	-0.055	0.823	-0.721	0.736	0.557
Hawaiian	0.720	-0.722	-0.682	0.733	-0.758	-0.691	-0.152	0.098	0.382	-0.221	0.422	0.355
Island Arc	-0.231	-0.324	0.483	0.388	-0.611	-0.237	-0.852	0.474	0.726	-0.429	0.509	0.465
Keweenawan	0.722	-0.613	-0.452	0.782	-0.493	-0.663	-0.403	-0.176	0.681	-0.686	0.657	0.076
Lunar Highlands	0.542	-0.327	-0.098	0.383	0.034	-0.526	0.294	-0.656	0.698	-0.506	0.711	0.132
Lunar Mare	0.434	-0.159	-0.124	0.513	-0.063	-0.357	-0.231	-0.637	0.817	-0.707	0.757	-0.090
Rio Grande	0.618	-0.599	-0.584	0.538	-0.507	-0.593	-0.569	0.337	0.709	-0.193	0.542	0.177

Table 7. Summary of relative importance of pyroxene "Others" - cation couples.

Group	1st most important	2nd most important
Archean	Ti-Al	Al-Al
Columbia Plateau	Fe^{3+}-Al	Ti-Al
Deep Sea	Fe^{3+}-Al	Ti-Al
Equilibrated Eucrites	(Cr^{3+}-Al)?	Ti-Al
Pasamonte	Ti-Al	Cr-Al
Shergotty	Ti-Al	Na-Al
Hawaiian	Ti-Al	Na-Fe^{3+}
Island Arc	Al-Al	Fe^{3+}-Al
Keweenawan	Fe^{3+}-Al	Ti-Al
Lunar Highlands	Ti-Al	Na-Ti
Lunar Mare	Ti-Al	Al-Al
Rio Grande	Fe^{3+}-Al	Ti-Al

with calcium in Hawaiian, island arc, Keweenawan, lunar, and Rio Grande basalts. The Table 5 correlation coefficients show clearly that the $^{VI}Fe^{3+}-^{IV}Al$ couple enters the pyroxene structure as $CaFe^{3+}SiAlO_6$ and thus is largely concentrated in the augitic pyroxenes.

Conclusions

1. The most magnesian pyroxenes are found in the lunar highland, island arc, Archean, deep sea, and Hawaiian suites. This is a reflection of the fact that these suites include some of the most primitive basalts.

2. The highest concentrations of pyroxene "Others" components (e.g. Cr, Al, Ti, Fe^{3+}, Na) are found in lunar mare and Archean basalts reflecting, among other things, rapid cooling histories.

3. The Cr content of pyroxenes from "equilibrated" eucrites shows anomalous trends indicating that the Cr now resides in finely dispersed phases formed by subsolidus annealing.

4. The most important "Others" substitutional couple for all of the planetary suites combined is $^{VI}Ti^{4+}-^{IV}Al_2^{3+}$. It enters pyroxenes mainly as the component $CaTiAl_2O_6$.

5. A very important component in pyroxenes from terrestrial basalts is $CaFe^{3+}SiAlO_6$. This emphasizes the importance of f_{O_2} in pyroxene-liquid elemental partitioning. At the relative high oxygen fugacities of terrestrial petrogenesis, iron enters the pyroxene structure both as Fe^{3+} and Fe^{2+}, whereas in lunar and most basaltic meteorite pyroxenes, iron is only present as Fe^{2+}.

In summary, inspection of a large number of high quality pyroxene analyses demonstrates conclusively that these phases carry a signature of the planetary body in which they evolved. Conversely, these phases are powerful planetary probes capable of yielding information on the T, P, f_{O_2} and compositional parameters that obtained during the petrogenesis of their host basaltic liquids.

ACKNOWLEDGMENTS

Many people were involved in the pyroxene data collection and their proper attribution will be documented in the book "Basaltic Volcanism on the Terrestrial Planets." Mr. C. White conducted the computer manipulation of the chemical data; without his help this presentation would not have been possible. Most of the figures concerning the chemistry of pyroxenes from the planetary basalt suite were drafted by the staff of the Lunar and Planetary Institute, Houston, Texas. To them and to Rosanna Krommer, who served as an interface, I owe a debt of gratitude for getting these figures to me in time to be included in this volume. Thanks go to Robin Spencer for typing the manuscript and to Pauline Papike for other technical assistance. This research was funded by NASA grant NSG 9044 which is greatly appreciated.

REFERENCES

Bence, A. E., Baylis, D., Bender, J. F., and Grove, T. L. (1978) Diversity of major and minor elements in mid-ocean ridge basalts. In *Implications of Deep Drilling Results in the Atlantic Ocean,* in press.

Bence, A. E., Grove, T. L., and Papike, J. J. (1980) Basalts as probes of planetary interiors: constraints on the chemistry and mineralogy of their source regions. Precambria Research, 10, 249-279.

Bender, J. F., Hodges, F. N., and Bence, A. E. (1978) Petrogenesis of basalts from the project FAMOUS area: experimental study from 0 to 15 kbars. Earth Planet. Sci. Letters, 41, 277-302.

Dixon, W. J. (Ed.) (1975) *BMDP Biomedical Computer Programs.* University of California Press, Berkeley, California.

Haggerty, S. E. (1978) The redox state of planetary basalts. Geophys. Res. Letters, 5, 443-446.

Head, J. W., Wood, C. A., and Mutch, T. A. (1977) Geologic evolution of the terrestrial planets, Am. Scientist, 65, 21-24.

Huebner, J. S., Lipin, B. R., and Wiggins, L. B. (1976) Partitioning of chromium between silicate crystals and melts. Proc. 7th Lunar Sci. Conf., 12, 1195-1220.

Kesson, S. E. and Lindsley, D. H. (1976) Mare basalt petrogenesis: a review of experimental studies. Rev. Geophys. Space Phys., 14, 361-373.

Papike, J. J. and Bence, A. E. (1978) Lunar mare versus terrestrial mid-ocean ridge basalts: planetary constraints on basaltic volcanism. Geophys. Res. Letters, 5, 803-806.

Papike, J. J. and Bence, A. E. (1979) Planetary basalts: chemistry and petrology. Rev. Geophys. Space Phys., 17, 1612-1641.

Papike, J. J. and Cameron, M. (1976) Crystal chemistry of silicate minerals of geophysical interest. Rev. Geophys. Space Phys., 14, 37-80.

Papike, J. J. and Vaniman, D. T. (1978) Lunar 24 ferrobasalts and the mare basalt suite: comparative chemistry, mineralogy and petrology. In Mare Crisium: The View from Luna 24, p. 371-401. Proc. Lunar and Planetary Inst. Conf. on Lunar 24.

Papike, J. J., Cameron, K. L., and Baldwin, K. (1974) Amphiboles and pyroxenes: characterization of Other than quadrilateral components and estimates of ferric iron from microprobe data. Geol. Soc. Am. Abstr. Programs, 6, 1053-1054.

Papike, J. J., Hodges, F. N., Bence, A. E., Cameron, M., and Rhodes, J. M. (1976) Mare basalts: crystal chemistry, mineralogy, and petrology. Rev. Geophys. Space Phys., 14, 475-540.

Schreiber, H. D. and Haskin, L. A. (1976) Chromium in basalts: experimental determination of redox states and partitioning among synthetic silicate phases. Proc. 7th Lunar Sci. Conf., 2, 1221-1259.

Schweitzer, E. L., Papike, J. J., and Bence, A. E. (1978) Clinopyroxenes from deep sea basalts: A statistical analysis. Geophys. Res. Letters, 5, 573-576.

Schweitzer, E. L., Papike, J. J., and Bence, A. E. (1979) Statistical analysis of clinopyroxenes from deep-sea basalts, Am. Mineral., 64, 501-513.

Takeda, H., Miyamoto, M., Ishii, T., and Reid, A. M. (1976) Characterization of crust formation on a parent body of achondrites and the moon by pyroxene crystallography and chemistry. Proc. 7th Lunar Sci. Conf., 3, 3535-3548.

Taylor, S. R. and Jakes, P. (1974) The geochemical evolution of the moon. Proc. 5th Lunar Sci. Conf., 2, 1287-1305.

Taylor, S. R. (1975) *Lunar Science: A Post-Apollo View.* Pergamon, New York.